Corrosion Engineering

Principles and Practice

Pierre R. Roberge, Ph.D., P.Eng.

McGraw Hill

New York Chicago San Francisco
Lisbon London Madrid Mexico City
Milan New Delhi San Juan
Seoul Singapore Sydney Toronto

The McGraw·Hill Companies

Library of Congress Cataloging-in-Publication Data

Roberge, Pierre R.
Corrosion engineering : principles and practice/Pierre R. Roberge.—1st ed.
p. cm.
Includes bibliographical references and index.
ISBN 978-0-07-148243-1 (alk. paper)
1. Corrosion and anti-corrosives. I. Title.
TA418.74.R62 2008
620.1'1223—dc22 2008005204

34567890 DOC/DOC 14321

ISBN 978-0-07-148243-1
MHID 0-07-148243-1

This book is printed on acid-free paper.

Sponsoring Editor
Larry S. Hager

Production Supervisor
Richard C. Ruzycka

Editing Supervisor
Stephen M. Smith

Project Manager
Preeti Longia Sinha, International Typesetting and Composition

Copy Editor
Ragini Pandey

Proofreader
Pam Simonic

Indexer
WordCo Indexing Services, Inc.

Art Director, Cover
Jeff Weeks

Composition
International Typesetting and Composition

Contents

Preface

When I carried out my first corrosion investigation, some 25 years ago, on what turned out to be a 90-10 copper-nickel tubing Type I pitting problem it never occurred to me that this was indeed to trigger an important transition in my career. Well, that seems to be how many corrosion engineers have stumbled onto what was later to become a central focus of their work. There are many reasons for this. One common factor that often attracts an investigator's attention is the drastic contrast that exists between the importance and seriousness of a corrosion problem and the size of the damage itself.

In my first corrosion investigation a metallurgical microscope of reasonable magnification was required to examine the tubing samples provided. Yet, these microscopic pits were causing a major havoc to the air-cooling system of a relatively modern facility where my laboratory and office were located. Eventually the whole air-conditioning system unit had to be replaced at a cost of over $200,000. The precise root cause of the problem still remains a mystery since a few other systems operating with a common water intake and of the same design and vintage are still in operation today and never suffered Type I pitting problems.

My first case also revealed another aspect of many corrosion investigations that is quite fascinating. It has to do with the complexity of the interactions that eventually culminate in a failure or a need to repair. The belief was widespread at the time that many of the corrosion problems could be alleviated with the help of well-designed and calibrated expert systems. In many countries the development of these systems was funded on the premise that these software tools would artificially improve the level of expertise of technical personnel. Of course, this optimistic view could not possibly consider many of the hidden factors that are behind many corrosion situations: unreported system changes, rapid and frequent changes in technical personnel and many other factors that may remain invisibly at work on a micro scale for years before giving the final blow to a system.

As with many of my predecessors and many colleagues, I have come to the conclusion that the main line of defense against the multi-headed foe we call corrosion is by increasing awareness through education and training. In our modern world some of that training can be provided by various routes that are readily accessible almost anywhere via the Internet or the Web. However, textbooks and reference documents remain as precious today as they were a century ago when they were the main source of distributing information.

As an educator I have always been looking for useful and educative corrosion documents. Hopefully the reader will find the present text instructive in several useful ways.

In conclusion, I would like to acknowledge the numerous contributors who have directly or indirectly provided many of the cases discussed and illustrated in *Corrosion Engineering: Principles and Practice*. Your work at combating corrosion on all fronts has been greatly inspiring.

Pierre R. Roberge, Ph.D., P.Eng.

About the Author

Pierre R. Roberge, Ph.D., P.Eng., is a professor at the Royal Military College of Canada, where he teaches materials engineering, corrosion engineering, and electrochemical power sources. He previously worked as a research scientist in industry, specializing in the performance of materials in service and the production of energy with electrochemical power sources. Dr. Roberge has written numerous journal articles and conference papers and is the author of several engineering titles, including McGraw-Hill's *Handbook of Corrosion Engineering*.

CHAPTER 1

The Study of Corrosion

1.1 Why Study Corrosion?

Most people are familiar with corrosion in some form or another, particularly the rusting of an iron fence and the degradation of steel pilings or boats and boat fixtures. Piping is another major type of equipment subject to corrosion. This includes water pipes in the home, where corrosion attacks mostly from the inside, as well as the underground water, gas, and oil pipelines that crisscross our land. Thus, it would appear safe to say that almost everyone is at least somewhat familiar with corrosion, which is defined in general terms as the degradation of a material, usually a metal, or its properties because of a reaction with its environment.

This definition indicates that properties, as well as the materials themselves, may and do deteriorate. In some forms of corrosion, there is almost no visible weight change or degradation, yet properties change and the material may fail unexpectedly because of certain changes within the material. Such changes may defy ordinary visual examination or weight change determinations.

In a modern business environment, successful enterprises cannot tolerate major corrosion failures, especially those involving personal injuries, fatalities, unscheduled shutdowns, and environmental contamination. For this reason considerable efforts are generally expended in corrosion control at the design stage and in the operational phase. This is particularly true for industries where harsh chemicals are handled routinely.

Corrosion can lead to failures in plant infrastructure and machines which are usually costly to repair, costly in terms of lost or contaminated product, in terms of environmental damage, and possibly costly in terms of human safety. Decisions regarding the future integrity of a structure or its components depend upon an accurate assessment of the conditions affecting its corrosion and rate of deterioration.

With this information an informed decision can be made as to the type, cost, and urgency of possible remedial measures.

Required levels of maintenance can vary greatly depending on the severity of the operating environments. While some of the infrastructure equipment might only require regular repainting and occasional inspection of electrical and plumbing lines, some chemical processing plants, power generation plants, and aircraft and marine equipment are operated with extensive maintenance schedules.

Even the best design cannot be expected to anticipate all conditions that may arise during the life of a system. Corrosion inspection and monitoring are used to determine the condition of a system and to determine how well corrosion control and maintenance programs are performing. Traditional corrosion inspection practices typically require planned periodic shutdowns or service interruptions to allow the inspection process. These scheduled interruptions may be costly in terms of productivity losses, restart energy, equipment availability, and material costs. However, accidental interruptions or shutdowns are potentially much more disruptive and expensive.

1.2 The Study of Corrosion

To the great majority of people, corrosion means rust, an almost universal object of hatred. *Rust* is, of course, the name which has more recently been specifically reserved for the corrosion of iron, while *corrosion* is the destructive phenomenon which affects almost all metals. Although iron was not the first metal used by man, it has certainly been the most used, and must have been one of the first on which serious corrosion problems were encountered [1].

Greek philosophers viewed the physical world as matter organized in the form of *bodies* having length, breadth, and depth that could act and be acted upon. They also believed that these bodies made up a material continuum unpunctuated by voids. Within such a universe, they speculated about the creation and destruction of bodies, their causes, the essence they consisted of, and the purpose they existed for. Surfaces did not fit easily into these ancient pictures of the world, even those painted by the atomists, who admitted to the existence of voids. The problem of defining the boundary or limit of a body or between two adjacent bodies led Aristotle (fourth century BC) and others to deny that a *surface* has any substance. Given Aristotle's dominance in ancient philosophy, his view of surfaces persisted for many centuries, and may have delayed serious theoretical speculation about the nature of solid surfaces [2].

Perhaps the only ancient scientific account of surfaces is to be found in some passages of the great Roman philosopher Pliny the Elder (23–79 AD) who wrote at length about *ferrum corrumpitur*, or spoiled iron. By his time the Roman Empire had been established as the world's foremost civilization, a distinction due partly to the extensive use of iron for weaponry and other artifacts that were, of course, highly subject to rust and corrosion.

Pliny described corrosion phenomena taking place at the surface of metals, as well as remedies for minimizing the effects of corrosion. His reference to the use of oil as a means of protecting bronze objects against corrosion, as well as of allowing the soldering of lead surfaces, has been unambiguously verified by modern chemical analysis of Roman artifacts. According to Pliny, surfaces act as bodies that interact with each other and external agents. Pliny also speculated on the causes of metal corrosion (air and fire).

Numerous scientific and engineering discoveries have been made since then and the general understanding of corrosion mechanisms has progressed with these. Some of the discoveries that have improved the field of corrosion are listed in App. A (Historical Perspective). By the turn of the twentieth century the basic processes behind the corrosion of iron and steel were relatively well understood. One of the first modern textbooks on corrosion prevention and control was published by McGraw-Hill in 1910 [3]. The following are some excerpts that illustrate the state of knowledge when this landmark text came out.

On the Theory of Corrosion

In order that rust should be formed iron must go into solution and hydrogen must be given off in the presence of oxygen or certain oxidizing agents. This presumes electrolytic action, as every iron ion that appears at a certain spot demands the disappearance of a hydrogen ion at another, with a consequent formation of gaseous hydrogen. The gaseous hydrogen is rarely visible in the process of rusting, owing to the rather high solubility and great diffusive power of this element. Substances which increase the concentration of hydrogen ions, such as acids and acid salts, stimulate corrosion, while substances which increase the concentration of hydroxyl ions inhibit it. Chromic acid and its salts inhibit corrosion by producing a polarizing or dampening effect which prevents the solution of iron and the separation of hydrogen.

Electrolytic Theory of Corrosion of Iron

From the standpoint of the electrolytic theory, the explanation of the corrosion of iron is not complicated, and so far has been found in accordance with all the facts. Briefly stated, the explanation is as follows: Iron has a certain solution tension, even when the iron is chemically pure and the solvent pure water. The solution tension is modified by impurities or additional substances contained in the metal and in the solvent. The effect of the slightest segregation in the metal, or even unequal stresses and strains in the surface, will throw the surface out of equilibrium, and the solution tension will be greater at some points than at others.

The points or nodes of maximum solution pressure will be electro-positive to those of minimum pressure, and a current will flow, provided the surface points are in contact, through a conducting film. If the film is water, or is in any way moist, the higher its conductivity the faster iron will pass into solution in the electro-positive areas, and the faster corrosion proceeds. Positive hydrogen ions migrate to the negative areas, negative hydroxyls to the positives.

On the Effects of Cold Work

A considerable body of evidence has been brought forward from time to time to show that in addition to the segregation of impurities in steel, the presence of scratches, sand pitholes, and, in fact, all indentations or wounds on the surface of steel, will stimulate rusting by becoming centers of corrosion. Such marks or indentations are almost invariably electropositive to surrounding areas, and the depolarization which results in the rapid disengagement of hydrogen at these spots leads to stimulated pitting. This effect can be very prettily shown by means of the ferroxyl indicator.*

On Puddle Iron† and Steel

Mr. J. P. Snow, Chief Engineer of the Boston and Maine Railroad, has called attention to a very significant case of corrosion in connection with the destruction of some railroad signal bridges erected in 1894, and removed and scrapped in 1902. These structures were built at the time that steel was fast displacing puddled iron as bridge material.

The result was that the bridges were built from stock material which was partly steel and partly wrought iron. The particular point of interest in this case lies in the fact that while some of the members of the bridge structures rusted to the point of destruction in eight years, others were in practically as good condition as on the day they were erected.

Moreover, the tonnage-craze, from which the quality of product in so many industries is today suffering, is causing to be placed on the market a great mass of material, only a small proportion of which is properly inspected, which is not in proper condition to do its work: rails and axles which fail in service and steel skeletons for high buildings which may carry in them the germs of destruction and death.

---

*The ferroxyl indicator is a mixture of two indicators used to reveal the nature of surface corrosion on steel. Phenolphthalein in the ferroxyl indicator reveals surface areas that are becoming basic and potassium ferricyanide which turns blue in the presence of the iron (II) ions produced during corrosion. The use of ferroxyl indicator will be discussed in more details in Chap. 7.

†Puddle iron is a type of wrought iron produced in a puddling furnace, a process invented at the end the eighteenth century. The process results in an iron that contains a slightly increased carbon content and a higher tensile strength compared to wrought iron. The puddling furnace also allows a better control of the chemical composition of the iron. The Eiffel Tower and many bridges were built with puddle iron.

That the old, largely hand-worked metal of about 30 years ago is superior in rust-resisting quality to the usual modern steel and iron is attested by the recorded evidence of a large number of observers.

On Paints and Corrosion Inhibitive Pigments

The many theories which have attempted to explain the rusting of iron during the last century have stimulated a large amount of original research on the relation of various pigments to the corrosion problems. In the course of the investigations undertaken, the subject of protective coatings for iron and steel was naturally brought into prominence and received a considerable amount of attention.

The study of protective coatings for iron has led many paint manufacturers, as well as scientific investigators, to make closer studies of the causes of corrosion. It is evident that the electrochemical explanation of corrosion must have a direct bearing on paint problems.

1.3 Needs for Corrosion Education

The specific needs for corrosion education vary greatly with the level of education required, the functions expected of the personnel, and of course the applications where corrosion is a concern. In order to indicate the suitability of the various teaching aids and texts for particular types of training, four categories of corrosion personnel based upon their particular activities have been identified by the European Federation of Corrosion (EFC).

- Group A: corrosion scientists and engineers
- Group B: technologists
- Group C: technicians
- Group D: operatives

In real-world situations, all corrosion personnel would, of course, work toward the solution of often very specific corrosion problems and there will be considerable overlap between the tasks assigned to individuals. Any person working primarily in one group will probably have interests and activities in other groups. The distinction between these groups is therefore more one of perspective, rather than level of skills. When designing a corrosion training course or program, it should be realized that while the training material should be broadly based and cover all major aspects of corrosion and protection, there will inevitably be some emphasis put on the lecturer's field of interest and expertise. There could also be some specific requirements depending on the trainee's type of work, for example, aeronautical, automotive, oil and gas, and medical.

Corrosion Scientists and Engineers

This group comprises persons who are going to work on the development of techniques and methods and need to have a good understanding of the mechanism of corrosion—personnel such as chemists, metallurgists, physicists, engineers, and so forth who are carrying out research and teaching in the field of corrosion and protection. A corrosion training program designed for this group should focus on the phenomena associated with corrosion and its prevention in a manner based upon the scientific principles involved. Besides specific courses and laboratories in corrosion prevention and control, additional courses in physical chemistry, electrochemistry, chemical thermodynamics, and physical metallurgy should be required as prerequisites for the corrosion education.

Corrosion Technologists and Technicians

Corrosion technologists, who must collaborate directly with the corrosion scientist and engineers, should also have a good understanding of scientific principles and be capable of applying these to practical problems. Corrosion technicians are typically qualified to implement decisions made by the corrosion technologists, or to carry out experimental work under supervision of a corrosion scientist or an engineer. Technicians will normally work under supervision, and will be concerned with design, surveys, inspection, commissioning plant, control, laboratory and field testing.

A common syllabus could satisfy both the technologists and technicians groups, but it is apparent that the depth of approach and emphasis would not be necessarily the same. Thus, in the case of the technologist a more fundamental approach may be required, and in addition courses in physical chemistry and physical metallurgy, which should precede the course on corrosion, will be necessary to enable the technologist to appreciate the electrochemical and metallurgical aspects of the subject. On the other hand, the technician will not be required to go so deeply into theory and emphasis of the general course should be on the practical aspects of corrosion protection and on corrosion monitoring and testing.

Operatives

Operatives are the personnel who carry out the actual work in the field under the supervision of corrosion engineers. For such groups the training objectives should focus on treatments of the principles sufficiently to provide a basic knowledge relevant to the special topics being taught. These courses will be highly specialized and directed to specific jobs. Special attention will be paid to carrying out the work effectively and the training supplemented with case studies.

For all active personnel, certification in the field of corrosion and corrosion prevention is an issue of growing importance because certification provides confidence in the quality of services provided.

Certification bodies exist on all continents which issue certificates of conformity using criteria established by professional societies and councils in various countries. One major provider of corrosion training for such certificates is the National Association of Corrosion Engineers (NACE) International, which offers professional certification programs with up to 10 different certification categories (Table 1.1).

Course Name	Duration (days)	What Is the Link to Certification?
NACE Basic Corrosion Course	5	Persons passing this course exam, who have 2 years of experience, may apply to become a certified NACE Corrosion Technician.
NACE Protective Coatings and Linings (Basic)	5	This course exam is on the parallel path to NACE Corrosion Technologist and NACE Sr. Corrosion Technologist certification.
NACE Marine Coating Inspector Course	3	A NACE Coating Inspector who passes this exam may receive a "Marine" endorsement on his or her NACE Coating Inspector card. Anyone may enroll in this course.
CIP 1-day Bridge Specialty Course	1	A NACE Coating Inspector who passes this exam may receive a "Bridge" endorsement on his or her NACE Coating Inspector card. NOTE: Only a person who is a NACE-recognized Coating Inspector Technician or Certified Coating Inspector may enroll in or attend this course.
NACE CP 3—Cathodic Protection Technologist	6	Persons passing this exam, who meet education and experience requirements for CP Technologist, may apply to be NACE CP Technologists.

Table 1.1 NACE International Certification Courses

Course Name	Duration (days)	What Is the Link to Certification?
NACE CP 4—Cathodic Protection Specialist	6	Persons passing this exam, who meet education and experience requirements requirements for CP Specialist, may apply to be NACE CP Specialists.
Successful Coating and Lining of Concrete	2	No link to any certification—this course has no exam.
Corrosion Control in the Refining Industry	4.5	No link to any certification—this course has no exam.
Internal Corrosion Technologist Course (Plan B Only)	5	Persons passing this exam, who meet education and experience requirements for Internal Corrosion Technologist, may apply to be NACE Internal Corrosion Technologists.
CP Tutorials (pre–CP Level 1)	1–1.5	The tutorials have no exams. Coordinate your offering of the tutorials with HQ's schedule of CP-1 classes; then offer the tutorials just prior to the CP-1 class in your area. The tutorials will help CP-1 students perform better.

Table 1.1 NACE International Certification Courses (*continued*)

1.4 The Functions and Roles of a Corrosion Engineer

Work associated with corrosion assessment, mitigation, and management encompasses a wide range of technical disciplines, from expert support and review, through laboratory studies and failure investigations, and from corrosion assessment to corrosion management reviews and risk-based management implementation. The corrosion engineer may be expected to provide a specialist corrosion consultancy, support, and management function for a larger group. The tasks of a corrosion engineer may also include product research and development work for customer applications, being responsible for the development of new products, and interfacing with customers and suppliers to provide solutions for technical challenges.

A corrosion engineer is often the member of a team with expertise in chemical and materials engineering, failure analysis, electrochemistry, biochemistry, and applied microbiology. In a large organization, the primary function of the corrosion team would be to ensure that adequate corrosion prevention and control requirements are being implemented during all phases of procurement and operations. The corrosion team would also be responsible for ensuring that relevant program documents are prepared and submitted in accordance with acquisition requirements and schedule. The work of a corrosion engineer may bring him into frequent contact with responsible people in many of the branches of his organization:

- With the engineering staff to work out new designs or modify existing ones in order to reduce the opportunity for corrosion.
- With the maintenance engineers so that corrosion problems and their probable causes are ascertained in order to cope with them by making repairs or avoid them altogether through preventive maintenance.
- With the production department to recognize their particular requirements and needs for improvement in order to increase the reliability and safe usage of equipment prone to be affected by corrosion.
- With the accounting department to establish the actual cost of corrosion in each case and the savings that may be expected by reducing losses from this source.
- With the purchasing department to advise on the choice of materials, to work out appropriate specifications and quality control for materials, equipment, and fabrication procedures.
- With the sales department to discover any deficiencies of the product that might be corrected by a better corrosion control and demonstrate the sales value of the improvements resulting from any corrective measure.
- With management to keep them abreast of particular needs and accomplishments in order to receive the support required to be fully effective in fighting corrosion.

As Francis L. LaQue pointed out in a paper published in 1952 and rerun in the August 1985 issue of *Materials Performance*, a corrosion engineer is for many organizations an engineer trained to recognize the nature of corrosion and understand the mechanics of corrosion processes [4]. With this knowledge, the corrosion engineer can make a faster and more accurate diagnosis or analysis of any corrosion related problem and be in a much better position to reason from one experience to another, appraise the information presented, plan research to uncover new information, and interpret and apply results of investigations when they have been completed.

In order to choose the proper material and overcome a corrosion problem the corrosion engineer is expected to know what materials are available and what are their corrosion resistant advantages and limitations. The environmental degradation of materials is often a critical and limiting factor in the development of virtually every advanced technology area, such as power generation, energy conversion, waste treatment, and communications and transportation. As new materials enter the marketplace and new engineering systems evolve to take advantage of their properties, it is of paramount importance for the corrosion engineer to understand the chemical limits of these materials and to develop corrosion control approaches that can be integrated into the design and operation of the systems.

The evolution of traditional and advanced engineering systems requires engineering materials capable of performing under increasingly hostile service environments. Unless these materials are chemically stable in such environments, their otherwise useful properties (strength, toughness, electrical and thermal conductivity, magnetic and optical characteristics, etc.) may be compromised. In our modern, high-technology society, this applies to all materials, including metals, ceramics, polymers, semiconductors, and glasses. It is therefore necessary to know a good deal about the corrosive characteristics of the chemical or chemicals involved and how these are affected by such factors as concentration, temperature, velocity, aeration, or the presence of oxidizing or reducing substances or special contaminants.

Regardless of how attractive a material may be from any other point of view, it is of no use for a particular purpose if it cannot be secured in the required form. Filter cloth cannot be woven from an alloy available only as castings. Several materials may possess the corrosion resistant and mechanical properties required for a job, but many of them may be too expensive to be considered. For example, silver might be somewhat better than nickel for tubes in an evaporator to concentrate caustic soda to 50 percent, but it would not be enough to justify the extra cost involved, and steel might be a better choice economically overall for handling dilute caustic under less stringent conditions [4].

For simple economical reasons, it is much more efficient to prevent corrosion than to explain why it occurred, suggest what should have been done to avoid it, or even prescribe how the damage might be repaired. However, corrosion engineers are often forced to work in these less than optimal scenarios.

Civil engineers are more concerned with designing and building bridges that do not collapse than with rebuilding them after failure. If given the opportunity at the proper time, a good corrosion engineer should be able to guide, design, specify materials, and know how they should be fabricated so that costly corrosion failures might become as rare as catastrophic failures in structural engineering. In addition, periodic inspection of existing equipment should be undertaken so that any corrosion may be detected in time to initiate corrective action or to avoid hazards or interruptions of production that would accompany an unexpected failure from corrosion. This inspection may be done by the corrosion engineer or by those scheduled by the engineer.

Whatever the corrosion engineer recommends to cope with corrosion must always take the economic details into consideration. The results will then be worth more than the cost, and the most economical of several possible solutions to a problem will be recognized and chosen in preference to the others. While corrosion has nothing but negative aspects from most points of view, money spent to enable a corrosion engineer to control corrosion will be returned manyfold and will represent one of the most profitable investments that can be made. From management's point of view, the results of an engineer's activities may be expected to show up in one of the following forms [4]:

1. Ensuring maximum life of new equipment.
2. Preservation of existing equipment.
3. Protecting or improving the quality of a product in order to maintain or improve a competitive position.
4. Avoiding costly interruptions of production.
5. Reducing or eliminating losses of valuable products by spillage or leaks.
6. Refitting of equipment withdrawn from service because of corrosion.
7. Reducing hazards to life and property that might be associated with corrosion: Explosions of pressure vessels or piping systems, release of poisonous or explosive gases or vapors are a few examples.

1.5 The Corrosion Engineer's Education

Universities, colleges, and technical schools do not typically offer specific programs in corrosion prevention or control. The subject is most often learned at the school of "hard knocks," as old timers would say. However, it is interesting to examine the type of academic education practicing corrosion engineers have received. A survey was carried out by inviting members of two active corrosion-focused Internet discussion lists* to answer simple questions concerning their educational background. Sixty respondents, with a total of more than 1000 years of corrosion experience among them, answered these questions. Some results of this survey are summarized in Fig. 1.1.

---

*The UMIST Corrosion Discussion List (http://www.cp.umist.ac.uk/corros-l/) and the NACE International Corrosion Network (http://www.nacecorrosionnetwork.com/read/all_forums/).

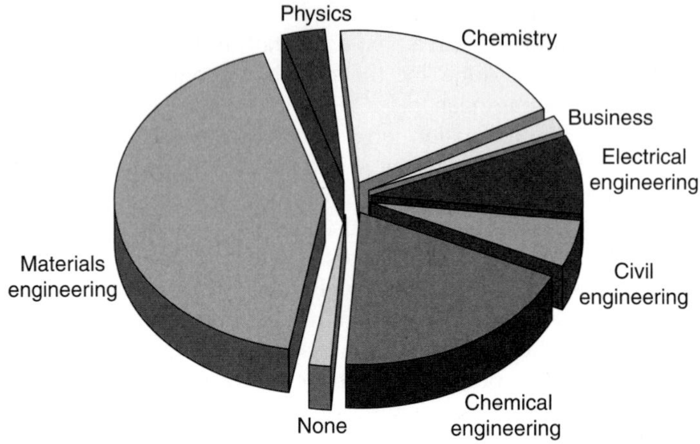

FIGURE 1.1 Distribution of disciplines in which active corrosion engineers have graduated.

As can be seen in Fig. 1.1, materials engineering was by far the main academic discipline taken by corrosion engineers. However, the wide spectrum of corrosion activities is also reflected in the breadth of expertise required of these engineers when they embark accidentally, coincidentally, or otherwise in a corrosion career. Interesting comments and observations were also gathered during the survey. The following are some of the comments collected during the Internet survey:

> I have found that corrosion is more the result of chemical and electrochemical interactions with the service environment than necessarily with the materials selected. The materials engineers I have worked with have an outstanding understanding of the manufacture of alloys, but not necessarily a good understanding of the effects of chemical attack and degradation on materials post-manufacture.

> Most metallurgical programs do not include electrochemistry, a must for a corrosion engineer. Chemical Engineers with some metallurgy classes would likely be the best equipped directly out of school.

> A corrosion engineer needs a broad background. When dealing with coatings, knowledge of chemistry is helpful. When dealing with cathodic protection, knowledge of electrical engineering is helpful. Material selection and high temperature corrosion is best left to metallurgists. Microbial influenced corrosion is certainly a biological process. All corrosion engineers deal with life cycle costs and risk. Corrosion is multidisciplinary so a corrosion engineer needs to know materials, chemical engineering, mechanical engineering, some chemistry, a bit of electricity and electronics and so, it is not easy to become a corrosion engineer in full.

> Understanding fundamental origins of corrosion, the electrochemical basis for much of it as well as how and why standard tests are designed is critical. Encyclopedic knowledge of facts available in databases is of less importance.

1.6 Strategic Impact and Cost of Corrosion Damage

It is the belief of many that corrosion is a universal foe that should be accepted as an inevitable process. Actually, somethings can and should be done to prolong the life of metallic structures and components exposed to the environments. As products and manufacturing processes have become more complex and the penalties of failures from corrosion, including safety hazards and interruptions in plant operations, have become more costly and more specifically recognized, the attention given to the control and prevention of corrosion has generally increased.

Since the first significant report by Uhlig in 1949 that the cost of corrosion to nations is indeed great [5], the conclusion of all subsequent studies has been that corrosion represents a constant charge to a nation's gross national product (GNP). The annual cost of corrosion to the United States was estimated in Uhlig's report to be $5.5 billion or 2.1 percent of the 1949 GNP. This study attempted to measure the total costs associated with corroding components by summing up costs for owners and operators (direct cost) as well as those of users (indirect cost).

Corrosion cost studies of various forms and importance have since been undertaken by several countries, including the United States, the United Kingdom, Japan, Australia, Kuwait, Germany, Finland, Sweden, India, and China [6]. A common finding of these studies has been that the annual corrosion costs range from approximately 1 to 5 percent of the GNP of each nation. Several studies separated the total corrosion costs into two parts:

1. The portion of the total corrosion cost that could be avoided if better corrosion control practices were used.
2. Costs where savings required new and advanced technology (currently unavoidable costs).

Estimates of avoidable corrosion costs in these studies have varied widely with a range from 10 to 40 percent of the total cost. Most studies have categorized corrosion costs according to industrial sectors or to types of corrosion control products and services. All studies have focused on direct costs even if it has been estimated that indirect costs due to corrosion damage were often significantly greater than direct costs. Indirect costs have been typically excluded from these studies simply because they are more difficult to estimate.

Potential savings and recommendations in terms of ways to realize savings from corrosion damage were included in most of the reports as formal results or as informal directions and discussion. Two of the most important and common findings were:

1. Major improvements could be provided by a better dissemination of the existing information through education and training, by technical advisory and consulting services, and by research and development activities.

2. There were many opportunities for large savings through more cost-effective uses of available technologies to reduce corrosion.

The most recent study resulted from discussions between NACE International representatives, members of the U.S. Congress, and the Department of Transportation (DOT). An amendment for the cost of corrosion was included in the Transportation Equity Act for the 21st Century, which was passed by the U.S. Congress in 1998. The amendment requested that a study be conducted in conjunction with an interdisciplinary team of experts from the fields of metallurgy, chemistry, economics, and others, as appropriate.

Two different approaches were taken in the ensuing study to estimate the cost of corrosion. The first approach followed a method where the cost was determined by summing the costs for corrosion control methods and contract services. The costs of materials were obtained from various sources, such as the U.S. Department of Commerce Census Bureau, existing industrial surveys, trade organizations, industry groups, and individual companies. Data on corrosion control services, such as engineering services, research and testing, and education and training, were obtained primarily from trade organizations, educational institutions, and individual experts. These services included only contract services and not service personnel within the owner/operator companies.

The second approach followed a method where the cost of corrosion was first determined for specific industry sectors and then extrapolated to calculate a national total corrosion cost. Data collection for the sector-specific analyses differed significantly from sector to sector, depending on the availability of data and the form in which data were available. In order to determine the annual corrosion costs for the reference year of 1998, data were obtained for various years in the surrounding decade, but mainly for the years 1996 to 1999.

Indirect costs were defined in this study as costs incurred by those other than just the owners or operators of a given plant, structure, or system. Measuring and determining the value of indirect costs are generally complex assessments; however, several methods, such as risk-based analyses, are available to evaluate these costs. Owners and operators may be forced to assume the corrosion costs through taxing, penalties, litigations, or paying for clean up of spilled product. In such cases, the costs become direct costs. However, there are some indirect costs, such as traffic delays due to bridge repairs and rehabilitation that are more difficult to turn over to the owner or operator of a structure. These become indirect costs to a user but may still have a significant impact on the overall economy due to lost productivity.

The total cost due to the impact of corrosion for the individual economic sectors was $137.9 billion per year (Table 1.2).

Category	Industry Sectors	Estimated Direct Cost of Corrosion per Sector	
		$billion	percent
Infrastructure (16.4% of total)	Highway Bridges	8.3	37
	Gas and Liquid Transmission Pipelines	7.0	27
	Waterways and Ports	0.3	1
	Hazardous Materials Storage	7.0	31
	Airports	—	—
	Railroads	—	—
	Subtotal	**$22.6**	**100%**
Utilities (34.7% of total)	Gas Distribution	5.0	10
	Drinking Water and Sewer Systems	36.0	75
	Electrical Utilities	6.9	14
	Telecommunications	—	—
	Subtotal	**$47.9**	**100%**
Transportation (21.5% of total)	Motor Vehicles	23.4	79
	Ships	2.7	9
	Aircraft	2.2	7

Table 1.2 Summary of Estimated Direct Cost of Corrosion for Industry Sectors Analyzed in the 2001 Study

Category	Industry Sectors	Estimated Direct Cost of Corrosion per Sector	
		$billion	percent
	Railroad Cars	0.5	2
	Hazardous Materials Transport	0.9	3
	Subtotal	**$29.7**	**100%**
Production and Manufacturing (12.8% of total)	Oil and Gas Exploration and Production	1.4	8
	Mining	0.1	1
	Petroleum Refining	3.7	21
	Chemical, Petrochemical, and Pharmaceutical	1.7	10
	Pulp and Paper	6.0	34
	Agricultural	1.1	6
	Food Processing	2.1	12
	Electronics	—	—
	Home Appliances	1.5	9
	Subtotal	**$17.6**	**100%**
Government (14.6% of total)	Defense	20.0	99.5
	Nuclear Waste Storage	0.1	0.5
	Subtotal	**$20.1**	**100%**
	Total	**$137.9**	

Table 1.2 Summary of Estimated Direct Cost of Corrosion for Industry Sectors Analyzed in the 2001 Study (*continued*)

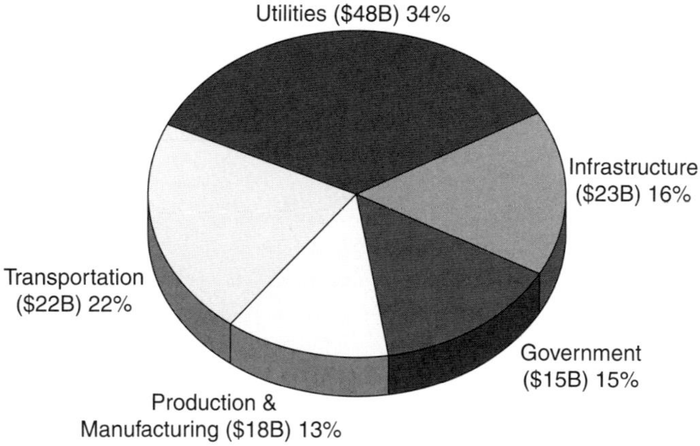

Figure 1.2 Corrosion costs breakdown across industrial sectors.

A breakdown of these costs by individual sectors is shown in Fig. 1.2. Since not all economic sectors were examined, the sum of the estimated costs for the analyzed sectors did not represent the total cost of corrosion for the entire U.S. economy.

By estimating the percentage of U.S. GNP for the sectors for which corrosion costs were determined and by extrapolating the figures to the entire U.S. economy, a total cost of corrosion of $276 billion was estimated. This value shows that the impact of corrosion is approximately 3.1 percent of GNP. This cost is considered to be a conservative estimate since only well-documented costs were used in the study. The indirect cost of corrosion was conservatively estimated to be equal to the direct cost, giving a total direct plus indirect cost of $552 billion or 6 percent of the GNP.

References

1. Trethewey KR, Chamberlain J. *Corrosion for Science and Engineering*. 2nd ed. Burnt Mill, UK: Longman Scientific & Technical, 1995.
2. Paparazzo E. Surfaces—lost and found. *Nature Materials* 2003; 2: 351–3.
3. Cushman AS, Gardner HA. *The Corrosion and Preservation of Iron and Steel*. New York: McGraw-Hill, 1910.
4. LaQue FL. What can management expect from a corrosion engineer? *Materials Performance* 1985; 25: 82–4.
5. Uhlig HH. The cost of corrosion in the United States. *Chemical and Engineering News* 1949; 27: 2764.
6. Koch GH, Brongers MPH, Thompson NG, Virmani YP, Payer JH. *Corrosion Costs and Preventive Strategies in the United States*. FHWA-RD-01-156. Springfield, Va.: National Technical Information Service, 2001.

CHAPTER 2
Corrosion Basics

2.1 Why Metals Corrode

The driving force that causes metals to corrode is a natural consequence of their temporary existence in metallic form. In order to produce metals starting from naturally occurring minerals and ores, it is necessary to provide a certain amount of energy. It is therefore only natural that when these metals are exposed to their environments they would revert back to the original state in which they were found. A typical cycle is illustrated by iron. The primary corrosion product of iron, for example, is $Fe(OH)_2$ (or more likely $FeO{\cdot}nH_2O$), but the action of oxygen and water can yield other products having different colors:

- $Fe_2O_3{\cdot}H_2O$ or *hydrous ferrous oxide,* sometimes written as $Fe(OH)_3$, is the principal component of red-brown rust. It can form a mineral called hematite, the most common iron ore.
- $Fe_3O_4{\cdot}H_2O$ or *hydrated magnetite,* also called ferrous ferrite ($Fe_2O_3{\cdot}FeO$), is most often green but can be deep blue in the presence of organic complexants.
- Fe_3O_4 or *magnetite* is black.

The energy required to convert iron ore to metallic iron is returned when the iron corrodes to form the original compound. Table 2.1 describes the results of x-ray diffraction of products found on specimens exposed to real environments where it can be seen that the metals often revert to naturally occurring mineral forms during the corrosion process [1]. The amount of energy required and stored in a metal or that is freed by its corrosion varies from metal to metal. It is relatively high for metals such as magnesium, aluminum, and iron, and relatively low for metals such as copper, silver, and gold. Table 2.2 lists a few metals in order of diminishing amounts of energy required to convert them from their oxides to metal.

The high reactivity of magnesium and aluminum expressed as energy in Table 2.2 is paralleled by the special efforts that were historically required to transform these metals from their respective ores. The industrial process to produce aluminum metal on a large

Sample Description	Chemical or Mineral Name*	Chemical Formula*
Product formed on magnesium during 3-month immersion in tap water	Nesquehonite *Calcium fluosilicate* *Beta silicon carbide* *Sodium sulfide* *Sodium fluoride* Magnesium carbonate Chloride hydroxide hydrate Magnesium pyrophosphate *Anorthoclase* *Alpha cristobalite* *Sodium hydroxide* *Calcium aluminum oxide sulfate*	$MgCO_3 \cdot 3H_2O$ $CaSiF_6$ β-SiC $Na2S$ NaF $MgCl_2 \cdot 2MgCO_3$ $Mg(OH)_2 \cdot 6H_2O$ $Mg_2P_2O_7$ $(Na,K)AlSi_3O_8$ SiO_2 $NaOH$ $Ca_4Al_6O_{12}SO_4$
Substance found on heat exchanger	*Halite*	$NaCl$
Substance found beneath paint on metal surface	*Alpha quartz*	SiO_2
Product formed on automobile bumper support during 3-year service	Lepidocrocite Goethite	γ-$Fe_2O_3 \cdot H_2O$ $Fe_2O_3 \cdot 2H_2O$
Product from conversion unit in marine environment	Zinc ferrite Cobalt ferrite Cobaltous ferrite *Halite* Chromic oxide Nickel, zinc ferrospinel *Sodium fluothorate* *Embolite* Magnesioferrite *Beryllium palladium* Magnetite *Nickel titanium*	$ZnO \cdot Fe_2O_3$ $CoO \cdot Fe_2O_3$ $CoO \cdot Fe_2O_3$ $NaCl$ Cr_2O_3 $(Ni,Zn)O \cdot Fe_2O_3$ $Na_3Th_2F_{11}$ $Ag(Cl,Br)$ $MgFe_2O_4$ $BePd$ Fe_3O_4 $NiTi$
Product formed on copper during 3-month immersion in tap water	Botallackite *Ilvaite*	$CuCl_2 \cdot 3Cu(OH)_2 \cdot 3H_2O$ $Ca(Fe,Mn,Mg)_2(Fe,Al)(SiO_4)_2OH$

Table 2.1 Results of X-Ray Diffraction of Products Found on Specimens Exposed to Real Environments [1]

Sample Description	Chemical or Mineral Name*	Chemical Formula*
Product from Al-Cu alloy exposed to deep-sea environment	Ammonium copper fluoride dihydrate *Potassium cyanide* Chi alumina Calcium aluminate *Alpha cadmium iodide*	$(NH_4)_2 \cdot CuF_4 \cdot 2H_2O$ *KCN* Al_2O_3 $3CaO \cdot Al_2O_3$ CdI_2
Product from Al-Zn-Mg-Cu alloy exposed to deep-sea environment	Chi alumina *Alpha cadmium iodide*	Al_2O_3 CdI_2
Product from Al-Mn alloy exposed to deep-sea environment	Ammonium copper fluoride dihydrate *Nobleite*	$(NH_4)_2CuF_4 \cdot 2H_2O$ $CaB_6O_{10} \cdot 4H_2O$

* Substances shown in italics are not corrosion products of the primary metals or alloys involved in the system.

Table 2.1 Results of X-Ray Diffraction of Products Found on Specimens Exposed to Real Environments [1] (*continued*)

	Metal	Oxide	Energy (MJ kg^{-1})
Highest Energy	Li	Li_2O	40.94
	Al	Al_2O_3	29.44
	Mg	MgO	23.52
	Ti	TiO_2	18.66
	Cr	Cr_2O_3	10.24
	Na	Na_2O	8.32
	Fe	Fe_2O_3	6.71
	Zn	ZnO	4.93
	K	K_2O	4.17
	Ni	NiO	3.65
	Cu	Cu_2O	1.18
	Pb	PbO	0.92
	Pt	PtO_2	0.44
	Ag	Ag_2O	0.06
Lowest Energy	Au	Au_2O_3	–0.18

Table 2.2 Positions of Some Metals in the Order of Energy Required to Convert Their Oxides to Produce 1 Kilogram of Metal

scale, for example, was only invented at the end of the nineteenth century and objects made of this metal were still considered to be a novelty when the 2.85-kg aluminum cap was set as the last piece of the Washington Monument in 1884. Aluminum was then considered to be a precious metal.

The energy difference between metals and their ores can be expressed in electrical terms that are in turn related to heats of formation of the compounds. The difficulty of extracting metals from their ores in terms of the energy required, and the consequent tendency to release this energy by corrosion, is reflected by the relative positions of pure metals in a list, which is discussed later as the electromotive series in Chap. 4.

2.2 Matter Building Blocks

Since metals are the principal materials which suffer corrosive deterioration, it is important to have an understanding of their atomic organization in order to fully understand corrosion.

Metals as well as all materials are made up of atoms; metals are also composed, of course, of those smaller particles which make up the atoms. These numerous particles arrange themselves so that those bearing positive charges or those which are neutral cluster together to form a nucleus around which negatively charged particles or electrons rotate in orbits or shells.

Chemical shorthand exists to express these atomic states. For example, Fe is the chemical shorthand for a neutral atom of iron, whereas Fe^{2+} denotes an iron atom that has been stripped of two electrons and is called a ferrous ion or Fe(II). Similarly, Fe^{3+} denotes an iron atom stripped of three electrons and is called a ferric ion or Fe(III). The process of stripping electrons from atoms is referred to by electrochemists as *oxidation*. Note that the term oxidation is not necessarily associated with oxygen.

An opposite process can also occur in which extra electrons are added to the neutral atom giving it a net negative charge. Any increase in negative charge (or decrease in positive charge) of an atom or ion is called *reduction*.

Many chemical compounds, such as salts, are made up of two or more ions of opposite charge. When these are dissolved in water, they can readily split into two or more separate ions which display equal but opposite charges. This process is also called *ionization*. It is these particles that are responsible for the conduction of electric currents in aqueous solutions.

For a non reacted atom, the negative particles exactly balance the positive charges present in the atomic nucleus. The electrons occupy shells in an orderly fashion to balance the positive charge of the nucleus. The electrons in the outermost shell are called valence electrons.

These electrons can participate in chemical reactions and be "stripped" from the atom, therefore drastically changing its properties. Thus, the charge of the nucleus is unbalanced and the atom that displays a positive charge is called an *ion*.

Nearly all metals and alloys exhibit a crystalline structure. The atoms which make up a crystal exist in an orderly three-dimensional array. Figure 2.1 is a schematic representation of the unit cells of the most common crystal structures found in metals and alloys. The unit cell is the smallest portion of the crystal structure which contains all of the geometric characteristics of the crystal.

Most metals fall in these three simple crystal structure categories. For example, V, Fe, Cr, Nb, and Mo have a body-centered cubic structure while Al, Ca, Ni, Cu, and Ag are face-centered cubic crystal systems and Ti, Zn, Co, and Mg are hexagonal close packed. The solubility of one metal into another to create alloys is greatly determined by the respective similarities between the crystal lattice of these metals and by other properties such as the size of the atoms. Noteworthy families of alloys made of iron (Fe, BCC), nickel (Ni, FCC), and chromium (Cr, BCC) are explained and described by their crystal structure as illustrated in Fig. 2.2.

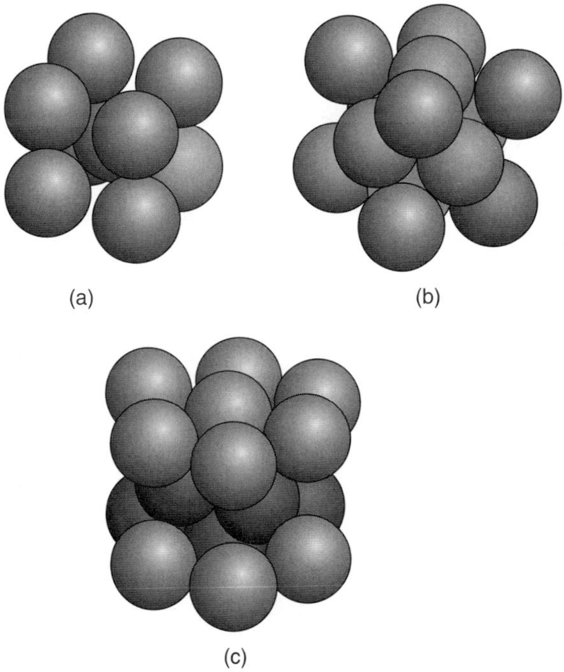

FIGURE 2.1 Schematic representation of the unit cells of the most common crystal structures found in metals: (*a*) body-centered cubic; (*b*) face-centered cubic; (*c*) hexagonal close packed.

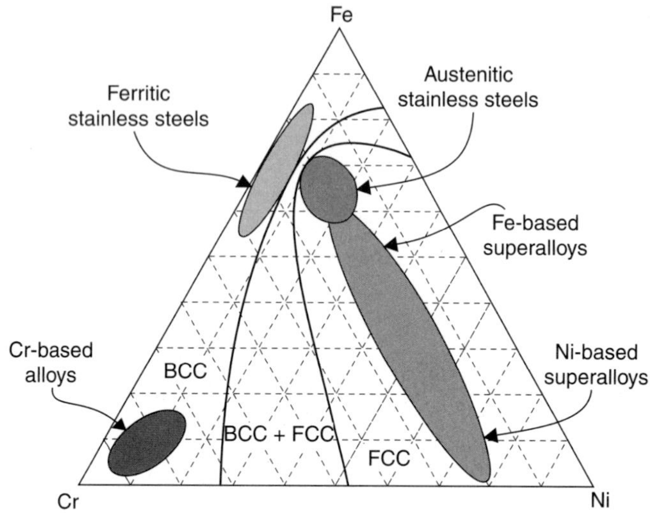

FIGURE 2.2 Cr-Fe-Ni ternary phase diagram showing body-centered and face-centered cubic crystal domains with examples of alloying families.

The crystals, or grains, of a metal are made up of unit cells repeated in a three-dimensional array. However, the crystalline nature of metals is not readily obvious because the metal surface usually conforms to the shape in which it has been cast or formed. In some instances, the crystal structure can be observed naturally. Figure 2.3 is a photograph of the surface of a hot-dip galvanized post with characteristic spangle

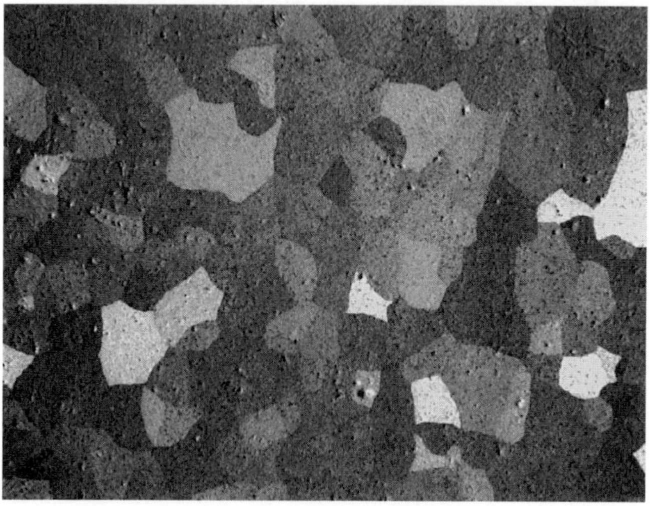

FIGURE 2.3 Hot-dip galvanized steel with spangle patterns that are a form of crystallized grains.

FIGURE 2.4 Brass door knob with evidence of crystal structure etched by corrosive perspiration. (Courtesy of Kingston Technical Software)

patterns that are, in fact, large grains visible without any etching. Brass door knobs are normally bright and shiny, however, after some time the corrosive perspiration from hands etches the crystalline features of the alloy on the surface as shown in this second example (Fig. 2.4).

The procedure to determine the grain size or microstructure of a metal or alloy is to first prepare a sample for microscopic study by grinding and polishing the surface of a specimen mounted in a plastic material. The polished surface is then corroded with a suitable etching reagent, such as those briefly described in Table 2.3, that attacks more readily the grain boundaries of a metallic microstructure to reveal its characteristic features.

The grains shown in Figs. 2.3 and 2.4 are extremely large for most metal crystals. Normally, metal grains are so small they can only be satisfactorily observed with a microscope. The general range of grain size usually runs from 25 to 250 μm in diameter. The American Society for Testing and Materials (ASTM) grain number (G), defined in Eq. (2.1) and illustrated in Fig. 2.5, is a convenient way to describe the size of grains in a material:

$$N = 2^{G-1} \tag{2.1}$$

where N is the number of grains per square inch at a magnification of 100 times.

Uses	Etchant	Comments
Iron & steel	Nital 3% or 5%	Most common etchant used. Reveals ferrite grain boundaries, pearlite, martensite, and bainite. Not as good as picral for high-resolution work with low-temperature transformation products.
Iron & steel	Picral	As above but more sensitive to carbide than other features. Good for revealing carbides in cold rolled and annealed strip. Gives superior resolution with fine pearlite, martensite, and bainite.
Iron & steel	Brauners reagent	Reveals austenite grain boundaries in tempered steels.
Iron & steel	Fry's reagent	Reveals deformed regions and flow lines. Temper at 150 to 200°C prior to etching.
Iron & steel	SASPA	Segregation sensitive etchant. Also good for flow lines.
Iron & steel	LePera's reagent	Good for quantitative assessment of martensite and bainite in dual-phase steels. Martensite appears white, bainite appears black, and ferrite appears tan.
Stainless steels	Aqua regia	Reveals general structure of stainless steels.
Stainless steels	Glycer regia	For austenitic iron-chromium based alloys.
Stainless steels	Acid ferric chloride	General stainless steel etchant.
Stainless steels	Vilella's reagent	Etches Fe-Cr, Fe-Ni, and Fe-Cr-Mn steels. Use hot to reveal austenite grain boundaries.
Stainless steels	Kalling's reagent	Develops dendritic pattern. Darkens ferrite and martensite, austenite light, does not attack carbides.
Copper alloys	Acid ferric chloride	General etch for copper, brasses, and bronzes.

Acid ferric chloride (10 g ferric chloride, 30 mL HCl, 120 mL water)
Aqua regia (3 parts HCl plus 1 part nitric acid)
Brauners reagent (2-naphthylamine 6-sulphonic acid)
Fry's reagent [mixture composed of HCl, copper (II) chloride, ethanol]
Glycer regia (45 mL glycerol, 15 mL nitric acid, 30 mL HCl)
Kalling's reagent (2 g cupric chloride, 40–80 mL methanol, 40 mL water, 40 mL HCl)
LePera's reagent (1% aqueous sodium metabisulfite + 4% picral solution)
Nital (1–10 mL nitric acid, 90–99 mL methanol)
Picral (2–4 g picric acid, 100 mL ethanol)
SASPA (saturated aqueous solution picric acid)
Vilella's reagent (1 g picric acid, 5 mL HCl, 100 mL ethanol)

TABLE 2.3 Chemical Etchants Used to Reveal the Microscopic Features of Some Common Alloys

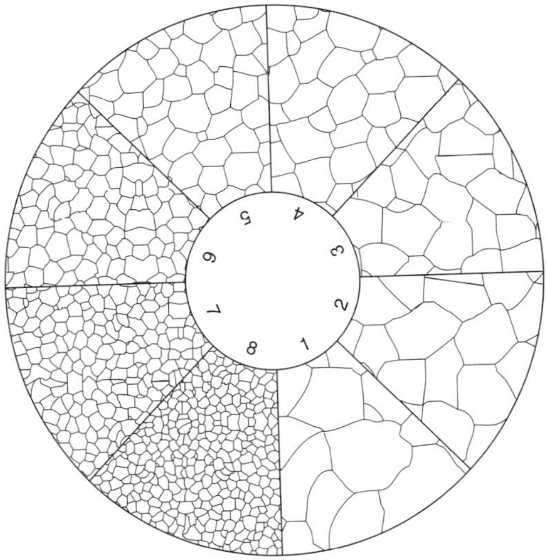

FIGURE 2.5 Visual description of various grain ASTM grain sizes.

Grain size has an important effect on physical properties. For service at ordinary temperatures it is generally considered that fine-grained steels give a better combination of strength and toughness, whereas coarse-grained steels have better machinability. The Hall-Petch relation in Eq. (2.2) describes the relation between the grain size and the yield strength of a material. This relation indicates that the smaller the grain size of a metal, the stronger it is and the higher its yield strength.

$$\sigma_y = \sigma_0 + \frac{k_y}{\sqrt{d}} \tag{2.2}$$

where k_y is a constant specific to a material, σ_0 is another constant related to the starting stress for dislocation movement, d is the grain diameter, and σ_y is the yield strength.

In addition to the influence of impurities, inclusions, and cold work, grain boundaries and differences in grain orientation may also result in significantly different electrochemical reactivity in many metals and alloys. When an alloy is cooled somewhat rapidly, the grain boundaries may end up having a composition significantly different than the grain core. This variation in chemical composition from the interior to the exterior of a grain is referred to as microsegregation or coring. The dendrites quite visible in Fig. 2.6(*a*) are typical to the microsegregation associated with nonequilibrium solidification. The same alloy, that is, 70-Ni 30-Cu (N04400) (also known as Monel 400), would normally have in its hot rolled form the crystalline grain structure revealed in Fig. 2.6(*b*).

(a)
(b)

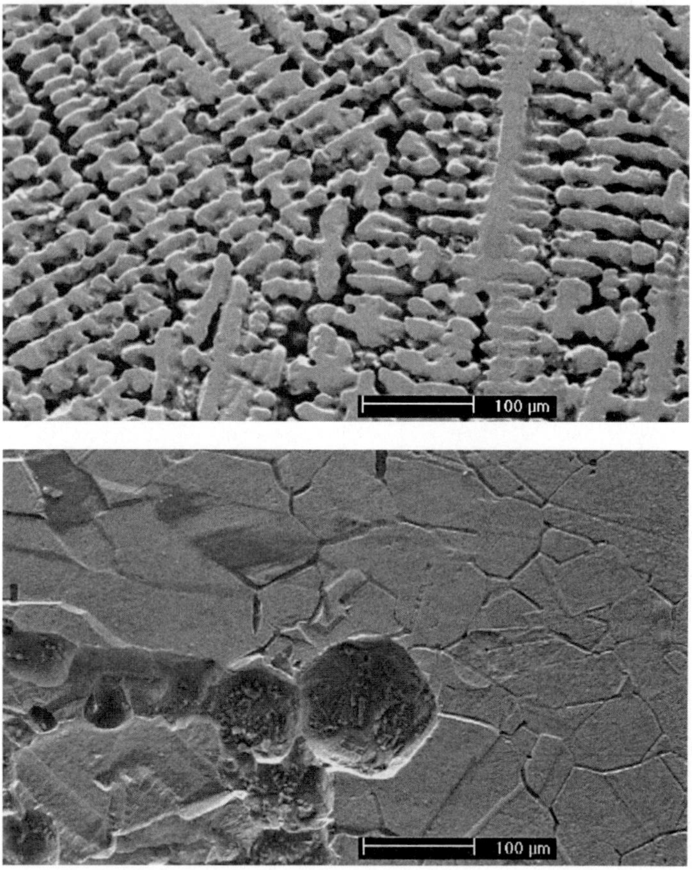

Figure 2.6 Surface corrosion of the same 70-Ni 30-Cu alloy (N04400) (*a*) when cast and left to cool and (*b*) after hot-rolling.

Grain boundary effects are of little or no consequence in most applications or uses of metals. If a metal corrodes, uniform attack results since grain boundaries are usually only slightly more reactive than the matrix. However, under certain conditions, grain interfaces are very reactive and intergranular corrosion results (see Chap. 6 for more details).

2.3 Acidity and Alkalinity (pH)

When discussing the ionic content of an aqueous medium, the question often arises as to how acid, or alkaline, is the solution. Quite simply, this refers to whether there is an excess of H^+ (hydrogen) or OH^- (hydroxyl) ions present. The H^+ ion is acid while the OH^- ion is alkaline or basic. The other ionic portion of an acid or alkali added to water can increase its conductivity or change other properties of the liquid, but does not increase or decrease its acidity. For instance, whether a given amount of H^+ ion is produced in water by introducing hydrochloric (HCl), sulfuric (H_2SO_4), or any other acid is immaterial. The pH of the solution will be the same for the same number of dissolved hydrogen atoms.

The pH may be measured with a meter or calculated if certain parameters are established. Water itself dissociates to a small extent to produce equal quantities of H^+ and OH^- ions displayed in the following equilibrium:

$$H_2O \rightleftarrows H^+ + OH^- \tag{2.3}$$

pH, originally defined by Danish biochemist Søren Peter Lauritz Sørensen in 1909, is a measure of the concentration of hydrogen ions. The term pH was derived from the manner in which the hydrogen ion concentration is calculated; it is the negative logarithm of the hydrogen ion (H^+) concentration:

$$pH = -\log_{10}(a_{H^+}) \tag{2.4}$$

where $\log_{10}$ is a base-10 logarithm and a_{H+} is the activity (related to concentration) of hydrogen ions. The "p" in Eq. (2.4) stands for the German word for power, *potenz*, so pH is an abbreviation for *power of hydrogen*.

A higher pH means there are fewer free hydrogen ions, and that a change of one pH unit reflects a tenfold change in the concentrations of the hydrogen ion. For example, there are 10 times as many hydrogen ions available at pH 7 than at pH 8. The pH scale commonly quoted ranges from 0 to 14 with a pH of 7 considered to be neutral.

Substances with a pH less that 7 are considered to be acidic, and substances with a pH equal to or greater than 7 are considered to be basic or alkaline. Thus, a pH of 2 is very acidic and a pH of 12 very alkaline. However, it is technically possible to have very acidic solutions with a pH lower than zero and concentrated caustic solutions with a pH greater than 14. Such solutions are in fact typical of many ore extracting processes that require the digestive power of caustics and acids.

Low-pH acid waters accelerate corrosion by supplying hydrogen ions to the corrosion process. Although even absolutely pure water contains some free hydrogen ions, dissolved carbon dioxide (CO_2) in the water can increase the hydrogen ion concentration. Dissolved CO_2 may react with water to form carbonic acid (H_2CO_3), as shown in Eq. (2.5):

$$CO_2 + H_2O \rightleftarrows H_2CO_3 \text{ with } K_{eq} = \frac{[H_2CO_3]}{[CO_2]} = 1.7 \times 10^{-3} \text{ at } 25°C \tag{2.5}$$

where K_{eq} is the reaction equilibrium expressed in equation.

As illustrated in the predominance diagram shown in Fig. 2.7, carbonic acid subsequently dissociates in bicarbonate and carbonate ions as expressed respectively in Eqs. (2.6) and (2.7):

$$H_2CO_3 \rightleftarrows HCO_3^- + H^+,\ pK_a = 3.6 \tag{2.6}$$

$$HCO_3^- \rightleftarrows CO_3^{2-} + H^+,\ pK_a = 10.3 \tag{2.7}$$

Care must be taken when quoting and using the dissociation constant in Eq. (2.6). This equilibrium value is correct for the H_2CO_3 molecule, and shows that it is a stronger acid than acetic acid or formic acid as might be expected from the influence of the electronegative oxygen substituent. However, carbonic acid only exists in solution in equilibrium with carbon dioxide, and so the concentration of H_2CO_3 is much lower than the concentration of CO_2, reducing the measured acidity. The equation may be rewritten as follows:

$$CO_2 + H_2O \rightleftarrows HCO_3^- + H^+, K_a = 4.3 \times 10^{-7} \text{ or } pK_a = 6.36 \tag{2.8}$$

Even more acidity is sometimes encountered in mine waters and in water contaminated by industrial wastes. Many salts added to an aqueous system also have a direct effect on the pH of that mixture through the following process of hydrolysis, shown here for the addition of ferric ions to water:

$$Fe^{3+} + 3H_2O \rightleftarrows Fe(OH)_3 + 3H^+ \tag{2.9}$$

In this particular example the equilibrium is established between ferric ions, water, ferric hydroxide or $Fe(OH)_3$, and the acidity of the water. This particular example is quite useful to explain the severity of a situation that can develop in confined areas such as pitting and crevices.

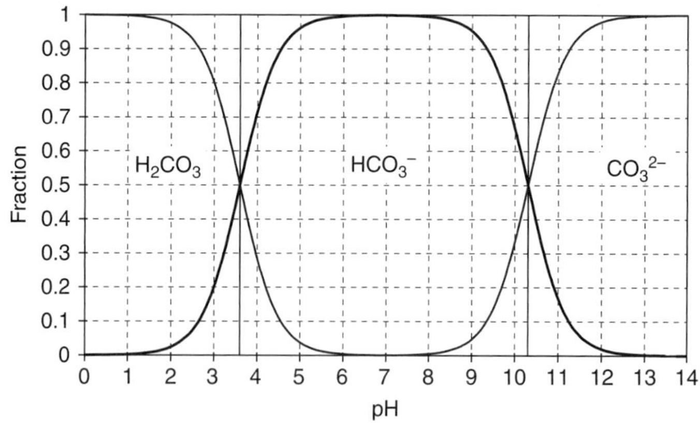

FIGURE 2.7 pH predominance diagram of the chemical species associated with the dissolution of carbon dioxide in water.

2.4 Corrosion as a Chemical Reaction

2.4.1 Corrosion in Acids

One of the common ways of generating hydrogen in a laboratory is to place zinc into a dilute acid, such as hydrochloric or sulfuric. When this is done, there is a rapid reaction in which the zinc is attacked or "dissolved" and hydrogen is evolved as a gas (Fig. 2.8). This is shown in Eqs. (2.10) to (2.14):

$$Zn + 2HCl \rightarrow ZnCl_2 + H_2^0(g) \tag{2.10}$$

$$Zn + 2H^+ + 2Cl^- \rightarrow Zn^{2+} + 2Cl^- + H_2^0(g) \tag{2.11}$$

Equations (2.10) and (2.11) are the chemical shorthand for the following statement: One zinc atom plus two hydrochloric acid molecules dissociate as ions H^+ and Cl^- and become one molecule of zinc chloride in Eq. (2.10) [written as a soluble salt in the form of Zn^{2+} and Cl^- ions in Eq. (2.11)] plus one molecule of hydrogen gas which is given off as indicated by the vertical arrow. It should be noted that the chloride ions do not participate directly in this reaction, although they could play an important role in real corrosion situations.

Similarly, zinc combines with sulfuric acid to form zinc sulfate (a salt) and hydrogen gas as shown in Eqs. (2.12) and (2.13):

$$Zn + H_2SO_4 \rightarrow ZnSO_4 + H_2^0(g) \tag{2.12}$$

$$Zn + 2H^+ + SO_4^{2-} \rightarrow Zn^{2+} + SO_4^{2-} + H_2^0(g) \tag{2.13}$$

FIGURE 2.8 Bubbling, or "plating out of hydrogen" on zinc immersed in a 0.1 M sulfuric acid solution. (Courtesy Kingston Technical Software)

Note that each atom of a substance that appears on the left-hand side of these equations must also appear on the right-hand side. There are also some rules that denote in what proportion different atoms combine with each other. As in the preceding reaction, the sulfate ions that are an integral part of sulfuric acid do not participate directly to the corrosion attack and therefore one could write both Eqs. (2.11) and (2.13) in a simpler form:

$$Zn + 2H^+ \rightarrow Zn^{2+} + H_2^0(g) \tag{2.14}$$

Many other metals are also corroded by acids, often yielding soluble salts and hydrogen gas, as shown in Eqs. (2.15) and (2.16) for iron and aluminum, respectively:

$$Fe + 2H^+ \rightarrow Fe^{2+} + H_2^0(g) \tag{2.15}$$

$$2Al + 6H^+ \rightarrow 2Al^{3+} + 3H_2^0(g) \tag{2.16}$$

Note that zinc and iron react with two H^+ ions, whereas aluminum reacts with three. This is due to the fact that both zinc and iron, when corroding, lose two electrons and display two positive charges in their ionic form. They are said to have a valence of 2, whereas aluminum loses three electrons when leaving an anodic surface and hence displays three positive charges and is said to have a valence of 3. Some metals have several common valences, others only one. Figure 2.9 shows some of the oxidation states found in compounds of the transition-metal elements.

2.4.2 Corrosion in Neutral and Alkaline Solutions

The corrosion of metals can also occur in fresh water, seawater, salt solutions, and alkaline or basic media. In almost all of these environments, corrosion occurs importantly only if dissolved oxygen

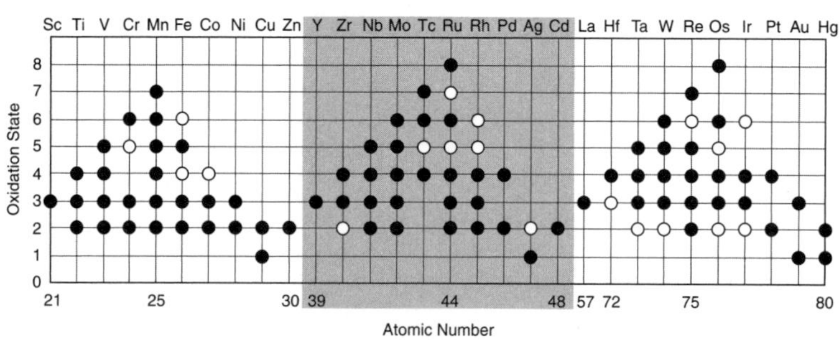

FIGURE 2.9 Oxidation states found in compounds of the transition-metal elements. A solid circle represents a common oxidation state, and a ring represents a less common (less energetically favorable) oxidation state.

is also present. Water solutions rapidly dissolve oxygen from the air, and this is the source of the oxygen required in the corrosion process. The most familiar corrosion of this type is the rusting of iron when exposed to a moist atmosphere.

$$4Fe + 6H_2O + 3O_2 \rightarrow 4Fe(OH)_3 \downarrow \tag{2.17}$$

In Eq. (2.17), iron combines with water and oxygen to produce an insoluble reddish-brown corrosion product that falls out of the solution, as shown by the downward pointing arrow.

During rusting in the atmosphere, there is an opportunity for drying, and this ferric hydroxide dehydrates and forms the familiar red-brown ferric oxide (rust) or Fe_2O_3, as shown in Eq. (2.18):

$$2Fe(OH)_3 \rightarrow Fe_2O_3 + 3H_2O \tag{2.18}$$

Similar reactions occur when zinc is exposed to water or moist air followed by natural drying:

$$2Zn + 2H_2O + O_2 \rightarrow Zn(OH)_2(s) \tag{2.19}$$

$$Zn(OH)_2 \rightarrow ZnO + H_2O \tag{2.20}$$

The resulting zinc oxide is the whitish deposit seen on galvanized pails, rain gutters, and imperfectly chrome-plated bathroom faucets.

As discussed previously, the iron that took part in the reaction with hydrochloric acid in Eq. (2.15) had a valence of 2, whereas the iron that takes part in the reaction shown in Eq. (2.17) has a valence of 3. The clue to this lies in the examination of the equation for the corrosion product $Fe(OH)_3$. Note that water ionized into H^+ and OH^-. It is further known that a hydrogen ion has a valence of 1 (it has only one electron to lose). It would require three hydrogen ions with the corresponding three positive charges to combine with the three OH^- ions held by the iron. It can thus be concluded that the iron ion must have been Fe^{3+} or a ferric ion.

Also note that there is no oxidation or reduction (electron transfer) during the reaction in either Eq. (2.18) or (2.20). In both cases the valences of the elements on the left of each reaction remain what it is on the right. The valences of iron, zinc, hydrogen, and oxygen elements remain unchanged throughout the course of these reactions, and it is consequently not possible to divide these reactions into individual oxidation and reduction reactions.

Reference

1. Mapes RS, Berkey WW. X-ray diffraction methods for the analysis of corrosion products. In: Ailor WH, ed. *Handbook on Corrosion Testing and Evaluation*. New York: John Wiley & Sons, 1971; 697–730.

CHAPTER 3
Corrosion Electrochemistry

3.1 Electrochemical Reactions

An electrochemical reaction is defined as a chemical reaction involving the transfer of electrons. It is also a chemical reaction which involves oxidation and reduction. Since metallic corrosion is almost always an electrochemical process, it is important to understand the basic nature of electrochemical reactions. The discoveries that gradually evolved in modern corrosion science have, in fact, played an important role in the development of a multitude of technologies we are enjoying today. Appendix A provides a list of some of these discoveries.

An important achievement early in the history of electrochemistry was the production of power sources, following the production of the first batteries by Alessandro Volta. Figure 3.1 illustrates the principle of a Daniell cell in which copper and zinc metals are immersed in solutions of their respective sulfates. The Daniell cell was the first truly practical and reliable electric battery that supported many nineteenth-century electrical innovations such as the telegraph. In the process of the reaction, electrons can be transferred from the corroding zinc to the copper through an electrically conducting path as a useful electric current. Zinc more readily loses electrons than copper, so placing zinc and copper metal in solutions of their salts can cause electrons to flow through an external wire which leads from the zinc to the copper.

$$\text{Zinc anode: } Zn(s) \rightarrow Zn^{2+} + 2e^- \tag{3.1}$$

$$\text{Copper cathode: } Cu^{2+} + 2e^- \rightarrow Cu(s) \tag{3.2}$$

The difference in the susceptibility of two metals to corrode can often cause a situation that is called galvanic corrosion named after Luigi Galvani, the discoverer of the effect. The purpose of the separator shown in Fig. 3.1 is to keep each metal in contact with its

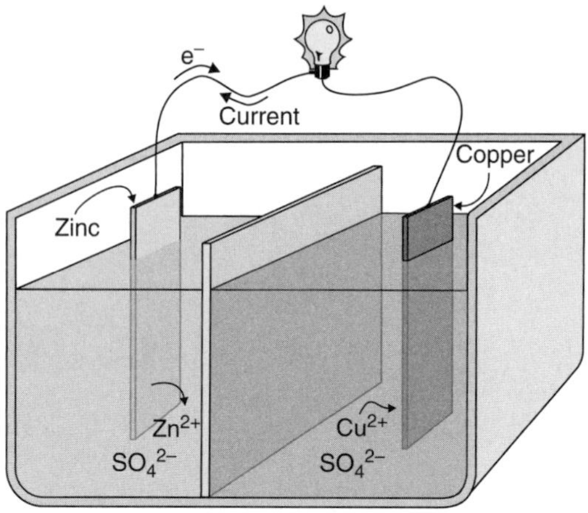

FIGURE 3.1 Schematic of a Daniell cell.

own soluble sulfates, a technical point that is critical in order to keep the voltage of a Daniell cell relatively constant [1]. The same goal can be achieved by using a salt bridge between two different beakers as shown in Fig. 3.2. The salt bridge, in that case, provides the electrolytic path that is necessary to complete an electrochemical cell circuit. This situation is common in natural corrosion cells where the environment serves as the electrolyte that completes the corrosion cell. The conductivity of an aqueous environment such as soils, concrete, or natural waters has often been related to its corrosivity.

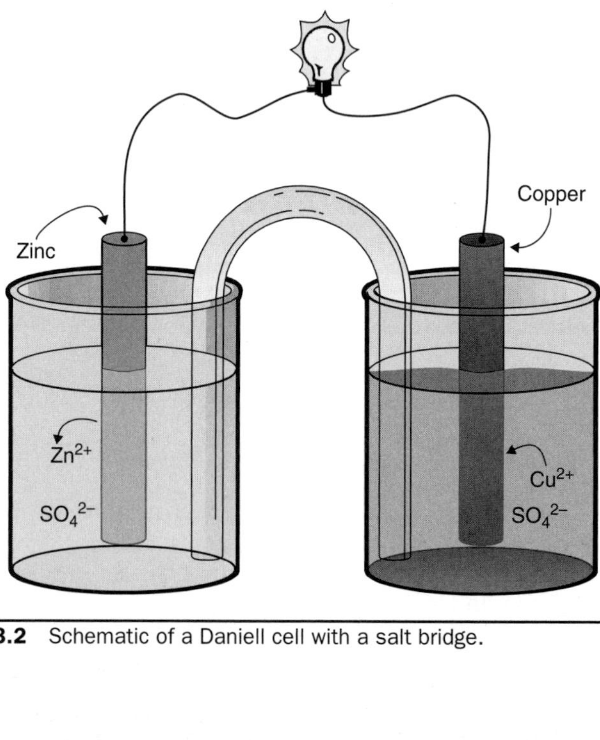

FIGURE 3.2 Schematic of a Daniell cell with a salt bridge.

The short-hand description in Eq. (3.3) is valid for both cells shown in Figs. 3.1 and 3.2. Such a description is often used to simplify textual reference to such cells.

$$(-)\mathrm{Zn}/\mathrm{Zn}^{2+}, \mathrm{SO}_{4\ (\mathrm{Conc}_1)}^{2-}//\mathrm{Cu}^{2+}, \mathrm{SO}_{4\ (\mathrm{Conc}_2)}^{2-}/\mathrm{Cu}(+) \quad (3.3)$$

$Conc_1$ and $Conc_2$ in Eq. (3.3) indicate respectively the concentration of zinc sulfate and copper sulfate that may differ in the two half-cells while the two slanted bars (//) describe the presence of a separator. The same short-hand description also identifies the zinc electrode as the anode that is negative in the case of a spontaneous reaction and the copper cathode as positive.

The fact that corrosion consists of at least one oxidation and one reduction reaction is not always as obvious as it is in chemical power cells and batteries. The two reactions are often combined on a single piece of metal as it is illustrated schematically in Fig. 3.3.

In Fig. 3.3, a piece of zinc immersed in hydrochloric acid solution is undergoing corrosion. At some point on the surface, zinc is transformed to zinc ions, according to Eq. (3.4). This reaction produces electrons and these pass through the solid conducting metal to other

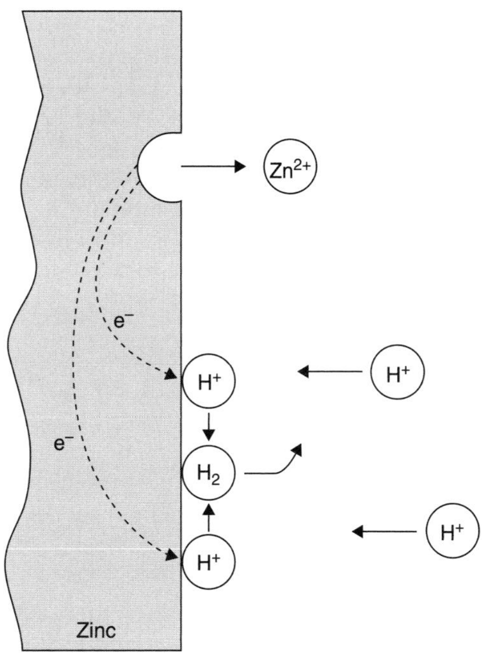

FIGURE 3.3 Electrochemical reactions occurring during the corrosion of zinc in air-free hydrochloric acid.

sites on the metal surface where hydrogen ions are reduced to hydrogen gas according to Eq. (3.5) (Fig. 2.7 in Chap. 2).

$$\text{Anodic reaction: Zn(s)} \rightarrow \text{Zn}^{2+} + 2\text{e}^- \tag{3.4}$$

$$\text{Cathodic reaction: 2H}^+ + 2\text{e}^- \rightarrow \text{H}_2\text{(g)} \tag{3.5}$$

Equations (3.4) and (3.5) illustrate the nature of an electrochemical reaction typically illustrated for zinc in Fig. 3.3. During such a reaction, electrons are transferred, or, viewing it another way, an oxidation process occurs together with a reduction process. The overall corrosion processes are summarized in Eq. (3.6):

$$\text{Overall corrosion reaction: Zn} + 2\text{H}^+ \rightarrow \text{Zn}^{2+} + \text{H}_2\text{(g)} \tag{3.6}$$

Briefly then, for corrosion to occur there must be a formation of ions and release of electrons at an anodic surface where oxidation or deterioration of the metal occurs. There must be a simultaneous reaction at the cathodic surface to consume the electrons generated at the anode. These electrons can serve to neutralize positive ions such as the hydrogen ions (H^+), or create negative ions. The anodic and cathodic reactions must go on at the same time and at equivalent rates. However, what is usually recognized as the corrosion process occurs only at the areas that serve as anodes.

3.2 Anodic Processes

Let us consider in greater detail what takes place at the anode when corrosion occurs. For instance, reconsider Eq. (3.6). This reaction involves the reduction of hydrogen ions to hydrogen gas, according to Eq. (3.5). This hydrogen evolution reaction occurs with a wide variety of metals and acids, including hydrochloric, sulfuric, perchloric, hydrofluoric, formic, and other strong acids. The individual anodic reactions for iron, nickel, and aluminum are listed as follows:

$$\text{Iron anodic reaction: Fe(s)} \rightarrow \text{Fe}^{2+} + 2\text{e}^- \tag{3.7}$$

$$\text{Nickel anodic reaction: Ni(s)} \rightarrow \text{Ni}^{2+} + 2\text{e}^- \tag{3.8}$$

$$\text{Aluminum anodic reaction: Al(s)} \rightarrow \text{Al}^{3+} + 3\text{e}^- \tag{3.9}$$

Examining the Eqs. (3.7) to (3.9) shows that the anodic reaction occurring during corrosion can be written in the general form:

$$\text{General anodic reaction: M(s)} \rightarrow \text{M}^{n+} + n\text{e}^- \tag{3.10}$$

That is, the corrosion of metal *M* results in the oxidation of metal *M* to an ion with a valence charge of n+ and the release of n electrons. The value of n, of course, depends primarily on the nature of the metal.

Some metals such as silver are univalent, while other metals such as iron, titanium, and uranium are multivalent and possess positive charges as high as 6 (see Fig. 2.8). Equation (3.10) is general and applies to all corrosion reactions.

3.3 Faraday's Law

If the current generated by one of the anodic reactions expressed earlier was known, it would be possible to convert this current to an equivalent mass loss or corrosion penetration rate with a very useful relation discovered by Michael Faraday, a nineteenth century pioneer in electrochemistry. Faraday's empirical laws of electrolysis relate the current of an electrochemical reaction to the number of moles of the element being reacted and the number of moles of electrons involved. Supposing that the charge required for such reaction was one electron per molecule, as is the case for the plating or the corrosion attack of silver described respectively in Eqs. (3.11) and (3.12):

$$Ag^+ + e^- \rightarrow Ag(s) \tag{3.11}$$

$$Ag(s) \rightarrow Ag^+ + e^- \tag{3.12}$$

According to Faraday's law, the reaction with 1 mol of silver would require 1 mol of electrons, or 1 Avogadro's number of electrons (6.022×10^{23}). The charge carried by 1 mol of electrons is known as 1 faraday (F). The faraday is related to other electrical units through the electronic charge; the electronic charge is 1.6×10^{-19} coulomb (C). Multiplying the electronic charge by the Avogadro number means that 1 F equals 96,485 C/(mol of electrons). Combining Faraday's principles with specific electrochemical reactions of known stoichiometry leads to Eq. (3.13) that relates the charge Q to chemical descriptors N and n:

$$Q = F \cdot \Delta N \cdot n \tag{3.13}$$

where N is the number of moles and ΔN the change in that amount
n is the number of electrons per molecule of the species being reacted

The charge Q can be defined in terms of electric current as in Eq. (3.14),

$$Q = \int_0^t I \cdot dt \tag{3.14}$$

where I is the total current in amperes (A)
t is the duration of the electrochemical process in seconds (s)

	mA cm^{-2}	mm y^{-1}	mpy	g m^{-2} day^{-1}
mA cm^{-2}	1	3.28 *M/nd*	129 *M/nd*	8.95 *M/n*
mm y^{-1}	0.306 *nd/M*	1	39.4	2.74 *d*
mpy	0.00777 *nd/M*	0.0254	1	0.0694 *d*
g m^{-2} day^{-1}	0.112 *n/M*	0.365 */d*	14.4 */d*	1

where mpy = milli-inch per year
n = number of electrons freed by the corrosion reaction
M = atomic mass
d = density

*Note: the table should be read from left to right, i.e.,
1 mA cm^{-2} = (3.28 *M/nd*) mm y^{-1} = (129 *M/nd*) mpy = (8.95 *M/n*) g m^{-2} day^{-1}

TABLE 3.1 Conversion between Current, Mass Loss, and Penetration Rates for all Metals*

	mA cm^{-2}	mm y^{-1}	mpy	g m^{-2} day^{-1}
mA cm^{-2}	1	11.6	456	249
mm y^{-1}	0.0863	1	39.4	21.6
mpy	0.00219	0.0254	1	0.547
g m^{-2} day^{-1}	0.00401	0.0463	1.83	1

*Note: the table should be read from left to right, i.e.,
1 mA cm^{-2} = 11.6 mm y^{-1} = 456 mpy = 249 g m^{-2} day^{-1}

TABLE 3.2 Conversion between Current, Mass Loss and Penetration Rates for Steel*

The corrosion current itself can be either estimated by using specialized electrochemical methods or by using weight-loss data and a conversion chart (Table 3.1) based on Faraday's principle. Table 3.1 provides the conversion factors between commonly used corrosion rate units for all metals and Table 3.2 describes these conversion factors adapted to iron or steel (Fe) for which $n = 2$, $M = 55.85$ g/mol and $d = 7.88$ g cm^{-3}.

3.4 Cathodic Processes

When hydrogen ions are reduced to their atomic form they often combine, as shown earlier, to produce hydrogen gas through reaction with electrons at a cathodic surface. This reduction of hydrogen ions at a cathodic surface will disturb the balance between the acidic hydrogen (H^+) ions and the alkaline hydroxyl (OH^-) ions and make the solution less acidic or more alkaline or basic at the corroding interface.

In neutral waters the anodic corrosion of some metals like aluminum, zinc, or magnesium develops enough energy to split water directly as illustrated in Fig. 3.4 and Eq. (3.15).

$$\text{Water splitting cathodic reaction: } 2H_2O(l) + 2e^- \rightarrow H_2 + 2OH^- \quad (3.15)$$

The change in the concentration of hydrogen ions or increase in hydroxyl ions can be shown by the use of pH indicators, which change color and thus can serve to demonstrate and locate the existence of surfaces on which the cathodic reactions in corrosion are taking place. There are several other cathodic reactions encountered during the corrosion of metals. These are listed below:

Oxygen reduction:

$$\text{(acid solutions) } O_2 + 4H^+ + 4e^- \rightarrow 2H_2O \quad (3.16)$$

$$\text{(neutral or basic solutions) } O_2 + 2H_2O + 4e^- \rightarrow 4OH^- \quad (3.17)$$

$$\text{Hydrogen evolution: } 2H^+ + 2e^- \rightarrow H_2(g) \quad (3.5)$$

$$\text{Metal ion reduction: } Fe^{3+} + e^- \rightarrow Fe^{2+} \quad (3.18)$$

$$\text{Metal deposition: } Cu^{2+} + 2e^- \rightarrow Cu(s) \quad (3.19)$$

Hydrogen ion reduction, or hydrogen evolution, has already been discussed. This is the cathodic reaction that occurs during

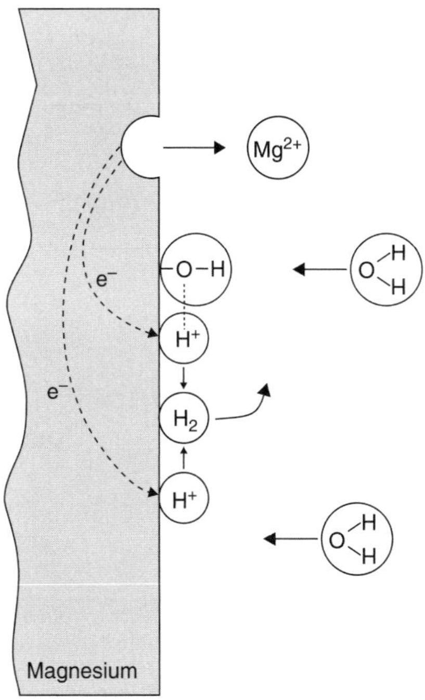

Figure 3.4 Electrochemical reactions occurring during the corrosion of magnesium in neutral water.

corrosion in acids. Oxygen reduction [Eqs. (3.16) and (3.17)] is a very common cathodic reaction, since oxygen is present in the atmosphere and in solutions exposed to the atmosphere. Although less common, metal ion reduction and metal deposition, can cause severe corrosion problems in special situations. One particular case worth mentioning here is the plating of copper ions, produced upstream in a water circuit, on the internal aluminum surface of a radiator, for example.

The plated nodules, which may form even at very low concentrations of copper ions, tend to be dispersed and are thus a good catalyst for the subsequent reduction of dissolved oxygen. It is therefore highly recommended to avoid using copper tubing in a water circuit where aluminum is also present. Such deposition corrosion can be avoided by preventing the pick-up of cathodic ions that will enter the equipment, or by scavenging them by passing the contaminated product through a tower packed with more anodic metal turnings (such as aluminum) on which the ions can deposit (Fig. 3.5).

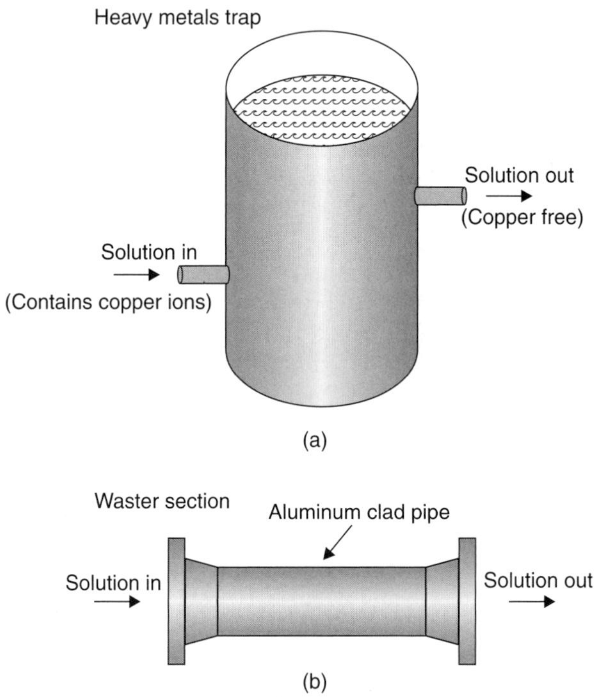

FIGURE 3.5 Method for removing troublesome ions from solution. (*a*) Heavy metal trap: Solutions containing copper ions enter barrel filled with aluminum shavings; (*b*) Waster section: Aluminum-clad pipe inserted in a system removes heavy metal ions. The section is replaced once corroded.

Note that all of the above reactions are similar in one respect—they consume electrons. All corrosion reactions are simply combinations of one or more of the above cathodic reactions, together with an anodic reaction similar to Eq. (3.10). Thus, almost every case of aqueous corrosion can be reduced to these equations, either singly or in combination.

Consider the corrosion of zinc by water or moist air. By multiplying the zinc oxidation reaction (3.4) by 2 and summing this with the oxygen reduction reaction, one obtains Eq. (3.21).

$$2Zn(s) \rightarrow 2Zn^{2+} + 4e^- \text{ (oxidation)} \tag{3.20}$$

$$+\, O_2 + 2H_2O + 4e^- \rightarrow 4OH^- \text{ (reduction)} \tag{3.17}$$

$$2Zn + 2H_2O + O_2 \rightarrow 2Zn^{2+} + 4OH^- \rightarrow 2Zn(OH)_2(p) \tag{3.21}$$

The products of this reaction are Zn^{2+} and OH^-, which immediately react to form insoluble $Zn(OH)_2$. Likewise, the corrosion of zinc by a solution containing copper ions is merely the summation of the oxidation reaction for zinc and the metal deposition reaction involving cupric ions in Eq. (3.19).

$$Zn(s) \rightarrow Zn^{2+} + 2e^- \text{ (oxidation)} \tag{3.4}$$

$$+\, Cu^{2+} + 2e^- \rightarrow Cu(s) \text{ (reduction)} \tag{3.19}$$

$$Zn + Cu^{2+} \rightarrow Zn^{2+} + Cu(p) \tag{3.22}$$

During corrosion, more than one oxidation and one reduction reaction may occur. For example, during the corrosion of an alloy, its component metal atoms go into solution as their respective ions. Thus, during the corrosion of a chromium-iron alloy, both chromium and iron are oxidized. Also, more than one cathodic reaction can occur on the surface of a metal.

Consider the corrosion of zinc in a hydrochloric acid solution containing dissolved oxygen. Two cathodic reactions are possible: the evolution of hydrogen and the reduction of oxygen (Fig. 3.6). Since there are two cathodic reactions or processes which consume electrons, the overall corrosion rate of zinc is increased. Thus, acid solutions which either contain dissolved oxygen or are exposed to air are generally more corrosive than air-free acids. Therefore, removing oxygen from acid solutions will often make these solutions less corrosive.

If a piece of mild steel is placed in a solution of hydrochloric acid, a vigorous formation of hydrogen bubbles is observed. Under such conditions, the metal corrodes very quickly. The dissolution of the metal occurs only at anodic surfaces. The hydrogen bubbles form only at the cathodic surfaces, even though it may appear they

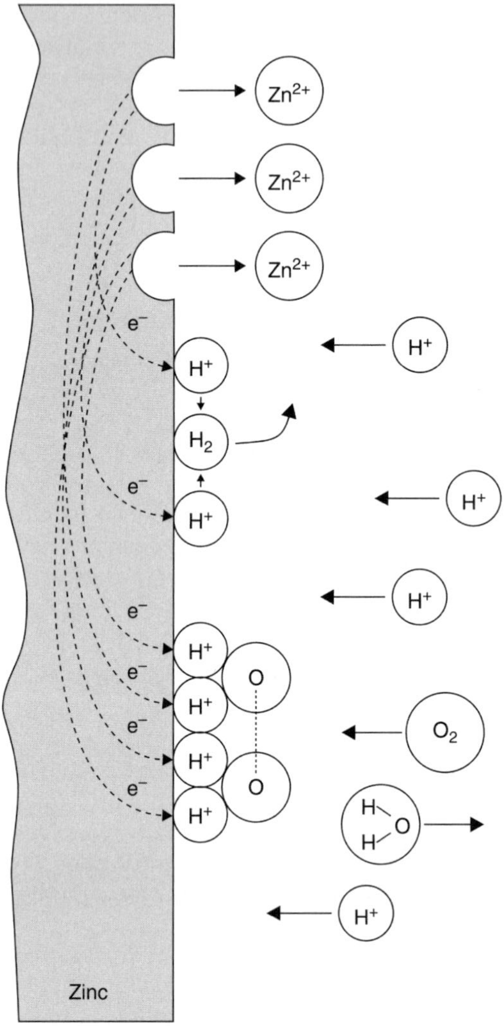

FIGURE 3.6 Electrochemical reactions occurring during the corrosion of zinc in aerated hydrochloric acid.

come from the entire surface of the metal rather than at well-defined cathodic areas (Fig. 3.7). The anodic and cathodic areas may shift from time to time so as to give the appearance of uniform corrosion.

If this action could be seen through a suitable microscope, many tiny anodic and cathodic areas would be observed shifting around on the surface of the metal. These areas, however, are often so small as to be invisible and so numerous as to be almost inseparable.

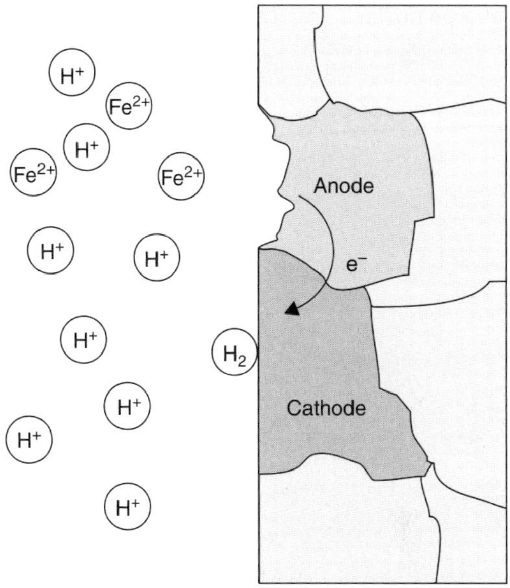

FIGURE 3.7 Formation of ions at an anodic area and release of hydrogen at a cathodic area in a local cell on an iron surface.

3.5 Surface Area Effect

When a piece of metal is freely corroding, the electrons generated at anodic areas flow through the metal to react at cathodic areas similarly exposed to the environment where they restore the electrical balance of the system. The fact that there is no net accumulation of charges on a corroding surface is quite important for understanding most corrosion processes and ways to mitigate them. However, the absolute equality between the anodic and cathodic currents expressed in Eq. (3.23) does not mean that the current densities for these currents are equal.

$$I_{\text{anodic}} = I_{\text{cathodic}} \tag{3.23}$$

When Eq. (3.23) is expressed in terms of current densities in Eq. (3.24) by considering the relative anodic (S_a) and cathodic (S_c) surface areas and their associated current densities i_a and i_c expressed in units of mA cm^{-2}, for example, it becomes clear that a difference in the surface areas occupied by each reaction will have to be compensated by inequalities in the current densities as expressed in Eq. (3.25).

$$I_{\text{anodic}} = i_a \times S_a = I_{\text{cathodic}} = i_c \times S_c \tag{3.24}$$

$$i_a = i_c \frac{S_c}{S_a} \tag{3.25}$$

The implications of the surface area ratio S_c/S_a in Eq. (3.25) are particularly important in association with various forms of local cell corrosion such as pitting and stress corrosion cracking for which a large surface area ratio is a serious aggravating factor. It is easy to understand that the effect of a certain amount of anodic current concentrated on a small area of metal surface will be much greater than when the effect of the same amount of current is dissipated over

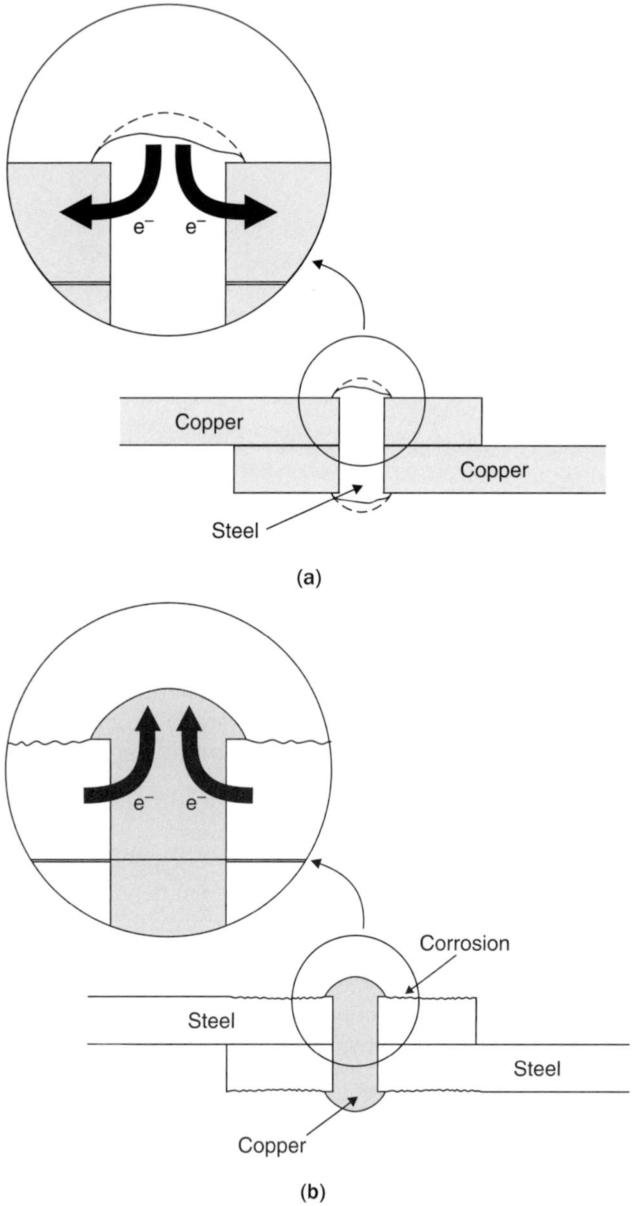

FIGURE 3.8 Galvanic coupling caused by riveting with dissimilar metals: (*a*) steel rivets on copper plates, (*b*) copper rivets on steel plates.

a much larger area. This factor is expressed in Eq. (3.25) which states that the ratio of cathodic to anodic surfaces is an important amplifying factor of the anodic current when S_c/S_a is >> 1 and a stifling factor when it is << 1.

This area effect in terms of current density is illustrated by combinations of steel and copper as either plates or the fasteners used to join them and immersed in a corrosive solution. If steel rivets are used to join copper plates, the current density on the relatively large cathodic copper plates will be low, cathodic polarization of the copper will be slight, and the voltage of the galvanic couple will maintain a value close to the open circuit potential. At the same time, the current density on the small anodic steel rivets will be high and the consequent corrosion quite severe, giving rise to a particularly vicious form of corrosion called galvanic corrosion [Fig. 3.8*(a)*].

With the opposite arrangement of copper rivets joining steel plates, the current density on the copper cathodes will be high, with consequently considerable cathodic polarization of the copper reducing the open circuit potential below its initial value. The diminished anodic current will be spread over the relatively large steel plates and the undesirable galvanic effect will hardly be noticeable [Fig. 3.8*(b)*].

Open circuit potential measurements are grossly inadequate for predicting the magnitude of galvanic effects since they do not take into account area and polarization effects. They are reliable only for predicting the direction of such effects.

Reference

1. Roberge PR. *Corrosion Basics—An Introduction.* 2nd ed. Houston, Tex: NACE International, 2006.

CHAPTER 4

Corrosion Thermodynamics

4.1 Free Energy

In electrical and electrochemical processes, electrical work is defined as the product of charges moved (Q) times the potential (E) through which it is moved. If this work is done in an electrochemical cell in which the potential difference between its two half-cells is E, and the charge is that of 1 mol of reactant in which n mol of electrons are transferred, then the electrical work (w) done by the cell must be $-nE$. In this relationship, the Faraday constant F* is required to convert coulombs from moles of electrons. In an electrochemical cell at equilibrium, no current flows and the energy change occurring in a reaction is expressed in Eq. (4.1).

$$w = \Delta G = -nFE \tag{4.1}$$

Under standard condition, the standard free energy of the cell reaction ΔG^0 is directly related to the standard potential difference across the cell, E^0:

$$\Delta G^0 = -nFE^0 \tag{4.2}$$

For solids, liquid compounds, or elements, standard condition is the pure compound or element; for gases it is 100 kPa pressure; and for solutes it is the ideal 1 M (mol/L) concentration.

Electrode potentials can be combined algebraically to give cell potential. For a galvanic cell, such as the Daniell cell shown previously in Chap. 3, a positive cell voltage will be obtained if the difference is taken in the way described in Eq. (4.3) and illustrated in Fig. 4.1.

$$E_{\text{cell}} = E_{\text{cathode}} - E_{\text{anode}} \tag{4.3}$$

* The Faraday constant is 96 485 C/mol.

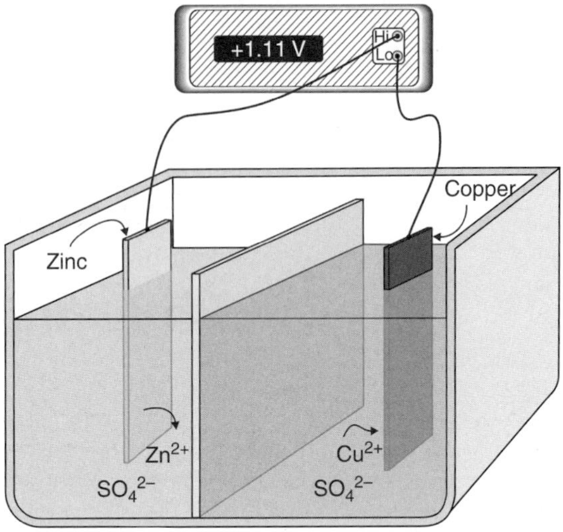

FIGURE 4.1 Daniell cell instrumented to read its equilibrium potential.

The free energy change in a galvanic cell, or in a spontaneous cell reaction, is negative and the positive cell voltage is a measure of available energy. The opposite is true in an electrolytic cell that requires the application of an external potential to drive the electrolysis reaction, in which case E_{cell} would be negative.

Other thermodynamic quantities can be derived from electrochemical measurements. For example, the entropy change (ΔS) in a cell reaction is given by the temperature dependence of ΔG:

$$\Delta S = -\left(\frac{\partial \Delta G}{\partial T}\right)_P \tag{4.4}$$

Hence

$$\Delta S = nF\left(\frac{\partial E}{\partial T}\right)_P \tag{4.5}$$

And

$$\Delta H = \Delta G + T\Delta S = nF\left[T\left(\frac{\partial E}{\partial T}\right)_P - E\right] \tag{4.6}$$

where ΔH is the enthalpy change and T the absolute temperature in degrees Kelvin (K).

The equilibrium constant (K_{eq}) for the same reaction can be obtained with Eq. (4.7)

$$RT \ln K_{eq} = -\Delta G^0 = nFE^0 \tag{4.7}$$

4.2 Standard Electrode Potentials

The potential difference across an electrochemical cell is the potential difference measured between two electronic conductors connected to the electrodes. In the external circuit, the electrons will flow from the most negative point to the most positive point and, by convention, the current will flow in the opposite direction. Since the electrode potential can be either positive or negative, the electrons in the external circuit can also be said to flow from the least positive electrode to the most positive electrode. A voltmeter may be used to measure the potential differences across electrochemical cells but cannot measure directly the actual potential of any single electrode. Nevertheless, it is convenient to assign part of the cell potential to one electrode and part to the other.

There are several potential benchmarks in common use, but the most ancient is the half-cell in which hydrogen gas is bubbled over a platinum electrode immersed in a solution having a known concentration of hydrogen ions. This historically important reference electrode is called the standard hydrogen electrode (SHE) if a standard solution of acid is used. By definition, the equilibrium potential of this electrode is zero at any temperature. However, the SHE can be somewhat inconvenient to use because of the need to supply hydrogen gas. Therefore, other reference electrodes are much preferred for practical considerations.

The potential difference across a reversible cell made up of any electrode and a SHE is called the reversible potential of that electrode, E. If this other electrode is also being operated under standard conditions of pressure and concentration, then the reversible potential difference across the cell is the standard electrode potential E^0 of that electrode.

Tables of standard electrode potentials such as Table 4.1 and Table 4.2 can be obtained if any one electrode, operated under standard conditions, is designated as the standard electrode or standard reference electrode with which other electrodes can be compared.

Since an electrochemical reaction can be written either as an oxidation or a reduction causing confusion in relation to the sign of the potential of that reaction, a convention was adopted in Stockholm in 1953 to write the standard potential of a reaction in reference to its reduction (E^0_{red}) as shown in Table 4.1 and Table 4.2.

Half-Reaction	E^0_{red}
$Ag^+ + e^- \rightleftharpoons Ag$	0.7996
$Al^{3+} + 3\ e^- \rightleftharpoons Al$	−1.706
$Au^+ + e^- \rightleftharpoons Au$	1.68
$Cl_2(g) + 2\ e^- \rightleftharpoons 2\ Cl^-$	1.3583
$Co^{2+} + 2\ e^- \rightleftharpoons Co$	−0.28
$Cr^{3+} + 3\ e^- \rightleftharpoons Cr$	−0.74
$Cu^+ + e^- \rightleftharpoons Cu$	0.522
$Cu^{2+} + 2\ e^- \rightleftharpoons Cu$	0.3402
$Fe^{2+} + 2\ e^- \rightleftharpoons Fe$	−0.409
$Fe^{3+} + 3\ e^- \rightleftharpoons Fe$	−0.036
$2\ H^+ + 2\ e^- \rightleftharpoons H_2$	0.0000...
$H_2O_2 + 2\ e^- \rightleftharpoons 2\ OH^-$	0.88
$H_2O_2 + 2\ H^+ + 2\ e^- \rightleftharpoons 2\ H_2O$	1.776
$Hg^{2+} + 2\ e^- \rightleftharpoons Hg$	0.851
$Hg_2^{2+} + 2\ e^- \rightleftharpoons Hg$	0.7961
$K^+ + e^- \rightleftharpoons K$	−2.924
$Mg^{2+} + 2\ e^- \rightleftharpoons Mg$	−2.375
$Mn^{2+} + 2\ e^- \rightleftharpoons Mn$	−1.04
$Na^+ + e^- \rightleftharpoons Na$	−2.7109
$Ni^{2+} + 2\ e^- \rightleftharpoons Ni$	−0.23
$O_2 + 2\ H_2O + 4\ e^- \rightleftharpoons 4\ OH^-$	0.401
$O_2 + 4\ H^+ + 4\ e^- \rightleftharpoons 2\ H_2O$	1.229
$O_3(g) + 2\ H^+ + 2\ e^- \rightleftharpoons O_2(g) + H_2O$	2.07
$Pb^{2+} + 2\ e^- \rightleftharpoons Pb$	−0.1263
$PbO_2 + 4\ H^+ + 2\ e^- \rightleftharpoons Pb^{2+} + 2\ H_2O$	1.467
$Pt^{2+} + 2\ e^- \rightleftharpoons Pt$	1.2
$Sn^{2+} + 2\ e^- \rightleftharpoons Sn$	−0.1364
$Zn^{2+} + 2\ e^- \rightleftharpoons Zn$	−0.7628

TABLE 4.1 Standard-State Reduction Half-Cell Potentials in Alphabetical Order

Half-Reaction	E^0_{red}
$O_3(g) + 2\ H^+ + 2\ e^- \rightleftharpoons O_2(g) + H_2O$	2.07
$H_2O_2 + 2\ H^+ + 2\ e^- \rightleftharpoons 2\ H_2O$	1.776
$Au^+ + e^- \rightleftharpoons Au$	1.68
$PbO_2 + 4\ H^+ + 2\ e^- \rightleftharpoons Pb^{2+} + 2\ H_2O$	1.467
$Cl_2(g) + 2\ e^- \rightleftharpoons 2\ Cl^-$	1.3583
$O_2 + 4\ H^+ + 4\ e^- \rightleftharpoons 2\ H_2O$	1.229
$Pt^{2+} + 2\ e^- \rightleftharpoons Pt$	1.2
$H_2O_2 + 2\ e^- \rightleftharpoons 2\ OH^-$	0.88
$Hg^{2+} + 2\ e^- \rightleftharpoons Hg$	0.851
$Ag^+ + e^- \rightleftharpoons Ag$	0.7996
$Hg_2^{2+} + 2\ e^- \rightleftharpoons Hg$	0.7961
$Cu^+ + e^- \rightleftharpoons Cu$	0.522
$O_2 + 2\ H_2O + 4\ e^- \rightleftharpoons 4\ OH^-$	0.401
$Cu^{2+} + 2\ e^- \rightleftharpoons Cu$	0.3402
$2\ H^+ + 2\ e^- \rightleftharpoons H_2$	0.0000...
$Fe^{3+} + 3\ e^- \rightleftharpoons Fe$	−0.036
$Pb^{2+} + 2\ e^- \rightleftharpoons Pb$	−0.1263
$Sn^{2+} + 2\ e^- \rightleftharpoons Sn$	−0.1364
$Ni^{2+} + 2\ e^- \rightleftharpoons Ni$	−0.23
$Co^{2+} + 2\ e^- \rightleftharpoons Co$	−0.28
$Fe^{2+} + 2\ e^- \rightleftharpoons Fe$	−0.409
$Cr^{3+} + 3\ e^- \rightleftharpoons Cr$	−0.74
$Zn^{2+} + 2\ e^- \rightleftharpoons Zn$	−0.7628
$Mn^{2+} + 2\ e^- \rightleftharpoons Mn$	−1.04
$Al^{3+} + 3\ e^- \rightleftharpoons Al$	−1.706
$Mg^{2+} + 2\ e^- \rightleftharpoons Mg$	−2.375
$Na^+ + e^- \rightleftharpoons Na$	−2.7109
$K^+ + e^- \rightleftharpoons K$	−2.924

TABLE 4.2 Standard-State Reduction Half-Cell Potentials by Decreasing Order of Potential

4.3 Nernst Equation

The Nernst equation was named after the German chemist Walther Nernst who established very useful relations between the energy and the potential of a cell to the concentrations of participating ions and other chemical species. Equation (4.8) can be derived from the equation linking free energy changes to the reaction quotient (Q_{reaction}):

$$\Delta G = \Delta G^0 + RT \ln Q_{\text{reaction}} \tag{4.8}$$

where Q_{reaction} is defined in Eq. (4.10) for a generalized equation of the form:

$$aA + bB + \ldots \rightarrow mM + nN + \ldots \tag{4.9}$$

The capital letters A, B, M, and N in Eq. (4.9) represent, respectively, the reactants and products of a given reaction while the small letters represent the coefficients required to balance the reaction.

$$Q_{\text{reaction}} = \frac{a_M^m \cdot a_N^n \cdots}{a_A^a \cdot a_B^b \cdots} \tag{4.10}$$

At equilibrium, $\Delta G = 0$ and Q_{reaction} corresponds to the equilibrium constant (K_{eq}) described earlier in Eq. (4.7).

In the case of an electrochemical reaction, substitution of the relationships $\Delta G = -nFE$ and $\Delta G^0 = -nFE^0$ into the expression of a reaction free energy and division of both sides by $-nF$ gives the Nernst expression for an electrode reaction described in Eq. (4.11):

$$E = E^0 - \frac{RT}{nF} \ln Q_{\text{reaction}} \tag{4.11}$$

Combining constants at 25°C (298.15 K) gives the simpler form of the Nernst equation for an electrode reaction at this temperature:

$$E = E^0 - \frac{0.059}{n} \log_{10} Q_{\text{reaction}} \tag{4.12}$$

In Eq. (4.12), the electrode potential (E) would be the actual potential difference across a cell containing this electrode as a half-cell and a standard hydrogen electrode as the other half-cell. Alternatively, the relationship in Eq. (4.3) can be used to combine two Nernst equations corresponding to two half-cell reactions into the Nernst equation for a cell reaction:

$$E_{\text{cell}} = (E^0_{\text{cathode}} - E^0_{\text{anode}}) - \frac{0.059}{n} \log_{10} Q_{\text{reaction}} \tag{4.13}$$

Some of the species that take part in these electrode reactions are pure solid compounds and pure liquid compounds. In dilute aqueous solutions, water can be treated as a pure liquid. For pure solid compounds or pure liquid compounds, activities are constant and their values are considered to be unity. The activities of gases are usually taken as their partial pressures and the activities (a_i) of solutes such as ions are the product of the molar concentration and the activity coefficient of each chemical species (i):

$$a_i = \gamma_i[i]_i \approx [i] \tag{4.14}$$

The activity coefficient (γ_i) in Eq. (4.14) can be a complex function highly dependent on a multitude of variables often difficult to even estimate. For this reason it is usually convenient to ignore (γ_i) and use the concentration term $[i]$ as an approximation of a_i.

4.4 Thermodynamic Calculations

The present section illustrates how calculations from basic thermodynamic data can lead to open-circuit cell potential in any condition of temperature and pressure. Chemical power sources, with the exception of fuel cells, are all based on the corrosion of a metal connected to the negative terminal. The aluminum-air power source, that owes its energy to the corrosion of aluminum in caustic, was chosen for this example because of the relative simple chemistry.

4.4.1 The Aluminum-Air Power Source

The high electrochemical potential and low equivalent weight of aluminum combine to produce a theoretical energy density* of 2.6 kWh/kg and make it an attractive candidate as an anode material in metal/air electrochemical cells. The development of aluminum-based cells dates back to 1855 when M. Hulot described a voltaic cell containing aluminum with an acid electrolyte. Since then, many attempts to substitute aluminum for zinc in zinc/carbon and zinc/manganese dioxide cells have been reported.

Figure 4.2 shows a general schematic of a typical aluminum-air system. Tables 4.3 and 4.4, respectively, contain thermodynamic data

FIGURE 4.2 Schematic of an aluminum-air power source.

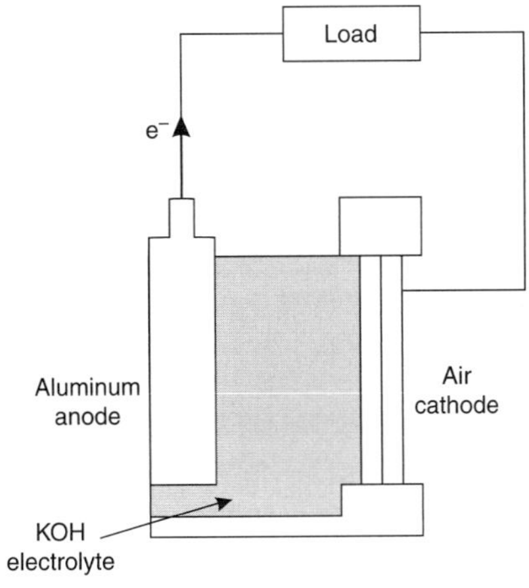

* Characteristic parameter of a battery indicating the amount of electrical energy stored per unit weight or volume.

Species	$G^0_{(298\ K)}$ (J mol^{-1})	$S^0_{(298\ K)}$ (J mol^{-1})	A	$B \times 10^3$	$C \times 10^{-5}$	C_p (J mol^{-1} K^{-1})	$G^0_{(333\ K)}$ (J mol^{-1})
O_2	0	205	29.96	4.184	−1.674	29.85	−7234.04
H_2	0	131	27.28	3.263	0.502	28.82	−4642.01
H_2O	−237000	69.9	10.669	42.284	−6.903	18.54	−239483
Al	0	28.325	20.67	12.38	0	24.79	−1040.43
$Al(OH)_3$	−1136542					0	−1136542
$Al_2O_3 \cdot H_2O$	−1825500	96.86	120.8	35.14	0	132.51	−1829152

TABLE 4.3 Thermodynamic Data of Pure Species Involved in the Operation of an Aluminum-Air Power Source

Species	$G^0_{(298\,K)}$ (J mol^{-1})	$S^0_{(298\,K)}$ (J mol^{-1})	$\check{S}^0_{(298\,K)}$ (J mol^{-1})	*A*	*B*	C_p (J mol^{-1} K^{-1})	$G^0_{(333\,K)}$ (J mol^{-1})
H^+	0	0	−20.9	0.065	−0.005	118.75	−234.9
OH^-	−157277	41.888	20.968	−0.37	0.0055	−452.03	−157849
Al^{3+}	−485400	−321.75	−384.45	0.13	−0.00166	372.84	−474876
$Al(OH)^{2+}$	−694100	−142.26	−184.06	0.13	−0.00166	267.95	−689651
$Al(OH)_2^+$	−900000	205.35	184.43	0.13	−0.00166	75.06	−907336
AlO_2^-	−838968	96.399	117.31	−0.37	0.0055	−284.94	−841778

TABLE 4.4 Thermodynamic Data of Soluble Species Involved in the Operation of an Aluminum-Air Power Source

for pure species and soluble species involved in the equilibria associated with aluminum, water, and oxygen. Table 4.5 contains the chemical and electrochemical reactions possibly occurring in a typical Al-air corrosion cell.

The overall anodic reaction of the aluminum-air battery is the corrosion of aluminum into a soluble form stable in a caustic environment, AlO_2^- in Eq. (4.15), that can subsequently precipitate as $Al_2O_3{\cdot}H_2O$, as shown in Eq. (4.16), depending on the concentration of ions in solution, pH, and temperature.

$$Al + 4OH^- \rightarrow AlO_2^- + 2H_2O + 3e^- \tag{4.15}$$

$$2AlO_2^- + 2H_2O \rightarrow Al_2O_3 \cdot H_2O + 2OH^- \tag{4.16}$$

Water equilibria
$2\ e^- + 2\ H^+ = H_2$
$4\ e^- + O_2 + 4\ H^+ = 2\ H_2O$
$OH^- + H^+ = H_2O$
Equilibria involving aluminum metal
$3\ e^- + Al^{3+} = Al$
$3\ e^- + Al(OH)_3 + 3\ H^+ = Al + 3\ H_2O$
$6e + Al_2O_3{\cdot}H_2O + 6\ H^+ = 2\ Al + 4\ H_2O$
$3\ e^- + AlO_2^- + 4\ H^+ = Al + 2\ H_2O$
$3\ e^- + Al(OH)^{2+} + H^+ = Al + H_2O$
$3\ e^- + Al(OH)_2^+ + 2\ H^+ = Al + 2\ H_2O$
Equilibria involving solid forms of oxidized aluminum
$Al(OH)_3 + H^+ = Al(OH)_2^+ + H_2O$
$Al_2O_3{\cdot}H_2O + 2\ H^+ = 2\ Al(OH)_2^+$
$Al(OH)_3 + 2\ H^+ = Al(OH)^{2+} + 2\ H_2O$
$Al_2O_3{\cdot}H_2O + 4\ H^+ = 2\ Al(OH)^{2+} + 2\ H_2O$
$Al(OH)_3 + 3\ H^+ = Al^{3+} + 3\ H_2O$
$Al_2O_3{\cdot}H_2O + 6\ H^+ = 2\ Al^{3+} + 4\ H_2O$
$Al(OH)_3 = AlO_2^- + H^+ + H_2O$
$Al_2O_3{\cdot}H_2O = 2\ AlO_2^- + 2\ H^+$
Equilibria involving only soluble forms of oxidized aluminum
$AlO_2^- + 4\ H^+ = Al^{3+} + 2\ H_2O$

TABLE 4.5 Reactions Describing the Chemistry of an Aluminum-Air Corrosion Cell

There is also a parasitic reaction at the aluminum anode that has to be considered because it has serious safety implications, that is, the production of hydrogen gas from the reduction of water described in Eq. (4.17):

$$2H_2O + 2e^- \rightarrow H_2 + 2OH^- \tag{4.17}$$

There is however only one reaction on the cathode, that is, the reduction of oxygen shown in Eq. (4.18):

$$O_2 + 2H_2O + 4e^- \rightarrow 4OH^- \tag{4.18}$$

The overall cell voltage can be calculated from thermodynamic data by computing Gibbs free energy for the individual species involved in the global reaction described in Eq. (4.19) using the coefficients expressed in that equation:

$$4Al + 4OH^- + 3O_2 \rightarrow 4AlO_2^- + 2H_2O \tag{4.19}$$

4.4.2 Detailed Calculations

Calculate G^0 for each species. The free energy of a substance, for which heat capacity data are available, can be calculated as a function of temperature using Eq. (4.20):

$$G^0_{(T2)} = G^0_{(T1)} - S^0_{(T1)}\left[T_2 - T_1\right] - T_2\int_{T1}^{T2}\frac{C_p^0}{T}dT + \int_{T1}^{T2}C_p^0 dT \tag{4.20}$$

For pure substances (i.e., solids, liquids, and gases) the heat capacity C_p^0 is often expressed, as in Table 4.3, as a function of the absolute temperature:

$$C_p^0 = A + BT + CT^{-2} \tag{4.21}$$

For ionic substances, one has to use another method, such as proposed by Criss and Cobble in 1964 [1], to obtain the heat capacity, provided the temperature does not rise above 200°C. The expression of the ionic capacity in equation (Eq. 4.22) makes use of absolute entropy values and the parameters a and b contained in Table 4.3.

$$C_p^0 = (4.186a + bS^0_{(298\ K)})(T_2 - 298.16)/\ln\left(\frac{298.16}{T_2}\right) \tag{4.22}$$

By combining Eqs. (4.21) or (4.22) with Eq. (4.20) one can obtain the free energy in Eq. (4.23) at any given temperature by using the fundamental data contained in Table 4.3 and Table 4.4:

$$G^0_{(T)} = G^0_{(298\ K)} + (C_p^0 - S^0_{(298\ K)})\cdot(T_2 - 298.16) - T_2\cdot\ln\left(\frac{T_2}{298.16}\right)\cdot C_p^0 \tag{4.23}$$

Although these equations appear slightly overwhelming, they can be computed relatively simply with the use of a modern spreadsheet.

Step One: Calculate *G* for Each Species

For species *O*, the free energy of 1 mol can be obtained from G^0 with Eq. (4.24):

$$G_{o(T)} = G^0_{o(T)} + 2.303\,RT\,\log_{10} a_O \tag{4.24}$$

For x mol of species *O* the free energy is expressed by Eq. (4.25):

$$xG_{o(T)} = x\left(G^0_{o(T)} + 2.303\,RT\,\log_{10} a_O\right) \tag{4.25}$$

For pure substances such as solids, a_O is equal to 1. For a gas, a_O is equal to its partial pressure (p_O), as a fraction of 1 atm. For soluble species, the activity of species *O* (a_O), is the product of the activity coefficient of that species (γ_O) with its molar concentration ([*O*]) (i.e., $a_O = \gamma O[O]$). The activity coefficient of a chemical species in solution is close to 1 at infinite dilution when there is no interference from other chemical species. For most other situations the activity coefficient is a complex function that varies with the concentration of the species and with the concentration of other species in solution. For the sake of simplicity the activity coefficient will be assumed to be of value 1; hence Eq. (4.25) can be rewritten as a function of [*O*]:

$$xG_{o(T)} = x\left(G^0_{o(T)} + 2.303\,RT\,\log_{10}[O]\right) \tag{4.26}$$

Taking the global reaction for the Al-O_2 system expressed in Eq. (4.19) and the G^0 values calculated for 60°C in Table 4.3 and Table 4.4, one can obtain thermodynamic values for the products and reactants, as is done in Table 4.6.

Step Two: Calculate Cell ΔG

The ΔG of a cell can be calculated by subtracting the *G* values of the reactants from the *G* values of the products in Table 4.6. Keeping the example of the global reaction at 60°C in mind, one would obtain

$$\Delta G = G_{\text{products}} - G_{\text{reactants}} = -3{,}846{,}087 - (-670{,}615) = -3{,}175{,}472\ \text{J} \tag{4.27}$$

The ΔG energy can be converted into potential:

$$E = \frac{-\Delta G}{nF} = \frac{3{,}188{,}818}{12 \times 96{,}485} = 2.74\ \text{V} \tag{4.28}$$

where $n = 12$ because each of the four Al in Eq. (4.19) gives off 3 e^-.

Step Three: Calculate the Specific Capacity (Ah kg^{-1})

The specific capacity relates the weight of active materials with the charge that can be produced, that is, a number of coulombs or ampere-hours (Ah). Because 1 A = 1 C s^{-1}, 1 Ah = 3600 C, and because 1 mol of e^- = 96,485 C (Faraday), 1 mol of e^- = 26.80 Ah.

Species	x	$G^0_{(333\ K)}$ (J mol^{-1})	a_0	$2.303 \cdot RT \cdot \log_{10} a_0$	$xG^0_{(333\ K)}$	$\Sigma G^0_{(333\ K)}$
Reactants						
Al	4	−1040.43	1	0.00	−4161.72	
OH^-	4	−157849	1	0.00	−631396.00	
O_2	3	−7234.04	0.2	−4452.02	−35058.17	−670615.89
Products						
AlO_2^-	4	−841778	0.1	−6369.39	−3392589.58	
H_2O	2	−239483	1	0.00	−478966.00	−3846087.21

Table 4.6 Calculated Free Energies for Species Involved in the Global Aluminum-Air Reactions at 60°C

By considering the global expression of the cell chemistry expressed in Eq. (4.19), one can relate the weight of the active materials to a certain energy and power. In the present case 12 mol of e^- are produced by using

4 mol of Al	4 × 26.98 g mol^{-1}, or 107.92 g
4 mol of OH^- as KOH	4 × 56.11 g mol^{-1}, or 224.44 92 g
3 mol of O_2 (as air)	0 g
Total	332.4 g or 0.3324 kg for 12 mol e^-

The theoretical specific capacity is thus 26.80 × 12/0.3324 = 967.5 Ah kg^{-1}.

Step Four: Calculate the Energy Density (Wh kg^{-1})

The energy density can then be obtained by multiplying the specific capacity obtained from calculating the specific capacity with the thermodynamic voltage calculated when translating AG into potentials: 2.74 × 967.5 = 2651 Wh kg^{-1}, or 2.651 kWh kg^{-1} if running on air and, because the voltage for running on pure oxygen is slightly higher (i.e., 2.78 V), 2.78 × 750.7 = 2087 Wh kg^{-1}, or 2.087 kWh kg^{-1} if running on compressed or cryogenic oxygen.

4.4.3 Reference Electrodes

The thermodynamic equilibrium of any other chemical or electrochemical reaction can be calculated in the same manner, provided the basic information is found. Table 4.7 contains the chemical description of most reference electrodes used in laboratories and field units, and Tables 4.8 and 4.9, respectively, contain the thermodynamic data associated with the solid and soluble chemical species making these electrodes. Table 4.10 presents the results of the calculations performed to obtain the potential of each electrode at 60°C.

4.5 Reference Half-Cells (Electrodes)

As mentioned earlier, the standard hydrogen half-cell is rather awkward to use under most circumstances. The other half-cells most frequently used in corrosion studies, along with their potentials relative to the standard hydrogen half-cell, are listed in Table 4.7.

Reference electrodes are commonly used with a saturated solution and an excess of salt crystals. The extra salt dissolves into the half-cell solution as some of the ions diffuse out of the reference cell body through the liquid junction during normal use. This extra buffer of salt extends the time before the reference cell starts to drift due to the depletion of ions as predicted by Nernst equation [Eq. (4.12)].

Name	Equilibrium reaction and Nernst equation	Conditions	Potential (V vs. SHE)	T coefficient (mV °C^{-1})
Standard hydrogen electrode (SHE)	$2\ H^+ + 2\ e^- = H_2$	pH = 0	0.00	
	$E^o - 0.059$ pH			
Silver chloride	$AgCl + e^- = Ag + Cl^-$	$a_{Cl^-} = 1$	0.2224	–0.6
	$E^o - 0.059 \log_{10} a_{Cl^-}$	0.1 M KCl	0.2881	..
		1.0 M KCl	0.235	..
		Saturated (KCl)	0.199	..
		Seawater	~0.250	..
Calomel	$Hg_2Cl_2 + 2\ e^- = 2\ Hg + 2\ Cl^-$	$a_{Cl^-} = 1$	0.268	
	$E^o - 0.059 \log_{10} a_{Cl^-}$	0.1 M KCl	0.3337	–0.06
		1.0 M KCl	0.280	–0.24
		Saturated	0.241	–0.65
Mercurous sulfate	$Hg_2SO_4 + 2\ e^- = 2\ Hg + SO_4^{-2}$	$a_{SO_4^{2-}} = 1$	0.6151	
	$E^o - 0.0295 \log_{10} a_{SO_4^{2-}}$			
Mercuric oxide	$HgO + 2\ e^- + 2\ H^+ = Hg + H_2O$		0.926	
	$E^o - 0.059$ pH			
Copper sulfate	$Cu^{+2} + 2\ e^- = Cu$ (sulfate solution)	$a_{CU^{2+}} = 1$	0.340	
	$E^o + 0.0295 \log_{10} a_{CU^{2+}}$	saturated	0.318	

TABLE 4.7 Equilibrium Potential at 25°C of Commonly Used Reference Electrodes

Species	$G^0_{(298\ K)}$ (J mol⁻¹)	$S^0_{(298\ K)}$ (J mol⁻¹)	*A*	*B*	*C*	$Cp_{(333K)}$ (J mol⁻¹ K⁻¹)	$G^0_{(333K)}$ (J mol⁻¹)
O_2	0	205	29.96	4.184	–1.674	29.85	–7234.04
H_2	0	131	27.28	3.263	0.502	28.8	–4642.01
H_2O	–237000	69.9	10.669	42.284	–6.903	18.5	–239483.00
Ag	0	42.55	21.297	8.535	1.506	25.5	–1539.69
Cu	0	33.2	22.635	6.276		24.7	–1210.91
Hg	0	76.02	26.94	0	0.795	27.7	–2715.41
AgCl	–109805	96.2	62.258	4.184	–11.297	53.5	–113277.00
Hg_2Cl_2	–210778	192.5	63.932	43.514	0	78.4	–217670.00
Hg_2SO_4	–625880	200.66	131.96			132	–633164.00
HgO	–58555	70.29	34.853	30.836	0	45.1	–61104.4

TABLE 4.8 Thermodynamic Data for Pure Species, the Free Energy, and Potential of Commonly Used Reference Electrodes at 60°C

Species	$G^0_{(298\ K)}$ (J mol^{-1})	$S^0_{(298\ K)}$ (J mol^{-1})	$\check{S}_{(298\ K)}$ (J mol^{-1})	*a*	*b*	$C_{p\ (333K)}$ (J mol^{-1} K^{-1})	$G^0_{(333K)}$ (J mol^{-1})
H^+	0	0	–20.9	0.065	–0.005	118.7525	–234.927
Cu^{2+}	65689	–207.2	–249.04	0.13	–0.00166	301.9618	72343.6
Cl^-	–131260	–12.6	8.32	–0.37	0.0055	–473.9694	–129881.
SO_4^{2-}	–744600	10.752	52.592	–0.37	0.0055	–397.1863	–744190.

TABLE 4.9 Thermodynamic Data for Soluble Species, the Free Energy, and Potential of Commonly Used Reference Electrodes at 60°C

Name	ΔG⁰ reactants J mol^{-1}	ΔG⁰ products J mol^{-1}	ΔG⁰ reaction J mol^{-1}	Potential V vs. SHE
Hydrogen	–470	–46,420	–4172	0.0216
Silver chloride	–113,277	–131,421	–18,144	0.1880
Calomel	–217,670	–265,193	–47,523	0.2463
Mercurous sulfate	–633,164	–749,621	–116,457	0.6035
Mercuric chloride	–61,574	–242,199	–180,624	0.9360
Copper sulfate	72,344	–1211	–73,555	0.3812

*Note: all species considered to be of activity = 1

Table 4.10 Detailed Calculations of the Equilibrium Associated with the Most Commonly Used Reference Electrodes at 60°C

4.5.1 Conversion between References

When reporting electrochemical potential measurements, it is always important to indicate which reference half-cell was used to carry out the work. This information is required to compare these measurements to similar data that could have been obtained using any other reference half-cell listed in Table 4.7. The scheme presented in Figure 4.3 provides a graphical representation to visualize some of the information listed in Table 4.7.

In the case of the potential measured between a steel pipe buried in the ground and a saturated copper-copper sulfate reference electrode (CCSRE) this might show a potential of –0.700 V. To convert this potential to a value on the scale in which the hydrogen electrode has a potential of zero, it is necessary to add 0.318 V to the potential that was measured, making it –0.382 V vs. SHE.

4.5.2 Silver/Silver Chloride Reference Electrode

The silver/silver chloride reference electrode is a widely used reference electrode because it is simple, relatively inexpensive, its potential is stable and it is nontoxic. As a laboratory electrode such as described in Figure 4.4, it is mainly used with saturated potassium chloride (KCl) electrolyte, but can be used with lower concentrations such as 1 M KCl and even directly in seawater. As indicated in Table 4.7, such changes in ionic concentrations also change the reference potential. Silver chloride is slightly soluble in strong potassium chloride solutions, so it is sometimes recommended that the potassium chloride be saturated with silver chloride to avoid stripping the silver chloride off the silver wire.

Typical laboratory electrodes use a silver wire that is coated with a thin layer of silver chloride either by electroplating or by dipping

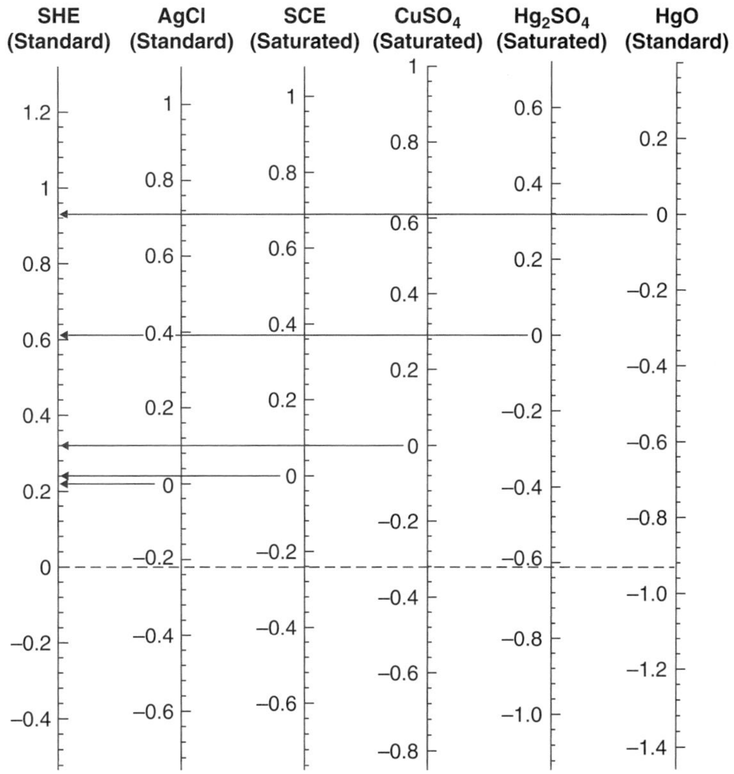

Figure 4.3 Graphical scheme to compare potentials of the most commonly used reference electrodes.

the wire in molten silver chloride. Industrial electrodes are fabricated along the same principle but using other geometries such as planar electrodes. When the electrode is placed in a saturated potassium chloride solution it develops a potential of 199 mV vs. SHE. The potential of the half-cell reaction shown in Eq. (4.29) is determined by the chloride concentration of the solution, as defined by the Nernst equation.

$$AgCl(s) + e^- \rightleftarrows Ag + Cl^- \quad E^0_{red} = 0.2224 \text{ vs. SHE} \tag{4.29}$$

$$E_{Ag/AgCl} = E^0_{Ag/AgCl} - 0.059 \log_{10} a_{Cl^-} \tag{4.30}$$

The silver-silver chloride half-cell electrode develops a potential proportional to the chloride concentration, whether it is sodium chloride, potassium chloride, ammonium chloride, or some other chloride salt, and remains constant as long as the chloride concentration remains constant.

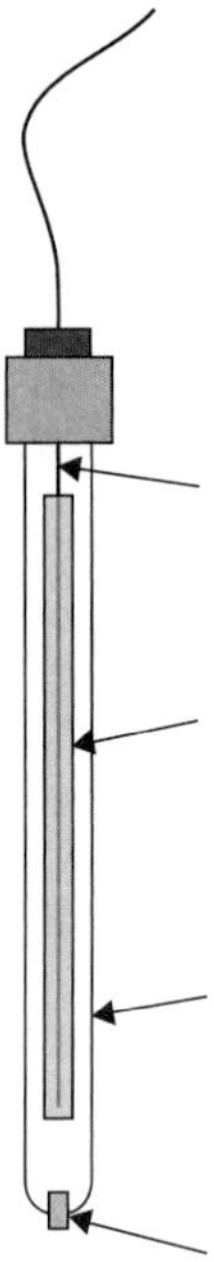

FIGURE 4.4 Schematic of a silver/silver chloride reference electrode.

The silver-silver chloride electrode simplicity of fabrication and fundamental ruggedness makes it a good candidate for many industrial applications where the electrochemical potential has to be measured or controlled. An important industrial example where this half-cell has become indispensable is to provide a measure of applied potential during the impressed current cathodic protection (ICCP) of sea-going ships.

4.5.3 Copper/Copper Sulfate Reference Electrode

Copper/copper sulfate half-cells are typically favored for potential measurements of systems buried in soils. Figure 4.5 illustrates the principle of construction of a copper/copper sulfate reference electrode (CCSRE) used for soil application and Fig. 4.6 shows a commercial CCSRE ready for field work. What is often referred to as a pipe-to-soil potential is actually the potential measured between the pipe and the reference electrode used to make the measurement. The soil itself has no standard value of potential against which the potential of a pipe can be measured independently.

The half-cell potential of a CCSRE is dependent only upon the electrochemical equilibrium established between Cu and its ions in solution as shown in Eq. (4.31) and in its corresponding Nernst equation.

$$Cu^{2+} + 2e^- \rightleftarrows Cu(s) \quad E^0_{red} = 0.340 \text{ vs. SHE} \tag{4.31}$$

$$E_{Cu/Cu^{2+}} = E^0_{Cu/Cu^{2+}} + 0.059 \log_{10} a_{Cu^{2+}} \tag{4.32}$$

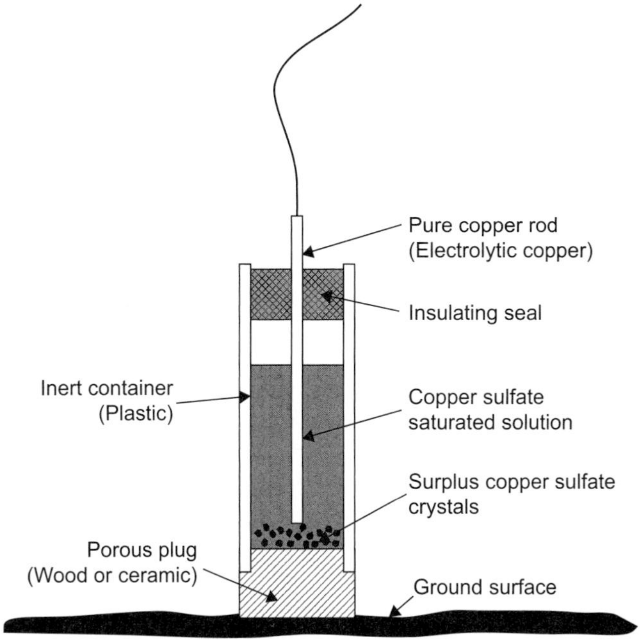

Figure 4.5 Schematic of a copper/copper sulfate reference electrode.

For Cu in a saturated Cu/$CuSO_4$ solution, this equilibrium is influenced modestly by temperature and not at all by other factors except for light. Therefore, this reference electrode has a relatively constant half-cell potential, making it reliable for field potential measurements.

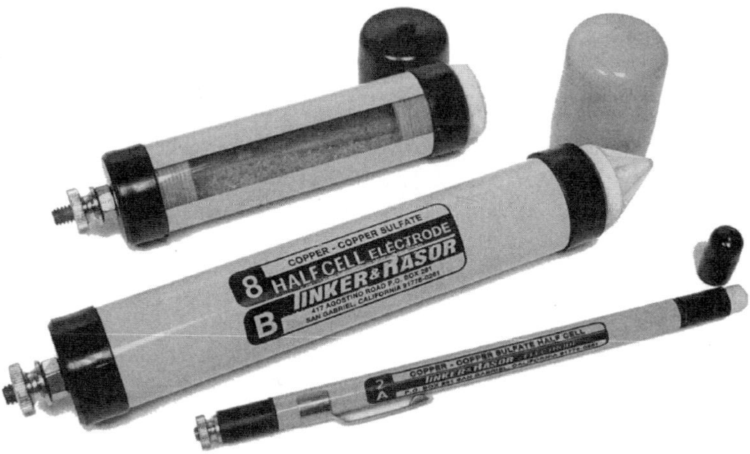

Figure 4.6 Commercial copper/copper sulfate reference electrode. (Photo Courtesy of Tinker & Rasor)

Calculation Example

A saturated CCSRE can be fabricated with a solution of copper sulfate made with 40 g of $CuSO_4{\cdot}5H_2O$ in 25 mL of distilled water. The saturated solution should contain approximately 260 g/L of $CuSO_4$ at 22°C. The following two sets of measurements were carried out between a saturated CCSRE maintained at 22°C and a second CCSRE for which the temperature (Table 4.11 Set 1) or the sulfate concentration (Table 4.11 Set 2) were changed [2].

To calculate the theoretical voltage of a saturated CCSRE at 22°C with the basic thermodynamic equations, assume that saturation is achieved with 260 g/L of $CuSO_4$ and that all activity coefficients are one. First use the thermodynamic data in Tables 4.8 and 4.9 to calculate the temperature corrected G^0 for each chemical species involved in the CCSRE. Then calculate the $G^0_{(reaction)}$ by summing up the G^0 of reactants and products following the aluminum-air example described earlier. Your answer should be 0.3372 V vs. SHE.

Set 1 - Temperature effects	
Temperature (°C)	**Measured Voltage Difference (mV)**
3	−16
22	0
36	+11
46	+24

Set 2 - Sulfate concentration effects	
Concentration (g/L)	**Measured Voltage Difference (mV)**
Saturated	0
230	−1
100	−6
77	−9
26	−17
10	−24

TABLE 4.11 Measured Voltage Differences between a Saturated CCSRE Maintained at 22°C and a CCSRE at Other Temperatures (Set 1) or Sulfate Concentrations (Set 2)

For calculating the shift in the CCSRE voltage due to the temperatures indicated in Set 1 and comparing it to the observed value see the table below.

Temperature (°C)	E^0	2.303*RT*/*nF**	E^0 corrected	Δ with 22°C
3	0.3119	0.0274	0.3124	−24.76
22	0.3367	0.0293	0.3372	0.00
36	0.3538	0.0307	0.3544	17.21
46	0.3656	0.0317	0.3661	28.95

* Correction due to the activity coefficient.

For calculating the shift in the CCSRE voltage due to the concentrations indicated in Set 2 and comparing it to the observed value see the table below.

Concentration (g/L)	Molarity	Corrected	Δ with 260 g/L
260	1.0417	0.3372	0.00
230	0.9215	0.3356	−1.56
100	0.4006	0.3250	−12.15
77	0.3085	0.3217	−15.48
26	0.1042	0.3079	−29.29
10	0.0401	0.2957	−41.44

4.6 Measuring the Corrosion Potential

The potential of a corroding metal, often termed E_{corr}, is probably the single most useful variable measured in corrosion studies or for corrosion monitoring. It is readily measured by determining the voltage difference between a metal in its environment and an appropriate reference electrode.

Figure 4.7 illustrates an experimental technique for measuring the corrosion potential of a metal M using a laboratory cell. This is accomplished by measuring the voltage difference between the reference electrode and the metal using a high impedance voltmeter capable to accurately measure small voltages without drawing any appreciable current. Note that in Fig. 4.7 the reference electrode is contained in a Luggin capillary to prevent any contamination of the reference electrode by the environment or the opposite, that is, leaking some corrosive agent in the environment being monitored, while making potential measurements very close to the surface of the metal being monitored.

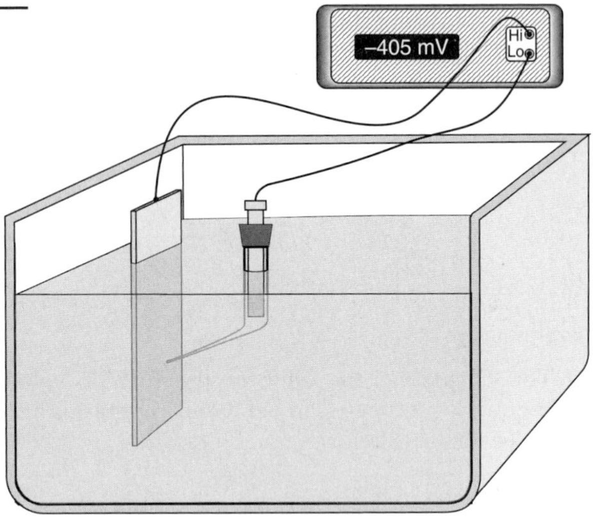

FIGURE 4.7 Experimental set-up to measure the corrosion potential of a specimen.

In measuring and reporting corrosion potentials, it is necessary to indicate the magnitude of the voltage and its sign. In the example shown in Fig. 4.7, the corrosion potential of metal M is –0.405 V. The minus sign indicates that the metal is negative with respect to the reference electrode. However, if the metal was connected to the low point (Lo) and the reference electrode to the high point (Hi) the reading would be +0.405 V. It is customary to connect the reference electrode to the low point or the instrumental ground to avoid any confusion in reporting. Nonetheless, some manufacturers of electrochemical equipment have done the opposite.

4.7 Measuring pH

Measuring pH involves either the use of indicators whose colors are dependent on pH or the use of pH measuring electrodes. pH indicators based on color changes are normally used in the form of pH papers. The paper is wetted with the solution being measured and the resulting color is compared with color standards to determine the pH.

A pH meter measures the difference in potential between a reference electrode insensitive to changes in pH and an electrode sensitive to such changes. A successful pH reading depends upon three components:

Electrodes: A pH electrode consists of two half-cells; an indicating electrode and a reference electrode. Most applications today use a combination electrode with both half-cells in one body. Over 90 percent of pH measurement problems are related to the improper use, storage, or selection of electrodes.

Meters: A pH meter is in reality a high-precision and high-impedance voltmeter capable of reading small millivolt changes from the pH electrode system. The meter is seldom the source of problems for pH measurements. Modern pH meters have temperature compensation (either automatic or manual) to correct for variations in slope caused by changes in temperature.

Buffers: These solutions of known pH value allow the user to calibrate the system to read accurate measurements.

4.7.1 Glass Electrodes

A glass electrode is a potentiometric sensor made from glass of a specific composition. All glass pH electrodes have extremely high electric resistance from 50 to 500 MΩ. There are different types of pH glass electrodes. Some of them have improved characteristics for working in very alkaline or acidic medium. But almost all electrodes can operate in the 1 to 12 pH range.

A typical pH probe is a combination electrode, which combines both the glass and reference electrodes into one body. The measuring part of the electrode, the glass bulb at the bottom of the pH probe (Figure 4.8), is coated both inside and out with a ~10nm layer of a hydrated gel. These two layers are separated by a layer of dry glass and the potential is created by the equilibrium in H^+ ions across the membrane.

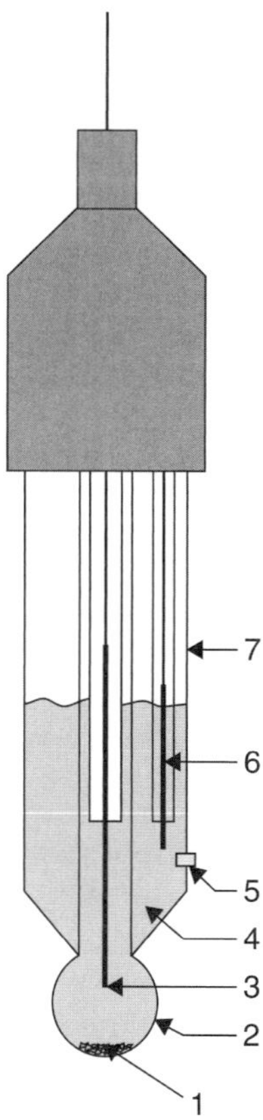

Figure 4.8 Schematic description of a typical pH glass electrode: 1) small amount of AgCl precipitate crystals, 2) pH sensing bulb made of special glass, 3) internal electrode, usually silver chloride electrode, 4) internal solution, usually 0.1 M HCl, 5) porous junction to provide an electrolytic contact with the solution being monitored, 6) reference electrode, usually the same type as 3, and 7) body of electrode, made from non-conductive glass or plastics.

FIGURE 4.9 Commercial antimony electrode with a voltage/pH conversion scale engraved on the body of the electrode. (Courtesy of Kingston Technical Software)

4.7.2 Antimony Electrode

Antimony is a unique metal in that it can provide a direct relationship between pH and its measured potential due to the formation of an oxide film on the metal surface. The potential difference or voltage developed between antimony and a copper/copper sulfate reference electrode is typically between 0.1 to 0.7 V for a variation of pH between 1 and 11.

Antimony electrodes must be cleaned with special cleaning procedures prior to use. Antimony is very brittle and must be treated carefully. The antimony tip must be kept smooth, and there must be no rough surface or pits. A scale for the conversion of voltage to pH is conveniently engraved on the side of commercial reference half-cells as illustrated in Fig. 4.9.

4.8 Potential-pH Diagram

The stability of a metal when exposed to a given environment depends on a multitude of factors that may vary greatly with the pH and oxidizing or reducing power of that environment. One useful concept to represent the effects of aqueous environments on metals became known as potential-pH (E-pH) diagrams, also called predominance or Pourbaix diagrams, which have been adopted universally since their introduction in late 1940s. E-pH diagrams are typically plotted for various equilibria on normal Cartesian coordinates with potential (E) as the ordinate (Y axis) and pH as the abscissa (X axis) [3].

Some interesting uses of such diagrams, which have been constructed for most metals and a few alloys, are to (1) predict whether or not corrosion can occur; (2) estimate the composition of the corrosion products formed; and (3), predict environmental changes which will prevent or reduce corrosion attack.

4.8.1 E-pH Diagram of Water

The following example illustrates how the stability or predominance diagram of water can be constructed from its basic thermodynamic information. Equation (4.33) describes the equilibrium between hydrogen ions and hydrogen gas in an aqueous environment:

$$2H^+ + 2e^- \rightleftarrows H_2(g) \tag{4.33}$$

Adding sufficient OH^- to both sides of reaction in Eq. (4.33) results in Eq. (4.34) in neutral or alkaline solutions:

$$2H_2O + 2e^- \rightleftarrows H_2(g) + 2OH^- \tag{4.34}$$

At higher pH than neutral, Eq. (4.34) is a more appropriate representation. However, since the concentrations of H^+ and OH^- ions are related by the dissociation constant of water expressed in Eq. (4.35), Eq. (4.33) and (4.34) basically represent the same reaction for which the thermodynamic behavior can be expressed by Nernst equation.

$$H_2O \rightleftarrows H^+ + OH^- \text{ with } K_{eq} = \frac{[H^+][OH^-]}{[H_2O]} = 10^{-14} \text{ at } 25^\circ\text{C} \tag{4.35}$$

$$E_{H^+/H_2} = E^0_{H^+/H_2} + \frac{RT}{nF} \ln \frac{[H^+]^2}{p_{H_2}} \tag{4.36}$$

that becomes Eq. (4.37) at 25°C and the partial pressure of hydrogen (p_{H_2}) of value unity.

$$E_{H^+/H_2} = E^0_{H^+/H_2} - 0.059 \text{ pH} \tag{4.37}$$

Equation (4.33) and its alkaline or basic form, Eq. (4.34), delineate the stability of water in a reducing environment and are represented in a graphical form by the sloping line (a) on the Pourbaix diagram in Fig. 4.10. Below the equilibrium reaction shown as line (a) in this figure, the decomposition of H_2O into hydrogen is favored while water is thermodynamically stable above the same line (a). As potential becomes more positive or noble, water can be decomposed into its other constituent, oxygen, as illustrated in Eqs. (4.38) and (4.39) for respectively the acidic form and neutral or basic form of the same process.

$$O_2 + 4H^+ + 4e^- \rightleftarrows 2\,H_2O \tag{4.38}$$

$$O_2 + 2H_2O + 4e^- \rightleftarrows 4OH^- \tag{4.39}$$

And again these equations are equivalent and only reflect the pH condition of the environment. The corresponding Nernst expression

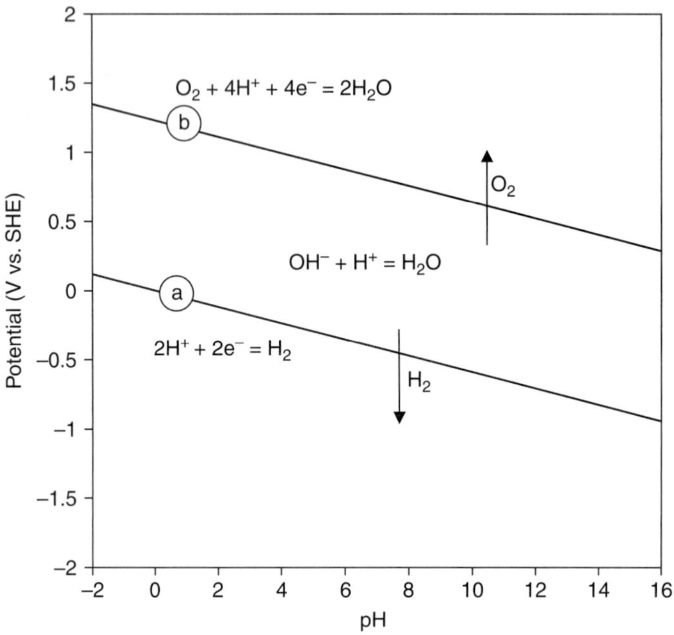

FIGURE 4.10 E-pH stability diagram of water at 25°C.

of the potential is described in Eq. (4.40) that simplifies as Eq. (4.41) at **25°C** and oxygen partial pressure, that is, p_{O_2} of value unity.

$$E_{O_2/H_2O} = E^0{}_{O_2/H_2O} + \frac{RT}{nF} \ln\left(p_{O_2} \times [H^+]^4\right) \tag{4.40}$$

$$E_{O_2/H_2O} = E^0{}_{O_2/H_2O} - 0.059\ \text{pH} \tag{4.41}$$

The line labeled (b) in Fig. 4.10 represents the behavior of E vs. pH for Eq. (4.41). The chemical behavior of water across all possible values of potential and pH as shown in Fig. 4.10 is divided into three regions. In the upper region, water can be oxidized to produce oxygen while in the lower region it can be reduced to form hydrogen gas. Water is therefore only thermodynamically stable between lines (a) and (b). It is common practice to superimpose these two lines (a) and (b) on all E-pH diagrams to mark the water stability boundaries.

4.8.2 E-pH Diagrams of Metals

Aluminum E-pH Diagram

The E-pH diagram of aluminum and zinc are quite similar and surely amongst the simplest E-pH diagrams of all metals. The Pourbaix diagram of aluminum will be used here to demonstrate how such diagrams are constructed from basic principles. In the following discussion, only four species containing the aluminum element will be considered, that is, two solid species (Al and $Al_2O_3{\cdot}H_2O$) and two ionic species (Al^{3+} and AlO_2^-). The first equilibrium to consider examines the possible presence of either Al^{3+} or AlO_2^- expressed in Eq. (4.42).

$$Al^{3+} + 2H_2O \rightleftarrows AlO_2^- + 4H^+ \tag{4.42}$$

Since there is no change in valence of the aluminum present in the two ionic species considered, the associated equilibrium is independent of the potential and the expression of that equilibrium can be derived in Eq. (4.43) for standard conditions.

$$RT \ln K_{eq} = RT \ln Q = -\Delta G^0_{reaction} \tag{4.43}$$

where Q is expressed in Eq. (4.44).

$$Q = \frac{a_{AlO_2^-} \times a^4_{H^+}}{a_{Al^{3+}} \times a^2_{H_2O}} \tag{4.44}$$

Assuming that the activity of H_2O is unity and that the activities of the two ionic species are equal, one can obtain a simpler expression of the equilibrium in Eq. (4.43) based purely on the activity of H^+, Eq. (4.45) and its logarithmic form, Eq. (4.46).

$$RT \ln [H^+]^4 = -\Delta G^0_{reaction} \tag{4.45}$$

$$\log_{10} [H^+] = -pH = \frac{-G^0_{reaction}}{4 \times 2.303 \times RT} \tag{4.46}$$

and if G^0 is expressed in joules and the temperature is 25°C or 298 K Eq. (4.46) is even further simplified in Eq. (4.47).

$$pH = 4.38 \times 10^{-5} \times G^0_{reaction} \tag{4.47}$$

By using the standard thermodynamic data from the literature [3], it is possible to calculate that the free energy of reaction in Eq. (4.42) is in fact equal to 120.44 kJ mol^{-1} when both $[Al^{3+}]$ and $[AlO_2^-]$ are equal. Equation (4.47) then becomes Eq. (4.48).

$$pH = 4.38 \times 10^{-5} \times 120{,}440 = 5.27 \tag{4.48}$$

This is represented, in the E-pH diagram shown in Fig. 4.11, by a dotted vertical line separating the dominant presence of Al^{3+} at low pH from the dominant presence of AlO_2^- at the higher end of the pH scale.

The next phase for constructing the aluminum E-pH diagram is to consider all possible reactions between the four chemical species containing aluminum retained for this exercise, that is, Al, $Al_2O_3{\cdot}H_2O$, Al^{3+}, and AlO_2^-. These reactions are summarized in Table 4.12. A computer program that would compare all possible interactions and rank the chemical species involved in terms of their thermodynamic

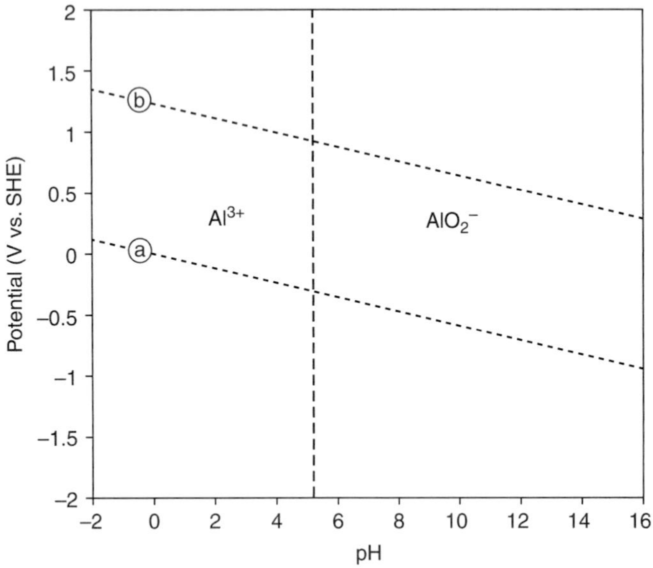

FIGURE 4.11 E-pH diagram showing the soluble species of aluminum in water at 25°C.

Equilibria involving aluminum metal
$3\ e^- + Al^{3+} = Al$
$3\ e^- + Al(OH)_3 + 3\ H^+ = Al + 3\ H_2O$
$6e + Al_2O_3{\cdot}H_2O + 6\ H^+ = 2\ Al + 4\ H_2O$
$3\ e^- + AlO_2^- + 4\ H^+ = Al + 2\ H_2O$
$3\ e^- + Al(OH)^{2+} + H^+ = Al + H_2O$
$3\ e^- + Al(OH)_2^+ + 2\ H^+ = Al + 2\ H_2O$
Equilibria involving solid forms of oxidized aluminum
$Al(OH)_3 + H^+ = Al(OH)_2^+ + H_2O$
$Al_2O_3{\cdot}H_2O + 2\ H^+ = 2\ Al(OH)_2^+$
$Al(OH)_3 + 2\ H^+ = Al(OH)^{2+} + 2\ H_2O$
$Al_2O_3{\cdot}H_2O + 4\ H^+ = 2\ Al(OH)^{2+} + 2\ H_2O$
$Al(OH)_3 + 3\ H^+ = Al^{3+} + 3\ H_2O$
$Al_2O_3{\cdot}H_2O + 6\ H^+ = 2\ Al^{3+} + 4\ H_2O$
$Al(OH)_3 = AlO_2^- + H^+ + H_2O$
$Al_2O_3{\cdot}H_2O = 2\ AlO_2^- + 2\ H^+$
Equilibria involving only soluble forms of oxidized aluminum
$AlO_2^- + 4\ H^+ = Al^{3+} + 2\ H_2O$

TABLE 4.12 Possible Reactions in the Al-H_2O System

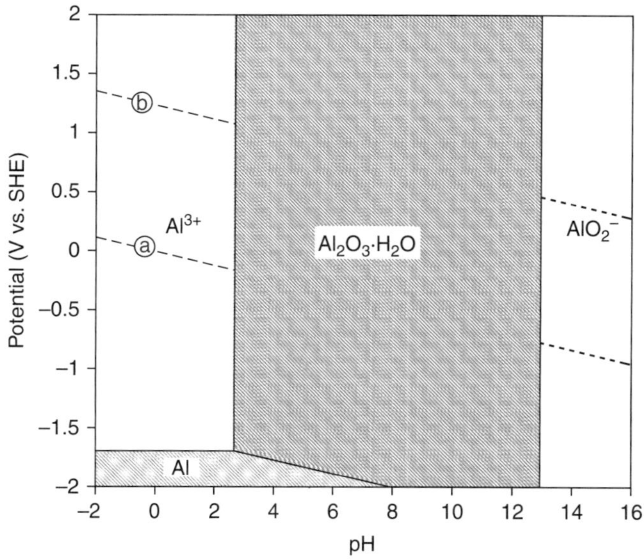

FIGURE 4.12 E-pH diagram of solid species of aluminum when the soluble species are at one molar concentration (25°C).

stability for all conditions of pH and potential was used to carry out this work [4]. Figure 4.12 illustrates the results of such computation for aluminum in the presence of water at 25°C when the activities of all species considered were set at value unity.

However, an additional consideration is necessary to make such diagrams useful for corrosion situations for which the presence of soluble species in the environment never reaches values of the order of 1 M. Figure 4.13 illustrates the results that were computed by setting the concentrations of soluble species at decreasing values of one (10^0), one hundredth (10^{-2}), one in ten thousand (10^{-4}), and one in a million (10^{-6}). The apparent stability of the solid species considered gradually recedes as lower values of soluble species are used in the calculations.

It is customary to use the lowest boundary (10^{-6}) as a practical indication of the corrosion stability of a metal and its solid products (Figure 4.14). The usefulness of this graphical representation of thermodynamic data for corrosion studies was discussed by Pourbaix who showed three possible states of a metallic material [3]:

- *Immunity region*: In the conditions of potential and pH of that region a metal is considered to be totally immune from corrosion attack and safe to use. Cathodic protection may be used to bring the potential of a metal closer to that region by forcing a cathodic shift, as shown for aluminum by the domain specified in Fig. 4.14 (or −1.0 to 1.2 V vs. CCSRE).

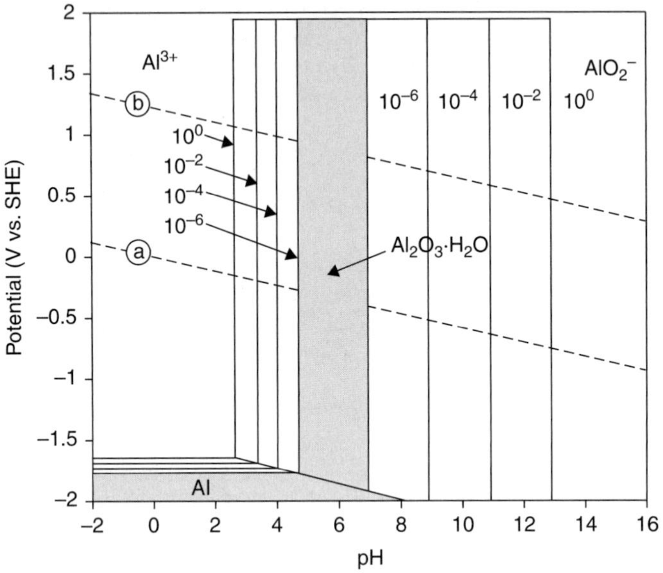

FIGURE 4.13 E-pH diagram of aluminum with four concentrations of soluble species (25°C).

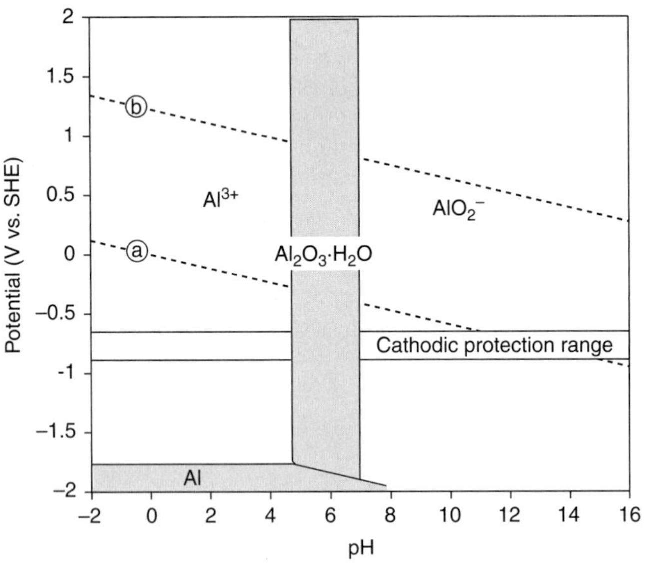

FIGURE 4.14 E-pH corrosion diagram of aluminum at 25°C with range where cathodic protection is useful.

- *Passive region*: In such region a metal tends to become coated with an oxide or hydroxide that may form on the metal either as a compact and adherent film practically preventing all direct contact between the metal itself and the environment, or as a porous deposit which only partially prevents contact between the metal and the environment.
- *Corrosion region*: Thermodynamic calculations indicate that, in such region of an E-pH diagram, a metal is stable as an ionic (soluble) product and therefore susceptible to corrosion attack. Experience is required to find out the extent and form of the corrosion attack that may occur in the corrosion region(s) of a Pourbaix diagram.

These three regions are indicated in Fig. 4.14 where the chemical reactivity of pure aluminum is noticeable by the small size of its immunity region. However, there is fortunately a band of stable oxy-hydroxide product at all potentials above the immunity region at neutral pH values. As mentioned earlier, the Pourbaix diagram of zinc is quite similar. Other metals such as beryllium, gallium, indium, and cadmium also have this column of corrosion product stable through the highest potential on a Pourbaix diagram.

The presence of a protective oxide layer on aluminum is the main reason why aluminum alloys are so broadly used with success in indoor and outdoor environments provided they fall within the passivation potential/pH boundaries shown in Fig. 4.14. The aluminum susceptibility to corrode in both acidic and basic environments is referred to as an amphoteric behavior. While the aluminum oxide will form naturally on aluminum, it is common practice to produce this oxide in a controlled process called anodization. As described in Chap. 5, the quality and properties of the protective oxide can thus be greatly enhanced, providing various finishes for a multitude of applications.

Iron E-pH Diagram

Figure 4.15 illustrates the E-pH diagram for iron at 25°C in the presence of water or humid environments. This diagram was calculated by considering all possible reactions associated with iron in wet or aqueous conditions listed in Table 4.13, excluding therefore drier forms of corrosion products such as magnetite (Fe_3O_4) or iron (ferric) oxide (Fe_2O_3). The various stability regions for these drier corrosion products are shown in Fig. 4.16 where the predominant compounds and ions are also indicated.

At potentials more positive than −0.6 and at pH values below 9, ferrous ion (Fe^{2+} or Fe II) is the stable substance. This indicates that iron will corrode under these conditions. In other regions of the iron E-pH diagram, it can be seen that the corrosion of iron

Equilibria
$2\,e^- + 2H^+ = 1H_2$
$4\,e^- + 1O_2 + 4H^+ = 2H_2O$
$2\,e^- + 1Fe(OH)_2 + 2H^+ = 1Fe + 2H_2O$
$2\,e^- + 1Fe^{2+} = 1Fe$
$2\,e^- + 1Fe(OH)_3^- + 3H^+ = 1Fe + 3H_2O$
$1\,e^- + 1Fe(OH)_3 + 1H^+ = 1Fe(OH)_2 + 1H_2O$
$1\,e^- + 1Fe(OH)_3 + 3H^+ = 1Fe^{2+} + 3H_2O$
$1Fe(OH)_3^- + 1H^+ = 1Fe(OH)_2 + 1H_2O$
$1\,e^- + 1Fe(OH)_3 = 1Fe(OH)_3^-$
$1Fe^{3+} + 3H_2O = 1Fe(OH)_3 + 3H^+$
$1Fe^{2+} + 2H_2O = 1Fe(OH)_2 + 2H^+$
$1\,e^- + 1Fe^{3+} = 1Fe^{2+}$
$1Fe^{2+} + 1H_2O = 1FeOH^+ + 1H^+$
$1FeOH^+ + 1H_2O = 1Fe(OH)_{2(sln)} + 1H^+$
$1Fe(OH)_{2(sln)} + 1H_2O = 1Fe(OH)_3^- + 1H^+$
$1Fe^{3+} + 1H_2O = 1FeOH^{2+} + 1H^+$
$1FeOH^{2+} + 1H_2O = 1Fe(OH)_2^+ + 1H^+$
$1Fe(OH)_2^+ + 1H_2O = 1Fe(OH)_{3(sln)} + 1H^+$
$1FeOH^{2+} + 1H^+ = 1Fe^{2+} + 1H_2O$
$1\,e^- + 1Fe(OH)_2^+ + 2H^+ = 1Fe^{2+} + 2H_2O$
$1\,e^- + 1Fe(OH)_{3(sln)} + 1H^+ = 1Fe(OH)_{2(sln)} + 1H_2O$
$1\,e^- + 1Fe(OH)_{3(sln)} + 2H^+ = 1FeOH^+ + 2H_2O$
$1\,e^- + 1Fe(OH)_{3(sln)} + 3H^+ = 1Fe^{2+} + 3H_2O$

Table 4.13 Possible Reactions in the Fe-H_2O System between the Species Most Stable in Wet Conditions

produces ferric ions (Fe^{3+} or Fe III), ferric hydroxide [$Fe(OH)_3$], ferrous hydroxide [$Fe(OH)_2$], and at very alkaline conditions, complex $HFeO_2^-$ ions. In Fig. 4.16, the solid corrosion products considered are ferric oxide (Fe_2O_3) and magnetite (Fe_3O_4), both important iron ore constituents.

The presence of a relatively large immunity region in Fig. 4.15 and Fig. 4.16, where corrosion products are solid and possibly protective, indicates that iron may corrode much less under these potential/ pH conditions.

These diagrams also indicate that if the potential of iron is made sufficiently negative or shifted cathodically below approximately −0.5 V vs. SHE in neutral or acidic environments, as indicated in

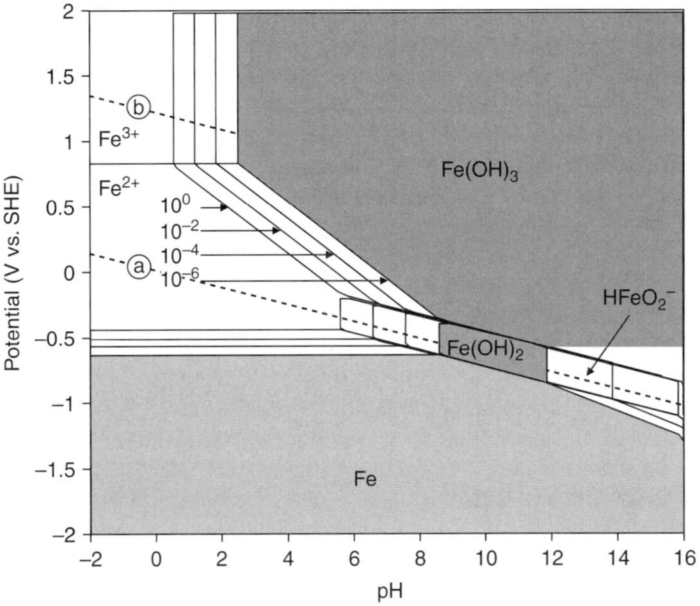

Figure 4.15 E-pH diagram of iron or steel with four concentrations of soluble species, three soluble species and two wet corrosion products (25°C).

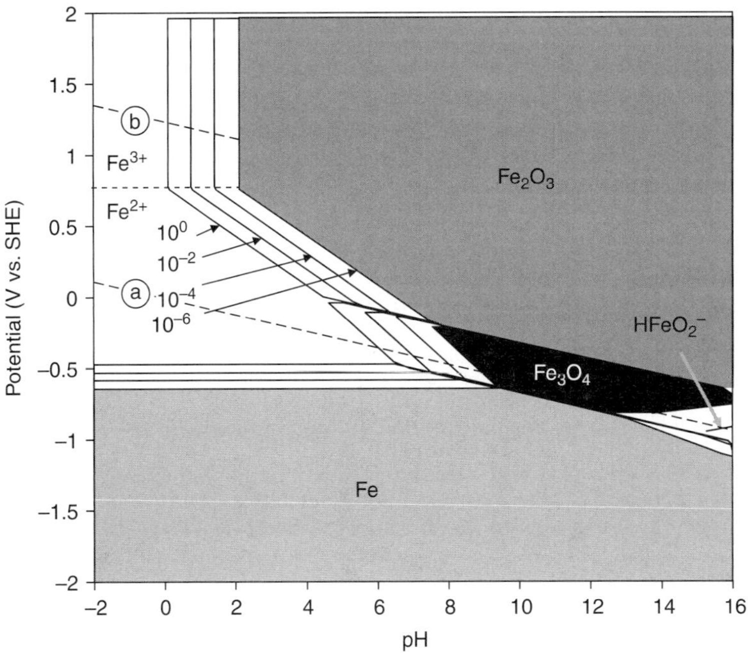

Figure 4.16 E-pH diagram of iron or steel with four concentrations of soluble species, three soluble species and two dry corrosion products (25°C).

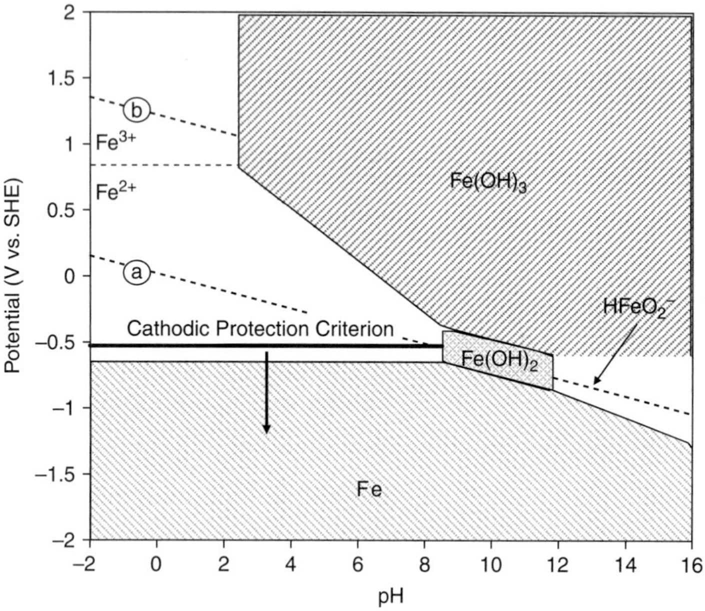

FIGURE 4.17 E-pH diagram of iron with the cathodic protection criterion at −053 V vs. SHE (−0.85 V vs. CCSRE).

Fig. 4.17, iron will corrode much less. This explains the generally accepted cathodic protection criterion of −0.85 V vs. CCSRE used across industries to protect steel assets buried in soils. The difference between this cathodic potential and line (a) is indicative that such potential will also tend to electrolyze water into hydrogen as indicated in Eqs. (4.33) and (4.34).

References

1. Criss CM, Cobble JW. The thermodynamic properties of high temperature aqueous solutions. *Journal of the American Chemical Society* 1964; 86: 5385–93.
2. Pawell SJ, Lopez RJ, Ondak E. Chemical and environmental influences on copper/copper sulfate reference electrode half cell potential. *Materials Performance* 1998; 38: 24–9.
3. Pourbaix M. *Atlas of Electrochemical Equilibria in Aqueous Solutions.* 2nd edn. Houston, Tex: NACE International, 1974.
4. KTS-Thermo. [2.1]. 2002. Kingston, Canada, Kingston Technical Software.

CHAPTER 5

Corrosion Kinetics and Applications of Electrochemistry to Corrosion

5.1 What Is Overpotential?

Thermodynamic principles can explain a corrosion situation in terms of the stability of chemical species and reactions associated with corrosion processes. However, thermodynamic principles cannot be used to predict corrosion currents or corrosion rates. In reality, polarization effects control the cathodic and anodic currents that are integral components of corrosion processes.

When two or more complementary processes such as those illustrated in Chap. 3 occur over a single metallic surface, the corrosion potential that results from such situations is a compromise between the various equilibrium potentials of all the anodic and cathodic reactions involved. The difference between the resultant potential (E) and each individual reaction equilibrium potential (E_{eq}) is called polarization and is quantified in terms of overpotential (η) described in Eq. (5.1):

$$\eta = E - E_{eq} \tag{5.1}$$

The polarization is said to be anodic when the anodic processes on the electrode are accelerated by moving the potential in the positive (noble) direction or cathodic when the cathodic processes are accelerated by moving the potential in the negative (active) direction. There are three distinct types of polarization and these are additive, as expressed in Eq. (5.2):

$$\eta_{total} = \eta_{act} + \eta_{conc} + iR \tag{5.2}$$

where η_{act} is the activation overpotential, a complex function describing the charge transfer kinetics of an electrochemical reaction. η_{act} is always present and the main polarization component at small polarization currents or voltages.

η_{conc} is the concentration overpotential, a function describing the mass transport limitations associated with electrochemical processes. η_{conc} is dominant at larger polarization currents or voltages.

iR is the ohmic drop. This function takes into account the electrolytic resistivity of an environment when the anodic and cathodic elements of a corrosion reaction are separated by this environment while still electrically coupled.

Activation polarization is usually the controlling factor during corrosion in strong acids since both η_{conc} and iR are relatively small. Concentration polarization usually predominates when the concentration of the active species is low; for example, in dilute acids or in aerated waters where the active component, dissolved oxygen, is only present at very low levels. The ohmic drop will become an extremely important factor when studying corrosion phenomena for which there is a clear separation of the anodic and cathodic corrosion sites, for example, crevice corrosion. The ohmic drop is also an important variable in the application of protective methods such as anodic and cathodic protection that forces a potential shift of the protected structure by passing a current in the environment.

Knowing the kind of polarization which is occurring can be very helpful, since it allows an assessment of the determining characteristics of a corroding system. For example, if corrosion is controlled by concentration polarization, then any change that increases the diffusion rate of the active species (e.g., oxygen) will also increase the corrosion rate. In such a system, it would therefore be expected that agitating the liquid or stirring it would tend to increase the corrosion rate of the metal. However, if a corrosion reaction is activation controlled then stirring or increasing the agitation will have no effect on the corrosion rate.

5.2 Activation Polarization

Activation polarization is due to retarding factors that are an inherent part of the kinetics of all electrochemical reactions. For example, consider the evolution of hydrogen gas illustrated previously in Chap. 3 and described by Eq. (5.3):

$$2H^+ + 2e^- \rightarrow H_2(g) \tag{5.3}$$

While this reaction seems to be relatively simple, the rate at which hydrogen ions are transformed into hydrogen gas is in reality a function of several factors, including the rate of electron transfer from a metal to hydrogen ions. In fact, there is a wide variability in this transfer rate of electrons on various metals and, as a result, the rate of hydrogen evolution from different metal surfaces can vary greatly.

The exchange current density (i_0) is surely the single most important variable that explains the large differences in the rate of hydrogen production on metallic surfaces. Table 5.1 contains the approximate exchange current density for the reduction of hydrogen ions on a range of materials. Note that the value for the exchange current density of hydrogen evolution on platinum is approximately 10^{-2} A/cm^2 whereas on mercury and lead it is 10^{-13} A/cm^2, eleven orders of magnitude difference for the rate of this particular reaction, or one hundred billion times easier on platinum than on mercury or lead!

This is the reason why mercury is often added to power cells such as the popular alkaline primary cells to stifle the thermodynamically favored production of gaseous hydrogen and prevent unpleasant incidents. This is also why lead acid batteries (car batteries) can provide power in a highly acidic environment in a relatively safe manner unless excessive charging currents are used.

Even so, the exchange current density remains an elusive parameter that may change rapidly with changing conditions at a metallic surface being naturally modified during its exposure in a given environment. One problem is that there is no simple method to estimate the exchange current density for a specific system. The exchange current density must be determined experimentally by scanning the potential with a laboratory setup such as shown in Fig. 5.1. In this experimental arrangement a potenstiostat/galvanostat power controller is used to pass current through the sample, or working electrode (W), and an auxiliary electrode (AUX) immersed in solution while monitoring the potential of the working electrode with a reference electrode and a Luggin capillary.

Metal	$\log_{10} i_0$ (A/cm^2)
Pb, Hg	−13
Zn	−11
Sn, Al, Be	−10
Ni, Ag, Cu, Cd	−7
Fe, Au, Mo	−6
W, Co, Ta	−5
Pd, Rh	−4
Pt	−2

Table 5.1 Approximate Exchange Current Density (i_0) for the Hydrogen Oxidation Reaction on Different Metals at 25°C

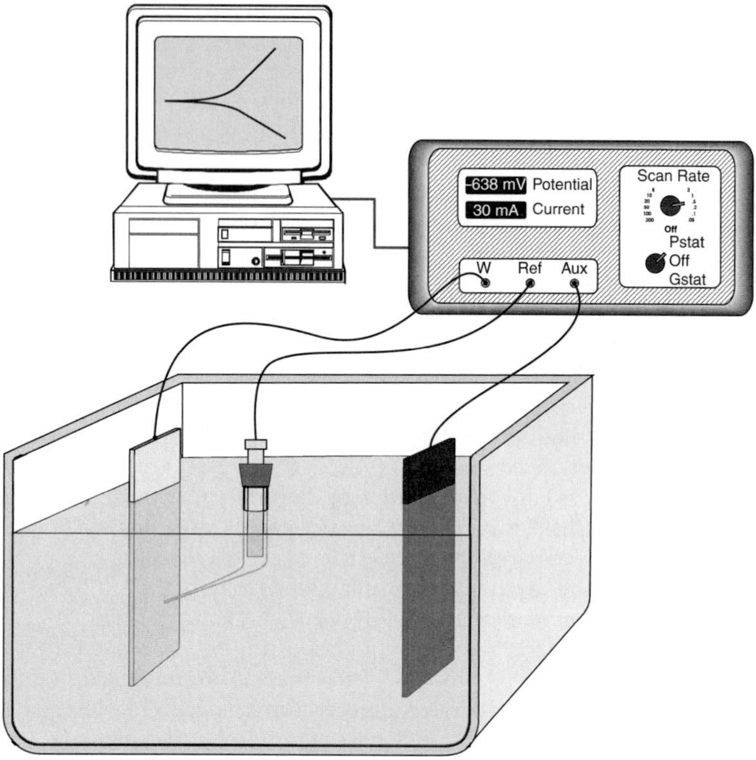

FIGURE 5.1 Electrochemical instrumentation to carry out potentiodynamic measurements in which a potenstiostat/galvanostat power controller is used to pass current through the sample or working electrode (W) and an auxiliary electrode (AUX), while monitoring the potential of the working electrode with a reference electrode.

The following theory explains the basic mathematics that may be used to extract the exchange current density from the results obtained. A general representation of the polarization of an electrode supporting one specific reaction is given in the Butler-Volmer equation (5.4):

$$i_{\text{reaction}} = i_0 \left\{ \exp\left(-\beta \frac{nF}{RT} \eta_{\text{reaction}} \right) - \exp\left((1-\beta) \frac{nF}{RT} \eta_{\text{reaction}} \right) \right\} \tag{5.4}$$

where i_{reaction} is the anodic or cathodic current

- β is the charge transfer barrier (symmetry coefficient) for the anodic or cathodic reaction, usually close to 0.5
- n is the number of participating electrons
- R is the gas constant, that is, 8.314 J mol^{-1} K^{-1}
- T is absolute temperature (K)
- F is 96,485 C/(mol of electrons)

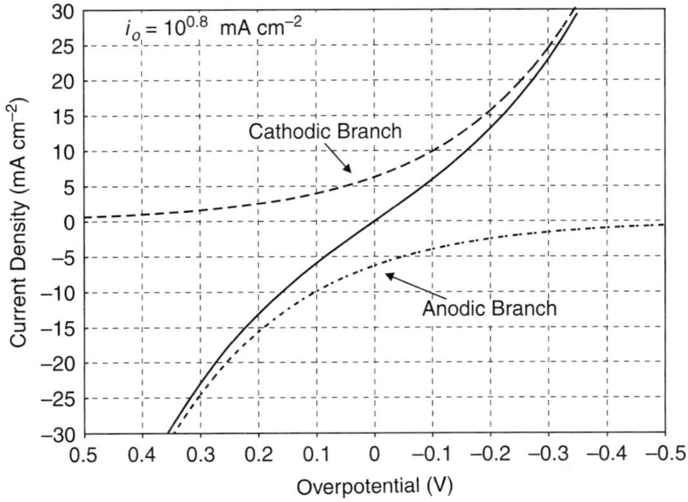

FIGURE 5.2 Current versus overpotential polarization plot of the ferric/ferrous ion reaction on palladium showing both the anodic and cathodic branches of the resultant current behavior.

The presence of two polarization branches in a single reaction expressed in Eq. (5.4) is illustrated in Fig. 5.2 for the polarization of a palladium electrode immersed in a solution containing similar concentrations of ferric (Fe^{3+}) and ferrous (Fe^{2+}) ions with a completely reversible reaction described in Eq. (5.5):

$$Fe^{3+} + e^- \rightleftharpoons Fe^{2+} \tag{5.5}$$

When η_{reaction} is cathodic, that is, negative, the second term in the Butler-Volmer equation becomes negligible and the cathodic current density (i_c) can be expressed by a simpler equation [Eq. (5.6)] and its logarithm [Eq. (5.7)]:

$$i_{\text{reaction}} = i_c = i_0 \exp\left(-\beta \frac{nF}{RT} \eta_{\text{reaction}}\right) \tag{5.6}$$

$$\eta_{\text{reaction}} = \eta_c = b_c \log_{10}\left(\frac{i_c}{i_0}\right) \tag{5.7}$$

where b_c is the cathodic Tafel coefficient described in Eq. (5.8) that can be obtained from the slope of a plot of η against log $|i|$, with the intercept yielding a value for i_0 as shown in Fig. 5.3.

$$b_c = -2.303 \frac{RT}{\beta nF} \tag{5.8}$$

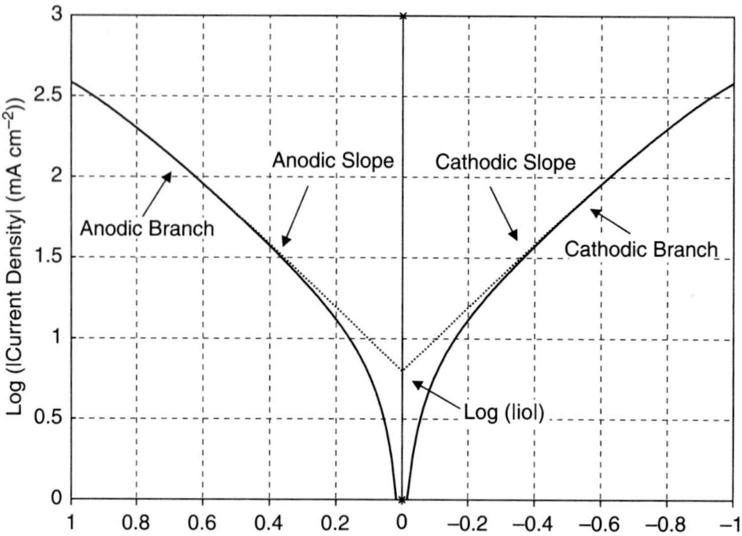

Figure 5.3 Plot of η against log $|i|$ or Tafel plot showing the exchange current density can be obtained with the intercept.

Similarly, when $\eta_{reaction}$ is anodic, that is, positive, the first term in the Butler-Volmer equation becomes negligible and the anodic current density (i_a) can be expressed by Eq. (5.9) and its logarithm in Eq. (5.10), with b_a obtained by plotting $\eta_{reaction}$ versus log $|i|$ [Eq. (5.11)]:

$$i_{reaction} = i_a = -i_0 \exp\left((1-\beta)\frac{nF}{RT}\eta_{reaction}\right) \tag{5.9}$$

$$\eta_a = b_a \log_{10}\left(\frac{|i_a|}{i_0}\right) \tag{5.10}$$

$$b_a = 2.303 \times \frac{RT}{\beta nF} \tag{5.11}$$

5.3 Concentration Polarization

Concentration polarization is the polarization component caused by concentration changes in the environment adjacent to the surface as illustrated in Fig. 5.4. When a chemical species participating in a corrosion process is in short supply, the mass transport of that species to the corroding surface can become rate controlling. A frequent case of concentration polarization occurs when the cathodic processes depend on the reduction of dissolved oxygen since it is usually in low concentration, that is, in parts per million (ppm) as shown in Table 5.2 [1].

Temperature (°C)	Volume (cm³)*	Concentration (ppm)	Concentration (M) (μmol L⁻¹)
0	10.2	14.58	455.5
5	8.9	12.72	397.4
10	7.9	11.29	352.8
15	7.0	10.00	312.6
20	6.4	9.15	285.8
25	5.8	8.29	259.0
30	5.3	7.57	236.7

* cm^3 at 0°C per kg of water.

Table 5.2 Solubility of Oxygen in Air-Saturated Water

As illustrated in Fig. 5.5, mass transport to a surface is governed by three forces, that is, diffusion, migration, and convection. In the absence of an electrical field the migration term is negligible since it only affects charged ionic species while the convection force disappears in stagnant conditions. For purely diffusion controlled

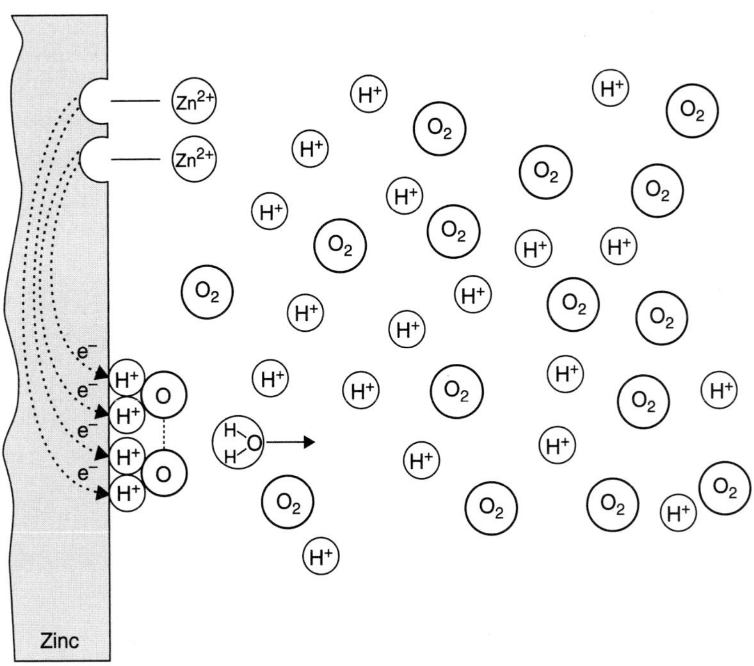

Figure 5.4 Concentration changes in the vicinity of an electrode causing a concentration polarization effect.

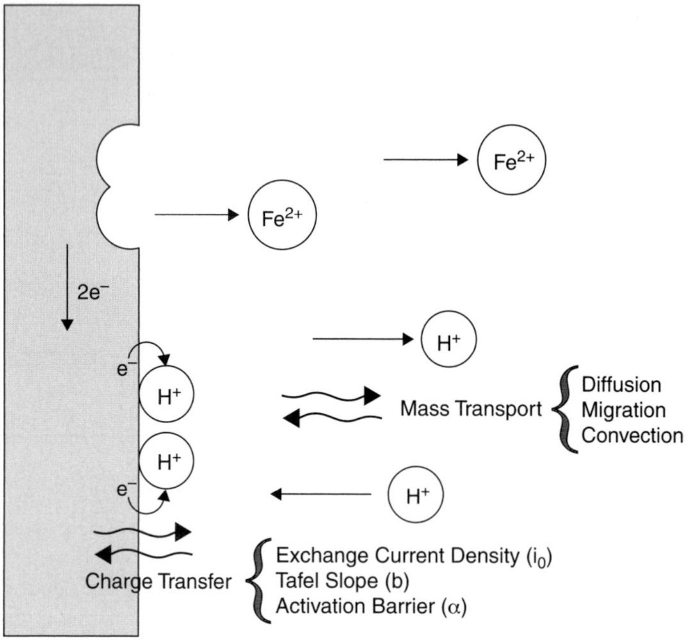

FIGURE 5.5 Graphical representation of the processes occurring at an electrochemical interface.

mass transport, the flux of a species O to a surface from the bulk is described with Fick's first law in Eq. (5.12):

$$J_O = -D_O\left(\frac{\delta C_O}{\delta x}\right) \tag{5.12}$$

where J_O is the flux of species O (mol s^{-1} cm^{-2})
D_O is the diffusion coefficient of species O (cm^2 s^{-1})
δC_O is the concentration gradient of species O across the interface between the metallic surface and the bulk environment (mol cm^{-3})
δx is the Nernst diffusion layer or diffuse layer (cm)

Table 5.3 contains values for D_O of some common ions. For more practical situations the diffusion coefficient can be approximated with the help of Eq. (5.13), that relates D_O to the viscosity of the solution (μ) and absolute temperature:

$$D_O = \frac{T \cdot A}{\mu} \tag{5.13}$$

where A is a constant for the system.

Cation	D × 10^5 (cm^2 s^{-1})	Anion	D × 10^5 (cm^2 s^{-1})
H^+	9.30	OH^-	5.25
Li^+	1.03	F^-	1.47
Na^+	1.33	Cl^-	2.03
K^+	1.95	NO_3^-	1.90
Ca^{2+}	0.79	ClO_4^-	1.79
Cu^{2+}	0.71	SO_4^{2-}	1.06
Zn^{2+}	0.70	CO_3^{2-}	0.92
O_2	2.26	HSO_4^-	1.33
H_2O	2.44	HCO_3^{-1}	1.11

Table 5.3 Diffusion Coefficients of Selected Ions at Infinite Dilution in Water at 25°C

Figure 5.6 illustrates the concentration-distance profile at the electrode surface approximated by a simple gradient. In this diagram the metallic surface is positioned at the ordinate axis while the x-axis expresses the distance away from the electrode and the y-axis the concentration of the chemical species being reacted. For well-mixed solutions, the concentration is constant in the bulk or convective region.

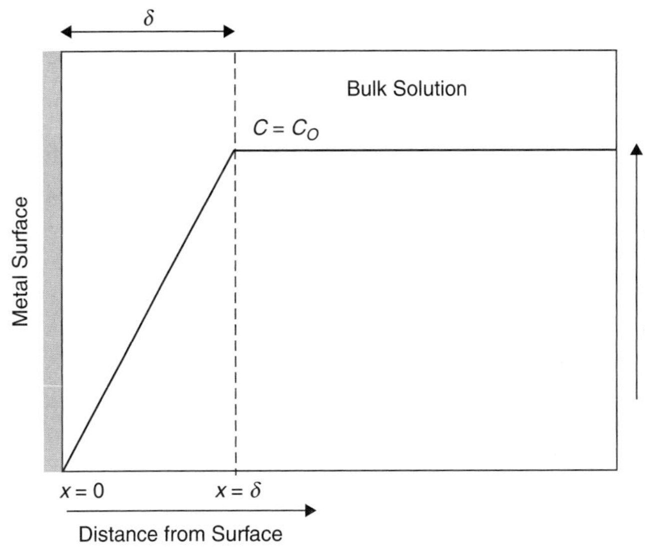

Figure 5.6 Nernst diffusion layer for a limiting current situation.

This is represented by the horizontal line where $C = C_O$. There is also a region where the concentration drops, falling to zero at the electrode surface. The Nernst diffusion layer associated with this drop has a specific thickness (δ) that depends upon the nature of the solution into which it extends. For stirred aqueous solutions the thickness of the diffusion layer varies between 0.01 and 0.001 mm.

For a chemical species O that is consumed by the cathodic reaction at the corroding surface, the concentration gradient ($\delta C_O/\delta x$) is greatest when the concentration of that species is completely depleted at the surface, that is, $C_O = 0$. It follows that the cathodic current is limited in that condition, as expressed by Eq. (5.14).

$$i_c = i_L = -nFD_O \frac{C_O^{\text{bulk}}}{\delta} \tag{5.14}$$

For intermediate cases, that is, when the cathodic current is smaller than i_L, η_{conc} can be evaluated using an expression [Eq. (5.15)] derived from Nernst equation:

$$\eta_{\text{conc}} = \frac{2.303 \times RT}{nF} \log_{10}\left(1 - \frac{i}{i_L}\right) \tag{5.15}$$

where $2.303 \times R \times T/F = 0.059$ V when $T = 298.16$ K.

5.4 Ohmic Drop

The ohmic overpotential appears in Eq. (5.2) as the simple product of a resistance and a current between the anodic and cathodic sites of a corrosion process. For many corrosion situations these sites are adjacent to each other and the ohmic drop is negligible, particularly so when the environment itself is a good electrolytic conductor, that is, seawater. However, there are special conditions where the separation of the anodic and cathodic sites can be an important factor in the corrosion progress, for example, galvanic corrosion, or even an integral part of a particular protection scheme, for example, anodic and cathodic protection.

5.4.1 Water Resistivity Measurements

The conductivity of an environment can itself be a complex function. When a salt dissociates, the resulting ions interact with surrounding water molecules to form charged clusters known as solvated ions. These ions can move through the solution under the influence of an externally applied electric field. Such motion of charge is known as ionic conduction and the resulting conductance is the reciprocal function of the resistance of an environment.

The dependence upon the size and shape of the conductor can be corrected by using the resistivity ρ rather than the resistance R, as

Water	ρ (Ω cm)
Pure water	20,000,000
Distilled water	500,000
Rain water	20,000
Tap water	1000–5000
River water (brackish)	200
Sea-water (coastal)	30
Sea-water (open sea)	20–25

TABLE 5.4 Resistivity of Some Typical Waters

expressed in Eq. (5.16) for the simple cell shown in Fig. 5.7. Table 5.4 lists some typical values of water resistivity [1].

$$\rho = \frac{R}{\left(\frac{\ell}{A}\right)} \tag{5.16}$$

where R is the measured resistance across the cell
A is the cross-sectional area of each electrode, provided that both electrodes have the same dimensions
ℓ is the gap separating the electrodes in Fig. 5.7

The ratio (ℓ/A) is also called the cell constant or *shape factor* and has units of cm^{-1} or m^{-1}. A variant of the electrochemical cell shown in Fig. 5.7 is commonly used to evaluate the conductivity of a solution between two electrodes by using an alternating current technique.

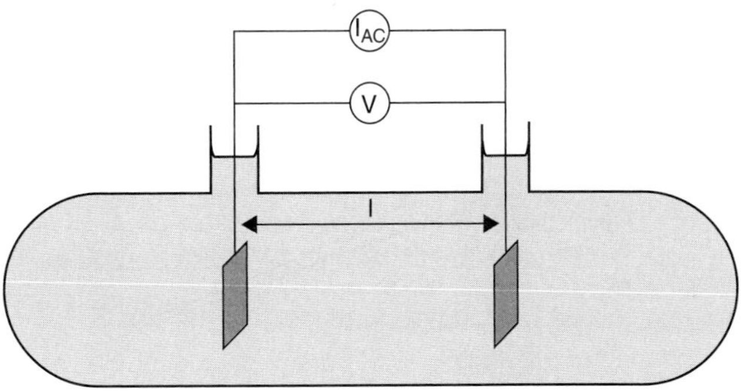

FIGURE 5.7 Schematic of a conductivity cell containing an electrolyte and two inert electrodes of surface A parallel to each other and separated by distance ℓ.

Other geometries would require the calculation of appropriate cell constants. The cell constant of an electrochemical cell with two concentric tubes as electrodes would be, for example, expressed by Eq. (5.17). Such an arrangement is a common design in the production of domestic water heaters in which a central sacrificial magnesium anode is inserted typically in the center of the hot-water tank to protect the surrounding tank material.

$$\text{Cell constant} = \frac{1}{2\pi h}\ln\left(\frac{r_2}{r_1}\right) \tag{5.17}$$

where h is the height of the cylindrical tank
r_2 is the internal tank radius
r_1 is the radius of the sacrificial anode

The ohmic drop can be minimized, when carrying out electrochemical tests, by placing the reference electrode in a Luggin capillary brought as close as possible to the surface being monitored (Fig. 5.8). Additionally, the Luggin capillary allows sensing of the solution potential close to the working electrode without the adverse shielding effects that may be caused when the reference electrode is positioned in front of the surface being monitored. A Luggin capillary

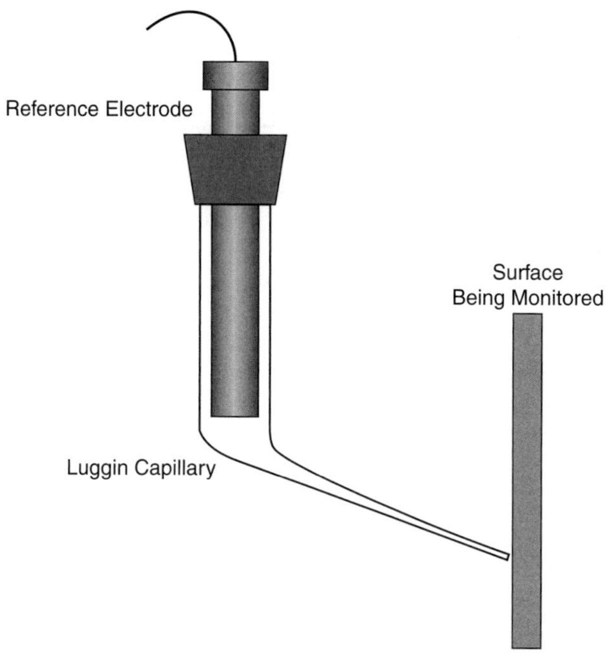

FIGURE 5.8 Schematic of a Luggin capillary to position a reference electrode in close proximity to an electrochemical cell's working electrode.

can be made of any material provided it is inert to the electrolytic environment. It basically consists of a bent tube generally filled with the test solution with a large enough opening to accommodate a reference electrode at one end and a usually much smaller opening at the other end to provide diffusional movement of the electrolyte.

5.4.2 Soil Resistivity Measurements

Soil resistivity is a function of soil moisture and the concentrations of ionic soluble salts and is considered to be the most comprehensive indicator of a soil's corrosivity. Typically, the lower the resistivity, the higher will be the corrosivity as discussed in more details in Chap. 10. Typically, soil resistivity decreases with increasing water content and the concentration of ionic species. Sandy soils, for example, are high up on the resistivity scale and therefore considered the least corrosive while clay soils are excellent at retaining water and at the opposite end of the corrosivity spectrum.

Four-Pin Method (Wenner Method)

Field soil resistivity measurements are most often conducted using the Wenner four-pin method and a soil resistance meter following the principles laid out by Wenner nearly one century ago [2]. The Wenner method requires the use of four metal probes or electrodes, driven into the ground along a straight line, equidistant from each other, as shown in Fig. 5.9 and Fig. 5.10. Soil resistivity is a relatively simple function derived from the voltage drop between the center pair of pins (P1 and P2 in Fig 5.9), with current flowing between the two outside pins (C1 and C2 in Fig 5.9) assuming that the measured resistivity is a measure of the hemispherical volume of earth probed by the central pins.

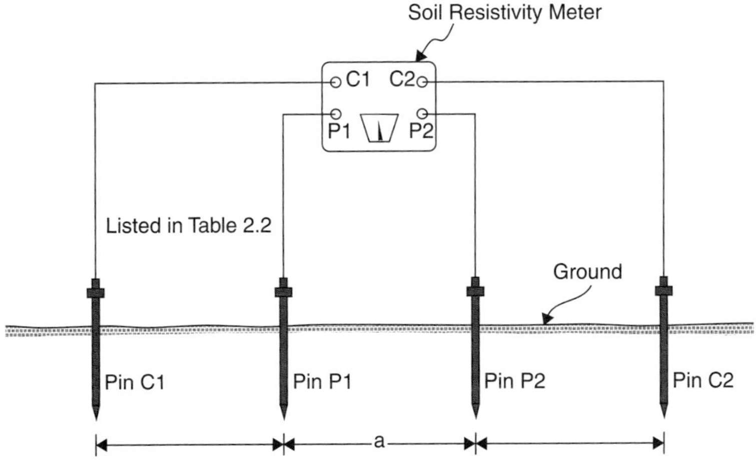

FIGURE 5.9 Wenner four-pin soil resistivity test setup.

(a)

(b)

FIGURE 5.10 Four-pin soil resistivity measurements being made; (*a*) field setup and (*b*) close-up on the instrument. (Courtesy of Tinker & Rasor)

An alternating current from the soil resistance meter causes current to flow through the soil, between pins C1 and C2 and the voltage or potential is measured between pins P1 and P2. Resistivity of the soil is then computed from the instrument reading, according to the following formula [1]:

$$\rho = 2\pi\, aR \tag{5.18}$$

where ρ is the soil resistivity (Ω cm)
a is the distance between probes (cm)
R is the soil resistance (Ω), instrument reading
π is 3.1416

The resistivity values obtained represent the average resistivity of the soil to a depth equal to the pin spacing. Resistance measurements are typically performed to a depth equal to that of the buried system (pipeline) being evaluated. Typical probe spacing is in increments of 0.5 to 1 m.

If the line of soil pins used when making four-pin resistivity measurements is closely parallel to a bare underground pipeline or other metallic structure, the presence of the bare metal may cause the indicated soil resistivity values to be lower than it actually is. Because a portion of the test current will flow along the metallic structure rather than through the soil, measurements along a line closely parallel to pipelines should therefore be avoided.

The recorded values from four-pin resistivity measurements can be misleading unless it is remembered that the soil resistivity encountered with each additional depth increment is averaged, in the test, with that of all the soil in the layers above. With experience, much can be learned about the soil structure by inspecting series of readings to increasing depths. The indicated resistivity to a depth equal to any given pipe spacing is a weighted average of the soils from the surface to that depth. Trends can be illustrated best by inspecting the sets of soil resistivity readings in Table 5.5.

Pin Spacing (m)	Pin Pacing (F)	Soil Resistivity (Ω cm)			
		Set A	Set B	Set C	Set D
0.76	2.5	960	1100	3300	760
1.5	5	965	1000	2200	810
2.3	7.5	950	1250	1150	1900
3.0	10	955	1500	980	3800
3.8	12.5	960	1610	840	6900
4.6	15	955	1710	780	12,500

Table 5.5 Examples of Soil Resistivity Readings Using Four-Pin Method

The first set of data in Table 5.5, Set A, represents uniform soil conditions. The average of the readings shown (~960 Ω cm) represents the effective resistivity that may be used for design purposes for impressed current groundbeds or galvanic anodes.

Data Set B represents low-resistivity soils in the first few feet. There may be a layer of somewhat less than 1000 Ω cm around the 1.5 m depth level. Below 1.5 m, however, higher-resistivity soils are encountered. Because of the averaging effect the actual resistivity at 2.3 m deep would be higher than the indicated 1250 Ω cm and might be in the order of 2500 Ω cm or more. Even if anodes were placed in the lower-resistivity soils, there would be resistance to the flow of current downward into the mass of the earth.

If designs are based on the resistivity of the soil in which the anodes are placed, the resistance of the completed installation will be higher than expected. The anodes will perform best if placed in the lower resistance soil. However, the effective resistivity used for design purposes should reflect the higher resistivity of the underlying areas. In this instance, where increase is gradual, using horizontal anodes in the low-resistivity area and a figure of effective resistivity of ~2500 Ω cm should result in a conservative design.

Data Set C represents an excellent location for anode location even though the surface soils have relatively high resistivity. It would appear from this set of data that anodes located >1.5 m deep, would be in low-resistivity soil of ~800 Ω cm, such a figure being conservative for design purposes. A lowering resistivity trend with depth, as illustrated by this set of data, can be relied upon to give excellent groundbed performance.

Data Set D is the least favorable of these sample sets of data. Low-resistivity soil is present at the surface but the upward trend of resistivity with depth is immediate and rapid. At the 2.3 m depth, for example, the resistivity could be tens of thousands of ohm-centimeters. One such situation could occur where a shallow swampy area overlies solid rock. Current discharged from anodes installed at such a location will be forced to flow for relatively long distances close to the surface before electrically remote earth is reached. As a result, potential gradients forming the area of influence around an impressed current groundbed can extend much farther than those surrounding a similarly sized groundbed operating at the same voltage in more favorable locations such as those represented by data Sets A and C.

Alternate Soil Resistivity Methods

In the two-pin (Shepard's Canes) method of soil resistivity measurement, the potential drop is measured between the same pair of electrodes used to supply the current [3]. As shown in Fig. 5.11, the probes are placed 0.3 m apart. If the soil is too hard for the probes to penetrate, the reading is taken at the bottom of two augured holes.

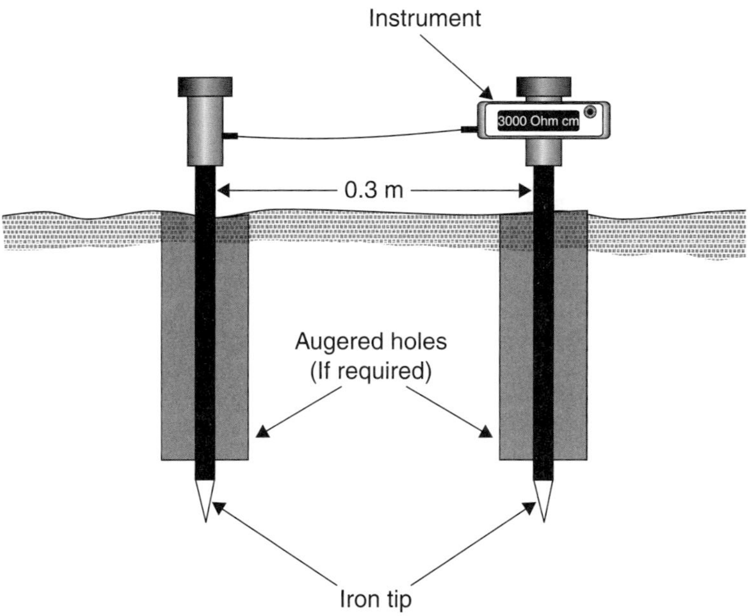

FIGURE 5.11 Two-pin (Shepard's Canes) method for soil resistivity measurements.

The instrument is calibrated for a probe spacing of 0.3 m and gives a reading directly in ohm cm. Although this method is less accurate than the four-pin method and measures the resistivity of the soil only near the surface, it is often used for preliminary surveys, as it is quicker than the four-pin method.

A soil rod is essentially a two-pin resistivity-measuring device where the electrodes are both mounted on a single rod, as shown in Fig. 5.12. As in the other two-pin method, the resistivity of the soil to a very shallow depth is measured. Also, the soil must be soft enough to allow penetration of the rod. Measurements using the soil rod, however, can be taken quickly when measuring in soft soil.

When it is impractical to make field measurements of soil resistivity, soil samples can be taken and the resistivity of the sample can be determined by using a soil box. As shown in Fig. 5.13, the method of measurement is essentially the four-pin method. Metal contacts in each end of the box pass current through the sample.

Potential drop is measured across probes inserted into the soil. The resistivity is calculated using constants provided with the particular geometry of soil box being used. Due to the disturbance of the soil during sampling and possible drying out of the soil during shipment, this method of soil resistivity measurement is less likely to represent true, in-place soil resistivity than an actual field test.

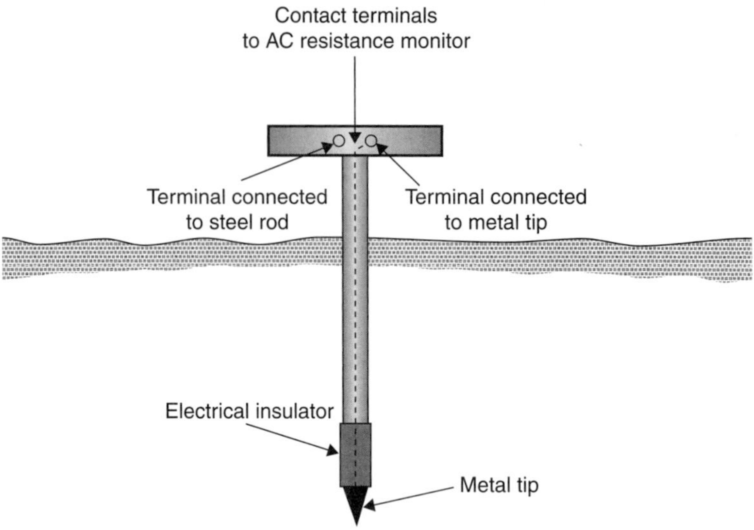

Figure 5.12 Soil rod method for soil resistivity measurements.

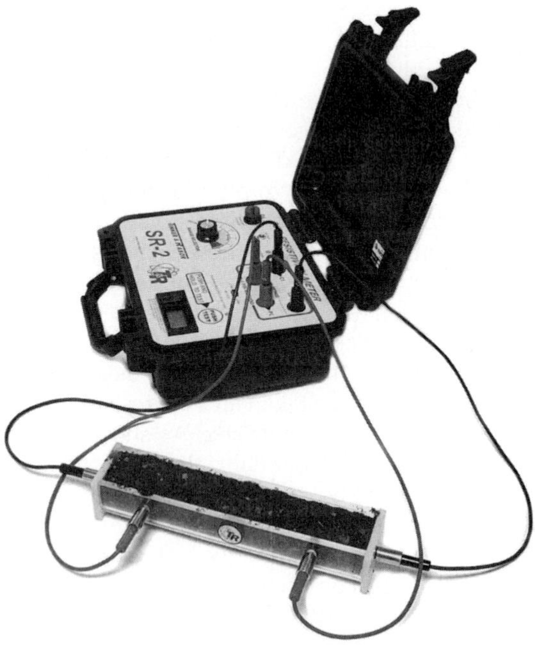

Figure 5.13 Soil box and resistivity equipment. (Courtesy of Tinker & Rasor)

5.5 Graphical Presentation of Kinetic Data (Evans Diagrams)

The use of polarization curves for the study of corrosion reactions can be traced back to the 1930s with the work of Wagner and Traud [4]. However the representation of the mixed potential behavior is often associated with Professor Evans who has popularized this representation of corrosion polarization measurements [5].

These polarization diagrams can be quite useful for describing or explaining parallel corrosion processes. According to the mixed-potential theory underlying these diagrams, any electrochemical reaction can be algebraically divided into separate oxidation and reduction reactions with no net accumulation of electrical charge. Under these circumstances the net measurable current is zero and the corroding metal is charge neutral, that is, all electrons produced by the corrosion of a metal have to be consumed by one or more cathodic processes.

In order to model a corrosion situation with mixed potential diagrams, one must first gather the information concerning the (1) activation overpotential for each corrosion process involved and (2) any additional information for processes that could be affected by concentration overpotential. The following sections present some examples that illustrate how the mixed potential theory may be used to explain simple cases where corrosion processes are purely activation controlled or cases where concentration controls at least one of the corrosion processes.

5.5.1 Activation Controlled Processes

For purely activation controlled processes, each reaction can be described by a straight line on an E versus Log i plot, with positive Tafel slopes for anodic processes and negative Tafel slopes for cathodic processes.

The following example illustrates the polarization behavior of carbon steel in a deaerated solution maintained at 25°C with a pH of zero. The solid line in Fig. 5.14 is the polarization plot itself and the dotted lines in this figure represent the anodic reaction in Eq. (5.19) and the cathodic reaction in Eq. (5.20) that describe the corrosion behavior of steel in these conditions. These lines are extrapolated from the linear sections of the plot on either the anodic or cathodic sides of the curve.

$$Fe_{(s)} \rightarrow Fe^{2+} + 2e^{+} \quad (5.19)$$

$$2H^{+} + 2e^{+} \rightarrow H_2(g) \quad (5.20)$$

While it is relatively easy to estimate the corrosion potential (E_{corr}) from the sharp peak observed at −0.221 V vs. standard

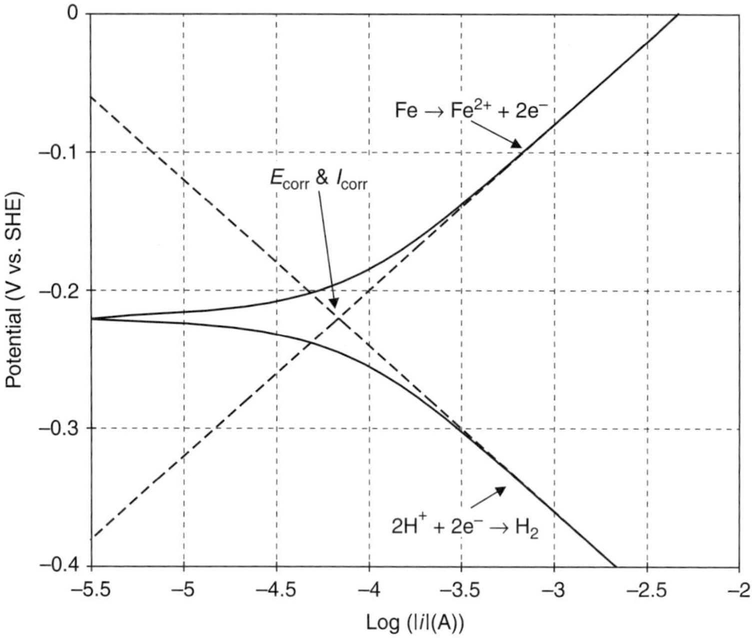

FIGURE 5.14 Polarization behavior of carbon steel in a deaerated solution maintained at 25°C and a pH of zero.

hydrogen electrode (SHE), when the current crosses zero (infinity on a log scale), the projected dotted lines are required to find the intercept indicating where the cathodic and anodic currents actually cancel each other. The corrosion current density (i_{corr}) can be obtained by dividing the anodic current by the surface area of the specimen, 1 cm^2 in the present case. According to the conversion table presented earlier in Chap. 3 (Table 3.2), the current density of 67 μA cm^{-2} evaluated in this example corresponds to a penetration rate of 0.8 mm y^{-1}.

The second example shows the carbon steel polarization behavior when exposed to a deaerated solution maintained at 25°C and pH of five. The mixed potential diagram of this system is shown in Fig. 5.15. The shift of the E_{corr} to a more negative value of −0.368 V vs SHE should be noted. The modeled projected lines provide an estimate of the corrosion current density of 4 μA cm^{-2} in this case and this current translates into a penetration rate of 0.05 mm y^{-1}.

5.5.2 Concentration Controlled Processes

When one of the reactions is limited by the rate of transport of the reactant to the metallic surface being corroded, the situation increases in complexity as illustrated in the polarization plot of the system in Fig. 5.16. The system represented here is similar to the previous one,

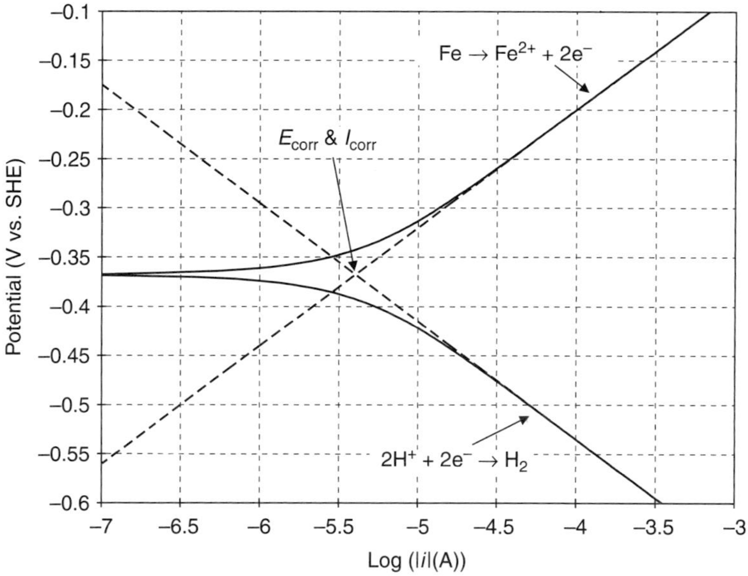

FIGURE 5.15 Polarization behavior of carbon steel in a deaerated solution maintained at 25°C and a pH of five.

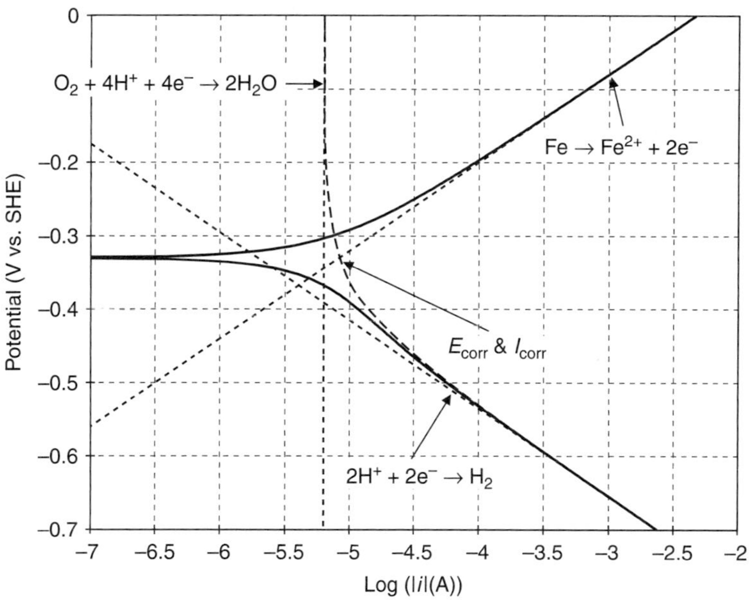

FIGURE 5.16 Polarization behavior of carbon steel in a stagnant aerated solution maintained at 25°C and a pH of five.

that is, pH of five at 25°C, with the exception that the environment is aerated and stagnant. In this situation the reduction of oxygen shown in Eq. (5.21) now becomes a possible second cathodic reaction.

$$O_2 + 4H^+ + 4e^- \rightarrow 2H_2O \tag{5.21}$$

In order to model the polarization plot, the total cathodic current corresponding to the sum of the currents of the hydrogen reaction and oxygen reduction has to be balanced by the single anodic current. The intercept where the opposing currents are balanced occurs at an E_{corr} of –0.33 V vs. SHE and, since the surface area is still 1 cm^2, an i_{corr} of 8.2 μA cm^{-2} or 0.1 mm y^{-1}. It should be noted that the current for the reduction of oxygen is constant across the potential range shown in Fig. 5.16 and has the value of 6.3 μA cm^{-2}.

The reduction of oxygen depends, among other factors, on the level of agitation of the environment. If the limiting current of this reaction is now increased through agitation by a tenfold factor to reach the value of 63 μA cm^{-2} a new situation emerges as depicted in Fig. 5.17. The marked positive shift of E_{corr}, now –0.224 V vs. SHE, is accompanied by a marked increase in current density, now of 63 μA cm^{-2} or 0.8 mm y^{-1}.

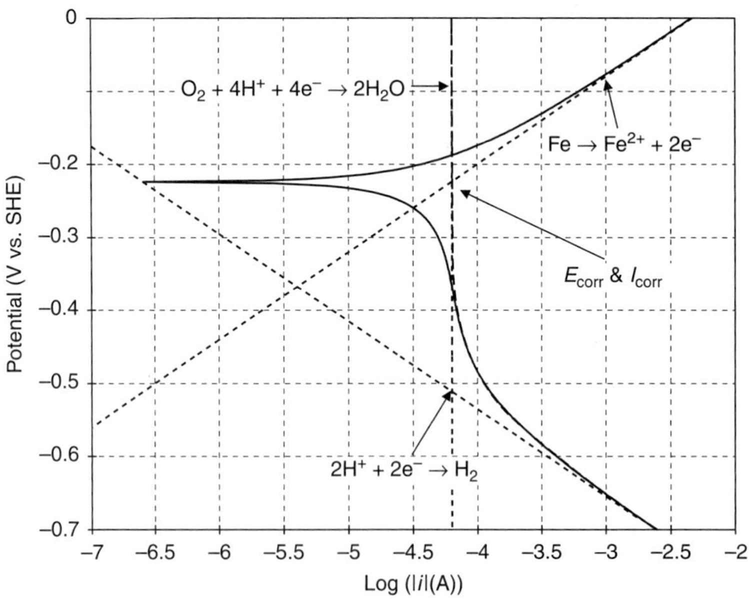

FIGURE 5.17 Polarization behavior of carbon steel in an agitated aerated solution maintained at 25°C and a pH of five.

5.6 Examples of Applied Electrochemistry to Corrosion

Given the electrochemical nature of corrosion processes it is not surprising to see that measurements and control methods based on electrochemical principles are so extensively used across the whole spectrum of corrosion science and engineering. The following sections provide some examples to illustrate how these principles are applied in practice.

5.6.1 Electrochemical Polarization Corrosion Testing

Corrosion testing with polarization methods basically consists in forcing potential or current changes on a sample under study while monitoring the resulting response in current or potential. This may be achieved by using either a direct current (DC) or an alternating current (AC) source. The instrumentation for carrying polarization testing as illustrated in Fig. 5.1 consists of

- A potentiostat is needed that will maintain the potential of the working electrode close to a preset value.
- A current measuring device for monitoring the current produced by the applied potential. The ability of the current measuring device to autorange or to change the scale automatically is also important.
- The ability of the data to be stored directly in a computer, or plotted out directly is also important.
- Polarization cells: Several test cells are available commercially for making polarization measurements. Polarization cells can have various configurations specific to the testing requirements from testing small coupons, to sheet materials, to testing inside autoclaves. In a plant environment the electrodes may be inserted directly into a process stream. Some of the features of a cell include [6]

    a) The working electrode, that is, the test sample, which may be accompanied by one or more auxiliary or counter electrodes.

    b) The reference electrode which is often separated from the solution by a solution bridge or a capillary Luggin probe (Fig. 5.1). This combination eliminates solution interchange with the reference electrode but allows it to be moved very close to the surface of the working electrode to minimize the effect of the solution resistance.

    c) A temperature monitoring device.

    d) An inlet and outlet for gas to allow deaeration, aeration, or the introduction of specific gases into the solution.

e) An electrical connection can be made directly with the working electrode, which will not be affected by the solution.

f) The working electrode must be introduced into the solution completely so as to eliminate any crevice at the solution interface, unless this is a desired effect.

g) The test cell itself must be composed of a material that will not corrode or deteriorate during the test, and which will not contaminate the test solution, the volume of the cell must be large enough to allow removal of the corroding ions from the surface of the working electrode without affecting the solution potential.

h) It may be necessary to include a mechanism for stirring the solution, such as a stirring bar, or bubbling gas to ensure uniformity of the solution chemistry.

DC Polarization Test Methods

DC polarization methods involve changing the potential of the working electrode and monitoring the current that is produced as a function of time or potential. For anodic polarization, the potential is changed in the anodic (or more positive) direction causing the working electrode to become the anode and forcing electrons to be withdrawn from the metal being tested. For cathodic polarization, the working electrode becomes more negative and electrons are added to the metal, sometimes causing electrodeposition. For cyclic polarization, both anodic and cathodic polarizations are performed in a cyclic manner [6]. There are several accepted methods to carry out DC polarization of specimens for corrosion testing.

Potentiodynamic Polarization. Potentiodynamic polarization refers to a polarization technique in which the potential of the electrode is varied over a relatively large potential domain at a selected rate by the application of a current through the electrolyte. Figure 5.18 is an example of a polarization plot obtained with a S43000 steel specimen in a 0.05 M H_2SO_4 solution.

A potentiodynamic polarization variant is cyclic voltammetry which involves sweeping the potential in a positive direction until a predetermined value of current or potential is reached, then the scan is reversed toward more negative values until the original value of potential is reached. In some cases, this scan may be done repeatedly to determine changes in the current-potential curve produced with scanning.

Another variation of potentiodynamic polarization is the potentiostaircase method which refers to a technique for polarizing an electrode in a series of potential steps where the time spent at each potential is constant, while the current is often allowed to stabilize prior to changing the potential to the next step. The step increase may be small, in which case, the technique resembles a potentiodynamic curve [6].

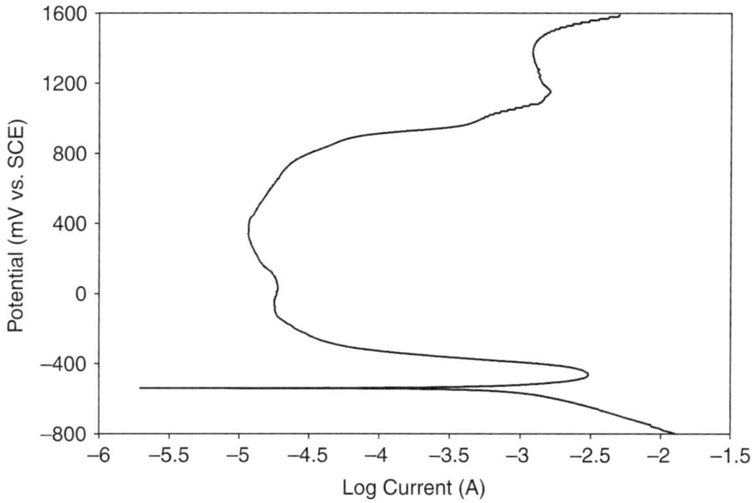

FIGURE 5.18 Typical anodic polarization plot for S43000 steel in a 0.05 M H_2SO_4 solution.

Electrochemical potentiodynamic reactivation (EPR) is another polarization method that evaluates the degree of sensitization of stainless steels such as S30400 and S30403 steels. This method uses a potentiodynamic sweep over a range of potentials from passive to active (called reactivation).

However, probably the most popular variant is the cyclic polarization test. This test is often used to evaluate the pitting susceptibility of a material. The potential is swept in a single cycle or slightly less than one cycle usually starting the scan at the corrosion potential. The voltage is first increased in the anodic or noble direction (forward scan). The voltage scan direction is reversed at some chosen current or voltage toward the cathodic or active direction (backward or reverse scan) and terminated at another chosen voltage. The presence of the hysteresis between the currents measured in the forward and backward scans is believed to indicate pitting, while the size of the hysteresis loop itself has been related to the amount of pitting that has occurred during the scan.

This technique has been especially useful to assess localized corrosion for passivating alloys such as S31600 stainless steel, nickel-based alloys containing chromium, and other alloys such as titanium and zirconium. Though the generation of the polarization scan is simple, its interpretation can be difficult [7].

In the following example, the polarization scans were generated after one and four days of exposure to a chemical product maintained at 49°C. The goal of these tests was to examine if S31600 steel could be used for short-term storage of a 50 percent commercial organic acid solution (aminotrimethylene phosphonic acid) in water.

A small amount of chloride ion (1 percent) was also present in this acidic chemical.

In this example, the potential scan rate was 0.5 mV s^{-1} and the scan direction was reversed at 0.1 mA cm^{-2}. Coupon immersion tests were run in the same environment for 840 hours. The S31600 steel specimens were exposed to the liquid, at the vapor/liquid interface, and in the vapor. The reason for the three exposures was that in most storage situations, the containment vessel would be exposed to a vapor/liquid interface and a vapor phase at least part of the time. Corrosion in these regions can be very different from liquid exposures. The specimens were also fitted with artificial crevice formers.

Figure 5.19 shows the polarization scan generated after one day and Figure 5.20 shows the polarization scan generated after four days of exposure. The important parameters considered were the position of the "anodic-to-cathodic" transition relative to the corrosion potential, the existence of the repassivation potential and its value relative to the corrosion potential, the existence of the pitting potential and its value relative to the corrosion potential, and the hysteresis (positive or negative). The interpretation of the results is summarized in Table 5.6.

The presence of the negative hysteresis would typically suggest that localized corrosion is possible depending on the value of the corrosion potential relative to the characteristic potentials present in these polarization plots. After the first day of exposure, pitting was not expected to be a problem because the pitting potential was far away from the corrosion potential. The currents generated were much

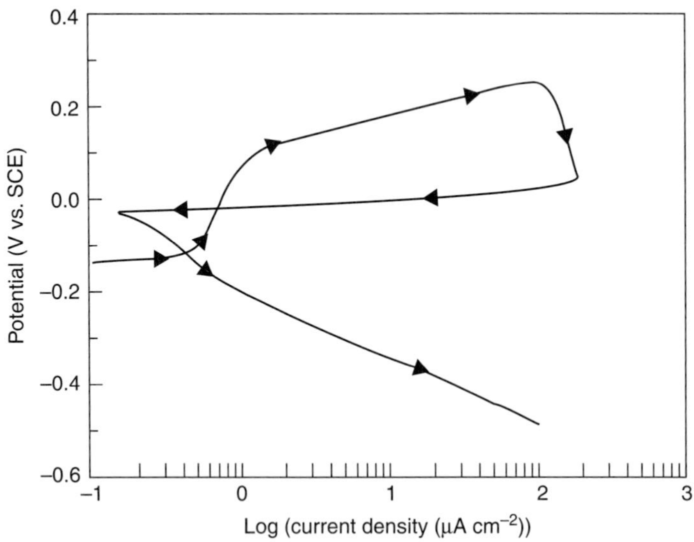

FIGURE 5.19 Polarization scan for S31600 steel in 50 percent aminotrimethylene phosphonic acid after one day of exposure (the arrow indicates scanning direction).

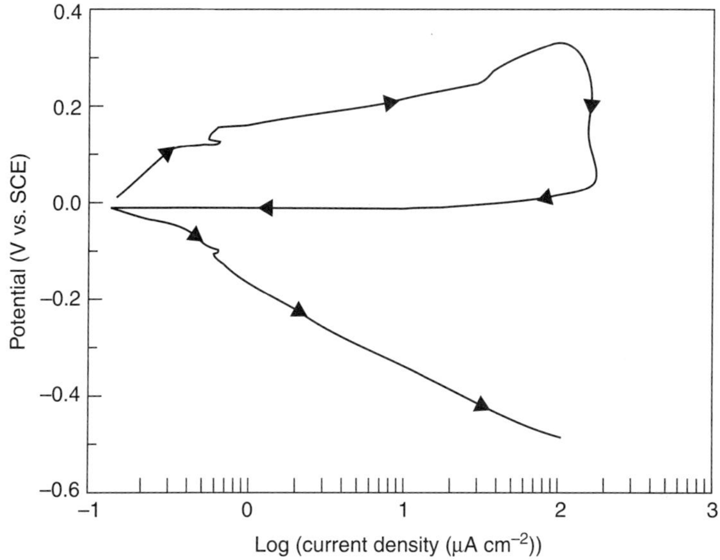

FIGURE 5.20 Polarization scan for S31600 steel in 50 percent aminotrimethylene phosphonic acid after four days of exposure (the arrow indicates scanning direction).

higher than those normally associated with S31600 steel in a passive state. These observations suggested that there was a risk of initiation of corrosion, particularly in localized areas where the pH can decrease drastically [7] (see crevice mechanism in Chap. 6).

After 4 days, the risk of localized corrosion increased. At this time, the repassivation potential and the potential of the change from anodic to cathodic current were equal to the corrosion potential. The pitting potential was only about 0.1 V more noble than the corrosion potential and the hysteresis still negative. The risk of pitting had increased enough to become a concern.

Feature	Value in Fig. 5.19	Value in Fig. 5.20
Repassivation potential–corrosion potential	0.12 V	0.0 V
Pitting potential–corrosion potential	0.22 V	0.12 V
Potential of anodic-to-cathodic transition–corrosion potential	0.12 V	0.0 V
Hysteresis	Negative	Negative
Active-to-passive transition	No	No

TABLE 5.6 Features and Values Used to Interpret Figs. 5.19 and 5.20.

Coupon immersion tests confirmed the long-term predictions. Slight attack was found under the artificial crevice formers in the complete liquid exposure. The practical conclusion of this in-service study was that, since localized corrosion often takes time to develop, a few days of exposure to this chemical product could be acceptable. However, it was recommended to avoid long-term exposure since both pitting and crevice corrosion would be expected for longer exposure periods.

Linear Polarization Resistance. Another widely used polarization method is linear polarization resistance (LPR). The polarization resistance of a material is defined as the $\Delta E/\Delta i$ slope of a potential-current density curve at the free corrosion potential (Fig. 5.21), yielding the polarization resistance R_p that can be itself related to the corrosion current (i_{corr}) with the help of the Stern-Geary approximation in Eq. (5.22) [8].

$$R_p = \frac{B}{i_{corr}} = \frac{(\Delta E)}{(\Delta i)}_{\Delta E \to 0} \tag{5.22}$$

where R_p is the polarization resistance
i_{corr} is the corrosion current
B is an empirical polarization resistance constant that can be related to the anodic (b_a) and cathodic (b_c) Tafel slopes with Eq. (5.23).

$$B = \frac{b_a \cdot b_c}{2.3(b_a + b_c)} \tag{5.23}$$

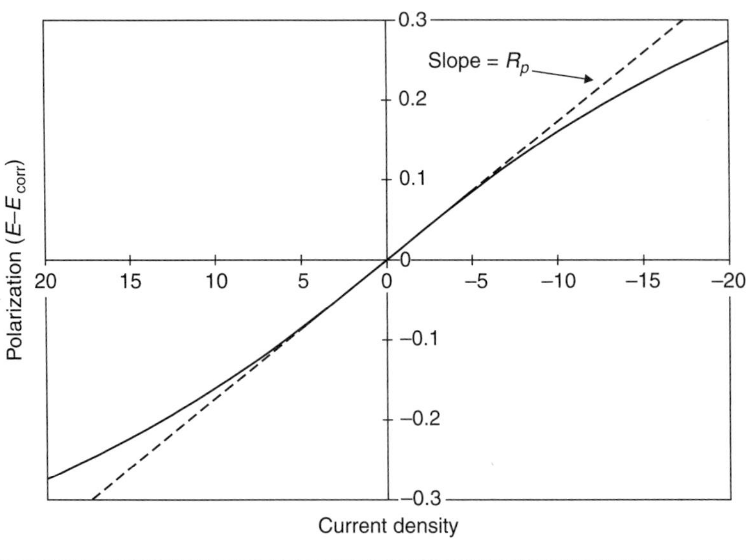

FIGURE 5.21 Hypothetical linear polarization plot.

The Tafel slopes in this equation can be evaluated experimentally using real polarization plots in the fashion described in Fig. 5.3 or obtained from the literature [8]. The corrosion currents may then be converted into other corrosion rate units using Faraday's law or, more simply, by using a conversion scheme provided in Chap. 3, that is, Table 3.1 for all metals, or Table 3.2 adapted to iron or steel.

For field measurements, it is necessary to use a probe that enters the vessel with a special probe retrieving device (Fig. 5.22). Several commercially available probes (Fig. 5.23) and analyzing systems can be directly interfaced with remote computer data-acquisition systems. Alarms can also be used to signal plant operators when high corrosion rates are experienced [9;10].

The following example illustrates how the corrosion efficiency of an inhibitor can be evaluated with LPR. Forty years ago, Hugel tested a variety of inhibitors for steel in 6 M HCl at 60°C and found that alkenyl and aromatic aldehydes were very effective [11]. Cinnamaldehyde was one of the best, providing almost 99 percent protection. Numerous patents have been issued since then on the use of aldehydes, and trans-cinnamaldehyde (TCA) in particular, as steel corrosion inhibitors in acid media have been used to reduce the corrosion of steel during pickling or oil field acidizing treatments.

The polarization curves presented in Fig. 5.24 were obtained with carbon steel exposed to a solution containing, respectively, (*a*) no inhibitor, (*b*) 250, (*c*) 1000, and (*d*) 2000 parts per million of

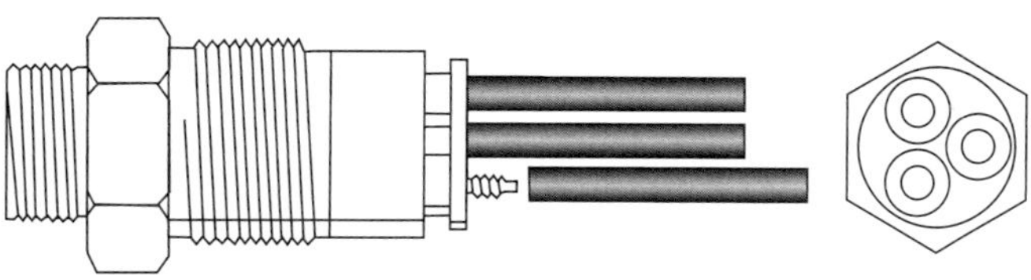

(a)

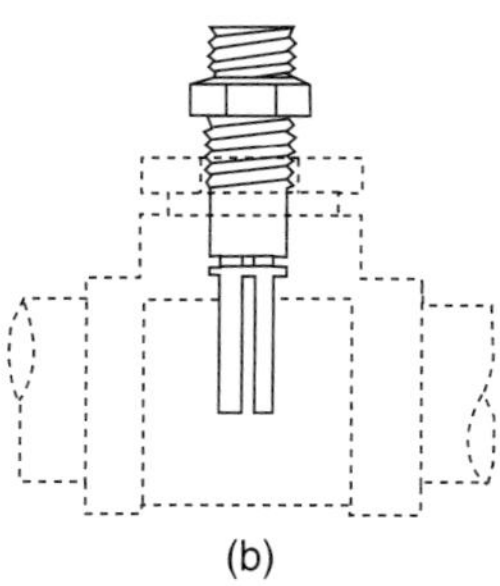

(b)

FIGURE 5.22 Typical linear polarization resistance probe (*a*) and probe in pipe tee (*b*) [10].

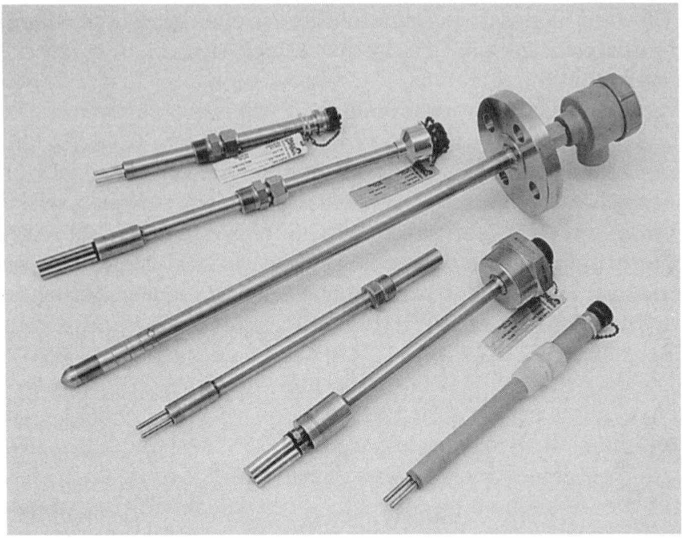

FIGURE 5.23 Commercial sensor elements to carry out linear polarization resistance (LPR) measurements. (Courtesy of Metal Samples Company)

TCA in a 6 M HCl solution. One can first estimate the R_p from the slopes of the polarization curves. Assuming that b_a and b_c have the same value, that is, 0.1 mV per decade of current, each R_p can then be converted to a corrosion current with the help of Eq. (5.22).

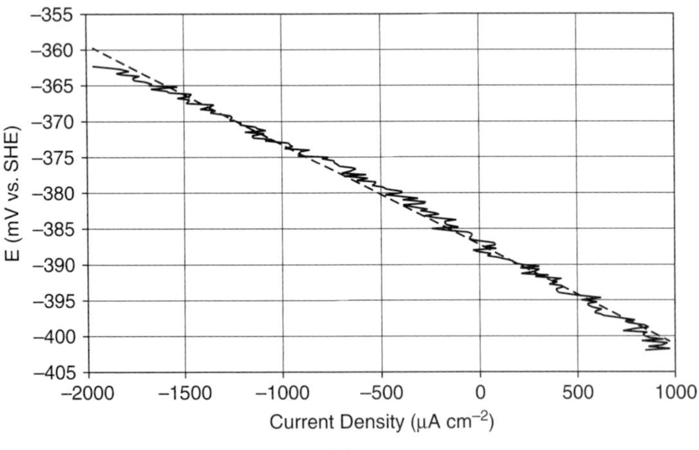

(a)

FIGURE 5.24 Corrosion of AISI 1018 carbon steel in 6 M HCl containing (*a*) no inhibitor, (*b*) 250 ppm trans-cinnamaldehyde (TCA), (*c*) 1000 ppm TCA, and (*d*) 5000 ppm TCA. The dotted line in these figures has been added to estimate the slope of the curves that can be related to the polarization resistance (R_p).

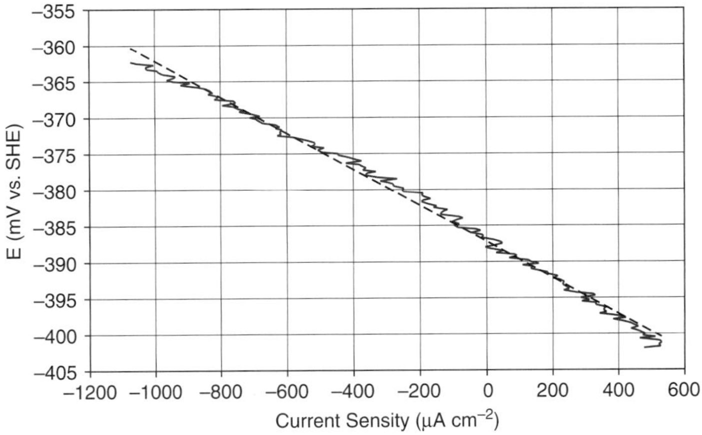

(b)

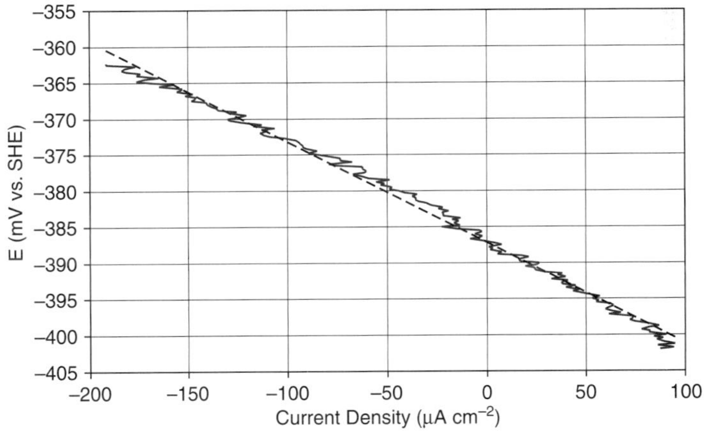

(c)

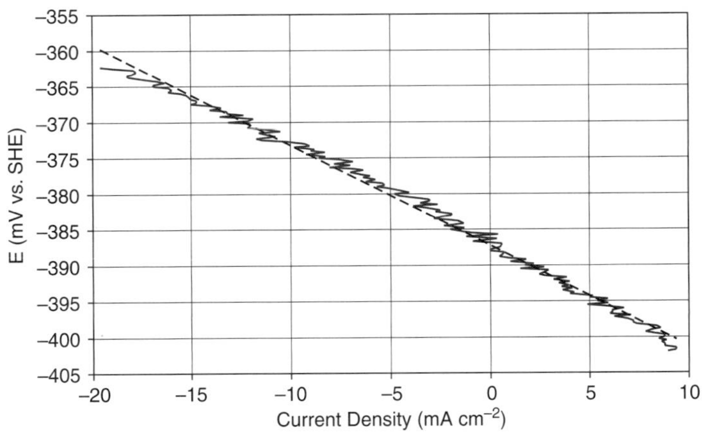

(d)

Figure 5.24 *(continued)*

The inhibitor efficiency of each solution can subsequently be calculated using Eq. (5.24).

$$\text{Inhibitor efficiency (\%)} = 100\,\frac{(\text{CR}_{\text{uninhibited}} - \text{CR}_{\text{inhibited}})}{\text{CR}_{\text{uninhibited}}} \tag{5.24}$$

where $CR_{\text{uninhibited}}$ is the corrosion rate of the uninhibited system
$CR_{\text{inhibited}}$ is the corrosion rate of the inhibited system

The results obtained with LPR on these inhibiting solutions are presented in Table 5.7 along with corrosion rates converted in mm/year with the help of Table 3.2 in Chap. 3.

Electrochemical Impedance Spectroscopy

Electrochemical impedance spectroscopy (EIS) has been successfully applied to the study of corrosion systems and proven to be a powerful and accurate method for measuring corrosion rates for a few decades. In this approach for determining the polarization resistance of a metal, a measure of the electrochemical impedance is made at a series of predetermined frequencies.

An important advantage of EIS over other electrochemical techniques is the possibility of using very small amplitude signals without significantly disturbing the properties being measured. However, in order to estimate the polarization resistance (R_p), that is proportional to the corrosion rate at the monitored interface according to Eq. (5.22), EIS results have to be interpreted with the help of a model of the interface.

Amongst the numerous equivalent circuits that have been proposed to model electrochemical interfaces only a few really apply in the context of a freely corroding system. The first circuit is the simplest equivalent circuit that can describe a metal/electrolyte interface (Fig. 5.25). Its behavior is described by Eq. (5.25).

$$Z(\omega) = R_s + \frac{R_p}{1+(j\omega R_p C_{\text{dl}})^{\beta}} \tag{5.25}$$

TCA (ppm)	R_p (Ω cm²)	Corrosion Current (mA cm⁻²)	Corrosion Rate (mm y⁻¹)	Efficiency (%)
0	14	1.55	18	0
250	25	0.87	10.1	44
1000	140	0.155	1.80	90
2000	1400	0.0155	0.18	99

Table 5.7 Inhibitor Efficiency of Trans-Cinnamaldehyde (TCA) to the Corrosion of Carbon Steel Exposed to a 6 M HCl Solution

where R_s is the solution resistance
R_p is the polarization resistance
ω is the frequency
C_{dl} is the double layer capacitance

The term Q, in Fig. 5.25, describes the "leaky capacitor" behavior corresponding to the presence of a constant-phase element (CPE) [12]. Figure 5.26(*a*) illustrates the complex plane presentation of the EIS model circuit in Fig. 5.25(*a*) where $R_s = 10\ \Omega$, $R_p = 100\ \text{k}\Omega$ and Q decomposes into $C_{dl} = 40\ \mu\text{F}$ and $n = 0.8$ and Fig. 5.26(*b*) shows how the same data would appear in a Bode plot format.

The second circuit [Fig. 5.25(*b*)] was proposed by Hladky et al. [13]. to take into account a diffusion-limited behavior corresponding to a Warburg component which can be described by Eq. (5.26). The exponent n in Eq. (5.26) can vary between 0.5 and 0.25 depending on the smoothness of the metallic surface that is 0.5 for highly polished surfaces and 0.25 for porous or very rough materials [14]. R and C in Eq. (5.26), are the resistance and capacitance associated with the distributed R-C line of infinite length.

$$Z(\omega) = (0.5 \bullet R/C)^{0.5} \bullet \omega^{-n} \tag{5.26}$$

Figure 5.27(*a*) illustrates the complex plane presentation of simulated data corresponding to model circuit in Fig. 5.25(*b*) where $R_s = 10\ \Omega$, $R_p = 100\ \text{k}\Omega$, $C_{dl} = 40\ \mu\text{F}$, and the exponent n of the Warburg component = 0.4. Figure 5.27(*b*) shows the same data in a Bode representation.

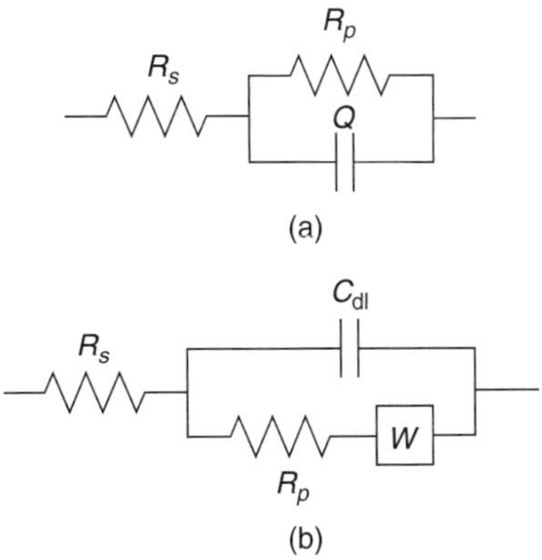

Figure 5.25 The equivalent circuit models proposed for the interpretation of EIS results measured in corroding systems: (*a*) simplest representation of an electrochemical interface, (*b*) one relaxation time constant with extended diffusion.

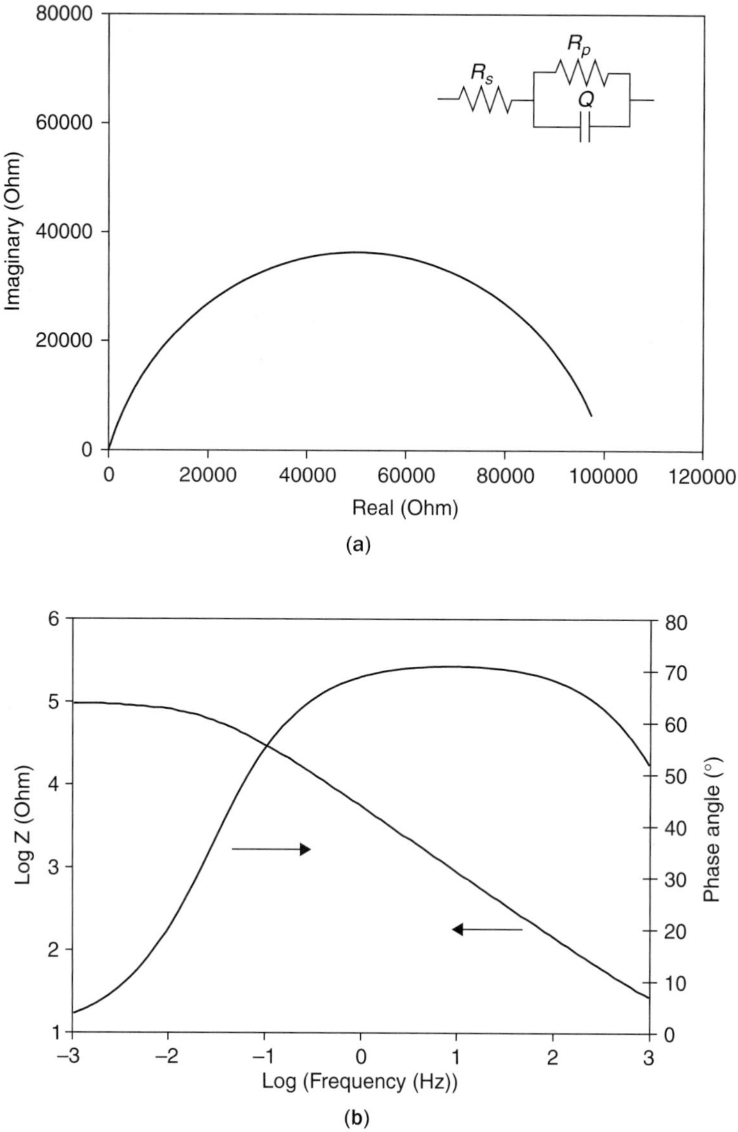

FIGURE 5.26 (*a*) Complex plane and (*b*) Bode plots of simulated data corresponding to the model circuit in Fig. 5.25(*a*) where $R_s = 10\ \Omega$, $R_p = 100\ \text{k}\Omega$, and Q decomposes into $C_{dl} = 40\ \mu\text{F}$ and $n = 0.8$.

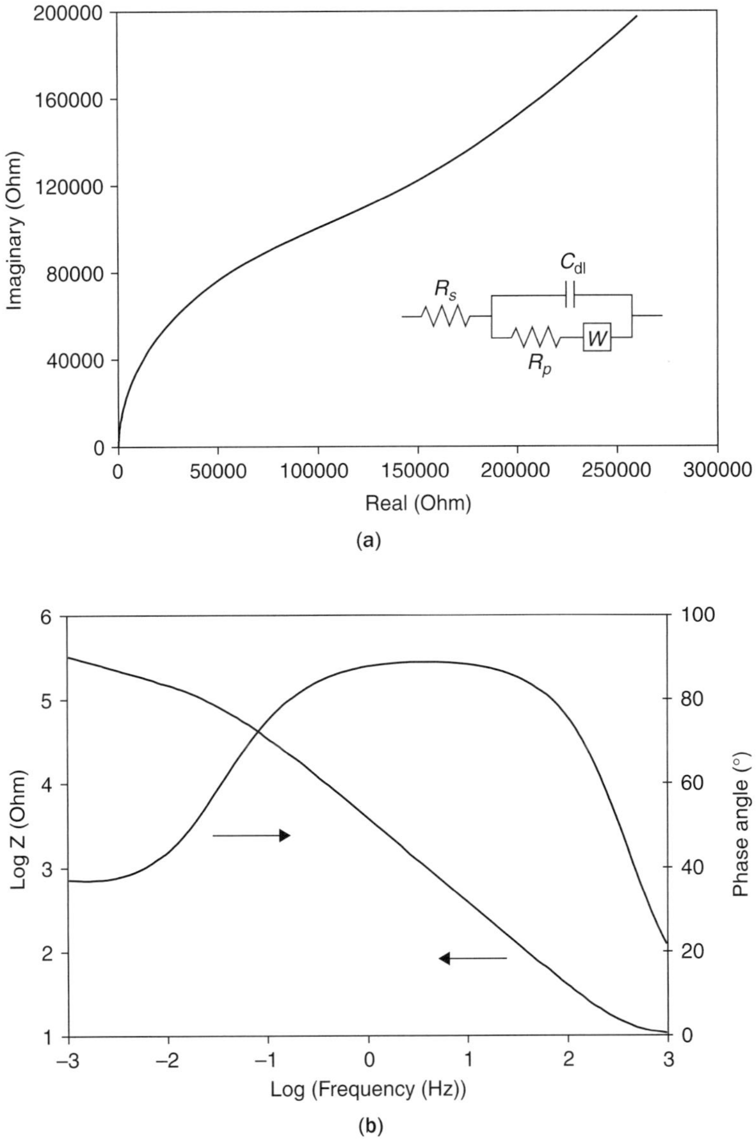

FIGURE 5.27 (*a*) Complex plane and (*b*) Bode plots of simulated data corresponding to model circuit in Fig. 5.25(*b*) where $R_s = 10\ \Omega$, $R_p = 100\ \text{k}\Omega$, $C_{dl} = 40\ \mu\text{F}$ and the exponent *n* of the Warburg component = 0.4.

Complications with Polarization Methods

The widespread use of DC or AC polarization methods in corrosion studies does not mean that they are without complications. Both linear polarization and Tafel extrapolation need special precautions for their results to be valid. The main complications or obstacles in performing polarization measurements can be summarized in the following categories:

- *Effect of scan rate:* The rate at which the potential is scanned may have a significant effect on the amount of current produced at all values of potential [6]. The rate at which the potential is changed is an experimental parameter over which the user has control. The goal is to set the polarization scan rate at a slow enough rate to minimize surface capacitance charging. If not, some of the current being generated serves to charge the surface capacitance in addition to the corrosion process with the result that the measured current can be greater than the current actually generated by the corrosion reactions alone.
- *Effect of solution resistance:* The distance between the Luggin probe and the working electrode is purposely minimized in most measurements to reduce the effect of the solution resistance. In solutions that have extremely high resistivity, for example, concrete, soils, and organic solutions, this can be an extremely significant effect.
- *Changing surface conditions:* Corrosion reactions take place at the surface of a metallic material. When the surface condition is modified, due to processing conditions, active corrosion, or any other reasons, the corrosion potential is usually also changed. This can have a strong effect on the polarization curves [6].
- *Determination of pitting potential:* In analyzing polarization curves the presence of a hysteresis loop between the forward and reverse scans often indicates that localized corrosion, for example, pitting or crevice corrosion, is in progress. This feature can be advantageously used to detect the susceptibility of an alloy to pitting in certain environments. However, it also means that the results obtained cannot be used with confidence to estimate general corrosion rates.

A critical problem in EIS, as well as any other scientific measurements, is the validation of the experimental data. The use of Kramers-Kronig (KK) transforms has been proposed to assess the quality of the measured impedance data [15]. These integral transforms were derived assuming four basic conditions:

- *Linearity:* A system is said to be linear if the response to a sum of individual inputs is equal to the sum of the individual responses. Practically, this implies that the impedance of a system does not vary with the magnitude of the probing signal.
- *Causality:* The response of a system should be solely due to the probing signal. Physically, this means that the system being tested responds fully and exclusively to the applied signal. This is an important consideration in electrochemical systems, because charge transfer interfaces are often active and do, in fact, generate noise in the absence of any external stimulus [15].
- *Stability:* A system is said to be stable if it comes back to its original state after a perturbation is removed. This condition ensures that there is no negative resistance in the system. This apparent restriction on the presence of negative resistance has often been associated with the presence of active-to-passive transitions in the system being investigated [15].
- *Finite value:* The real and imaginary components of a complex impedance must be finite over the entire frequency range sampled.

If a system satisfies the conditions of linearity, stability, and causality, it will *a priori* satisfy the KK transforms, provided that the frequency range is sufficiently broad for the integrals to be evaluated. However, a simple validity test is often sufficient to ascertain that the EIS measurements are indeed meaningful. For measurements made in real time, that is, by changing the frequency for each frequency point, the validity of a forward frequency scan can be verified by scanning in the opposite direction. Repeated measurements during two successive scans in opposite direction should generally satisfy KK transforms and represent the actual impedance of the system being tested.

5.6.2 Corrosion Monitoring

Electrochemical monitoring methods involve the determination of specific interface properties divided into three broad categories:

- *Corrosion potential measurements:* The potential at a corroding surface arises from the mutual polarization of the anodic and cathodic half-reactions resulting in the overall corrosion reaction. Corrosion potential is intrinsically the most readily observable parameter and understanding its behavior may provide very useful information on the thermodynamic state of a system.

- *Reaction rate as current density:* Partial anodic and cathodic current densities cannot be measured directly unless they are purposefully separated into a bimetallic couple. However, by polarizing a metal immersed in an aqueous environment, it is possible, with the use of simple assumptions and models of the underlying electrochemical behavior, to estimate net currents for both the anodic and cathodic polarizations from which a corrosion current density can be deduced.
- *Surface impedance:* A corroding interface can also be modeled for all its impedance characteristics, therefore revealing subtle mechanisms not visible by other means. EIS is now well established as a powerful technique for investigating corrosion processes and other electrochemical systems.

Corrosion potential or current produced by naturally occurring or externally imposed conditions can be measured with a variety of electrochemical techniques. Conversions of the measurements into corrosion rates or other meaningful data use equations or algorithms that are specific to each technique.

Limits of operation for field work are more serious than those experienced in a laboratory environment, mostly for reasons of practical probe geometry. For example, capillary salt bridges (e.g., Luggin capillary) commonly used in laboratory setups to reduce the solution resistance interference are definitively too delicate or cumbersome for field use [16].

Zero Resistance Ammetry

With this electrochemical technique galvanic currents between dissimilar electrode materials are measured with a zero resistance ammeter*. The design of dissimilarities between sensor elements may be made to target a feature of interest in the system being monitored (e.g., different compositions, heat treatments, stress levels, or surface conditions). Zero resistance ammetry (ZRA) may also be applied to nominally identical electrodes in order to reveal changes occurring in the corrosivity of the environment.

The main principle of the technique is that differences in the electrochemical behavior of two electrodes exposed to a process stream give rise to differences in the redox potential at these electrodes. Once the two electrodes are externally electrically connected, the more noble electrode becomes predominantly cathodic, while the more active electrode becomes predominantly anodic and sacrificial. When the anodic reaction is relatively stable the galvanic current

* A zero resistance ammeter (ZRA) is a current to voltage converter that produces a voltage output proportional to the current flowing between its input terminals while imposing a "zero" voltage drop to the external circuit.

monitors the response of the cathodic reaction to the process stream conditions. When the cathodic reaction is stable, it monitors the response of the anodic reaction to process fluctuations [16].

This technique has been found particularly useful to study depolarization effects of the cathode of a galvanic pair of electrodes to obtain feedback of low levels of dissolved gases, particularly oxygen, or the presence of bacteria, which depolarize the cathode of the galvanic pair and increase the coupling current. When used for detection of low levels of oxygen, other dissolved gases may interfere. Calibration against a dissolved oxygen meter is usually required if quantitative values are needed.

Electrochemical Impedance Spectroscopy

The measurement cycle time with EIS is quite critical and depends on the frequency range used, especially the low frequencies. A single frequency cycle at 1 mHz, for example, takes 15 minutes. A high-to low-frequency scan going to such a low frequency would take more than two hours. In order to make routine corrosion monitoring with EIS certain simplifications are needed to maximize the use of high frequency data and drastically shorten the measurement time. It is also important to simplify the data processing and analysis to make the technique user friendly for field corrosion monitoring.

In order to simplify the analysis of field EIS results, a method was developed which consists of finding the geometric center of an arc formed by three successive data points on a complex impedance diagram (Fig. 5.28) [17;18]. This technique was designed as an improvement over the two-point method based on the comparison of high and low frequency data points for which the impedance would be proportional to the R_s at the high frequency point and the summation of R_s and R_p at the low-frequency point [19]. In real world situations, one difficult assumption to satisfy with the two-point method is that data points should contain negligible imaginary components (i.e., 0 phase shift), a condition usually hard to achieve in a meaningful manner at low-measuring frequencies.

The three-point analysis technique was further developed by permuting the data points involved in the projection of centers in order to obtain a population of projected centers. This improvement has permitted to automate the data analysis while providing some information concerning the adherence of the results with the resistive-capacitive (RC) behavior described in Eq. (5.26) that is assumed for the evaluation of the parameters associated with uniform corrosion [18].

More recently, a full-spectrum, relatively low-cost EIS corrosion monitoring system has been developed, which is wireless, small (5 cm diameter, 1.2 cm height), requires nominally 10-mW power during its 200-second measurement period, and has an electronic identifier,

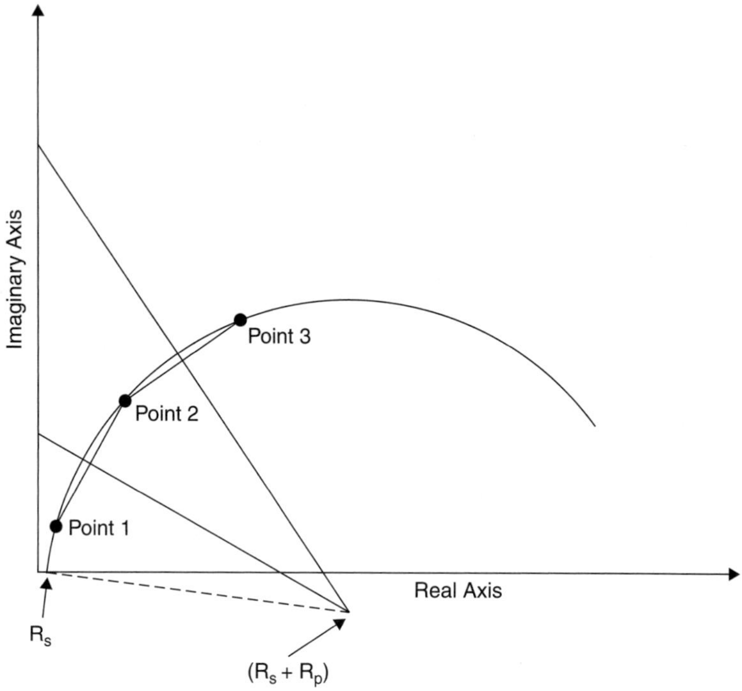

FIGURE 5.28 Schematic of the extrapolation method to obtain the polarization resistance from EIS data.

which allows for a single data logger to monitor multiple devices in the same general vicinity [Fig. 5.29(*a*) and (*b*)] [20].

The wireless EIS sensor determines the impedance at 15 to 20 independent frequencies, by measuring amplitude and phase at each frequency (Fig. 5.30). It computes corrosion rate, conductivity, and coating impedance and transmits the result wirelessly to a data logger. The miniature and wireless features make it suitable for embedding in concrete or placing in hidden and inaccessible locations, for example, in HVAC systems. Its minimal power consuming aspect lends itself useful for long-term monitoring of coating integrity. The miniature EIS system has been tested in various environments, namely concrete, water, and under coatings [Fig. 5.31(*a*) and 5.31(*b*)].

Electrochemical Noise Analysis

Fluctuations of the corrosion potential of a corroding metallic specimen are a well known and easily observable phenomenon. Electrochemical noise analysis (ENA) as a corrosion tool has increased steadily since Iverson's paper in 1968 [21]. The study of

(a)

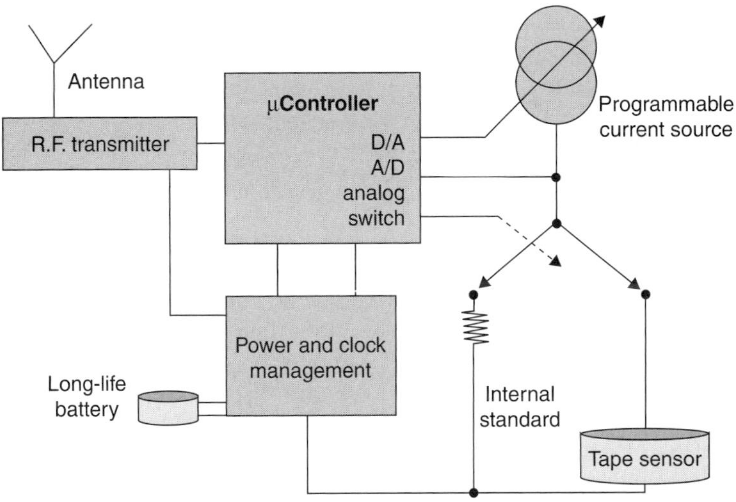

(b)

Figure 5.29 (*a*) Test and demonstration of breadboard coating health monitor, including wireless communication. Inset shows transceiver unit attached to laptop computer; (*b*) Block diagram of the electronics. (Courtesy of Guy D. Davis, DACCO SCI, Inc.)

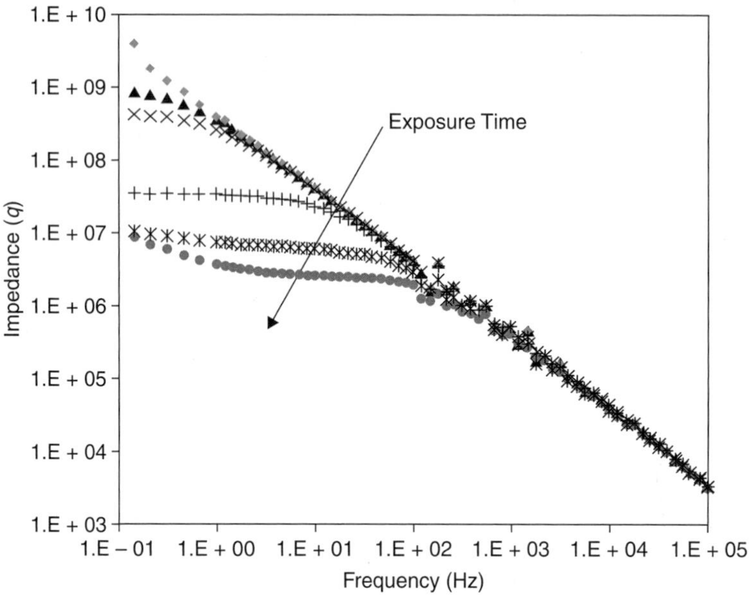

FIGURE 5.30 Magnitude of impedance of a coating versus frequency. Low frequency impedance values show good correlation with long-term exposure behavior. (Courtesy of Guy D. Davis, DACCO SCI, Inc.)

electrochemical noise (EN) has been found to be uniquely appropriate for monitoring the onset of events leading to localized corrosion and understanding the chronology of the initial events typical of this type of corrosion.

The EN technique differs in many ways from other electrochemical techniques used in corrosion. One important difference is that ENA does not require that the sensing element be polarized in order to generate a signal. However, it is also possible to measure current noise under an applied potential, or measure potential noise under an applied current. The potential and current between freely corroding electrodes (in many cases < 1 μV and < 1 nA) are measured with sensitive instrumentation. A measurement frequency of 1 Hz is usually appropriate to provide meaningful data. For simultaneous measurement of electrochemical potential and current noise, a three-electrode sensor is required. In field corrosion monitoring, the three-sensor elements are usually made of the same material.

Whilst the measurement of electrochemical noise is relatively straightforward, the data analysis can be complex and inconclusive. Even if ENA was first applied in field corrosion monitoring in the late 1960s, an understanding of the method of analysis is still evolving, partly because the technique has been used to look at several types of corrosion. The relationships between potential and current noise are inherently complex to analyze quantitatively because the naturally

FIGURE 5.31 (*a*) Schematic of the coating health monitor (CHM) with a tape sensing element mounted on a coated metal and (*b*) actual sensor tape electrode and electronics housing mounted on the frame of a commercial vehicle. (Courtesy of Guy D. Davis, DACCO SCI, Inc.)

occurring fluctuations do not have controlled frequencies as are applied, for example, in EIS.

There are basically three categories of ENA: visual examination, sequence-independent methods that treat the collection of voltage or current values without regard to their position in the sequence of readings (moments, mean, variance, standard deviation, skewness, and kurtosis), and those that take the sequence into account (autocorrelation, power spectra, fractal analysis, stochastic process analysis) [22].

Visual examination of the time record trace may provide indications as to the type of corrosion processes that are occurring. The following example illustrates how a simple examination of EN measurements could reveal the corrosivity of various points of an industrial gas scrubbing system where highly corrosive thin-film electrolytes are known to form [23]. These conditions arise when gas streams are cooled to a temperature below the dewpoint. The resulting thin electrolyte layer (moisture) is often highly concentrated in corrosive species.

The corrosion probe used in this example is illustrated in Figs. 5.32 and 5.33. A retractable probe with flexible depth was selected, in order to mount the sensor surface flush with the internal scrubber wall surface. The close spacing of the carbon steel sensor elements was designed to work with a discontinuous thin surface electrolyte film. This corrosion sensor was connected to a handheld multichannel data recorder by shielded multistrand cabling Fig. 5.34. As the ducting of the gas scrubbing tower was heavily insulated, no special precautions were taken to cool the corrosion sensor's surface.

Potential noise and current signals recorded during the first hour of exposure at the conical base of the gas scrubbing tower are presented in Fig. 5.35. According to the operational history of the plant, condensate had a tendency to accumulate at this location where highly corrosive conditions had been noted. The high levels of potential noise and current noise in Fig. 5.35 are indicative of a massive pitting attack which is consistent with the operational experience. It should be noted that the current noise is actually off-scale for most of the monitoring period, in excess of 10 mA. The high corrosivity indicated by the electrochemical noise data from this sensor location was confirmed by direct evidence of severe

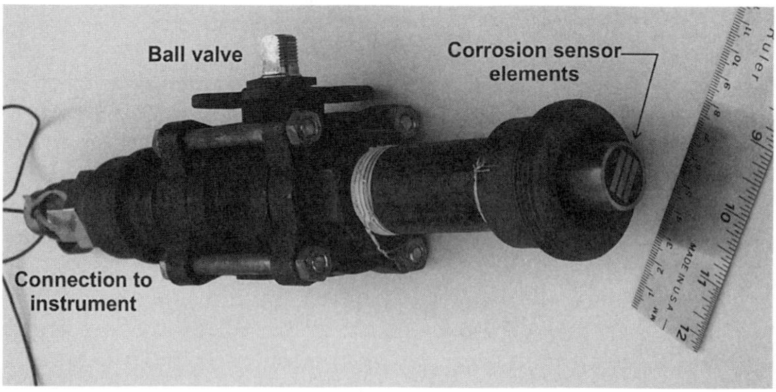

FIGURE 5.32 Corrosion sensor and access fitting used for thin film corrosion monitoring. (Courtesy of Kingston Technical Software)

Figure 5.33 Close-up of corrosion sensing elements used for thin film corrosion monitoring. (Courtesy of Kingston Technical Software)

pitting attack on the sensor elements, revealed by scanning electron microscopy (Fig. 5.36). In contrast, both current and voltage signals remained relatively constantly small at a position higher up in the tower, where the sensor surface remained mostly dry.

Figure 5.34 Field corrosion monitoring using electrochemical noise recorded with a handheld data logger. (Courtesy of Kingston Technical Software)

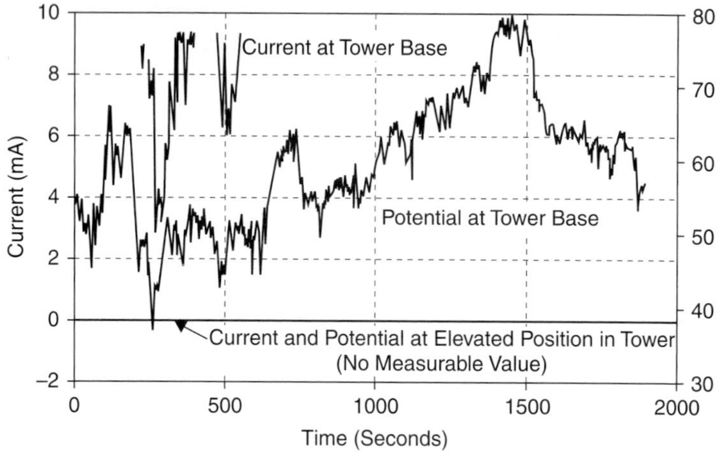

FIGURE 5.35 Potential and current noise records at two locations in a gas scrubbing tower.

An improvement over this simple analysis is commonly practiced in industry by tracking the width of the potential and current signals as an indication of corrosion activity in the system being monitored. Fig. 5.37 illustrates how the decrease in the current band obtained with a monitoring system was interpreted as a reduction in general corrosion activity in a debutanizer overhead piping where the interaction between operational changes and the corrosion mechanism were being investigated.

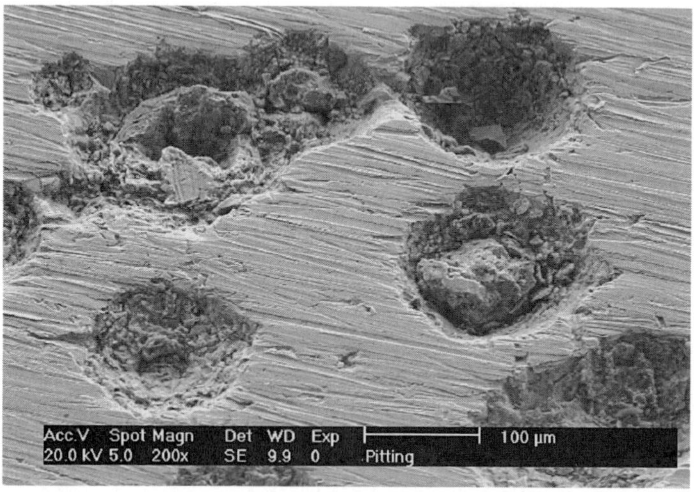

FIGURE 5.36 Scanning electron microscope image of a sensor element surface after exposure at the base of the scrubbing tower clearly showing corrosion pits.

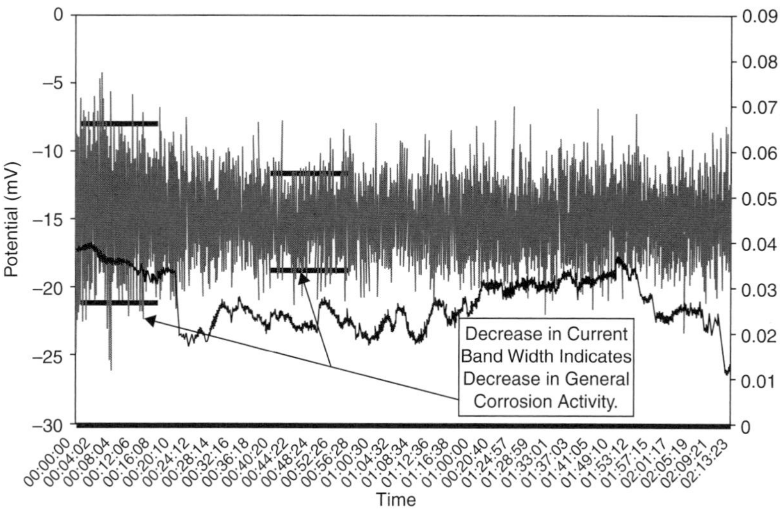

FIGURE 5.37 Electrochemical current noise (large band) and potential noise (lower signal) in a debutanizer overhead piping obtained with the Concerto VT noise system. (Courtesy of CAPCIS Ltd.)

The analysis of EN results obtained between two working electrodes (WEs) with an imposed or naturally developing asymmetry can be carried out by considering Eq. (5.27), which indicates that the noise impedance of a cell (Z_n) depends on the impedances of the two WEs, as well as their noise levels, represented by the power density spectra (Ψi_1 and Ψi_2) obtained by performing the analysis of noise signals with either fast Fourier transform (FFT) or with the maximum entropy method (MEM).

$$Z_n(f) = \sqrt{\frac{\Psi_V(f)}{\Psi_I(f)}} = |Z_1(f)Z_2(f)| \sqrt{\frac{\Psi_{i_1}(f) + \Psi_{i_2}(f)}{|Z_1(f)|^2 \, \Psi_{i_1}(f) + |Z_2(f)|^2 \, \Psi_{i_2}(f)}} \tag{5.27}$$

For the simplest case of two WEs with the same impedance ($Z_1 = Z_2$) the noise impedance is equal to the modulus of the electrode impedance $|Z(f)|$. This result is valid independently of the origin of the noise signals (localized or uniform corrosion, bubble evolution due to the cathodic reaction) and the shape of the impedance plot, even if the noise levels of the two electrodes are different. In such case, noise measurements are equivalent to impedance measurements for which the external signal perturbation has been replaced by the internal noise generated by the corrosion processes [24].

However, when the two WEs do not have the same impedance, the noise impedance analysis requires a more cautious interpretation. Depending on the source of the current noise the measured impedance may either be that of a quiet cathode or that of a quiet anode while in intermediate cases the results of the measurement of Z_n would show a mixed behavior and be more difficult to interpret. For example, if hydrogen bubbles are evolving on the cathode while the anode undergoes generalized corrosion, the noise of the cathode is orders of magnitude larger than that of the anode, so that Z_n becomes equal to the impedance modulus of the anode, $|Z_a|$. In these conditions, while the time records appear to show only the cathodic processes, the impedance measured is that of the anode, using the noise of the cathode as input signal.

An opposite case would be a cell where the anode is undergoing pitting, while the cathodic reaction is the reduction of dissolved oxygen or an imposed galvanic situation. Since the anodic noise is preponderant, Eq. (5.27) shows that Z_n is equal to the impedance modulus of the cathode, $|Z_c|$. The anodic noise is the internal signal source utilized for the measurement of the impedance of the cathode.

Coupled Multielectrode Array Systems and Sensors

The use of multielectrode array systems (CMAS) for corrosion monitoring is relatively new. The advantages of using multiple electrodes include the ability to obtain greater statistical sampling of current fluctuations, a greater ratio of cathode-to-anode areas in order to enhance the growth of localized corrosion once initiated. CMAS also provide the ability to estimate the pit penetration rate and obtain macroscopic spatial distribution of localized corrosion [25].

Figure 5.38 shows the principle of the CMAS in which a resistor is positioned between each electrode and the common coupling point [26]. Electrons from a corroding or a relatively more corroding electrode flow through the resistor connected to the electrode and produce a small potential drop usually of the order of a few microvolts. This potential drop is measured by the high-resolution voltage-measuring instrument and used to derive the current of each electrode. CMAS probes can be made in several configurations and sizes, depending on the applications. Figure 5.39 shows some of the typical probes that were reported for real-time corrosion monitoring.

Because the electrode surface area is usually between 1 and 0.03 mm^2, which is approximately 2 to 4 orders of magnitude less than that of a typical LPR probe or a typical electrochemical noise (EN) probe, the prediction of penetration rate or localized corrosion rate by assuming uniform corrosion on the small electrode is realistic in most applications. CMAS probes have been used for monitoring localized corrosion of a variety of metals and alloys in the following environments and conditions:

- Deposits of sulfate-reducing bacteria
- Deposits of salt in air
- High pressure simulated natural gas systems

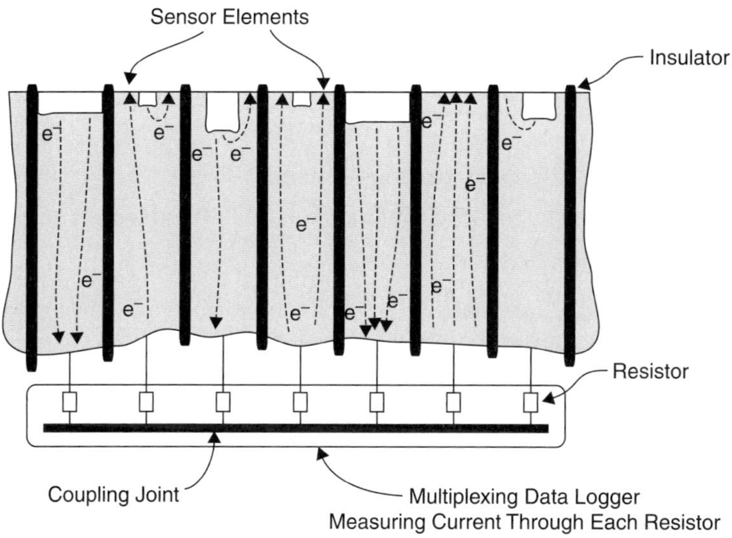

FIGURE 5.38 Multielectrode array multiplexed current and/or potential measurement (adapted from [26]).

- H_2S systems
- Oil/water mixtures
- Cathodically protected systems
- Cooling water
- Simulated crevices in seawater

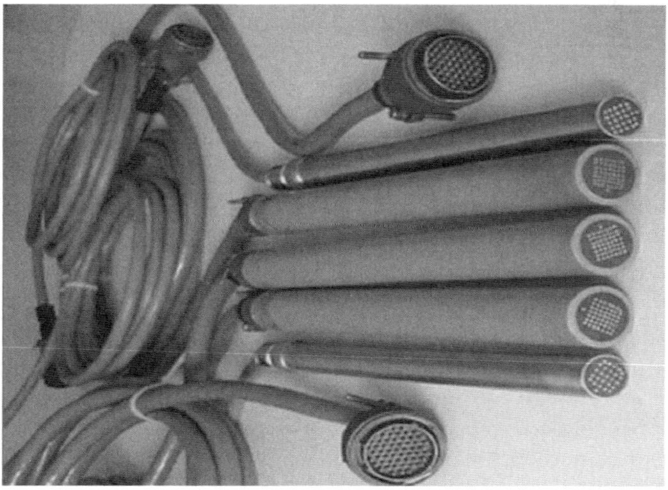

FIGURE 5.39 Typical CMAS probes used for real-time corrosion monitoring. (Courtesy of Corr Instruments, LLC)

- Salt-saturated aqueous solutions
- Concentrated chloride solutions
- Concrete
- Soil
- Low-conductivity drinking water
- Process streams of chemical plants at elevated temperatures
- Coatings

When CMAS are used in a relatively corrosive environment, produced data may be reduced to a single parameter so that the probe can be conveniently used for real-time and online monitoring purposes. The most anodic current has been used as a one-parameter signal for the CMAS probes. Because the anodic electrodes in a CMAS probe simulate the anodic sites on a metal surface, the most anodic current may be considered as the corrosion current from the most corroding site on the metal.

In a less corrosive environment or with a more corrosion-resistant alloy, the most anodic electrode may not be fully covered by anodic sites until the electrode is fully corroded. Therefore, the most anodic electrode may still have cathodic sites available, and the electrons from the anodic sites may flow internally to the cathodic sites within the same electrode. The total anodic corrosion current, I_{corr}, and the measured anodic current, I_a^{ex} may be related by Eq. (5.28).

$$I_a^{ex} = \varepsilon I_{corr} \tag{5.28}$$

where ε is a current distribution factor that represents the fraction of electrons resulting from corrosion that flows through the external circuit. The value of ε may vary between 0 and 1, depending on parameters such as surface heterogeneities on the metal, the environment, the electrode size, and the number of sensing electrodes. If an electrode is severely corroded and significantly more anodic than the other electrodes in the probe, the ε value for this corroding electrode would be close to 1, and the measured external current would be equal to the localized corrosion current.

5.6.3 Cathodic Protection

Cathodic protection is widely used on small to extremely large structures to protect metals and particularly steel against corrosion. This can often be accomplished by using a protective current that is either generated by a power supply in what is called impressed current cathodic protection (ICCP) or by using another metal that corrodes more readily than the metal being protected and therefore is sacrificed is the process. Chapter 13 provides a much detailed discussion of this important technique.

5.6.4 Anodic Protection

In contrast to cathodic protection, anodic protection is relatively new. The feasibility of anodic protection was first demonstrated in 1954 and tested on a small-scale stainless steel boiler designed to handle sulfuric acid [23]. Anodic protection refers to the corrosion protection achieved by maintaining an active-passive metal or alloy in its passive state by applying an external anodic current. The basic principle for this type of protection is explained by the behavior shown in Fig. 5.40.

When the potential of a metallic component is controlled and shifted in the anodic (positive) direction, the current required to cause that shift will vary. If the current required for the shift has the general polarization behavior illustrated in Fig. 5.40, the metal has an active-passive transition and can be anodically protected. Only a few systems exhibit this behavior in an appreciable and usable way. The corrosion rate of a metal with an active-passive behavior can be significantly reduced by shifting the potential of the metal so that it is at a value in the passive range shown in Fig. 5.40.

The current required to shift the potential in the anodic direction from the corrosion potential E_{corr} can be several orders of magnitude greater than the current necessary to maintain the potential at a passive value. The current will peak at the passivation potential value shown as E_{pp} (Fig. 5.40). To produce passivation the critical current density (i_{cc}) must be exceeded. The anodic potential must then be maintained in the passive region without allowing it to fall back in the active region or getting into the transpassive region, where the

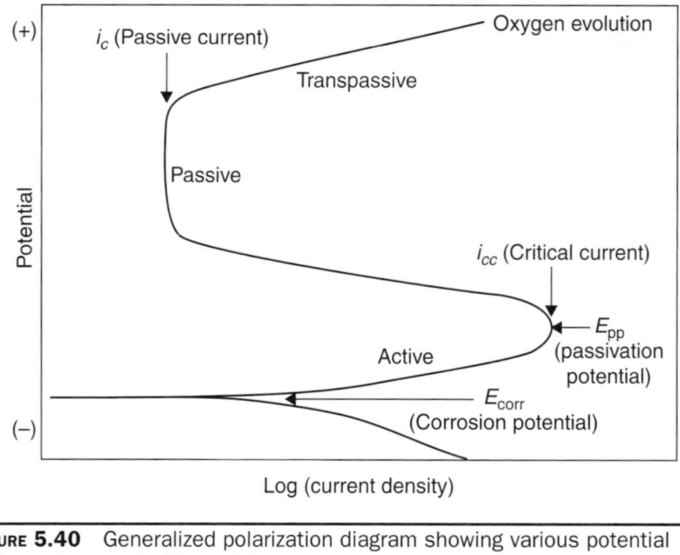

FIGURE 5.40 Generalized polarization diagram showing various potential regions of a passivable metal.

protective anodic film can be damaged and even break down completely. It follows that although a high current density may be required to cause passivation (> i_{cc}), only a small current density is required to maintain it, and that in the passive region the corrosion rate corresponds to the passive current density (i_p).

Anodic protection possesses unique features. For example, the applied current is usually equal to the corrosion rate of the protected system. Thus, anodic protection not only protects but also offers a direct means for monitoring the corrosion rate of a system. The main advantages of anodic protection are (1) low current requirements; (2) large reductions in corrosion rate (typically 10,000-fold or more); and (3) applicability to certain strong, hot acids and other highly corrosive media. It is important to emphasize that anodic protection can only be applied to metals and alloys possessing active-passive characteristics such as titanium, stainless steels, steel, and nickel-base alloys.

Since the potential must be maintained within the passive region, it is necessary to use a special device called a potential controller capable of maintaining a constant electrode potential by controlling the anodic current (Fig. 5.41). The potential controller acts as a potentiostat with its three terminals connected to the proper electrodes.

Furthermore, anodic protection can only be used in relatively well-behaved or understood environments since the composition of the electrolyte and its temperature can greatly influence the passive/active behavior of a metal or an alloy. The relative tendency for passivation is strongly dependent on the interactions between a metal and its environment and it may vary extensively with changes in either.

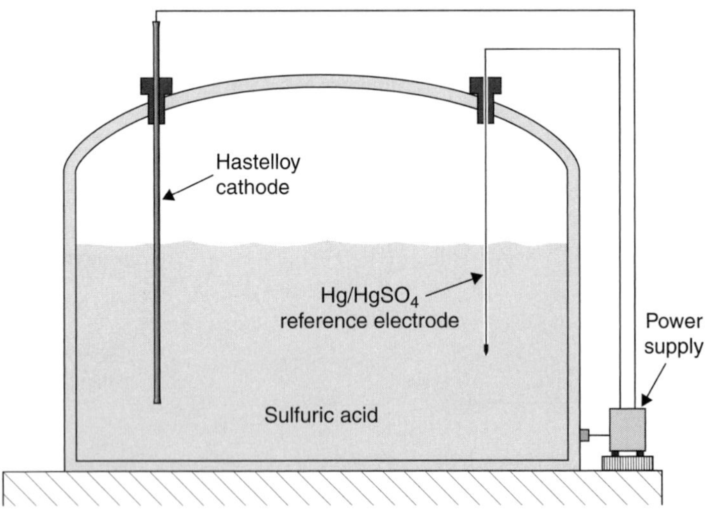

FIGURE 5.41 Schematic of an anodic protection system for a sulfuric acid storage vessel.

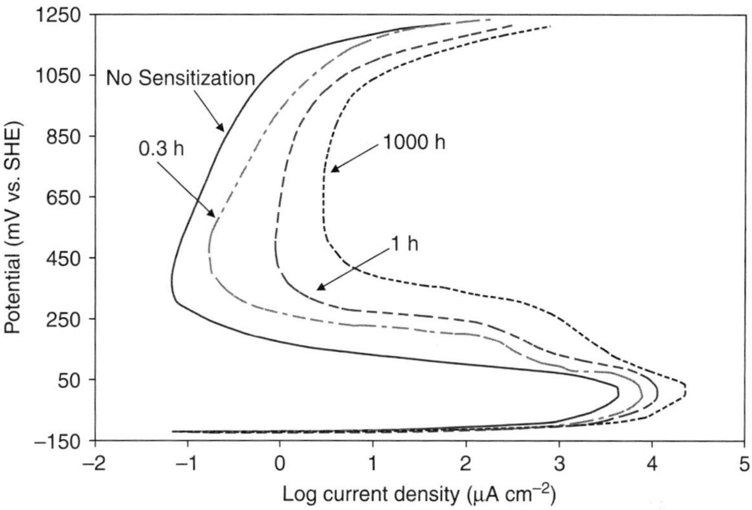

FIGURE 5.42 Anodic polarization curves of S30400 steel in a 1 M H_2SO_4 at 90°C after sensitization for various times.

The presence in the environment of impurities that retard the formation of a passive film or accelerate its degradation is often detrimental.

In this context, chloride ions can be quite aggressive for many alloys and particularly for steels and stainless steels. As an example, the addition of 3 percent hydrochloric acid to 67 percent sulfuric acid raises the critical current density for the passivation of a S31600 stainless steel from 0.7 to 40 mA cm^{-2} and the current density to maintain passivity from 0.1 to 60 μA cm^{-2}. Changes in the material may also affect its passivation behavior. Figure 5.42 illustrates how the sensitization of a S30400 stainless steel, for example, can affect its passivation behavior when exposed to sulfuric acid [27].

Anodic protection has been most extensively used to protect equipment for handling sulfuric acid. Sales of anodically protected heat exchangers used to cool sulfuric acid manufacturing plants have represented one of the more successful ventures for this technology. These heat exchangers are sold complete with the anodic protection systems installed and have a commercial advantage in that less costly materials can be used.

5.6.5 Aluminum Anodizing

As explained in Chap. 14, anodizing is a widely use technique to produce a protective inorganic coating of some engineering materials such as aluminum, magnesium, titanium and a few other metals and alloys by the application of an anodic potential that would be normally quite corrosive if it was not for the barrier created by the process itself. Of all metals that are routinely anodized, aluminum alloys are by far the most widely used for the production of a great variety of applications. There are many reasons and processes to anodize a part. Following are a few considerations that may help choose the type and process:

- *Appearance:* Products look finished, cleaner, better and last longer.
- *Corrosion resistance:* A smooth surface is retained while weathering is retarded. Useful for food handling and marine products.
- *Ease in cleaning:* Any anodized product will stay cleaner longer and are easier to clean when needed.
- *Abrasion resistance:* The treated metal is tough. It is harder than many abrasives and is ideal for tooling and air cylinder applications;
- *Non-galling:* Screws and other moving parts will not seize, drag, or jam while wear in these areas is diminished. Gun sights, instruments, and screw threads are typical applications.
- *Heat absorption:* Anodizing can provide uniform or selective heat absorption properties to aluminum for the food-processing industry.
- *Heat radiation:* Anodizing is used as a method to finish electronic heat sinks and radiators.

Anodizing Process

The anodizing process illustrated in Fig. 5.43 consists of a few steps through which a part is submitted sequentially by being dipped in baths to achieve the following goals:

- *Pretreatment:* Cleaning is done in a nonetching, alkaline detergent heated to approximately 80°C. This process removes accumulated contaminants and light oils.
- *Rinsing:* Multiple rinses, some using strictly deionized water, follow each process step.
- *Etching (chemical milling):* Etching in a sodium hydroxide solution to prepare the aluminum for anodizing by chemically removing a thin layer of aluminum. This alkaline bath gives the aluminum surface a matte appearance.
- *Desmutting:* Rinsing in an acidic solution to remove unwanted surface alloy constituent particles not removed by the etching process.
- *Anodizing:* Aluminum is immersed in a tank containing an acid electrolyte in which direct electric current is passed through the electrolytic cell between the aluminum part serving as the anode and the tank polarized as the cathode.

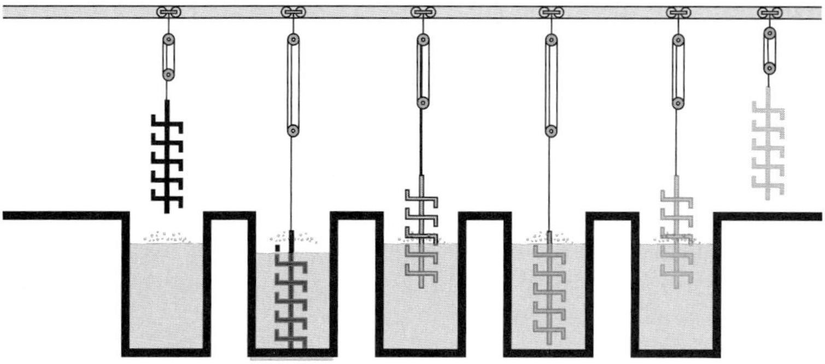

FIGURE 5.43 Sequential steps through which a part is submitted by being dipped in baths in order to clean, etch, anodize, and seal an aluminum part being anodized.

- *Coloring:* Anodic films are well suited to a variety of coloring methods including absorptive dyeing, both organic and inorganic dyestuffs, and electrolytic coloring.
- *Sealing:* In all the anodizing process, the proper sealing of the porous oxide coating is absolutely essential to the satisfactory performance of the coating. The pores must be rendered nonabsorbent to provide maximum resistance to corrosion and stains.

Many acidic solutions can be used for anodizing aluminum, but sulfuric acid solutions are surely the most common. Chromic, oxalic, and phosphoric acids are also used relatively often for specific applications [28]. The "standard" sulfuric acid anodizing bath (Type II) produces the best oxides for coloring. The solution consists of approximately 15 percent sulfuric acid and the anodizing bath is maintained at 20°C. As the anodizing temperature is increased, the oxide becomes more porous and improves in its ability to absorb color; however, it also loses its hardness and its luster, due to the dissolution action of the acid on the oxide surface.

Hardcoating (Type III) refers to a process that usually employs higher sulfuric acid concentrations, lower temperatures (between −1 and 4°C), and higher voltages, and current densities. Hardcoating imparts a very hard, dense, abrasion-resistant oxide on the surface of the aluminum. The dense oxide is formed due to the cooling effect of the cold electrolyte. At these temperatures, the sulfuric acid does not attack the oxide as fast as at elevated temperatures. Because of the lower temperature, the voltages needed to maintain the higher current densities also help form smaller, more dense pores, thus accounting for the hardness and excellent abrasion resistance.

Chromic acid anodizing (Type I) is used for applications in marine environments, on aircraft as a prepaint treatment, and in some cases when finishing assemblies where acid may be entrapped.

Although the film produced is extremely thin, it has excellent corrosion resistance and can be colored if desired. A typical bath might contain from 5 to 10 percent chromic acid, and be run at about 35 to 40°C. There are two main processes, one using 40 V and a newer process using 20 V.

Oxalic acid is sometimes used as an anodizing electrolyte using similar equipment. This bath will produce films as thick as 50 μm without the use of very low temperatures and usually gives a gold or golden bronze color on most alloys. The typical concentration is from 3 to 10 percent oxalic acid at about 27 to 32°C, using a DC voltage of about 50 V.

Phosphoric acid baths are used in the aircraft industry as a pretreatment for adhesive bonding. They are also very good treatments before plating onto aluminum. A typical bath might contain from 3 to 20 percent of phosphoric acid at about 32°C, with voltages as high as 60 V.

Properties of the Oxide Film

Anodized coatings are typically 2 to 25 μm thick, and consist of a thin nonporous barrier layer next to the metal with a porous outer layer that can be sealed by hydrothermal treatment in steam or hot water for several minutes. The resultant oxide is also nonconductive. This particular property of the anodic oxide is useful in the production of electrolytic capacitors using a special bath of boric and/or tartaric acids.

By balancing the conditions used in the anodizing process, oxides with almost any desired properties can be produced, from the thin oxides used in decorative applications to the extremely hard, wear-resistant oxides used in engineering applications. The utilization of electropolishing or chemical bright dipping in conjunction with a thin anodic oxide produces a finish whose appeal cannot be duplicated by other means. Matte finishes produced by etching the aluminum surface provides a "pewter" look that is the finish of choice of many architects.

If the anodic oxide is slightly soluble in the electrolyte, then porous oxides are formed. As the oxide grows under the influence of the applied DC current, it also dissolves, and pores develop. It is this property that provides the ability to color the oxide using organic dyes, pigment impregnation, or electrolytic deposition of various metals into the pores of the coating.

The anodizing process conditions have a great influence on the properties of the oxide formed. The use of low temperatures and acid concentration yield less porous, harder films. As the anodizing temperature is increased, the oxide becomes more porous and improves in its ability to absorb color. However, it also loses its hardness and its luster, due to the dissolution action of the acid on the oxide surface. As the pore size increases, sealing becomes more difficult and a greater amount of color is leached out into the sealing bath.

Sealing of Anodic Coatings

To achieve the maximum protective qualities and corrosion resistance required for finished articles, the anodic oxide must be sealed after it is formed and/or colored. This produces a hydrated oxide layer with improved protective properties. Figure 5.44 illustrates how the initially porous anodic film evolves during the sealing process.

The hydrothermal sealing (90 to 100°C) process consists of immersing the anodized parts in a solution of boiling water or other solution such as nickel acetate, wherein the aluminum oxide is hydrated. The hydrated form of the oxide has greater volume than the unhydrated form and thus the pores of the coating are filled or

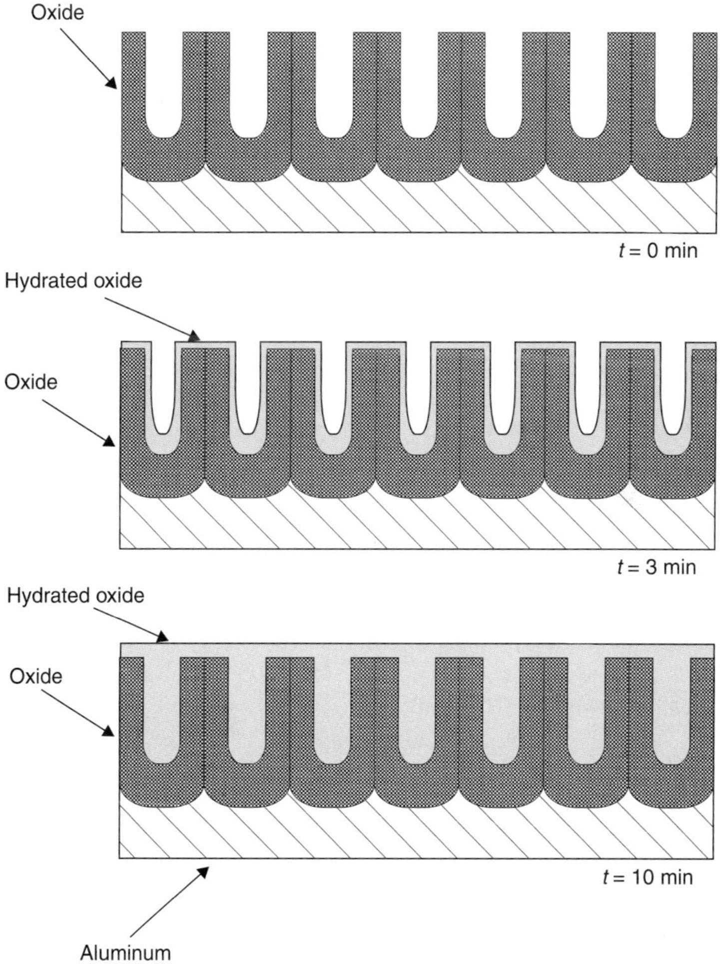

FIGURE 5.44 Schematic description of the evolution of a porous anodic film on aluminum as a function of the sealing time at 85°C.

plugged and the coating becomes resistant to further staining and corrosion. Nickel containing seals will, in most cases, prevent leaching of dyes during the sealing operation.

Due to the higher energy costs inherent in hydrothermal sealing, chemical manufacturers have developed mid-temperature seals (70 to 90°C). These seals, which contain metal salts such as nickel, magnesium, lithium, and others, have become very popular due to the lower energy costs and their ease of operation. One disadvantage of the lower temperature is the tendency of organically dyed parts to leach during sealing. This can be compensated for by a slight increase in the bath concentration and by operating the solution at the upper temperature limits (90°C).

Nickel-free environment-friendly seals are fast becoming the seal of choice where clear or electrolytically colored parts are concerned. Because there is nothing to leach, these mid-temperature seals accomplish hydration of the oxide without the use of the heavy metal ions. When the seals become contaminated or are no longer effective, they can be discharged to the sewer without subsequent treatment (except possible pH adjustment). This offers the finisher a safer alternative to the effluent treating necessary with heavy metal containing seals.

A significant modification in the sealing of anodized aluminum was also developed for room temperature sealing (20 to 35°C). Unlike the high temperature and mid-temperature seals, which depend on hydration for sealing, the cold seals rely on a chemical reaction between the aluminum oxide and the nickel fluoride contained in the seal solution. Unfortunately, this reaction is slow at ambient temperatures and the sealing process can require up to 24 hours. However, it has been found that a warm-water rinse (70°C) after the cold seal immersion will accelerate the sealing process, allowing for handling and packing of the sealed parts. The sealing of organically dyed parts in cold seals has been found to be advantageous. Light stability testing (fade resistance) has shown that parts sealed in cold seals gain additional lightfastness.

5.6.6 Chloride Extraction

Electrochemical chloride extraction is a method particularly appreciated by civil engineers who want to remediate existing concrete structures contaminated with chlorides from deicing salts or marine mist. The hardware involved is similar to that involved in cathodic protection. Electrochemical extraction of chloride ions is achieved by positioning an anode in contact with a caustic electrolyte spread on the external concrete surface, and impressing a direct current between the anode and the reinforcing steel acting as the cathode (Fig. 5.45). Under the application of this electrical field, chloride ions migrate away from the negatively charged steel and toward the positively charged external anode.

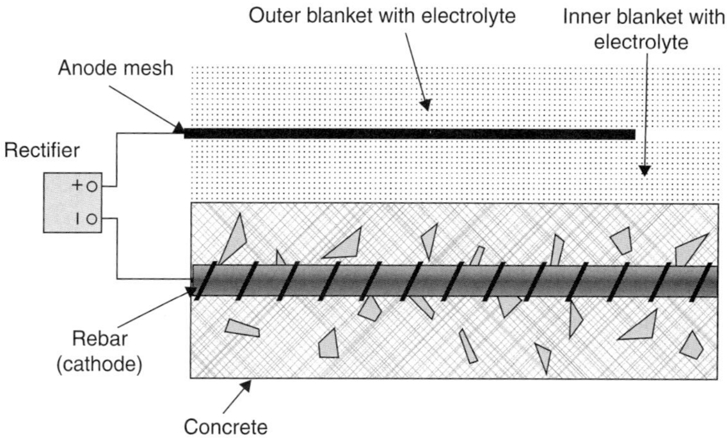

FIGURE 5.45 Schematic of electrochemical chloride extraction and re-alkalization treatments.

Chloride extraction has been recommended for structures that do not contain pre- or post-tensioned steel and have little damage to the concrete itself. The current densities and applied voltages involved are significantly higher than those used in cathodic protection, therefore increasing the risk of hydrogen evolution and subsequent hydrogen embrittlement (Chap. 6). The unsuitability of the technique to prestressed concrete is thus not surprising. Further requirements are a high degree of rebar electrical continuity and preferably low concrete resistance. Since several days or even weeks are involved in the extraction processes using suitable current densities, the technique is more realistically applicable to highway substructures rather than bridge decks, where it could cause serious traffic disruptions as illustrated in Fig. 5.46.

In practice, the chloride extraction process does not remove the chloride ions from the concrete completely. Rather, a certain percentage is removed and the balance is redistributed away from the reinforcing bars. There is also increasing evidence that the protective layer around the reinforcing steel is considerably enhanced, elevating the threshold for corrosion.

As with ICCP, the applied current density has to be controlled. Under excessive current magnitude several problems can arise, such as reduction in bond strength between rebar and concrete, softening of the cement paste around the rebar steel, and cracking of the concrete. Concrete containing alkali-reactive aggregates is not considered a suitable candidate for the process as the expansive reactions leading to cracking and spalling associated with these aggregates tend to be aggravated [29].

Figure 5.46 Bridge deck energized for electrochemical chloride extraction. (Courtesy of Vector Corrosion Technologies)

References

1. Shreir LL, Jarman RA, Burstein G.T. *Corrosion Control*. 3rd edn. Oxford, U.K.: Butterworths Heinemann, 1994.
2. Wenner F. A method of measuring earth resistivity. *Bulletin of the Bureau of Standards* 1915; 12: 469–78.
3. Shepard ER. Pipe line currents and soil resistivity as indicators of local corrosive soil areas. *National Bureau of Standards Journal of Research* 1931; 6: 683–708.
4. Wagner C, Traud W. Uber die deutung von korrosionvorgangen durch uberlagerung von elektrochemischen teilvorgangen und uber die potentialbildung an mischellektroden. *Zeitschrift fur Elektrochemie* 1938; 44: 391–402.
5. Evans UR. *An Introduction to Metallic Corrosion*. London, U.K.: Edward Arnold, 1948.
6. Van Orden AC. Applications and problem solving using the polarization technique. *CORROSION 98*, Paper # 301. Houston, Tex.: NACE International, 1998.
7. Silverman DC. Tutorial on cyclic potentiodynamic polarization technique. *CORROSION 98*, Paper # 299. Houston, Tex.: NACE International, 1998.
8. Grauer R, Moreland PJ, Pini G. *A Literature Review of Polarisation Resistance Constant (B) Values for the Measurement of Corrosion Rate*. Houston, Tex.: NACE International, 1982.
9. Standard Guide for on-line monitoring of corrosion in plant equipment (Electrical and electrochemical methods). *Annual Book of ASTM Standards*. Philadelphia, Pa.: American Society for Testing of Materials, 2001; 03(02): G 96–90.
10. Dean SW. Corrosion monitoring for industrial processes. In: Cramer DS, Covino BS, eds. Vol. 13A: *Corrosion: Fundamentals, Testing, and Protection*. Metals Park, Ohio: ASM International, 2003; 533–41.
11. Hugel G. *Corrosion Inhibitors—Study of their Activity Mechanism*. 1st European Symposium on Corrosion Inhibitors. Ferrara, Italy: University of Ferrara, 1960.

12. Boukamp BA. *Equivalent Circuit (Equivcrt.PAS) Users Manual.* Report CT89/214/128. The Netherlands: University of Twente, 1989.
13. Hladky K, Callow LM, Dawson JL. Corrosion rates from impedance measurements: an introduction. *British Corrosion Journal* 1980; 15: 20.
14. de Levie R. *Electrochemical Response of Porous and Rough Electrodes.* New York, N.Y.: Wiley-Interscience, 1967.
15. Urquidi-MacDonald M, Egan PC. Validation and extrapolation of electrochemical impedance spectroscopy data analysis. *Corrosion Reviews* 1997; 15.
16. *Techniques for Monitoring Corrosion and Related Parameters in Field Applications.* NACE 3T199. Houston, Tex.: NACE International, 1999.
17. Roberge PR. Analyzing electrochemical impedance corrosion measurements by the systematic permutation of data points. In: Munn RS. ed. *Computer Modeling in Corrosion* [STP 1154]. Philadelphia, Pa.: American Society for Testing and Materials, 1992; 197–211.
18. Roberge PR, Sastri VS. On-line corrosion monitoring with electrochemical impedance spectroscopy. *Corrosion* 1994; 50: 744–54.
19. Haruyama S, Tsuru TA. Corrosion monitor based on impedance method electrochemical corrosion testing. In: Mansfeld F, Bertocci U. eds. *Computer Modeling in Corrosion* [STP 727], Philadelphia, Pa.: American Society for Testing and Materials. 1981; 167–186.
20. Davis GD, Raghu S, Carkhuff BG, Garra F, Srinivasan R, Phillips TE. *Corrosion Health Monitor for Ground Vehicles.* Tri-Service Corrosion Conference, Paper # 103, November 14–18, Orlando, Fla: NACE International, 2005.
21. Iverson WP. Transient voltage changes produced in corroding metals and alloys. *Journal of the Electrochemical Society* 1968; 115: 617–8.
22. Cottis RA, Al-Awadhi MAA, Al-Mazeedi H, Turgoose S. Measures for the detection of localized corrosion with electrochemical noise. *Electrochimica Acta* 2001; 46: 3665–74.
23. Roberge PR. *Handbook of Corrosion Engineering.* New York, N.Y.: McGraw-Hill, 2000.
24. Huet F, Bautista A, Bertocci U. Listening to corrosion. *The Electrochemical Society Interface* 2001; 10: 40–3.
25. Yang L, Sridhar N, Pensado O, Dunn DS. An in-situ galvanically coupled multielectrode array sensor for localized corrosion. *Corrosion 2002*; 58: 1004–14.
26. Yang.L, Sridhar N. Coupled multielectrode online corrosion sensor. *Materials Performance* 2003; 42: 48–52.
27. Sedriks AJ. *Corrosion of Stainless Steels.* New York, N.Y.: John Wiley & Sons, 1979.
28. Grubbs CA. Anodizing of aluminum. *Metal Finishing* 2002; 100: 463–78.
29. *Electrochemical Chloride Extraction from Concrete Bridge Components.* Technical Brief #2. Toronto, Canada: Canadian Strategic Highway Research Program (C-SHRP), 1995.

CHAPTER 6

Recognizing the Forms of Corrosion

6.1 Recognizing Corrosion

The previous chapters provide an introduction to the general science of corrosion processes with some practical applications. In reality, the principles that govern these scientific concepts are rarely of interest to people facing corrosion problems. The main questions generally asked by most people facing a corrosion problem are

- How serious is this problem?
- How can it be fixed and how much will it cost?
- What caused the problem in the first place?

The present chapter will focus on answering the first of these questions and Chap. 7 the last. The following quotation from Fontana and Greene's classic textbook on corrosion engineering originally published in 1967 [1] summarizes a training principle that has been reused extensively by many instructors and that is central in all modern training manuals on the subject.

> It is convenient to classify corrosion by the forms in which it manifests itself, the basis for this classification being the appearance of the corroded metal. Each form can be identified by mere visual observation. In most cases the naked eye is sufficient, but sometimes magnification is helpful or required. Valuable information for the solution of a corrosion problem can often be obtained through careful observation of the corroded test specimens or failed equipment.

It is now widely accepted that much can be deduced from examination of materials which have failed in service and that it is often possible by visual examination to decide which corrosion mechanisms have been at work and what corrective measures are required. In another widely used NACE document, Paul Dillon and

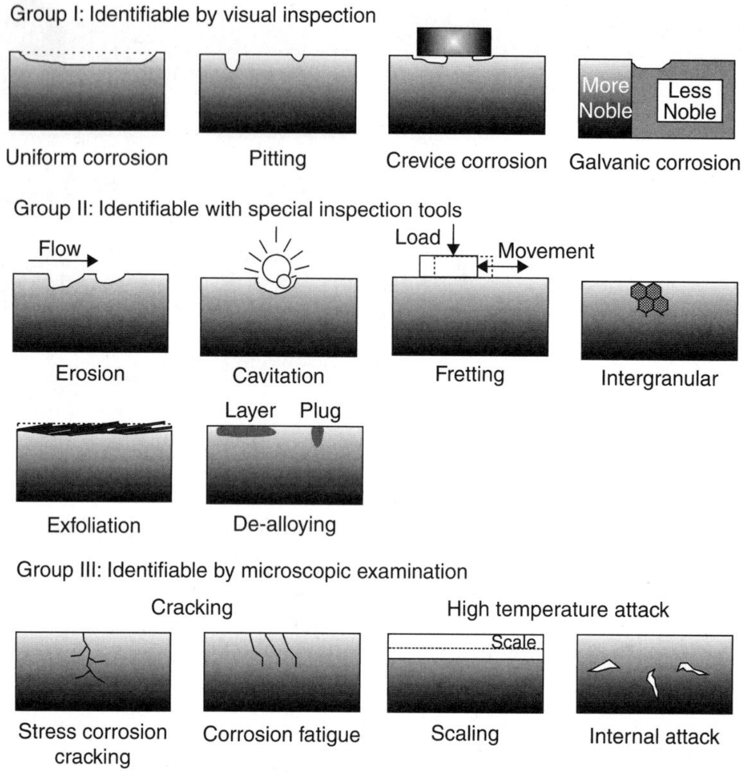

FIGURE 6.1 Main forms of corrosion attack regrouped by ease of identification.

his coauthors have grouped the eight forms of corrosion depicted graphically in Fig. 6.1 in the following three categories [2]:

Group I: Corrosion problems readily identifiable by visual examination.

1. Uniform corrosion is characterized by an even, regular loss of metal from the corroding surface.
2. Localized corrosion during which all or most of the metal loss occurs at discrete areas. In this scheme crevice corrosion is said to be a particular form of pitting usually due to localized differences in the environment (*pitting, crevice*).
3. Galvanic corrosion occasioned by electrical contact between dissimilar conductors in an electrolyte.

Group II: Corrosion damage that may require supplementary means of examination for identification.

4. Velocity effects include erosion–corrosion, a form of attack caused by high velocity flow; cavitation caused at even higher flow by the collapse of bubbles formed at areas of low pressure in a flowing stream; and fretting that is caused by vibratory relative motion of two surfaces in close contact under load (*erosion–corrosion, cavitation, fretting*).
5. Intergranular corrosion at the grain boundaries in the metal structure *(intergranular, exfoliation).*
6. Dealloying corrosion due to the selective dissolution of one component of an alloy.

Group III: Corrosion specimens for these types should usually be verified by microscopy of one kind or another.

7. Cracking phenomena includes corrosion fatigue, a mechanical phenomenon enhanced by nonspecific corrosive environments, and environmental cracking, in which a brittle failure is induced in an otherwise ductile material under tensile stress in an environment specific for the alloy system (*stress corrosion cracking, fatigue*).
8. High-temperature corrosion (*scaling, internal attack*).
9. Microbial effects caused by certain types of bacteria or microbes when their metabolism produces corrosive species in an otherwise innocuous environment, or when they produce deposits which can lead to corrosion attack.

In this widely distributed guide, Paul Dillon acknowledged microbial effects as a *ninth* broad type of corrosion attack. It could also be argued that many of the forms in the previous list are more families or multiple forms of corrosion damage. It is obvious that pitting and crevice corrosion, for example, are quite distinct in how they occur, have very dissimilar triggering mechanisms, and would be prevented by totally different methods. Pitting and crevice corrosion were indeed treated as two distinct forms of corrosion by Fontana and Greene in their 1967 manual and by many others since then.

Similarly the velocity effects group (erosion–corrosion, cavitation, and fretting corrosion) is a relatively artificial arrangement of corrosion types that are in fact quite different in appearance. So, if we count the actual forms of corrosion in Dillon's document, we have a total closer to fifteen than to eight! If we add to these forms the ninth form of corrosion discussed in Fontana and Greene, that is, hydrogen damage either as blistering or embrittlement, we now have a total of *seventeen corrosion forms*!

As Fontana indicated in 1967, "This listing is arbitrary but covers practically all corrosion failures and problems." As this statement clearly acknowledged, the listing popularized by Fontana is an attempt to summarize all known corrosion types in recognizable categories.

It is therefore a good yardstick to define the state of knowledge at the time such a list is created.

These corrosion forms could also have been organized on the basis of other factors than their visible appearance or inspectability. The degree of localization, for example, would be a way to organize corrosion problems as a function of the surface selectivity of attack. Another method to organize corrosion problems would be by using metallurgical features of importance such as the grain structure of a metallic material. A common type of corrosion attack for which the grain structure is important is intergranular or intercrystalline corrosion during which a small volume of metal is preferentially removed along paths that follow the grain boundaries to produce what might appear to be fissures or cracks. Intergranular and transgranular corrosion sometimes are accelerated by tensile stress. In extreme cases, the cracks proceed entirely through the metal, causing rupture or perforation. This condition is known as stress corrosion cracking (SCC).

While the types of corrosion identified in Fig. 6.1 are described individually in the following sections it should be recognized at the onset that during any damaging corrosion process these types often act in synergy. The unfolding of a crevice situation, for example, will typically create an environment favorable for pitting, intergranular attack, and even cracking.

The actual importance of each corrosion type will also differ between systems, environments, and other operational variables. However, there are surprising similarities in the corrosion failure distributions within the same industries as can be seen by comparing Figs. 6.2(*a*) and(*b*). Both corrosion failure distribution charts represent

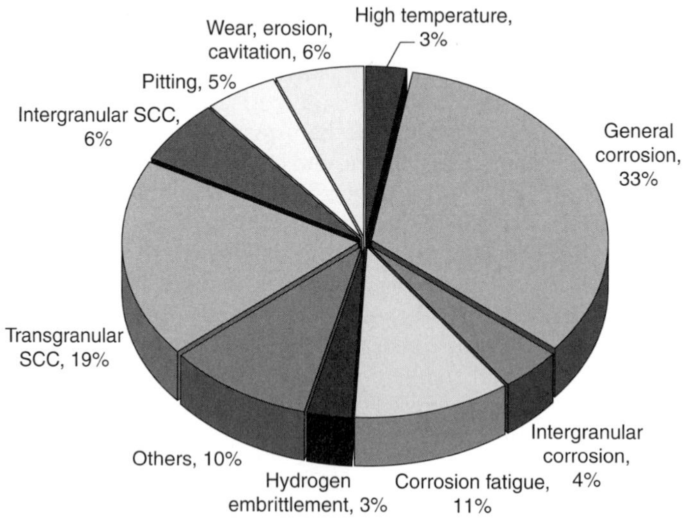

Figure 6.2 Failure-statistics of large chemical process plant in Germany (*a*), and in the United States (*b*).

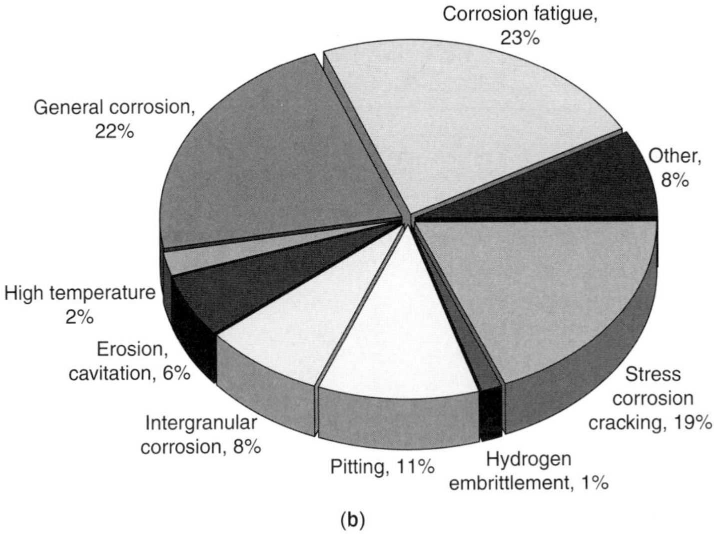

FIGURE 6.2 *(continued)*

data for two large chemical plants on two different continents, one located in Germany [Fig. 6.2(*a*)] and the other in the United States [Fig. 6.2(*b*)] [3].

6.2 General or Uniform Attack

Uniform corrosion corresponds to the corrosion attack with the greatest metal weight loss and is a common sight where steel structures are abandoned to rust (Fig. 6.3). In fact, the rich hues produced by the corrosion of some metals have been put to use in notable outdoor applications, for example, copper as a long-lasting roofing material and weathering steel in buildings and sculptures (Fig. 6.4). From a corrosion inspection point of view, uniform attack is relatively detectable and its effects predictable hence it is deemed to be less troublesome than other forms of corrosion unless the corroding material is hidden from sight. The internal corrosion of pipeline, for example, or the corrosion of hidden components and that of any other buried or immerged structures are good examples that even the simplest corrosion process needs to be monitored.

Designing in a system a corrosion allowance based on the possible loss of a material thickness is one of the simplest methods for dealing with uniform attack. Ultrasonic inspection has been used for decades to measure the thickness of solid objects. A piezoelectric crystal serves as a transducer to oscillate at high frequencies, coupled directly or indirectly to one surface of the object whose thickness is to be measured. The time the wave of known velocity takes to travel through the material is used to determine its thickness. Since the late 1970s,

Figure 6.3 Underground fuel tanks abandoned to rust in a field.

Figure 6.4 Outdoor sculpture besides City Hall in Kingston, Ontario.

ultrasonic equipment has been enhanced greatly by combining the basic electronics with computers. However, many instruments still in use today are for single-point thickness measurements, which do not provide the capability of the more sophisticated systems.

Rugged instruments based on portable computers are now available from many vendors. These systems, complete with motor-driven robotic devices to manipulate the transducer(s), have created the ability to measure wall thickness of corroded components at tens of thousands of points over 0.1 m^2, which can be converted into mass loss and pitting rates. This capability, coupled with increased precision of field measurements achievable with computer-controlled systems, has made these automated ultrasonic systems well suited for online corrosion monitoring [4].

Developments are now being made with individual transducers or transducer arrays that are left in place to provide continuous monitoring. Permanently attached transducers improve accuracy by removing errors in relocating a transducer to exactly the same point with exactly the same couplant thickness. With proper transducer selection, equipment setup, and controlled temperature conditions, the accuracy of controlled ultrasonic inspection can exceed ±0.025 mm in a laboratory setting. Field inspections are typically to within ±0.1 mm.

Uncoated components having a smooth external surface after cleaning off any biomass or debris can be inspected for internal corrosion or erosion wall losses with the traditional single backwall echo approach (Fig. 6.5). Through-coating measurements allow coated systems to be inspected without removal of the coating after applying echo-to-echo technique and A-scan imaging provided the coating is well-bonded to the metal surface and its thickness less than 6 mm (Fig. 6.6).

The thickness of the metal substrate is determined simply by the time of flight for the ultrasonic signal to reach the back surface and

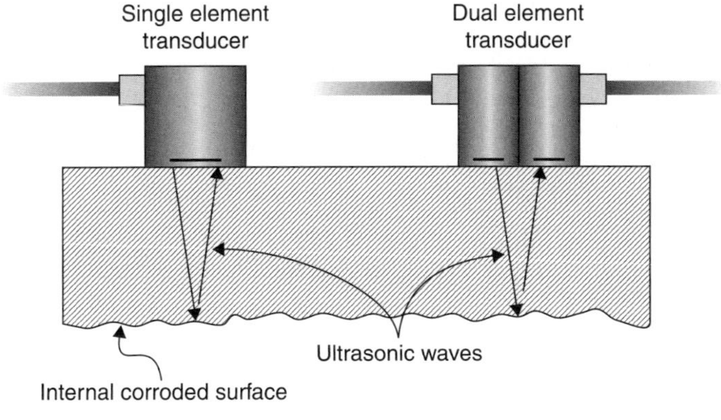

FIGURE 6.5 Thickness measurements taken using two types of UT probes (adapted from [2]).

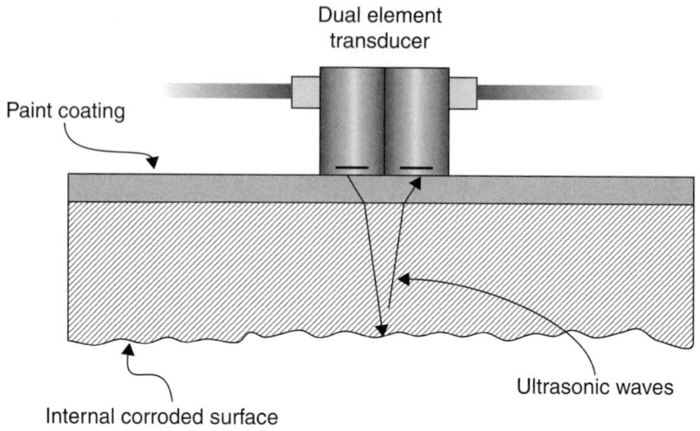

FIGURE 6.6 Thickness measurements obtained through a paint coating using a dual UT probe (adapted from [2]).

return to the transducer measured using either signal T_2 or T_3 as shown in Fig. 6.7. Ultrasonic testing (UT) digital gauges with 4- to 5-MHz, dual-element transducers are able to inspect carbon steel or austenitic stainless steel for walls thicker than 1 mm [5]. Dual-element transducers can focus the ultrasonic beam at a specific depth range thus enabling optimum sensitivity on corroded, eroded, or irregular internal riser surfaces (Fig. 6.6). These transducers are highly sensitive to small pits in their optimum thickness range.

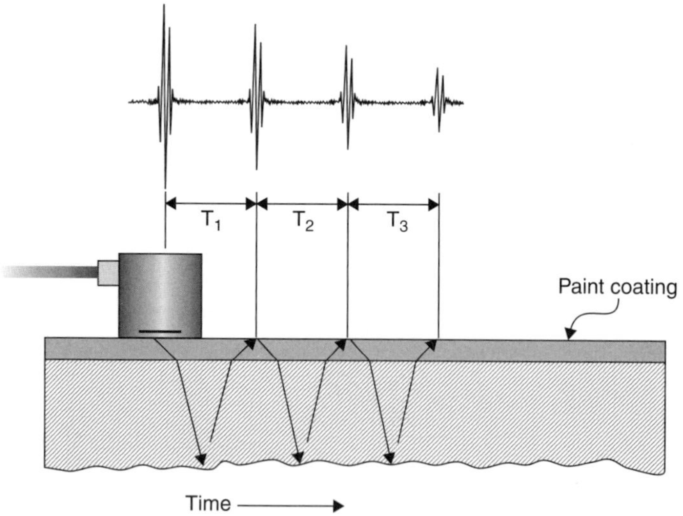

FIGURE 6.7 Thickness testing using longitudinal waves (wave propagation is actually perpendicular to the surface, but is spread out in the image to show the source of ultrasonic probing signals) (adapted from [2]).

6.3 Localized Corrosion

6.3.1 Pitting Corrosion

Probably the most common type of localized corrosion is pitting, in which small volumes of metal are removed by corrosion from certain areas on the surface to produce craters or pits that may culminate in complete perforation of a pipe or vessel wall (Fig. 6.8). Pitting corrosion may occur on a metal surface in a stagnant or slow-moving liquid. It may also be the first step in crevice corrosion, poultice corrosion, and many of the corrosion cells described in Chap. 7.

Pitting is considered to be more dangerous than uniform corrosion damage because it is more difficult to detect, predict, and design against. A small, narrow pit with minimal overall metal loss can lead to the failure of an entire engineering system. Only a small amount of metal is corroded, but perforations can lead to costly repair of expensive equipment.

One spectacular catastrophe resulting from a single pit has been described in the television series called *Seconds from Disaster*. The sewer explosion that killed 215 people in Guadalajara, Mexico, in

FIGURE 6.8 Water freely flowing out of a silver-plated water jug with pit through corrosion.

FIGURE 6.9 Pits in the galvanized water pipe that contributed to the erosion–corrosion of the gas line and the subsequent leak in a sewer main that caused the Guadalajara 1992 explosion [6]. (Courtesy of Dr. Jose M. Malo, Electric Research Institute, Mexico)

April 1992, also caused a series of blasts that damaged 1600 buildings and injured 1500 people. At least nine separate explosions were heard, starting at approximately 10:30 a.m., ripping a jagged trench that ran almost 2 km. The trench was contiguous with the city sewer system and the open holes at least 6 m deep and 3 m across. In several locations, much larger craters of 50 m in diameter were evident with numerous vehicles buried or toppled into them. An eyewitness said that a bus was "swallowed up by the hole." Damage costs were estimated at 75 million U.S. dollars.

The sewer explosion was traced to the installation of a water pipe by a contractor several years before the explosion. This water pipe leaked water on a gasoline line lying underneath. The cathodically protected gasoline pipeline had a hole within a cavity (Fig. 6.9) and an eroded area, all in a longitudinal direction. A second hole did not perforate the internal wall. The galvanized water pipe obviously had suffered stray current corrosion effects which were visible in pits of different sizes [6]. The subsequent corrosion of the gasoline pipeline, in turn, caused leakage of gasoline into a main sewer line.

Pitting cavities may fill with corrosion products and form caps over the pit cavities sometimes creating nodules or tubercles (Fig. 6.10). While the shapes of pits vary widely (Fig. 6.11) they are usually roughly saucer-shaped, conical, or hemispherical for steel and many associated alloys. The following are some factors contributing to initiation and propagation of pitting corrosion:

- Localized chemical or mechanical damage to a protective oxide film

FIGURE 6.10 Corrosion products forming nodules or tubercles over pit cavities. (Courtesy of Drinking Water Services, City of Ottawa)

- Water chemistry factors that can cause breakdown of a passive film such as acidity, low dissolved oxygen concentrations which tend to render a protective oxide film less stable and high chloride concentrations
- Localized damage to or poor application of a protective coating
- The presence of nonuniformities in the metal structure of the component, for example, nonmetallic inclusions.

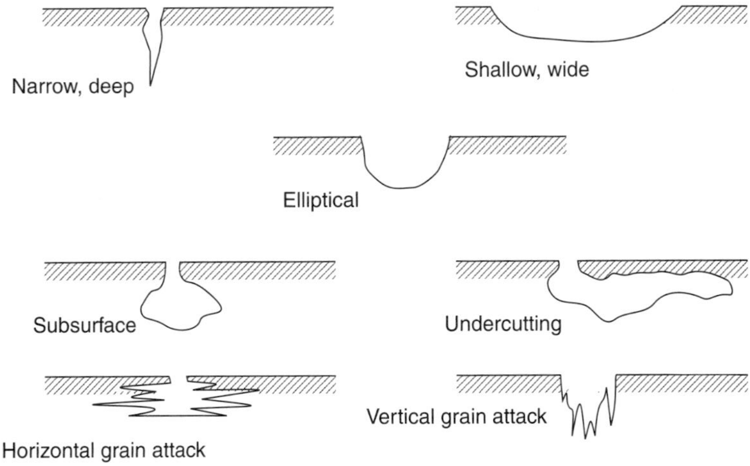

FIGURE 6.11 Typical cross-sectional shapes of corrosion pits.

The complex interactions between these factors may cause major differences on how pitting corrosion will initiate and develop in real situations. Copper, for example, a relatively simple material in terms of its metallurgy, can suffer three well-documented types of pitting corrosion depending on specific conditions in the water it carries:

Type I pitting is associated with hard or moderately hard waters with a pH between 7 and 7.8, and it is most likely to occur in cold water. The pitting is deep and narrow, and results in pipe failure (Fig. 6.12).

Type II pitting occurs only in certain soft waters, with a pH below 7.2 and occurs rarely in temperatures below 60°C. The pitting that occurs is narrower than Type I, but still may result in pipe failure (Fig. 6.13).

Type III pitting occurs in cold soft waters having a pH above 8.0. It is a more generalized form of pitting, which tends to be wide and shallow and results in blue water, byproduct releases, or pipe blockage (Fig. 6.14).

The practical importance of pitting corrosion depends on the thickness of the metal and on the penetration rate. In general, the rate of penetration decreases if the number of pits increases. This is because adjacent pits have to share the available adjacent cathodic area, which controls the corrosion current that can flow. A pit may go through four separate stages: (1) initiation, (2) propagation, (3) termination, and (4) reinitiation.

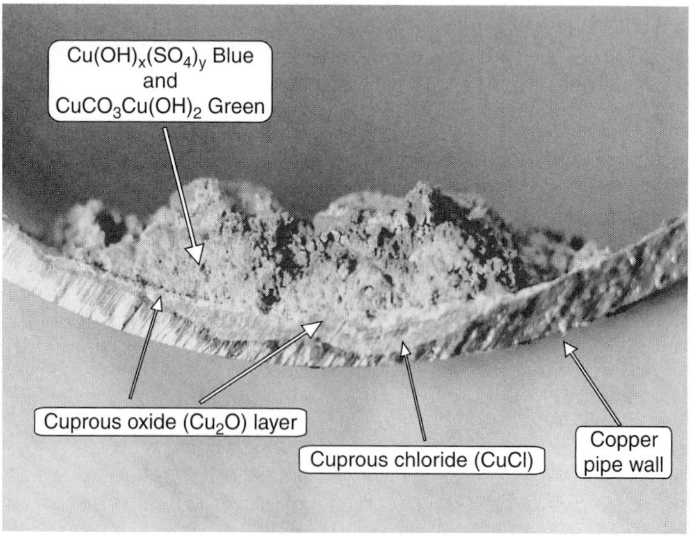

FIGURE 6.12 Type I copper pitting corrosion. (Courtesy of Russ Green, TMI)

FIGURE 6.13 Type II copper pitting corrosion. (Courtesy of Russ Green, TMI)

Typically, a local cell will lead to the initiation of a pit due to the presence of an abnormal anodic site surrounded by normal surface which acts as a cathode, or by the presence of an abnormal cathodic site surrounded by a normal surface in which a pit will have

FIGURE 6.14 Type III copper pitting corrosion. (Courtesy of Russ Green, TMI)

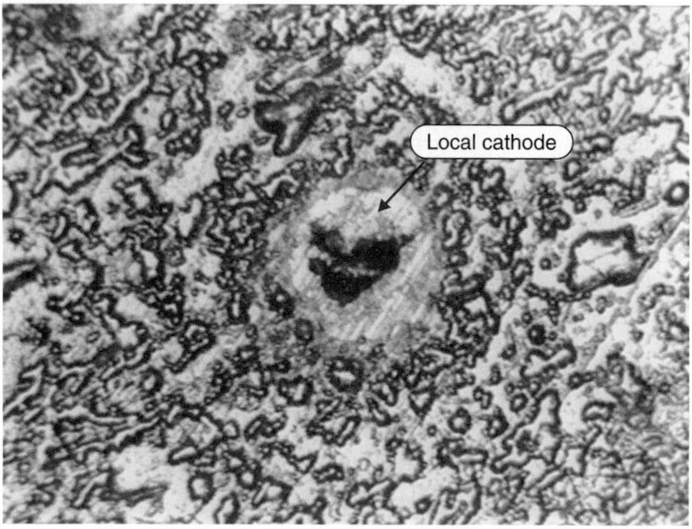

FIGURE 6.15 Pitted surface of aluminum A92519 exposed to a 3.5 percent NaCl solution for seven days (×100).

disappeared due to corrosion. In the second case, postexamination should reveal the local cathode, since it will remain unattacked (Fig. 6.15). Most cases of pitting are believed to be caused by local cathodic sites in an otherwise normal surface.

In the propagation stage, the rate increases due to changes in the anodic and cathodic environment which become respectively more acidic and alkaline. A pit may terminate due to increased internal resistance of the local cell caused by either filling with corrosion products, filming of the cathode, and so forth. If a pitted surface is dried out, pitting corrosion will be stifled, at least temporarily. When rewetted, some of the pits may reinitiate. Movement of the solution over a metal surface often reduces and may even prevent pitting that otherwise would occur if the liquid was stagnant.

As mentioned earlier, pitting is often associated with other forms of corrosion. Intergranular corrosion and cracks, for example, may progress from the main pit cavity further into the metal. In the example shown in Fig. 6.16, pitting at the edge of an aluminum-lithium sheet has progressed as intergranular corrosion at the root of the pits. In other cases intergranular corrosion is the precursor to the formation of cavernous pits as shown in Fig. 6.17. And crevice corrosion described later can be considered to be an aggravated case of pitting corrosion. Stray current corrosion that occurs when an electric current leaves a metal surface and flows into the environment can cause a very characteristic form of macroscopic pits as explained and illustrated in Chap. 7.

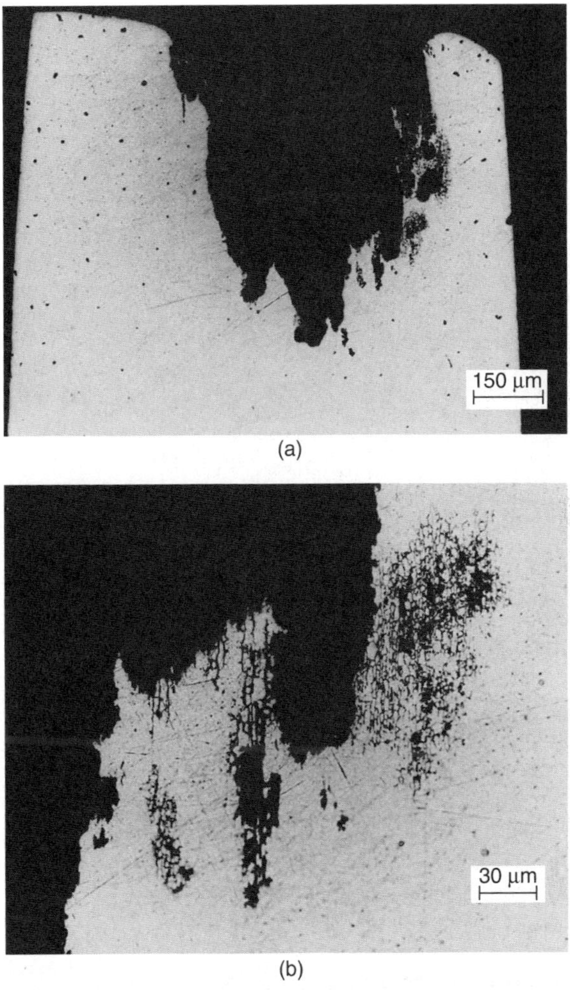

Figure 6.16 Photomicrograph of a section through an edge of the 8090-T851 panel immersed in seawater during four months; (*a*) at 64× and (*b*) at 320× to illustrate the intergranular nature of the corrosion attack.

Pitting corrosion can be assessed by various methods, including simple visual examination of a corroded specimen or monitoring coupon. When the sites of attack are numerous, the tedious job of counting pits can be facilitated by using rating charts, as shown in Fig. 6.18 [7].

Engineers concerned with soil corrosion and underground steel piping are aware that the maximum pit depth found on a buried structure is somehow related to the percentage of the structure inspected. Finding the deepest actual pit requires a detailed inspection of the whole structure. As the area of the structure inspected decreases, so does the probability of finding the deepest actual pit.

FIGURE 6.17 Stainless-steel cavernous pits formed on S30400 stainless steel following an intergranular corrosion attack. (Courtesy of Russ Green, TMI)

FIGURE 6.18 Rating charts that eliminate tedious counting of pitting sites.

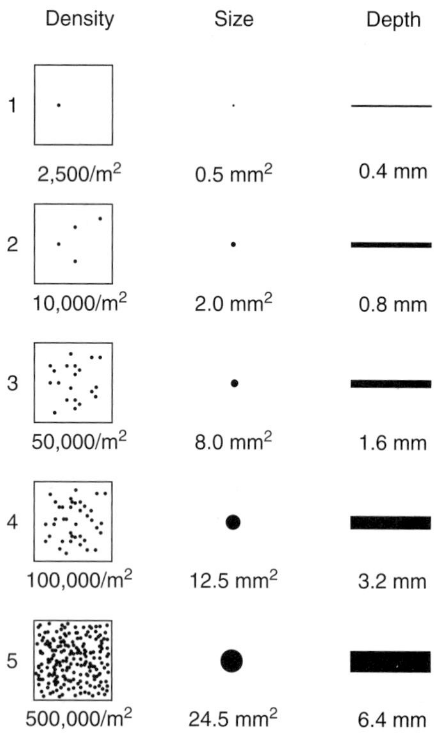

The concept of using probabilistic statistics to describe corrosion processes was first applied in the 1930s [8;9]. In their landmark paper, U.R. Evans and R.B. Mears described a simple arrangement of metallic coupons subdivided in a multitude of small squares by masking with paraffin (Fig. 6.19). The influence of sixteen basic variables on the probability and relative velocity of corrosion attack was estimated by carrying out a series of experiments with this arrangement in which all the factors except one were kept constant, thus isolating the effect of varying this one factor. The external factors studied were

1. Time of duration of the experiment
2. Area of metal exposed to the liquid
3. Oxygen-concentration in the atmosphere during the experiment
4. Quality of distilled water
5. Time of preexposure (to dried air)
6. Temperature of the experiment
7. Temperature of preexposure
8. Concentration of salt used (potassium chloride)
9. Effect of anion (comparison with other potassium salts)
10. Concentration of inhibitor (potassium carbonate in mixtures with potassium chloride)
11. Effect of other inhibitors
12. Effect of additions of acid or alkali
13. Effect of sulfur dioxide or carbon dioxide in the gas-phase

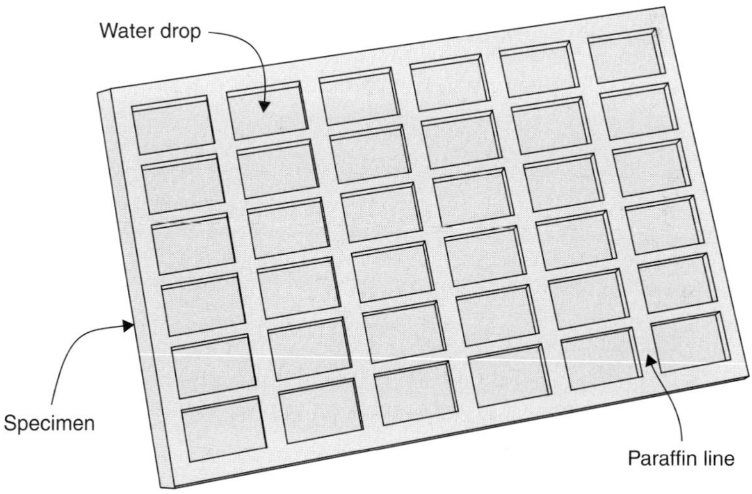

Figure 6.19 Schematic representation of the test specimen used to demonstrate the probabilistic nature of corrosion phenomena.

14. Nature of abrasive treatment and preliminary washing with various liquids
15. Character of the scratch line
16. Effect of corrosion in the neighborhood (protective action of a recent scratch-line for a less recent one)

A number of statistical transformations have since then been proposed to quantify the distributions in pitting variables. Gumbel is given the credit for the original development of extreme value statistics (EVS) for the characterization of pit depth distribution [10]. The EVS procedure is to measure maximum pit depths on several replicate specimens that have pitted, then arrange the pit depth values in order of increasing rank. The Gumbel or extreme value cumulative probability function [$F(x)$] is shown in Eq (6.1), where λ and α are the location and scale parameters, respectively. This probability function can be used to characterize data sets and estimate the extreme pit depth that possibly can affect a system.

$$F(x) = \exp\left(-\exp\left[-\frac{x-\lambda}{\alpha}\right]\right) \tag{6.1}$$

6.3.2 Crevice Corrosion

Crevice corrosion occurs in cracks or crevices formed between mating surfaces of metal assemblies, and usually takes the form of pitting or etched patches. Both surfaces may be of the same metal or of dissimilar metals, or one surface may be a nonmetal as shown in Fig. 6.20. It can also occur under scale and surface deposits and under loose fitting

FIGURE 6.20 Large holes created in a bridge structural steel by the presence of rocks. (Courtesy Kingston Technical Software)

washers and gaskets that do not prevent the entry of liquid between them and the metal surface. Crevices may proceed inward from a surface exposed to air, or may exist in an immersed structure.

The series of events leading to the formation of a severe crevice can be summarized in the following three stages. Firstly, crevice corrosion is believed to initiate as the result of the differential aeration mechanism mentioned earlier. Dissolved oxygen in the liquid which is deep in the crevice is consumed by reaction with the metal [Fig. 6.21(*a*)]. Secondly, as oxygen diffusion into the crevice is restricted, a differential aeration cell tends to be set up between the

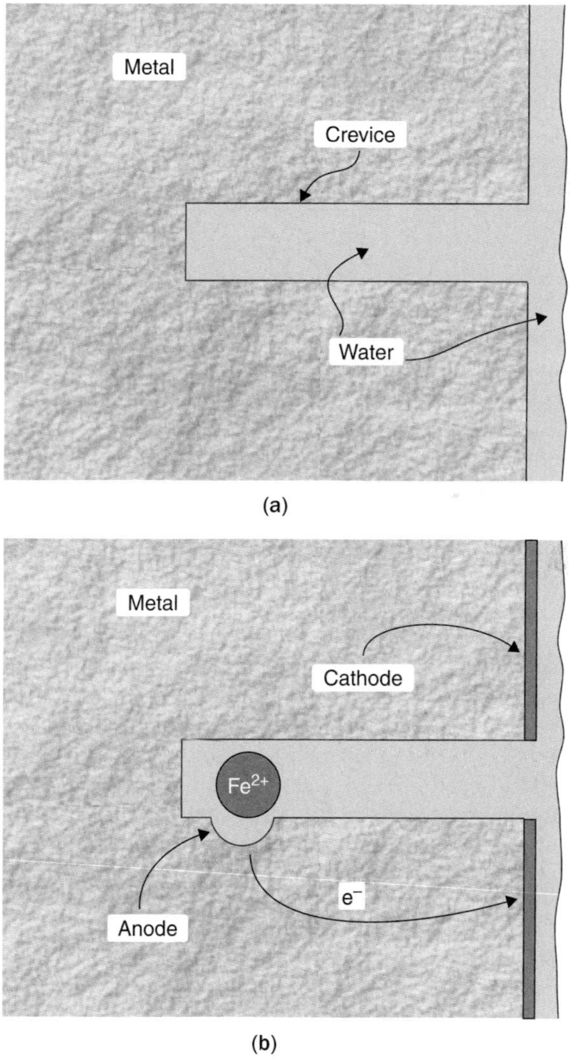

FIGURE 6.21 Schematic description of the stages of a crevice formation: (*a*) first stage; (*b*) second stage; (*c*) third stage.

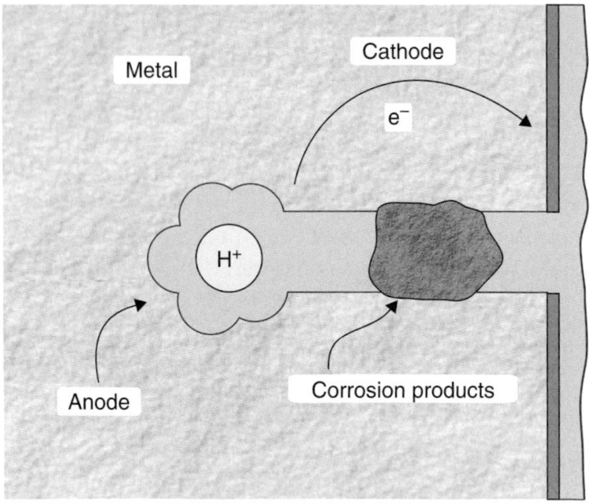

(c)

Figure 6.21 (*continued*)

crevice microenvironment and the external surface. The corrosion reactions now specialize in the crevice (anodic) and on the surface more accessible to ambient air (cathodic). The large cathodic surface (S_c) versus anodic surface (S_a) ratio (S_c/S_a) that forms in these conditions is a definitive aggravating factor of the anodic (corrosion) reaction [Fig. 6.21(*b*)].

The cathodic oxygen reduction reaction cannot be sustained in the crevice area, making it the anode of a differential aeration cell. This anodic imbalance may lead to the creation of highly corrosive microenvironmental conditions in the crevice, conducive to further metal dissolution. It is also thought that subsequent pH changes at anodic and cathodic sites further stimulate local cell action [Fig. 6.21(*c*)]. The aggravating factors present in a fully developed crevice can be summarized in the following points:

- Metal ions produced by the anodic corrosion reaction readily hydrolyze giving off protons (acid) and forming corrosion products. The pH in a crevice can reach very acidic values. For example, a pH of zero may develop in crevices involving noble alloys exposed to a neutral environment.
- The acidification of the local environment can produce a serious increase in the corrosion rate of most metals.
- The corrosion products seal even further the crevice environment.
- The accumulation of positive charge in the crevice becomes a strong attractor to negative ions in the environment, such as chlorides and sulfates that can be corrosive in their own right.

An Ideal Crevice

Many mathematical models have been developed to simulate the initiation and propagation of crevice corrosion processes as a function of external electrolyte composition and potential. Such models are deemed to be quite important for predicting the behavior of otherwise benign situations that can progress into serious corrosion situations. One such model was applied to several experimental datasets, including crevice corrosion initiation on stainless steel and active corrosion of iron in several electrolytes [11]. The model was said to break new ground by

- Using equations for moderately concentrated solutions and including individual ion-activity coefficients. Transport by chemical potential gradients was used rather than equations for dilute solutions.
- Being capable of handling passive corrosion, active corrosion, and active/passive transitions in transient systems.
- Being generic and permitting the evaluation of the importance of different species, chemical reactions, metals, and types of kinetics at the metal/solution interface.

Solution of the model for a particular problem requires specification of the chemical species considered, their respective possible reactions, supporting thermodynamic data, grid geometry, and kinetics at the metal/solution interface. The simulation domain is then broken into a set of calculation nodes that may be spaced more closely where gradients are highest (Fig. 6.22). Fundamental equations describing the many aspects of chemical interactions and species movement are finally made discrete in readily computable forms. The model was tested by comparing its output with the results of several experiments with three systems:

- Crevice corrosion of UNS 30400 stainless steel in a pH-neutral chloride solution

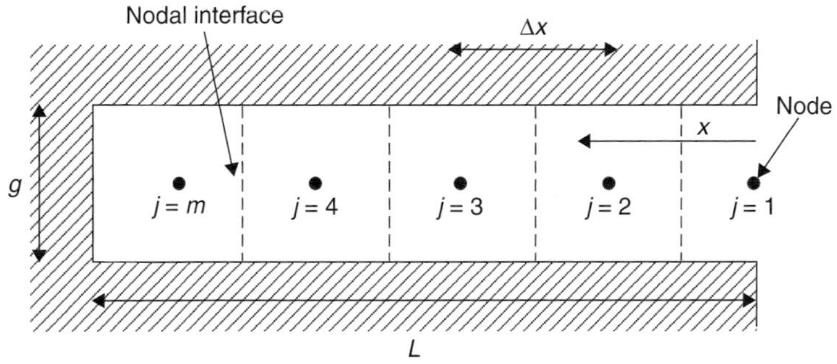

FIGURE 6.22 Schematic of crevice model geometry.

- Crevice corrosion of iron in various electrolyte solutions
- Crevice corrosion of iron in sulfuric acid

Comparison of modeled and experimental data for these systems gave agreement ranging from approximate to very good.

Real-World Crevices

There are many examples of crevice chemistries at work in all sorts of designs and applications. The following section describes a few of these. Other examples can be found throughout the text and particularly in Chap. 7, where pack rust of steel components and pillowing of aluminum fayed structures are described.

Crevice and Poultice Corrosion in Cars. The average life of a car was about 10 to 12 years in 1990, with a rather large scatter between manufacturers. Generally speaking, the areas which corrode within a car are those relatively small areas which are badly designed with regards to corrosion (Fig. 6.23). These areas could be considerable improved by better design, and greater attention to detail, and therefore would represent a minimal cost to the first and subsequent owners. Some examples of these areas are [12]

- Interweld gaps and narrow capillaries which are, by design, hardly accessible to electrocoat paint;
- Any protruding member at right angle to the direction of motion will generate spray
- Ledges and vertical panels which are out of the direct wheel trajectory can suffer from aggravated poultice corrosion
- Raw edged on the formed steel sheets can only be covered by a thin layer of protective coating

Poultice corrosion is a special form of crevice corrosion which forms when absorptive debris such as paper, wood, asbestos, sacking, cloth, and so forth., are in contact with a metal surface that becomes wetted periodically. No action occurs while the entire assembly is

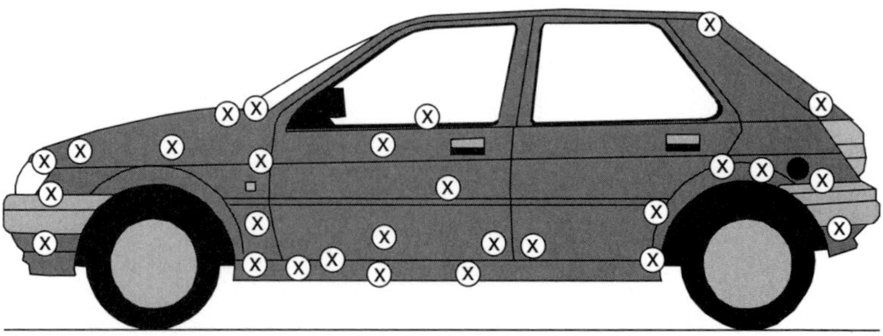

FIGURE 6.23 Diagrammatic view of corrosion prone areas in cars (adapted from [12]).

FIGURE 6.24 Sand trapped behind trim can wear away the protective paint coating and expose the steel to the environment and subsequent severe corrosion damage.

wet, but during the drying period, adjacent wet and dry areas develop. A differential aeration cell develops near the edges of the wet zones leading to corrosion.

Areas on automobiles where poultice has occurred, for example, include the hem flange within doors, wheel wells, and inside frames. Sand trapped behind trim can wear away the protective paint coating and expose the steel to the environment and subsequent severe corrosion damage (Fig. 6.24). These areas remain wet almost continuously with a highly corrosive liquid due to the moisture entrapment effect of the poultice. The aggravation caused by deicing salts can be quite serious in these areas due to wet dry cycling and accumulation that can reach saturation. A detailed analysis of poultices collected on 50 cars driven in four major North American cities has revealed the presence of large quantities of ions such as sodium, calcium, sulfate, and chloride (Table 6.1) These ions were thought to be most probably coming from road deicing and dust control agents [13].

Corrosion under Insulation. Corrosion under insulation (CUI), for example, is a particularly severe form of localized corrosion that has been plaguing chemical process industries ever since the energy crisis of the 1970s forced plant designers to include much more insulation in their designs [14].

	Chloride	Sulfate	Calcium	Sodium
Boston	33	150	63	42
Dallas	23	133	348	82
Detroit	679	531	431	140
Montreal	63	3390	1404	77

Table 6.1 Average Composition of Poultice (ppm) Found in Fifty Cars

Intruding water is the key problem in CUI. Water entering an insulation material and diffusing inward will eventually reach a region of dryout at the hot pipe or equipment wall. Next to this dryout region is a zone in which the pores of the insulation are filled with a saturated salt solution. When a shutdown or process change occurs and the metal-wall temperature falls, the zone of saturated salt solution moves unto the metal wall. Upon reheating, the wall will temporarily be in contact with the saturated solution, a very corrosive situation.

Corrosion may attack the jacketing, the insulation hardware, or the underlying piping or equipment. The drying or wetting cycles associated with CUI problems are strong accelerator of corrosion damage since they provoke the formation of an increasingly aggressive chemistry that may lead to the worst corrosion problems possible. Insulated pipes often carry high-pressure gas, oil, hydrocarbon, and many other highly dangerous chemicals such as used in the petrochemicals processing. Equipment failures in such applications mean leakages often causing fires, massive explosions, and fatalities.

Steel Deck Corrosion under Phenolic Roof Insulation. Phenolic foam roof insulation (PFRI) was hailed as the next panacea of roofing in the 1980s, as it had all the desired qualities of ideal roof insulation. It exhibited exceptional insulation with no thermal drift, great dimensional stability and easily passed fire resistance tests [1]. However, severe steel deck corrosion was observed as soon as three years after commercial PFRI was installed on some steel decks.

According to a study performed by the National Roofing Contractors Association (NRCA), there are two characteristics of phenolic insulation that appear to influence rapid steel-deck corrosion when the insulation becomes moist. The first is related to the leachate solution that percolates from wet insulation. Phenolic leachates were found to have a very low pH (1.8 to 2.1) for phenolic foams with fiberglass facers and slightly higher (2.4) for fiberglass facers with an inhibiting agent.

Phenolic foam roof insulation is manufactured with a catalyst blend of organic sulfonic acids. The pH of PFRI, when ground up in water, is 2.5 to 3.0. When moisture contacts PFRI, a very acidic

FIGURE 6.25 Section of galvanized roof panel after five years under PFRI.

environment is produced mostly due to residual formic acid from the manufacturing process. Steel roof decking commonly has either a prime coat of organic paint or is galvanized. The organic primer is unable to withstand such acidic environment and quickly breaks down, exposing bare steel and so does the galvanized coating after a longer exposure to PFRI as illustrated in Fig. 6.25 showing a corrugated galvanized deck material after five years under PFRI. Although the amount of formic acid leaching from the PFRI is small, it is recycled by the autocatalytic corrosion process.

The second characteristic that explains the rapid corrosion of steel decks in contact with PFRI is the affinity of the insulating material for water. PFRI readily absorbs water and retains it for long periods of time during which the substrate is kept wet and corroding. Laboratory analysis of PFRI taken from test cuts indicates that the moisture content by weight can average 250 percent. The high water absorption rate increases the amount of acid containing leachate that may come in contact with steel decks.

This high level of water absorption by PFRI can be revealed by performing a simple experiment in which a small piece of foam material is placed on distilled water. Figure 6.26 shows a 4 × 4 cm piece of PFRI that is sinking in water on its own weight only after 24 hours. Figure 6.27 shows how an alternate insulating foam material made with isocyanurate was still floating six weeks after the start of a similar experiment. The hydrophobic behavior of the isocyanurate foam is obviously an asset as a roofing material.

Filiform Corrosion. Filiform corrosion is a special form of crevice or oxygen cell corrosion occurring beneath organic or metallic coatings on steel, zinc, aluminum, or magnesium. Filiform corrosion normally starts at small, sometimes microscopic, defects in the coating. Lacquers and "quick-dry" paints are most susceptible to the problem.

FIGURE 6.26 Sample of PFRI sinking in water on its own weight twenty-four hours after it was left to float.

FIGURE 6.27 Sample of isocyanurate foam still floating on water after six weeks.

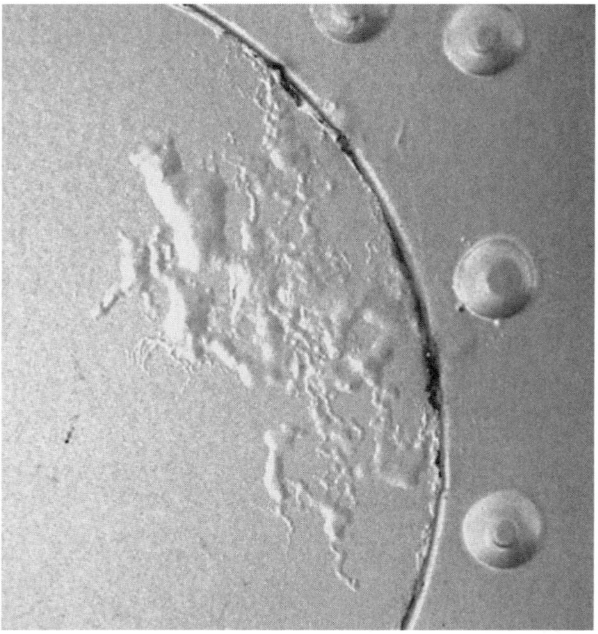

FIGURE 6.28 Filiform corrosion on an aircraft. (Courtesy of Kingston Technical Software)

The attack results in a fine network of random "threads" of corrosion product that develop as a shallow grooving of the metal surface beneath the coating material (Fig. 6.28).

The microenvironments produced by filiform corrosion on a steel surface are illustrated in Fig. 6.29 [15]. Essentially, a differential aeration cell is set up under the coating, with the lowest concentration of oxygen at the head of the filament since oxygen has to diffuse through the porous tail to the head region. A characteristic feature of such a differential aeration cell is the acidification of the environment with a low level of dissolved oxygen. This leads to the formation of an anodic metal dissolution site at the front of the head of the corrosion filament (Fig. 6.29).

For steel, pH values at the front of the head of one to four have been reported with a corrosion potential around –0.44V (SHE). In contrast, at the back of the head where the cathodic reaction dominates, the prevailing pH is around 12. For aluminum, an electrochemical potential at the front of the head of –0.73 V (SHE) has been reported, together with a 90 mV difference between the front and the back of the head [16]. Acidic pH values close to 1 have been reported at the head and pH values higher than 3.5 have been associated with the tail. Such low pH environments naturally encourage the formation of gaseous hydrogen that itself becomes a powerful disbounding force acting on a coating already damaged.

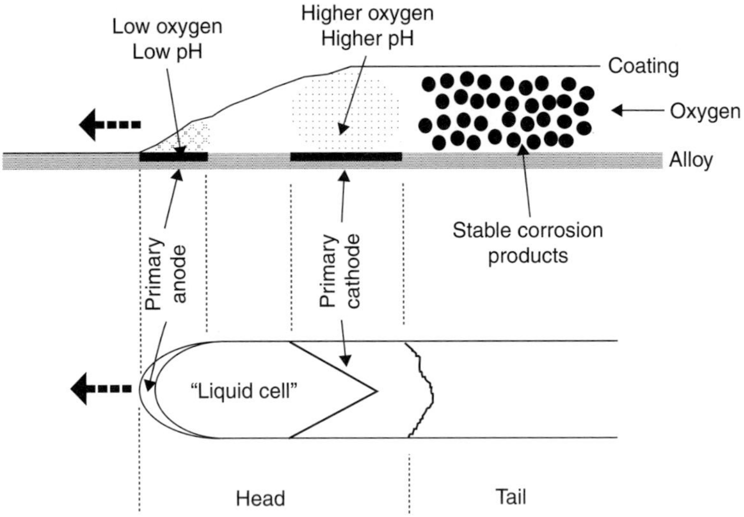

Figure 6.29 Graphical representation of the micro-environments created by filiform corrosion [7].

Corrosion of Double-Pane Windows. Glass is much more resistant to corrosion than most materials, so much so that it is easy to think of it as corrosion-proof. Glass windows after several years of exposure to the elements remain clear and apparently unaffected. Glass bottles hold a wide range of liquids that would dissolve other materials. In the laboratory, reactions are carried out in glass beakers and flasks without damage to the beakers or contamination of the solutions reacting. But, in spite of these indications that glass is indestructible by chemical attack, under certain conditions it will corrode, even dissolve.

One particular case of glass corrosion is quite similar to the crevice corrosion of metals with the main difference that the presence of condensed water in the crevice produces an alkaline environment instead of the acidic environment in the case of metals. However, that alkaline environment is particularly corrosive to glass. Such crevices are commonly produced when the seal of a double-pane window is broken.

There are two distinct stages to the corrosion process, occurring together or separately. The first stage is aqueous corrosion, caused by moisture. It refers to an ion exchange or alkali extraction (leaching) between sodium ions from the glass silicate network and hydrogen ions from the condensed water. The remaining components of the glass are not altered, but the effective surface area in contact with the solution is increased. This increase in surface area leads to extraction or leaching of the alkali ions from the glass as shown in Eq. (6.2), leaving a silica-rich layer on the surface. The pH of the solution in contact with the glass will be greatly affected by this ionic exchange due to the production of hydroxyl ions (OH^-).

$$2H_2O + Na_2SiO_{3(glass)} \rightarrow 2Na^+ + 2OH^- + H_2SiO_{3(surface)} \qquad (6.2)$$

The second stage of corrosion is a process of destruction of the leached surface layers of glass. Glass is resistant to most acids but is highly susceptible to attack by alkaline materials, especially a concentration of OH^- ions giving a pH greater than 9.0. The result is an attack of the network forming silica-oxygen (Si-O) bonds, leading to dissolution and reprecipitation of the glass components resulting in complete "frosting" of the normally transparent glass material (Fig. 6.30).

6.3.3 Galvanic Corrosion

Galvanic corrosion (also called "dissimilar metal corrosion" or wrongly "electrolysis") refers to corrosion damage induced when two dissimilar materials are coupled in a corrosive electrolyte. In a bimetallic couple, the less noble material becomes the anode and tends to corrode at an accelerated rate, compared with the uncoupled condition and the more noble material will act as the cathode in the corrosion cell.

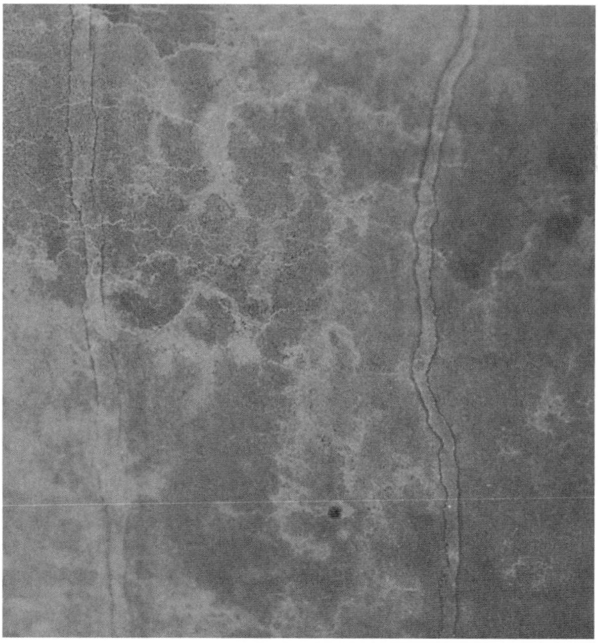

FIGURE 6.30 Double-pane window completely "frosted" due to a broken seal and subsequent corrosion of the glass material.

The bimetallic driving force was discovered in the late part of the eighteenth century by Luigi Galvani in a series of experiments with the exposed muscles and nerves of a frog that contracted when connected to a bimetallic conductor. The principle was later put into a practical application by Alessandro Volta who built, in 1800, the first electrical cell, or battery: a series of metal disks of two kinds, separated by cardboard disks soaked with acid or salt solutions. The principle was also engineered into the useful protection of metallic structures by Sir Humphry Davy in the early part of the nineteenth century. The sacrificial corrosion of one metal such as zinc, magnesium, or aluminum has become a widespread method of cathodically protecting metallic structures.

Galvanic Series

The potential of a metal in a solution is related to the energy that is released when the metal corrodes. Differences in corrosion potentials of dissimilar metals can be measured in specific environments by measuring the direction of the current that is generated by the galvanic action of these metals when exposed in a given environment. An arrangement of metals in a galvanic series based on observations in seawater, as shown in Fig. 6.31 [17], is frequently used as a first approximation of the probable direction of the galvanic effects in other environments.

In a galvanic couple involving any two metals in a galvanic series, corrosion of the metal higher in the list is likely to be accelerated, while corrosion of the metal lower in the list is likely to be reduced. Metals with more positive corrosion potentials are called noble or cathodic, and those with more negative corrosion potentials are referred to as active or anodic. Note that several metals in Fig. 6.31 are grouped. The potential differences within a group are not likely to be great and the metals can be combined without substantial galvanic effects under many circumstances.

Values of potential can change from one solution to another or in any solution when influenced by such factors as temperature, aeration, and velocity of movement. Consequently, there is no way, other than by direct potential measurements in the exact environment of interest, to predict the potentials of the metals and the consequent direction of a galvanic effect in that environment. As an example, zinc is normally very negative or anodic to steel at ambient temperatures, as indicated in the galvanic series shown in Fig. 6.31. However, the potential difference decreases with an increase in temperature until the potential difference may be zero or actually be reversed at 60°C [18;19].

Area Effects

Another important factor in galvanic corrosion is the area effect or the ratio of cathodic to anodic area. The larger the cathode compared

Noble	Platinum
or	Gold
Cathodic	Graphite
	Titanium
	Silver
	Chlorimet 3
	Hastelloy C
	18-8 Mo Stainless Steel (Passive)
	18-8 Stainless Steel (Passive)
	Chromium Steel >11% Cr (Passive)
	Inconel (Passive)
	Nickel (Passive)
	Silver Solder
	Monel
	Bronzes
	Copper
	Brasses
	Chlorimet 2
	Hastelloy B
	Inconel (Active)
	Nickel (Active)
	Tin
	Lead
	Lead-tin Solders
	18-8 Mo Stainless Steel (Active)
	18-8 Stainless Steel (Active)
	Ni-resist
	Chromium Steel >11% Cr (Active)
	Cast Iron
	Steel or Iron
	2024 Aluminum
	Cadmium
Active	Commercially Pure Aluminium
or	Zinc
Anodic	Magnesium and Its Alloys

Figure 6.31 Galvanic series of some commercial metals and alloys in seawater.

with the anode, the more oxygen reduction, or other cathodic reaction, can occur and, hence, the greater the galvanic current.

From the standpoint of practical corrosion resistance, the most unfavorable ratio is a very large cathode connected to a very small anode. This effect is illustrated in Figs. 6.32 and 6.33. Figure 6.31 indicates that iron is anodic with respect to copper and therefore is more rapidly corroded when placed in contact with it. This effect is greatly accelerated if the area of the iron is small in comparison to the area of the copper, as shown in Fig. 6.32. However, under the reverse condition, when the area of the iron is very large compared

Figure 6.32 Steel rivets on a copper bar: (*a*) at the start of the experiment; (*b*) six months after being submerged in 3 percent sodium chloride solution; and (*c*) after ten months in the same solution.

to the copper, the corrosion of the iron is only slightly accelerated as in Fig. 6.33.

If the electrical conductance of the electrolyte bridging the galvanic contact is low, either because the bulk conductivity is low or because the electrolyte is present only as a thin film as is the case in atmospheric exposure to humid environments, the effective areas taking part in galvanic cell reactions are small and the total amount of

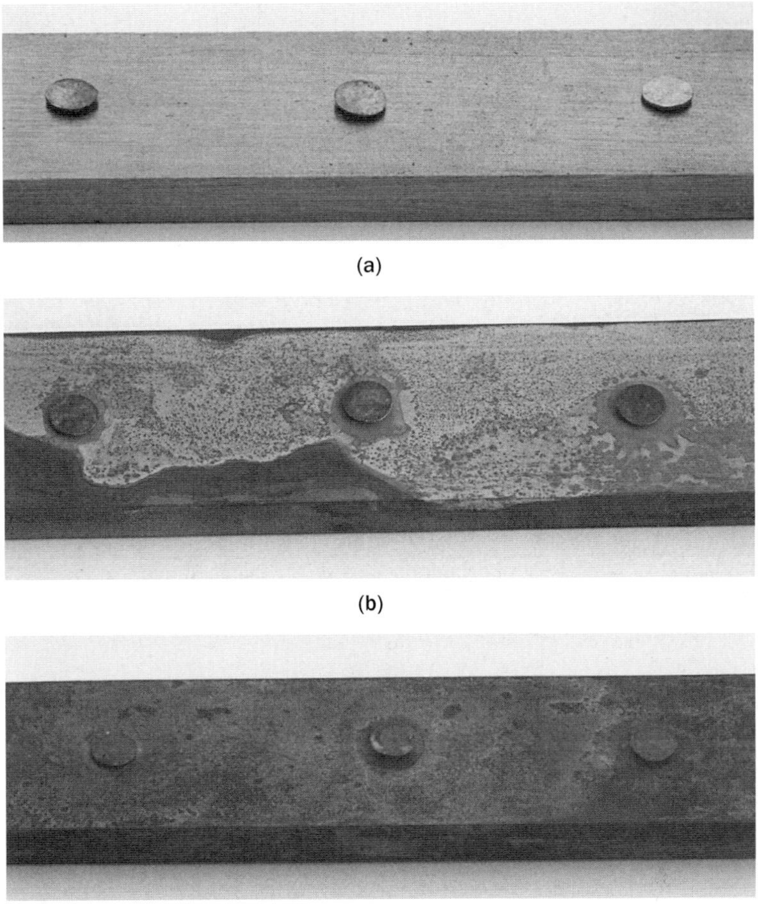

(a)

(b)

(c)

FIGURE 6.33 Copper rivets on a steel bar: (*a*) at the start of the experiment; (*b*) six months after being submerged in 3 percent sodium chloride solution; and (*c*) after ten months in the same solution.

corrosion is generally small or negligible, except maybe in the immediate vicinity of the metal junction.

Under immersed conditions in many supply waters, which generally have a relatively low electrical conductivity, adverse effects are uncommon if the contacting metals are of similar area. Thus galvanized steel pipes can be used with brass connectors, but serious corrosion to the pipe end is likely to result if the contact is made directly to a large area of copper, such as a tank or cylinder. Similarly, stainless steel and copper tubes can usually be joined without causing problems, but accelerated corrosion of the copper tube is likely to occur if it is attached to a stainless-steel tank. Under immersed

Figure 6.34 Plating out of copper nodules on an aluminum rod immersed in a 0.2 M copper sulfate solution.

conditions in a highly conducting electrolyte, such as seawater, effective areas will be greater and severe corrosion may be encountered on small anodic areas of many metals.

Deposition Corrosion

Deposition corrosion is a subtle form of galvanic corrosion in which pitting occurs in a liquid environment following the plating out of a more cathodic metallic element in solution onto a metal surface (Fig. 6.34). For example, soft water passing through a copper water pipe will accumulate some copper ions. If this water then goes in a galvanized steel or aluminum vessel, particles of metallic copper may plate out and their deposit on the vessel's internal surface will probably initiate pitting corrosion by local cell action.

This plating out action, or deposition corrosion, may be an important factor in the corrosion of the most active metals such as magnesium, zinc, or aluminum when they come into contact with solutions containing ions of more noble metals such as copper. Copper ions in concentrations less than one part per million have been observed to have a significant effect on the corrosion of aluminum by water.

6.3.4 Intergranular Corrosion

Intergranular corrosion is a form of localized attack in which a narrow path is corroded out preferentially along the grain boundaries of a metal. It often initiates on the surface and proceeds by local cell action in the immediate vicinity of a grain boundary. Although the detailed mechanism of intergranular corrosion varies with each metal system, its physical appearance at the microscopic level is quite similar for most systems (Figs. 6.16 and 6.17). The effects of this form of attack on mechanical properties may be extremely harmful.

In these cases, the driving force is a difference in corrosion potential that develops between the grain boundary and the bulk material in adjacent grains. The difference in potential may be caused by a difference in chemical composition between the two zones. This can develop as a result of migration of impurity or alloying elements in an alloy to the grain boundaries. If the concentration of alloying elements in the grain boundary region becomes sufficient, a second phase or constituent may separate or precipitate. This may have a corrosion potential different from that of the grains (or matrix) and cause a local cell to form.

The constituent may be anodic, cathodic, or neutral to the base metal or adjacent zone. Examples of anodic constituents are the intermetallic phases Mg_5Al_8 and $MgZn_2$ in aluminum alloys and Fe_4N in iron alloys. Examples of cathodic constituents are $FeAl_3$ and $CuAl_2$ in aluminum alloys and Fe_3C in iron alloys. Examples of neutral constituents are Mg_2Si and $MnAl_6$ in aluminum alloys and Mo_6C and W_6C in wrought Ni-Cr-Mo alloys.

6.3.5 Dealloying

Another type of localized corrosion involves the selective removal by corrosion of one of the elements of an alloy by either preferential attack or by dissolution of the matrix material. The various kinds of selective dissolution have been named after the alloy family that has been affected, usually on the basis of the element dissolved (except in the case of graphitic corrosion).

Dezincification

Dezincification refers to the selective leaching of the zinc phase in alloys such as brasses that contain more than 15 percent Zn. The gross appearance and size of a part that has suffered dezincification is often unchanged except for the formation of a copper hue. The part, however, will have become weak and embrittled, and therefore subject to failure without warning. To the trained observer, dezincification is readily recognized under the microscope, and even with the unaided eye, because the red copper color is easily distinguished from the yellow of brass.

Conditions generally conducive to dezincification include the presence of an electrolyte (e.g., seawater), slightly acidic conditions, presence of carbon dioxide or ammonia species, and appreciable oxygen. There are two general types of dezincification. The most common is layer dezincification that proceeds uniformly, as shown in Fig. 6.35. The second type is referred to as the plug dezincification and occurs at localized areas. The site of plugs may be recognized by the presence above them of a deposit of brownish-white zinc-rich corrosion product.

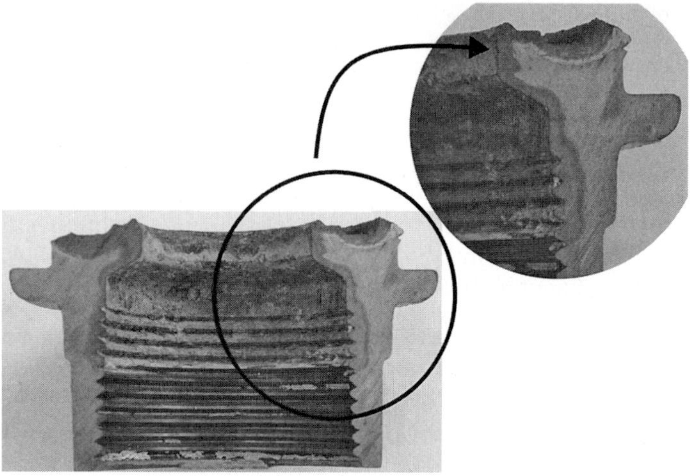

FIGURE 6.35 Layer dezincification of a brass fitting: parent material Cu 59.3 percent, Zn 35.7 percent, Pb 4.9 percent, leached area: Cu 95.0 percent, Zn 0.7 percent Pb 4.1 percent. (Courtesy of Defence R&D Canada-Atlantic)

Graphitic Corrosion

Graphitic corrosion refers to a form of selective leaching specific to gray cast iron which leaves only the graphite phase of the material. The term *graphitization* is commonly used to identify this form of corrosion but is not recommended because of its use in metallurgy for the decomposition of carbide to graphite. Gray cast iron pipe may suffer graphitic corrosion as a result of the selective dissolution of the ferrite in the alloy leaving a porous matrix made of the remaining 4 to 4.5 percent graphite. Graphitic corrosion occurs in salt waters, acidic mine waters, dilute acids, and soils, especially those containing chlorides from deicing salts or seawater and sulfate reducing bacteria (SRB).

Graphitic corrosion is quite common on water pipes made of gray cast iron, a typical material of choice of water utilities until the availability of ductile iron in the 1950s or plastic pipes later in the century. The resultant pipe surface looks normal, often exhibiting the original mill markings. When struck with a hammer or similar metal object, however, the pipe responds with a rather dull "thunk" rather than the sharp "clang" usually heard from sound metal. Graphitized pipe is also relatively soft, and one can carve into it with a chisel or other sharp object [20].

Graphitic corrosion usually occurs in three ways. Sometimes just the surface of the pipe graphitizes, forming a graphitic coating on the exterior. Frequently, this protects the pipe very well, leading to long service. At other times, a plug of graphite develops in the pipe wall. This pipe may serve for years, but if a pressure surge or water hammer occurs, the plug may blow out. The hole left by the plug is often greatly enlarged by the out-rushing water, creating a large penetration. Finally, the whole wall may graphitize, and if the pipe is subjected to a heavy earth load or perhaps a washout under a joint, a circumferential break occurs; hence the common term "water main break."

6.3.6 Hydrogen-Induced Cracking

Atomic hydrogen, and not the molecule, is the smallest atom of the periodic table and as such it is small enough to diffuse readily through a metallic structure. When the crystal lattice is in contact or is saturated with atomic hydrogen, the mechanical properties of many metals and alloys are diminished. Nascent atomic hydrogen can be produced as a cathodic reaction, either during natural corrosion processes or forced by cathodic protection, when certain chemical species are present which act as negative catalysts (i.e., poisons) for the recombination of atomic to molecular hydrogen as shown in Eq. (6.3).

$$2H^{o} \rightarrow H_2(g) \tag{6.3}$$

If the formation of molecular hydrogen is suppressed, nascent atomic hydrogen may diffuse into the interstices of the metal instead of being harmlessly evolved as a gaseous reaction product. There are many chemical species which poison this recombination (e.g., cyanides, arsenic, antimony, or selenium compounds). However, the most commonly encountered species is hydrogen sulfide (H_2S), which is formed in many natural decompositions, and in many petrochemical processes [21].

Processes or conditions involving wet hydrogen sulfide, that is, sour services, and the high incidence of sulfide-induced HIC may result in sulfide stress cracking (SSC), which has been a continuing source of trouble in the exploration and exploitation of oil and gas fields, and the subject of many international standards [22]. However, similar problems are encountered wherever wet hydrogen sulfide is encountered (e.g., acid gas scrubbing systems, heavy water plants, and waste water treatment).

Failures have occurred in the field when storage tank roofs have become saturated with hydrogen by corrosion and then subjected to a surge in pressure, resulting in the brittle failure of circumferential welds. In rare instances, even copper and Monel 400 (N04400) have suffered of HIC. More resistant materials, such as Inconels and Hastelloys often employed to combat HIC, can become susceptible under the combined influence of severe cold work, the presence of hydrogen recombination poisons, and their presence in a galvanic couple with a more anodic metal or alloy.

The HIC mechanism has not yet fully established. Various factors are believed to contribute to unlocking the lattice of the metal, such as hydrogen pressure at the crack tip, the competition of hydrogen atoms for the lattice-bonding electrons, easier plastic flow of dislocations in the metal at the crack tip in the presence of hydrogen, and the formation of certain metal hydrides in the alloy.

6.3.7 Hydrogen Blistering

Hydrogen blistering occurs predominantly in low-strength steel alloys when atomic hydrogen diffuses to internal defects such as laminations or nonmetallic inclusions where it may recombine as molecular hydrogen (H_2). When this happens, tremendous internal pressures can be generated and cause splits, fissures, and even blisters on the metal surface (Fig. 6.36). Typically 3 cm or so in diameter, blisters larger than 1 m in diameter have also been observed on some occasions. The tendency of steel alloys to blister can be prevented by using sound steels containing few inclusions and defects.

(a)

(b)

FIGURE 6.36 (*a*) Hydrogen induced cracking with mid-wall cracks running parallel to the pipeline wall; (*b*) Surface blisters may also contain cracks. (Courtesy of *MACAW's Pipeline Defects*, published by Yellow Pencil Marketing Co.)

6.4 Velocity Induced Corrosion

Velocity induced corrosion refers to problems caused or accelerated by the relative motion of the environment and the metal surface. It is characterized by surface features with a directional pattern which are a direct result of the flowing media.

With the exception of cavitation, flow-induced corrosion problems are generally termed erosion–corrosion, encompassing flow enhanced dissolution and impingement attack. The fluid can be aqueous or gaseous, single or multiphase [23]. There are several mechanisms described by the conjoint action of flow and corrosion that result in flow-accelerated corrosion (FAC) [24;25]:

Mass transport–control: Mass transport–controlled corrosion implies that the rate of corrosion is dependent on the convective mass-transfer processes at the metal/fluid interface. When steel is exposed to oxygenated water, the initial corrosion rate will be closely related to the convective flux of dissolved oxygen toward the surface, and later by the oxygen diffusion through the iron oxide layer. Corrosion by mass transport will often be streamlined and smooth.

Phase transport–control: Phase transport–controlled corrosion refers to conditions when wetting of the metal surface by a corrosive phase is flow-dependent. This may occur because one liquid phase separates from another or because a second phase forms from a liquid. An example of the second mechanism is the formation of discrete bubbles or a vapor phase from boiler water in horizontal or inclined tubes in high heat-flux areas under low flow conditions. The corroded sites will frequently display rough, irregular surfaces and be coated with or contain thick, porous corrosion deposits.

Erosion–corrosion: Erosion–corrosion has been associated with mechanical removal of the protective surface film resulting in a subsequent corrosion rate increase via either electrochemical or chemical processes. It is often accepted that a critical fluid velocity must be exceeded for a given material. The mechanical damage by the impacting fluid imposes disruptive shear stresses or pressure variations on the material surface and/or the protective surface film. Erosion–corrosion may be enhanced by particles (solids or gas bubbles) and impacted by multi-phase flows. The morphology of surfaces affected by erosion–corrosion and FAC may be in the form of shallow pits or horseshoes or other local phenomena related to the flow direction (Fig. 6.37).

Cavitation: Cavitation sometimes is caused by the formation and collapse of vapor bubbles in a liquid near a metal surface. Cavitation removes protective surface scales by the implosion of

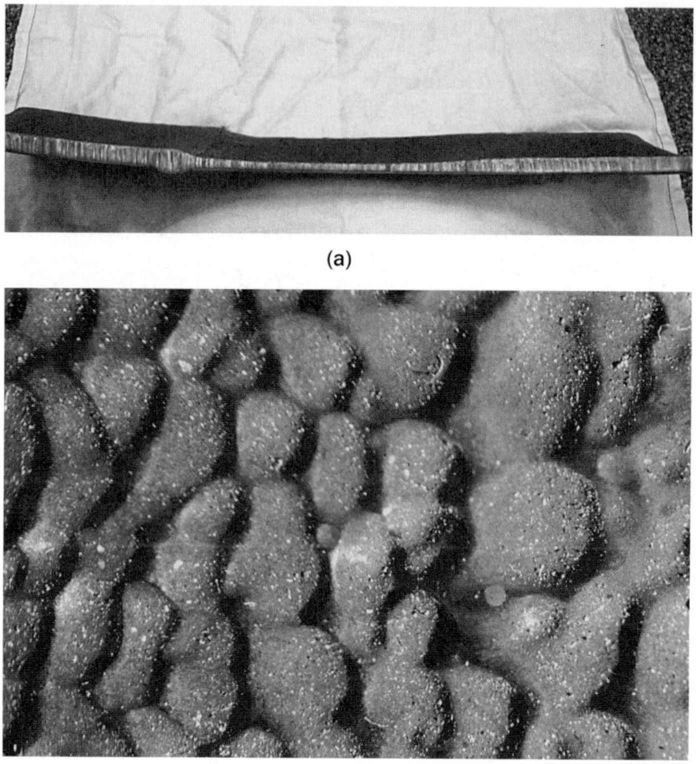

(a)

(b)

Figure 6.37 Pipe wall thinning (*a*) and corrosion patterns (*b*) caused by flow accelerated corrosion (FAC) inside a steam line. (Courtesy of Russ Green, TMI)

gas bubbles in a fluid. Calculations have shown that the implosions produce shock waves with pressures approaching 420 MPa. The subsequent corrosion attack is the result of hydromechanical effects from liquids in regions of low pressure where flow-velocity changes, disruptions, or alterations in flow direction have occurred. Cavitation damage often appears as a collection of closely spaced, sharp-edged pits or craters on the surface (Figure 6.38).

The destruction of a protective film on a metallic surface exposed to high flow rates can have a major impact on the acceleration of corrosion damage. Carbon steel pipe carrying water, for example, is usually protected by a film of rust that slows down the rate of mass transfer of dissolved oxygen to the pipe wall. The resulting corrosion rates are typically less than 1 mm/y. The removal of the film by flowing sand slurry has been shown to raise corrosion rates tenfold to approximately 10 mm/y [26].

When corrosion is controlled by dissolved oxygen mass transfer the corrosion rate can be estimated with the Sherwood number (Sh),

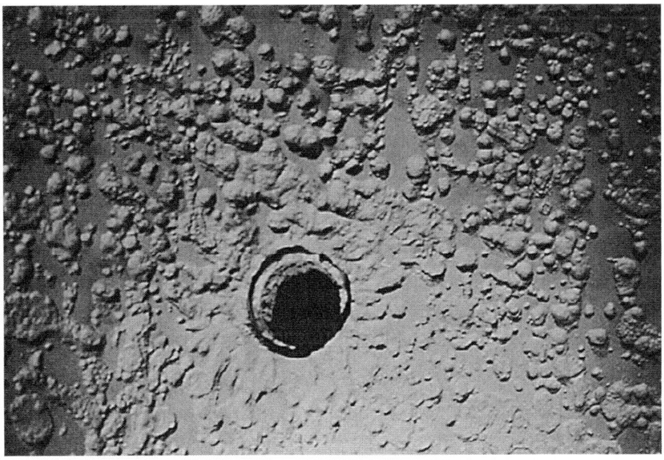

FIGURE 6.38 Cavitation corrosion inside a deaerator unit. (Courtesy of Defence R&D Canada-Atlantic)

which itself is a function of dimensionless Reynolds and Schmidt numbers in Eq. (6.4):

$$\mathrm{Sh} = \alpha\, \mathrm{Re}^{\beta}\, \mathrm{Sc}^{\gamma} \tag{6.4}$$

where α, β, and γ are experimental constants.

One of the most accepted mass transfer correlations gives values of 0.0165, 0.86, and 0.33 for respectively, α, β, and γ for fully developed turbulent flow in smooth pipes [27;28]. The Sherwood number represents the ratio of total mass transport to diffusion mass transport (D). Sh can therefore be directly related to corrosion rates ([29]) and it can be expressed in terms of the mass transfer coefficient k_m with Eq. (6.5):

$$\mathrm{Sh} = \frac{k_m L}{D} \tag{6.5}$$

where L is a characteristic length (m) describing the system. In the case of a tube L is the internal diameter of that tube.

The acceleration of corrosion may sometimes be accompanied by erosion of the underlying metal while in other cases erosion of the base metal is not a factor [23]. The relative roles of corrosion and erosion following damage to the protective film have been expressed in the relationship based on measurable bulk flow [30] expressed as Eq. (6.6):

$$\text{Erosion rate} + \text{Corrosion rate} \propto V^n \tag{6.6}$$

The exponent n in Eq. (6.6) depends on the relative contributions of corrosion and erosion to the total metal loss as illustrated in Table 6.2.

Mechanism of Metal Loss	Velocity Exponent, *n*
Corrosion	
• Liquid-phase mass transfer control	0.8–1
• Charge-transfer (activation) control	0
• Mixed (charge/mass transfer) control	0–1
• Activation/repassivation (passive films)	1
Erosion	
• Solid-particle impingement	2–3
• Liquid droplet impingement in high-speed gas flow	5–8
• Cavitation attack	5–8

TABLE 6.2 Flow Velocity as a Diagnostic Tool for Erosion–Corrosion Rates Following Damage to the Protective Film

Sources of various mechanical forces involved in the erosion of protective films and underlying metal are listed here and illustrated in Fig. 6.39 [29]:

- Turbulent flow, fluctuating shear stresses, and pressure impacts
- Impact of suspended solid particles
- Impact of suspended liquid droplets in high-speed gas flow
- Impact of suspended gas bubbles in aqueous flow
- The violent collapse of vapor bubbles following cavitation

For single-phase turbulent flow in pipes distinct "breakaway" velocities above which damage occurs in copper tubes have been observed, giving rise to the concept of a critical shear stress [31]. The largest values are obtained during quasicyclic bursting events close to the wall, which are said to be responsible for most of the turbulent energy production in the entire wall-bounded shear flow [32]. This is true of both nondisturbed and disturbed pipe flows. In practice, film removal in single-phase aqueous flow is invariably associated with the vortices created under disturbed flow conditions produced by sudden macro or microscale changes in the flow geometry. Table 6.3 summarizes the critical flow parameters that should be considered when designing a copper alloy tubing system for carrying seawater [28].

6.4.1 Erosion–Corrosion

The properties of surface films that naturally form on metals and alloys are important elements to understand the resistance of metallic materials to erosion–corrosion. Most metals and alloys used in

Alloy	CSS* (N m^{-2})	CV† (m s^{-1})	DV‡ (m s^{-1})	MDV§ (m s^{-1})
Cupro nickel with Cr	297	12.6	8.6	9
70–30 Cupro nickel	48	4.6	3.1	4.5–4.6
90–10 Cupro nickel	43	4.3	2.9	3–3.6
Aluminum bronzes			2.7	
Arsenical Al brass	19	2.7	1.9	2.4
Inhib. Admiralty				1.2–1.8
Low Si bronze				0.9
P deoxidized copper	9.6	1.9	1.3	0.6–0.9

* Critical shear stress
† Critical velocity in a 25 mm tube
‡ Design velocity based on 50 percent critical shear stress
§ Accepted or recommended maximum tubular design velocity

Table 6.3 Critical Flow Parameters for Copper Alloy Tubing in Seawater

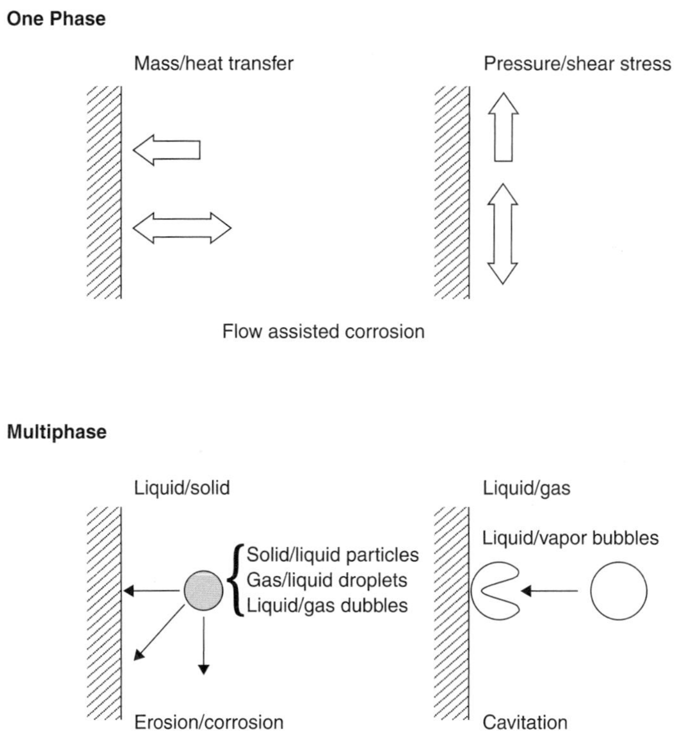

Figure 6.39 Interaction of flowing fluid with a solid boundary [9].

industry owe their corrosion resistance to the formation and retention of a protective film. Protective films fall into two categories [28]:

- Relatively thick and porous diffusion barriers such as formed on carbon steel as red rust and copper oxide on copper
- Thin invisible passive films such as formed on stainless steels, nickel alloys, and other passive metals such as titanium.

However, if the flow of liquid becomes turbulent, the random liquid motion impinges on the surface to remove this protective film. Additional oxidation then occurs by reaction with the liquid. This alternate oxidation and removal of the film will accelerate the rate of corrosion. The resulting erosive attack may be uniform, but quite often produces pitted areas over the surface that can result in full perforation (Fig. 6.40).

Obviously, the presence of solid particles or gaseous bubbles in the liquid can accentuate the attack. Also, if the fluid dynamics are such that impingement or cavitation attack is developed, even more severe corrosion can occur.

Chromium has proven to be most beneficial toward improving the properties of the passive film of ferrous and nickel-based alloys while molybdenum, when added to these alloys, improves their pitting resistance. Oxide passive films that contain insufficient molybdenum, such as in many nickel-based alloys and stainless steels, are susceptible to pitting in stagnant and low-flowing seawater, but perform well on boldly exposed surfaces at intermediate and high flow velocities. In oilfield conditions, fluid velocity acts in

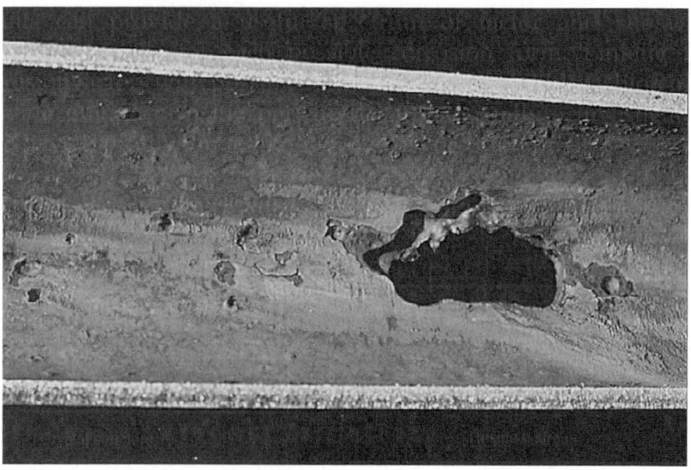

FIGURE 6.40 Erosion–corrosion of a brass tube carrying out seawater. (Courtesy of Defence R&D Canada-Atlantic)

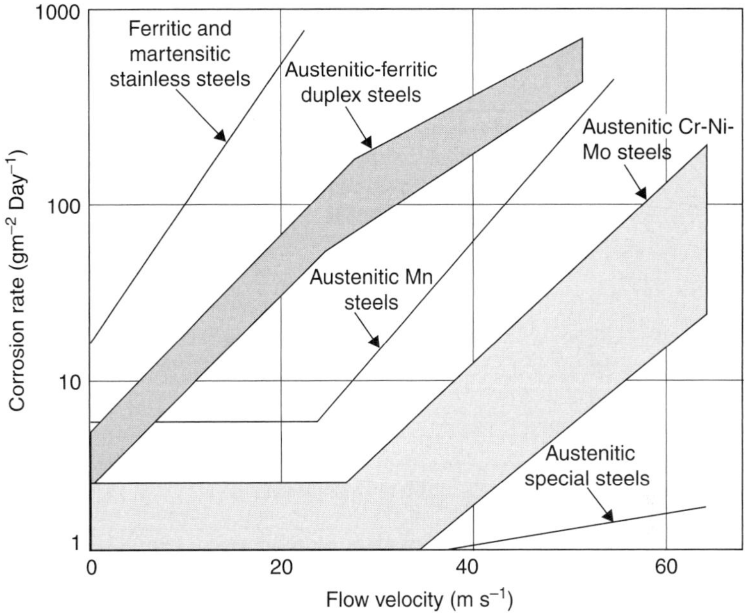

FIGURE 6.41 Influence of flow velocity on loss of material in media with huge levels of hydrogen sulfide and low pH [6].

synergy with corrosive constituents to produce a very demanding environment. Figure 6.41 illustrates the erosion–corrosion behavior of various materials exposed to flowing sour brine for a range of fluid velocities typical of oil and gas exploration [33].

Titanium forms a tenacious TiO_2 oxide film which is resistant in most oxidizing and reducing environments. Titanium and some nickel-chromium-molybdenum alloys perform well in low, intermediate, and high flow velocities. Small alloy additions to carbon steels can have a marked influence on the resistance to FAC. Carbon steels are generally susceptible to FAC, while low alloy chromium steels such as 1.25 Cr-0.5 Mo and 2.25 Cr-1 Mo are very resistant to FAC. Figure 6.42 shows the effect of steel composition on FAC developed by Huibregts from tests carried out at 960 m s^{-1} using wet steam containing 10 percent water [34]. Figure 6.42 was created using the Eq. (6.7):

$$\frac{\text{FACrate}}{\text{FACrate}_{\text{max}}} = \frac{1}{0.61 + 2.43[\text{Cr\%}] + 1.64[\text{Cu\%}] + 0.3[\text{Mo\%}]} \tag{6.7}$$

The addition of a minor element can also improve the resistance of copper-nickel alloys to erosion–corrosion. The effect of iron content on the corrosion and impingement resistance of 90/10 copper-nickel is maximized with the addition of about 2 percent

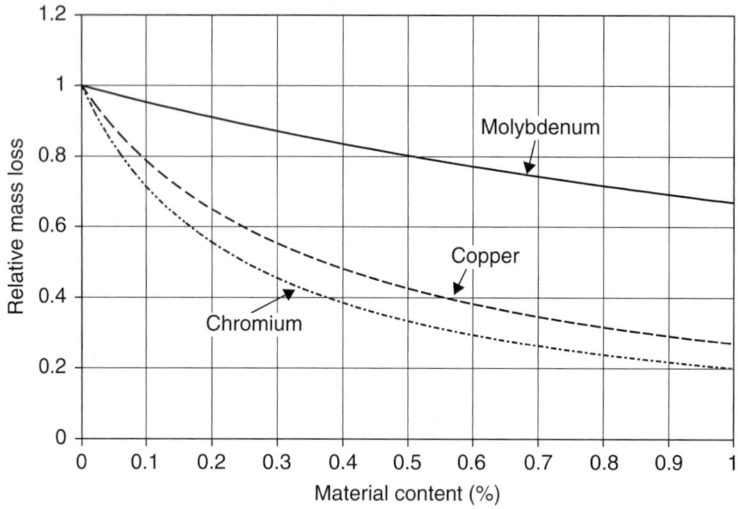

Figure 6.42 Effect of alloy content on erosion–corrosion rates.

iron to the cupronickel alloy [24]. Solutionized iron is beneficial and responsible for increased erosion–corrosion resistance through the incorporation of iron oxides into the corrosion product film. Nickel and iron incorporate into the corrosion product film in flowing seawater. If iron is precipitated in cupronickel alloys to form iron and nickel-iron second phases, deterioration of the passive film can occur and alter the corrosion behavior of the cupronickel alloy.

6.4.2 Cavitation

Cavitation damage (sometimes referred to as cavitation corrosion or cavitation erosion) is a form of localized corrosion combined with mechanical damage that occurs in turbulent or rapidly moving liquids and takes the form of areas or patches of pitted or roughened surface. Cavitation has been defined as the *deterioration of a surface caused by the sudden formation and collapse of bubbles in a liquid.* It has been similarly defined as the *localized attack that results from the collapse of voids or cavities due to turbulence in a liquid at a metal surface.* Cavitation also occurs in areas of high vibration such as on engine pistons and piston liners (Fig. 6.43).

In some instances, cathodic protection has been successful in reducing or preventing cavitation damage, but because cavitation damage usually involves physical as well as electrochemical processes, it cannot always be prevented by this means. In some cases, inhibitors have been used successfully to limit cavitation corrosion, as in the water side of diesel engine cylinder liners. Cavitation is a problem with ship propellers, hydraulic pumps and turbines, valves, orifice

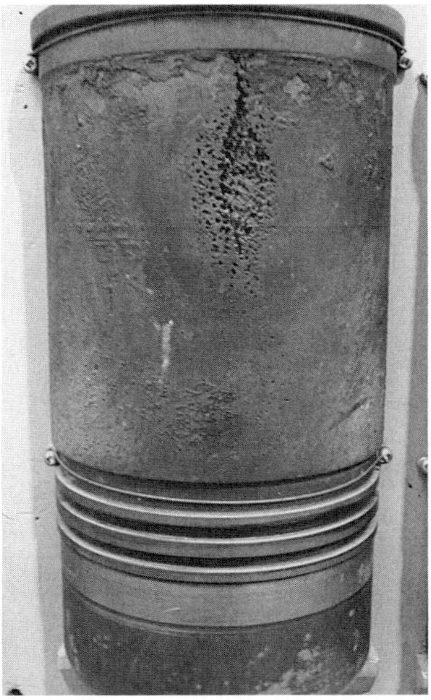

FIGURE 6.43 Cavitation damage of a diesel engine piston liner on the return stroke. (Courtesy of Defence R&D Canada-Atlantic)

plates, and all places where the static pressure varies very abruptly following the Bernoulli principle Eq. (6.8):

$$P + \frac{\rho}{2}V^2 + \rho g h = \text{Constant} \tag{6.8}$$

where P is the absolute static pressure
h is the elevation
g is the gravitational acceleration

Strictly speaking, the Bernoulli equation applies to flow along a streamline, however, in turbulent flow the bulk flow velocity can be used with little error. Thus the increase in the velocity as the liquid is accelerated through an orifice or over an impeller can result in a drop of the local static pressure. As the liquid slows down, after it passes a vena contracta or approaches the volute in a pump, the pressure rises again, leading to the collapse of the cavities formed by the previous drop in pressure (Fig. 6.39). Five different types of cavitation can be observed depending on the flow conditions and geometry [35,36].

- Traveling cavitation is a type of cavitation in which individual transient cavities or bubbles form in the liquid and move with it as they expand or shrink during their life cycles. To the naked eye traveling cavitation may appear as sheet cavitation.
- Cloud cavitation is caused by vorticity shed into the flow field. It causes strong vibration, noise and erosion. The shedding of cloud cavitation is periodic and the reentrant jet is the basic mechanism to generate cloud cavitation.

- Sheet cavitation is also known as fixed, attached, cavity or pocket cavitation. Sheet cavitation is stable in a quasi-steady sense. The liquid vapor interface becomes wavy and breaks down in the closure region of the cavity. Downstream flow, which contains large scale eddies, is dominated by bubble clusters.
- Supercavitation is when the sheet cavity grows in such a way it envelopes the whole solid body. Ventilation can be used to create or to enhance a supercavity. Supercavitation is desirable to achieve viscous drag reduction on underwater vehicles operating at high speeds. High speeds in excess of 1500 m/s are reported for supersonic operation of underwater projectiles.
- Vortex cavitation occurring on the tips of rotating blades is known as tip vortex cavitation. Cavities form in the cores of vortices in regions of high shear. This type of cavitation is not restricted to rotating blades. It can also occur in the separation zones of bluff bodies.

The cavitation number (σ) is a dimensionless number that provides an estimate of cavitation tendency in a flowing stream as described in Eq. (6.9):

$$\sigma = \frac{2\cdot(P - P_v)}{\rho V^2} \tag{6.9}$$

where P_v is the liquid vapor pressure.

When the cavitation number is null, the pressure is reduced to the vapor pressure and cavitation will occur. The cavitation number and the net positive suction head (NPSH) are related according to Eq. (6.10):

$$\text{NPSH} = \frac{(\sigma + 1)V^2}{2g} \tag{6.10}$$

6.5 Mechanically Assisted Corrosion

Mechanical forces (e.g., tensile or compressive forces) will usually have minimal effects on the general corrosion of metals and compressive stresses even reduce the susceptibility of metals to crack. In fact, shot peening is often used to reduce the susceptibility of metallic materials to fatigue, stress corrosion cracking (SCC), and other forms of cracking. However, a combination of tensile stresses and a specific corrosive environment is one of the most important causes of catastrophic cracking of metal structures.

SCC and other types of environmental cracking are also the most insidious forms of corrosion because environmental cracks are microscopic in their early stages of development. In many cases, they are not evident on the exposed surface by normal visual examination, and can be detected only by microscopic examination with an optical or a scanning electron microscope. As the cracking penetrates farther into the material, it eventually reduces the effective supporting cross section to the point where the structure fails by overload or, in the case of vessels and piping, escape (seepage) of the contained liquid or gas occurs.

Cracking is usually either intergranular (intercrystalline) as shown in Fig. 6.44 or transgranular (transcrystalline) as illustrated in Fig. 6.45. Occasionally, both types of cracking are observed in a failure. Intergranular cracks follow the grain boundaries in the metal. Transgranular cracks cross the grains without regard for the grain boundaries. The morphology of the cracks may change when the same material is exposed to different environments.

Failures are not necessarily the result of ordinarily applied engineering loads or stresses. However, these loads have to be added to invisible residual stresses already present in a structure due to various sources such as fabricating processes (e.g., deep drawing, punching, rolling of tubes into tubesheets, mismatch in riveting, spinning, welding, and so forth).

Residual stresses will remain in a structure unless it is annealed or otherwise thermally stress relieved following fabrication, a practice that becomes increasingly impractical as a system gets larger

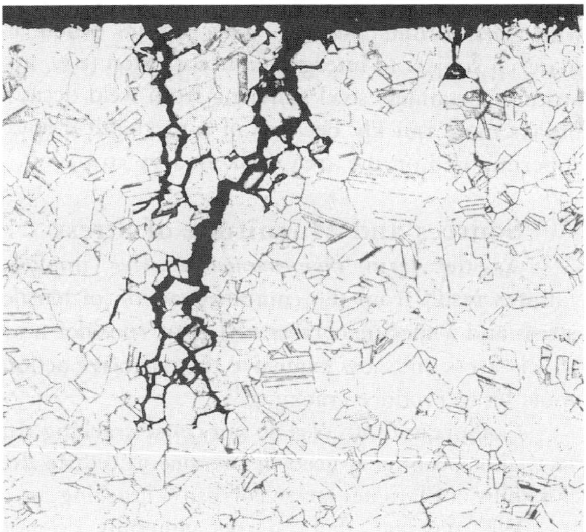

FIGURE 6.44 Typical intergranular stress corrosion cracks in cartridge brass (70 Cu, 30 Zn). Etched 30 percent H_2O_2, 30 percent NH_4OH, 40 percent H_2O (×75) (*Corrosion Basics: An Introduction*, 2nd edn., NACE International, by permission)

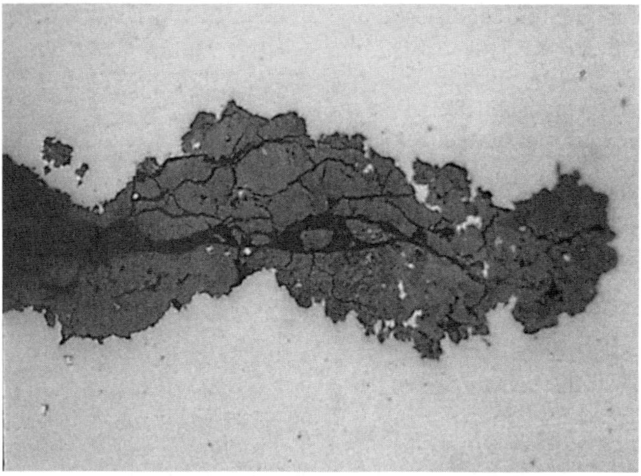

(a)

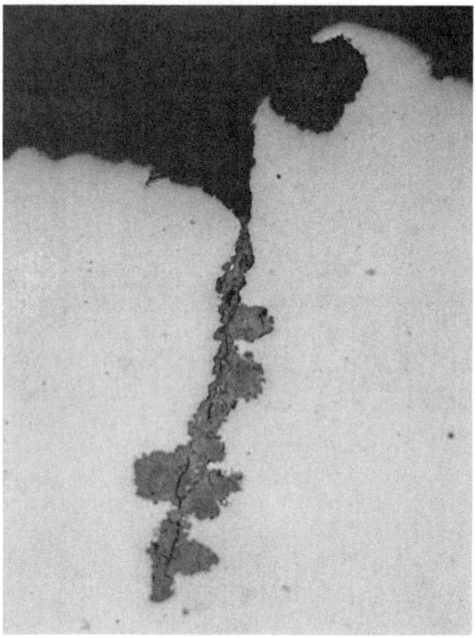

(b)

Figure 6.45 Axial or transgranular cracks filled with oxide or scale (*a*) observed on the external surface of a pipeline. These cracks may occur in association with pits and general corrosion (*b*). (Courtesy of *MACAW's Pipeline Defects*, published by Yellow Pencil Marketing Co.)

or more complex. Cooling from the high temperatures required may also induce internal stresses because of nonuniform cooling. In fact, very slow, controlled cooling is a prerequisite for effective stress relief.

Another source of stress may come from corrosion products from general corrosion or other forms of corrosion may build up between

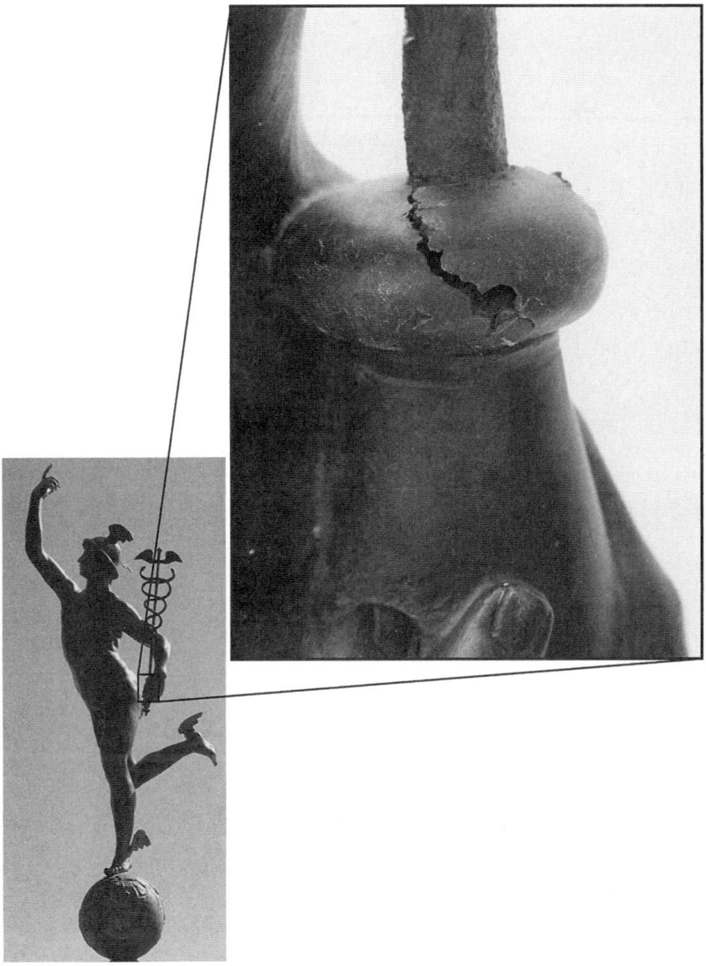

FIGURE 6.46 Stress corrosion crack in a bronze monument caused by buildup of rust around an iron rod. (Courtesy of Kingston Technical Software)

mating surfaces and, because they occupy so much more volume than the metal from which they are produced, generate sufficient stresses to cause SCC. In the example shown in Fig. 6.46, moisture working down the iron rod in combination with the galvanic corrosion due to the contact with the bronze support caused enough rust buildup to generate high stresses and induce SCC.

6.5.1 Stress Corrosion Cracking

It is not uncommon for growing cracks to be considered acceptable (Fig. 6.47) so long as they are substantially less than a critical crack size [Fig. 6.47(*a*) and Eq. (6.11)] and can be repaired at the next

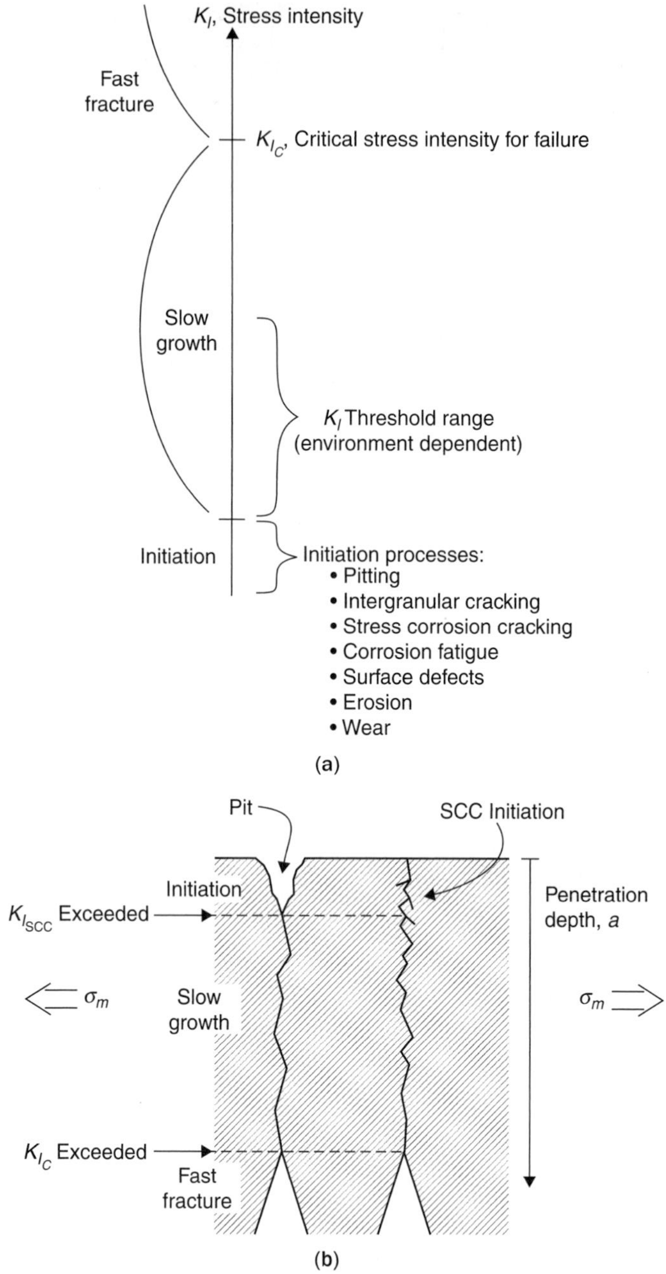

Figure 6.47 (*a*) Schematic view of relative magnitudes of K_I shown with transitions and principal dependencies. The K_I parameter for threshold range is shown to be variable, (*b*) Morphological regimes, (*c*) The $\Delta a/\Delta t$ versus K for SCC and similar environmentally dependent slow growth (e.g., hydrogen), (*d*) $\Delta a/\Delta n$ versus ΔK for fatigue crack growth showing possible effect of SCC in increasing the slow growth [37].

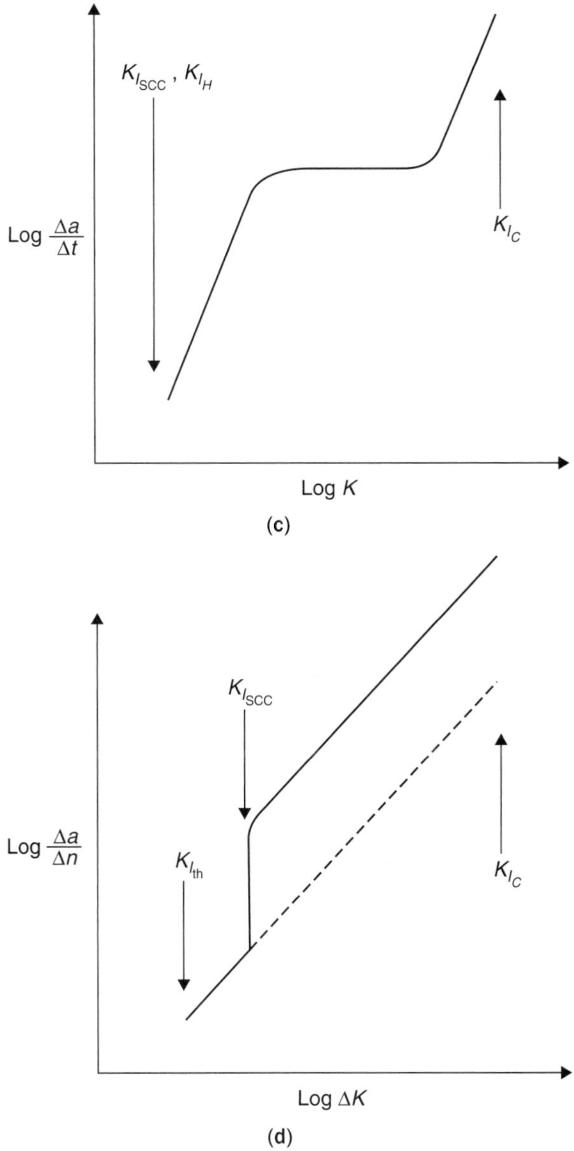

Figure 6.47 (*continued*)

outage or shutdown. Unless it is found during an inspection that the crack growth rate might end up in a failure before the next outage (with some safety factor), remedial actions are usually not taken [37].

$$K_{I_c} = \sigma_m \sqrt{C\pi a} \tag{6.11}$$

where σ_m is mean stress
a is depth of defect
C is geometric constant
K_{I_c} is critical stress intensity at which catastrophic fracture occurs

SCC is a mechanical–chemical process leading to the cracking of certain alloys at stresses below their tensile strengths. A susceptible alloy, the proper chemical environment, plus an enduring tensile stress are required. It is likely that there are no alloy systems which are completely immune to SCC in all environments. Usually, there is an induction period, during which cracking nucleates at a microscopic level. This period of latency may be quite long (e.g., many months or years) before progressing to the propagation stage.

Table 6.4 lists some environments in which SCC has been observed for at least some alloys of the systems listed. This listing does not imply that all alloys of a given material will be equally susceptible, or that there are none in the class which may be immune to the environments listed.

SCC is an anodic process, a fact which can be verified by using cathodic protection as an effective remedial measure. SCC may occasionally lead to fatigue corrosion, or the opposite. Usually, the true nature of the cracking can be identified by the morphology of the observed cracks. In a failure by SCC there is usually little metal loss due to general corrosion. Thus, the failure of a stress bolt rusted away until it eventually cannot sustain the applied load is not classified as SCC.

Aluminum alloys	NaCl-H_2O, NaCl solutions, seawater, mercury
Copper alloys	Ammonia vapors and solutions, mercury
Gold alloys	$FeCl_3$ solutions, acetic acid-salt solutions
Inconel	Caustic soda solutions
Lead	Lead acetate solutions
Magnesium alloys	NaCl-$K_2Cr_2O_7$ solutions, rural and coastal atmospheres, distilled water
Nickel	Fused caustic soda
Carbon steels	NaOH solutions; NaOH-$NaSiO_3$ solutions; $CaNO_3$, NH_4NO_3, and $NaNO_3$ solutions; mixed acids, (H_2SO_4-HNO_3); HCN; H_2S; seawater; NaPb alloy
Stainless steels	$BaCl_2$, $MgCl_2$ solutions; NaCl-H_2O_2 solutions; seawater, H_2S, NaOH-H_2S solutions
Titanium	Red fuming nitric acid

TABLE 6.4 Some Stress Corrosion Cracking Environments for Metals

However, if products from general corrosion are trapped so as to exert stress in a structure, they can cause SCC.

The idea, once prevalent, that only alloys and not pure metals are susceptible to SCC is quite possibly correct. The question is, "How pure is pure?" Copper containing 0.004 percent phosphorous or 0.01 percent antimony is reported to be susceptible to SCC in environments containing ammonia-based ions. Cracking has been produced in a decarburized steel containing less than 0.01 percent carbon, but still containing small amounts of manganese, sulfur, and silicon in a boiling ammonium nitrate solution. SCC has been produced in commercial titanium containing, among other constituents, 600 ppm of oxygen and 100 ppm of hydrogen. Hence, the idea that a given material cannot fail by SCC because it is commercially pure is not correct [21].

Transgranular SCC progresses by localized subsurface attack in which a narrow path is corroded randomly across grains without any apparent effect of the grain boundary on the crack direction. Transgranular SCC may occur during the SCC of austenitic stainless steels and less commonly during the SCC of low-alloy steels. It can also occur in the SCC of copper alloys in certain media (e.g., ammonia). It seldom occurs in aluminum alloys.

6.5.2 Corrosion Fatigue

Fatigue is the failure of a metal by cracking when it is subjected to cyclic stress. The usual case involves rapidly fluctuating stresses that may be well below the tensile strength. As stress is increased, the number of cycles required to cause fracture decreases. For steels, there is usually a stress level below which no failure will occur, even with an infinite number of cycles, and this is called the endurance limit. In practice, the endurance limit is defined as the stress level below which no failure occurs in one million cycles. A typical S-N curve fatigue curve, commonly known as an S-N curve, is obtained by plotting the number of cycles required to cause failure against the maximum applied cyclic stress.

When a metal is subjected to cyclic stress in a corrosive environment, the number of cycles required to cause failure at a given stress may be reduced well below the dotted line obtained for the same metal in air shown in Fig. 6.48. This acceleration of fatigue called "corrosion fatigue", is revealed by comparing the solid line in Fig. 6.48 with the dotted line reference. The solid curve indicates that metal life under such conditions can be much lower than the reference curve established in air. The S-N curve with corrosion tends to keep dropping, even at low stresses, and thus does not level off, as will the ordinary fatigue curve.

A marked drop in or elimination of the endurance limit may occur even in a mildly corrosive environment, especially in the case of a film-protected alloy. For example, deionized water, which ordinarily

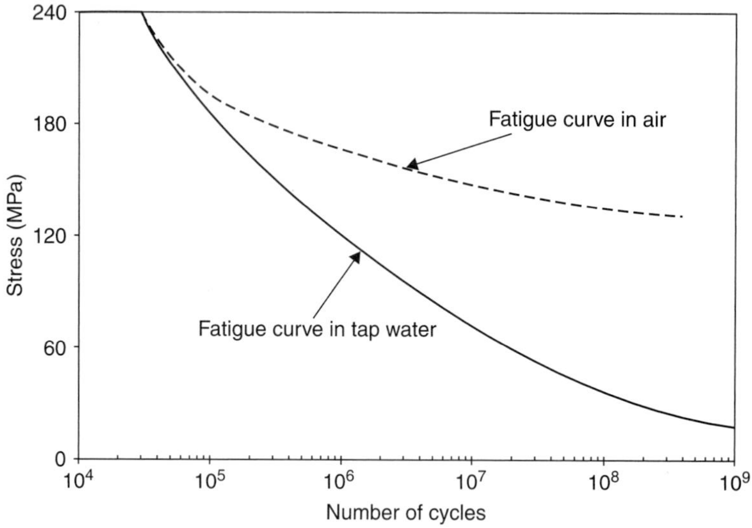

FIGURE 6.48 Fatigue and corrosion fatigue curves for an aluminum alloy [21].

will produce only film growth on an aluminum alloy in which it is immersed, will appreciably reduce the endurance limit of the same alloy subjected to cyclic stressing. This is because stress reversals cause repeated cracking of the otherwise protective surface film, and this allows access of the water to the unprotected metal below with resultant corrosion.

Failures that occur on vibrating structures (e.g., taut wires or stranded cables) exposed to the weather under stresses below the endurance limit are usually caused by corrosion fatigue. Corrosion fatigue also has been observed in steam boilers, due to alternating stresses caused by thermal cycling (Fig. 6.49).

The petroleum industry regularly encounters major trouble with corrosion fatigue in the production of oil. The exposure of drill pipe and of sucker rods to brines and sour crudes encountered in many producing areas results in failures which are expensive both from the standpoint of replacing equipment and from loss of production during the time required for "fishing" and rerigging.

For uniaxial stress systems, there will be an array of parallel cracks which are perpendicular to the direction of principal stress. Torsion loadings tend to produce a system of crisscross cracks at roughly 45° from the torsion axis. Corrosion fatigue cracks found in pipes subjected to thermal cycling will usually show a pattern made up of both circumferential and longitudinal cracks.

6.5.3 Fretting Corrosion

Fretting corrosion refers to corrosion damage at the asperities of contact surfaces (Fig. 6.50). This damage is induced under load and in the presence of repeated relative surface motion, as induced, for example, by vibration. Pits or grooves and oxide debris characterize

(a)

(b)

FIGURE 6.49 (*a*) As-received appearance of boiler tube section with crack failure. Note remnants of tack weld to strap on the back of the tube and that the top of tube is marked for establishing orientation; (*b*) Close-up of ID surface of failed tube section showing band of numerous parallel partial through-wall cracks in line with through-wall crack (arrow). (2× original magnification); (*c*) Photomicrograph showing as polished longitudinal metallographic specimen through cracked area on failed tube section. Cracks originate on ID surface. Note wedge- or needle-shaped, nonbranching cracks typical of corrosion fatigue. Faint black line near bottom of photo is an inclusion not related to the failure. (125× original magnification) (Courtesy of Corrosion Testing Laboratories, Newark, Delaware, USA).

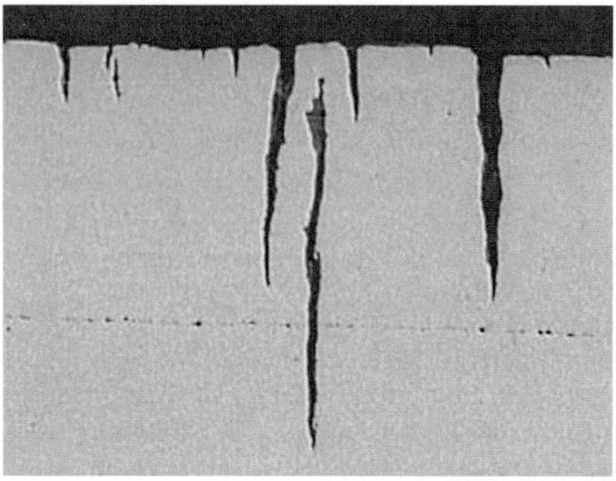

(c)

Figure 6.49 *(continued)*

this damage, typically found in machinery, bolted assemblies, and ball or roller bearings. Contact surfaces exposed to vibration during transportation are exposed to the risk of fretting corrosion.

Damage can occur at the interface of two highly loaded surfaces which are not designed to move against each other. The most common type of fretting is caused by vibration. The protective film on the metal surfaces is removed by the rubbing action and exposes fresh,

Figure 6.50 Fretting corrosion at the interface of two highly loaded surfaces which are not designed to move against each other. (Courtesy of Pacific Corrosion Control Corp)

active metal to the corrosive action of the atmosphere. Conditions necessary for the occurrence of fretting are (1) the interface must be under load, and (2) vibratory or oscillatory motion of small amplitude must result in the surfaces striking or rubbing together. The results of fretting are as follows:

1. Metal loss in the area of contact
2. Production of oxide debris
3. Galling, seizing, fatiguing, or cracking of the metal

References

1. Fontana MG, Greene ND. *Corrosion Engineering*. New York, N.Y.: McGraw Hill, 1967.
2. Dillon CP. *Forms of Corrosion: Recognition and Prevention*. Houston, Tex: NACE International, 1982.
3. Speidel MO, Fourt PM. Stress corrosion cracking and hydrogen embrittlement in industrial circumstances. In: Staehle RW, Hochmann J, McCright RD, Slater JE, eds. *Stess Corrosion Cracking and Hydrogen Embrittlement of Iron Base Alloys*. Houston, Tex.: NACE International, 1977; 57–60.
4. Techniques for Monitoring Corrosion and Related Parameters in Field Applications. NACE 3T199. Houston, Tex.: NACE International, 1999.
5. Lozev M, Grimmett B, Shell E, Spencer R. *Evaluation of Methods for Detecting and Monitoring of Corrosion and Fatigue Damage in Risers*. Project No. 45891GTH. Washington, Wash.: Minerals Management Service, U.S. Department of the Interior, 2003.
6. Malo JM, Salinas V, Uruchurtu J. Stray current corrosion causes gasoline pipeline failure. *Materials Performance*, 1994; 33: 63.
7. *Standard Guide for Examination and Evaluation of Pitting Corrosion*. ASTM G46–94. West Conshohocken, Pa.: American Society for Testing of Materials, 1999.
8. Evans UR, Mears RB, Queneau PE. Corrosion probability and corrosion velocity. *Engineering*, 1933; 136: 689.
9. Mears RB, Evans UR. The "probability" of corrosion. *Transactions of the Faraday Society*, 1935; 31: 527–42.
10. Gumbel EJ. *Statistical Theory of Extreme Values and Some Practical Applications*. Mathematics Series. 33. Washington, D.C.: National Bureau of Standards, 1954.
11. Walton JC, Cragnolino G, Kalandros SK. A numerical model of crevice corrosion for passive and active metals. *Corrosion*, 1996; 38: 1–18.
12. McArthur H. *Corrosion Prediction and Prevention in Motor Vehicles*. Chichester, U.K.: Ellis Horwood, 1988.
13. Baboian R. *Automotive Corrosion and Prevention*. Houston Tex.: NACE International, 1991.
14. Liss VM. Preventing corrosion under insulation. *Chemical Engineering*, 1987; 97–100.
15. Ruggeri RT, Beck TR. An analysis of mass transfer in filiform corrosion. *Corrosion*, 1983; 39: 452–65.
16. Slabaugh WH, DeJager W, Hoover SE, Hutchinson LL. Filiform corrosion of aluminum. *Journal of Paint Technology*, 1972; 44: 76–83.
17. Roberge PR. *Handbook of Corrosion Engineering*. New York, N.Y.: McGraw-Hill, 2000.
18. Hoxeng RB, Prutton CF. Electrochemical behavior of zinc and steel in aqueous media. *Corrosion*, 1949; 5: 330–8.
19. Haney EG. The zinc-steel potential reversal in cathodic protection. *Materials Performance*, 1982; 21: 44–51.

20. Fitzgerald III, JH. Longevity of a graphitized cast iron water main. *Materials Performance*, 2007; 46: 30.
21. Roberge PR. *Corrosion Basics—An Introduction.* 2nd edn. Houston, Tex.: NACE International, 2005.
22. Petroleum and natural gas industries—Materials for use in H_2S-containing environments in oil and gas production, NACE MR0175/ISO 15156. Houston, Tex.: NACE International, 2001.
23. Poulson BS. Erosion corrosion. In: Shreir LL, Jarman RA, Burstein GT, eds. *Corrosion Control.* Oxford, U.K.: Butterworths Heinemann, 1994; 1:293–1:303.
24. Shifler DA. Environmental effects in flow assisted corrosion of naval systems. *CORROSION 99*, Paper # 619. Houston, Tex.: NACE International, 1999.
25. *State-of-the-Art Report on Controlled-Flow Laboratory Corrosion Tests.* NACE Publication 5A195. Houston, Tex.: NACE International, 1995.
26. Postlethwaite J, Dobbin MH, Bergevin K. The role of oxygen mass transfer in the erosion-corrosion of slurry pipelines. *Corrosion*, 1986; 42: 514–21.
27. Berger FP, Hau K-FFL. Mass transfer in turbulent pipe flow measured by the electrochemical method. *International Journal of Heat and Mass Transfer*, 1977; 20: 1185–94.
28. Postlethwaite J, Nesic S. Erosion-corrosion in single and multiphase flow. In: Revie RW, ed. *Uhlig's Corrosion Handbook.* NewYork, N.Y.: Wiley-Interscience, 2000; 249–72.
29. Heitz E. Chemo-mechanical effects of flow on corrosion. In: Kennelley KJ, Hausler RH, Silverman DC, eds. *Flow-Induced Corrosion: Fundamental Studies and Industry Experience.* Houston, Tex.: NACE International, 1991; 1–29.
30. Lotz U. Velocity effects in flow induced corrosion. In: Kennelley KJ, Hausler RH, Silverman DC, eds. *Flow-Induced Corrosion: Fundamental Studies and Industry Experience.* Houston, Tex.: NACE International, 1991; 8-1–8-22.
31. Syrett BC. Erosion-corrosion of copper-nickel alloys in sea water and other aqueous environments—A literature review. *Corrosion*, 1976; 32: 242–52.
32. Dawson JL, Shih CC. Corrosion under flowing conditions—An overview and model. In: Kennelley KJ, Hausler RH, Silverman DC, eds. *Flow-Induced Corrosion: Fundamental Studies and Industry Experience.* Houston, Tex.: NACE International, 1991; 2-1–2-12.
33. Weber J. Flow induced corrosion: 25 years of industrial research. *British Corrosion Journal*, 1992; 27: 193–9.
34. Chexal B, Horowitz J, Dooley B, Millett P, Wood C, Jones R. *Flow-Accelerated Corrosion in Power Plants*, Revision 1. EPRI TR-106611-R1. Palo Alto, Calif.: Electric Power Research Institute, 1998.
35. Knapp RT, Daily JW, Hammitt FG. *Cavitation.* New York, N.Y.: McGraw-Hill, 1970.
36. Senocak I. *Computational Methodology for the Simulation of Turbulent Cavitating Flows.* Gainesville, FL:University of Florida, 2002.
37. Staehle RW. Lifetime Prediction of materials in environments. In: Revie RW, ed. *Uhlig's Corrosion Handbook.* New York, N.Y.: Wiley-Interscience, 2000; 27–84.

CHAPTER 7

Corrosion Failures, Factors, and Cells

7.1 Introduction

As described in the Chapter 6, corrosion damage may take various forms triggered themselves by apparently different compounding factors. The initiation and progression of corrosion processes indeed depend on the complex interaction of a multitude of factors such as

- Nature of the metal or alloy
- Presence of inclusions or other foreign matter at the surface
- Homogeneity of the metallic structure
- Nature of the corrosive environment
- Incidental environmental factors such as variations in the presence of dissolved oxygen, of temperature, and in the velocity of movement either of the environment or of the system itself
- Other factors such as stress (residual or applied, steady or cyclic)
- Oxide scales (continuous or broken)
- Presence of deposits on surfaces
- Fayed surfaces and the possibility of corrosion crevices
- Galvanic effects between dissimilar metals
- Occasional presence of stray electrical currents from external sources

The nature of the driving forces behind these factors has been the subject of scientific studies by many scientists in the early part of the twentieth century. In a landmark paper, Mears and Brown have summarized 18 mechanisms, listed in Table 7.1, by which differences in potential may develop on metal surfaces [1].

Causes of Corrosion Currents
Impurities in the metal
Orientation of grains
Grain boundaries
Differential grain size
Differential thermal treatment
Surface roughness
Local scratches or abrasions
Difference in shape
Differential strain
Differential pre-exposure to air or oxygen
Differential concentration or composition of solution
Differential aeration
Differential heating
Differential illumination
Differential agitation
Contact with dissimilar metals
Externally applied potentials
Complex cells

TABLE 7.1 Causes of Corrosion Currents

There is a vast body of information relating practically all the previously listed factors to actual field observations and subsequent analysis of failed components. These failure investigations are typically carried out in a detailed mechanistic "bottom-up" manner whereby a failed component would be sent to the laboratory where analytical techniques would then be used following well-established protocols. Chemical analysis, hardness testing, metallography, optical and electron microscopy, fractography, x-ray diffraction, and surface analysis are specialized tools used in such investigations.

However, this approach alone provides little or no insight into the real causes of failure. Underlying causes of serious corrosion damage often include human factors such as lack of corrosion awareness and inadequate training and poor communication. Further underlying causes may include weak maintenance management systems, insufficient repairs due to short-term profit motives, a poor organizational "safety culture," defective supplier's products, or an incorrect material selection.

It is thus apparent that there can be multiple causes associated with a single corrosion mechanism. Clearly, a comprehensive failure investigation providing information on the root cause of failure is much more valuable than one merely establishing the corrosion mechanism(s). Establishing the real causes of corrosion failures (often related to human behavior) is a much more difficult task than merely identifying the failure mechanisms.

In contrast to the traditional scientific mechanistic approach, systems engineers prefer the "top-down" approach that broadens the definition of the system and is more likely to include causes of corrosion failures such as human behavior. This is consistent with the lessons to be learned from the U.K. Hoar Report, which stated that corrosion control of even small components could result in major cost savings because of the effect on systems rather than just the components [2].

7.2 Information to Look For

It is not always possible to anticipate the actual environment in which a metallic structure will operate. Even if the initial conditions were known completely, there is often no assurance that operating temperatures, pressures, or even chemical compositions will remain constant over the expected equipment lifespan. The complexity of corrosion processes and their impact on equipment often buried or out-of-sight complicates many situations, rendering simple life prediction difficult.

Even corrosion tests carried out in controlled conditions often yield results with more scatter than many other types of materials test results because of a variety of factors, an important one being the effect on corrosion rates due to minor impurities in the materials themselves or in the testing environments [3].

7.2.1 Temperature Effects

In most chemical reactions, an increase in temperature is accompanied by an increase in reaction rate. A rough rule-of-thumb suggests that the reaction rate doubles for each 10°C rise in temperature. Although this rule has many exceptions, it is important to take into consideration the influence of temperature when analyzing why materials fail.

Changing the temperature of an environment can influence its corrosivity. Many household hot-water heater tanks, for example, were historically made of galvanized steel. The zinc coating offered a certain amount of cathodic protection to the underlying steel, and the service life was considered adequate. Water tanks seldom were operated above 60°C. With the development of automatic dishwashers and automatic laundry equipment, the average water temperature was increased so that temperatures of about 80°C have now become common in household hot water tanks.

Coinciding with the widespread use of automatic dishwashers and laundry equipment was a sudden upsurge of complaints of short life of galvanized steel water heater tanks. Electrochemical measurements showed that in many cases, iron was anodic to zinc above 75°C, whereas zinc was anodic to iron at temperatures below 60°C. This explained why zinc offered no cathodic protection above 75°C, and why red water and premature perforation of galvanized water tanks occurred so readily at higher temperatures. This particular problem was partly solved by using magnesium sacrificial anodes or protective coatings, and by the replacement with new alloys.

7.2.2 Fluid Velocity Effects

Unless otherwise protected, metals generally owe their corrosion resistance to a tightly adherent, protective film that builds up on the metal surface by corrosion processes. This film may consist of reaction products, adsorbed gases, or a combination of these. Any mechanical disturbance of this protective film can stimulate attack of the underlying metals until either the protective film is reestablished, or the metal has been corroded away. The mechanical disturbance itself can be caused by abrasion, impingement, turbulence, or cavitation.

Carbon steel pipe carrying water, for example, is usually protected by a film of rust that slows down the rate of mass transfer of dissolved oxygen to the pipe wall. The resulting corrosion rates are typically less than 1 mm/y. The removal of the film by flowing sand slurry has been shown to raise corrosion rates tenfold to approximately 10 mm/y [4]. Figure 7.1 illustrates the various states of an oxide-surface film behavior as liquid velocity or surface shear stresses are increased [5;6].

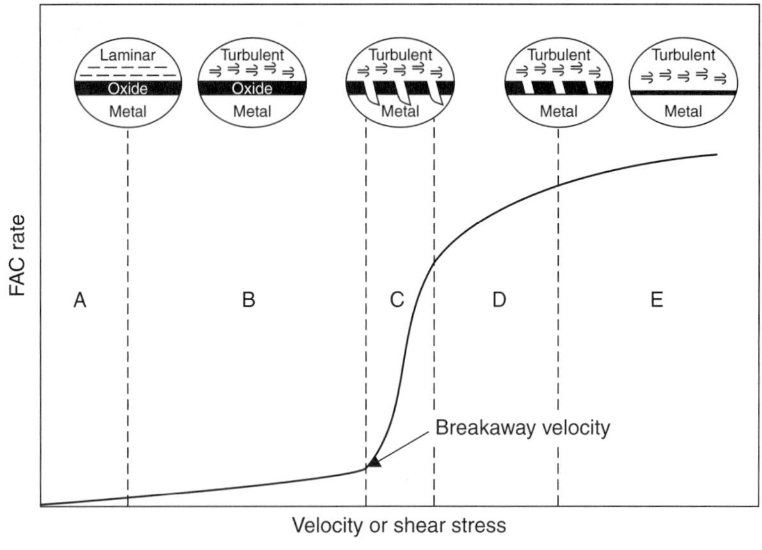

FIGURE 7.1 Changes in the corrosion and erosion mechanisms as a function of liquid velocity [5].

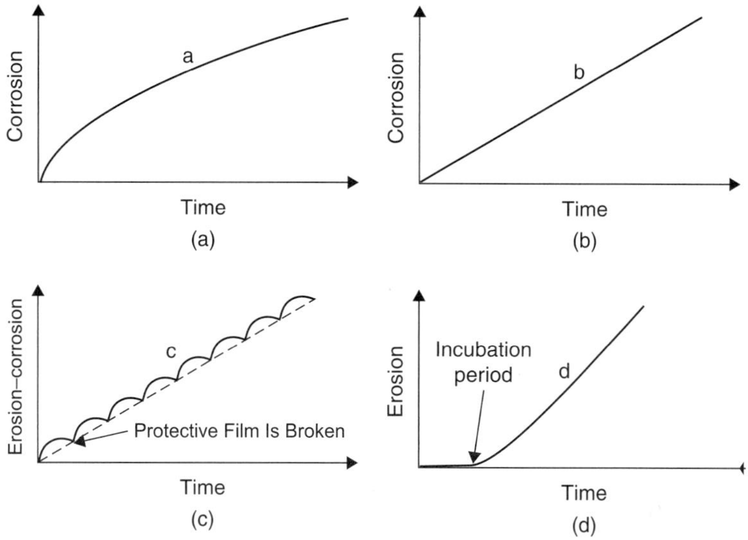

FIGURE 7.2 Various time dependent corrosion/erosion behaviors and processes: (*a*) corrosion follows a parabolic time law, (*b*) FAC follows a linear time law, (*c*) erosion and corrosion follows a quasi-linear time law with repeated breaks in the protective surface film, and (*d*) erosion linear time dependency after an initial incubation period [5].

Changes in the corrosion and erosion mechanisms associated with flow accelerated corrosion (FAC) are summarized in Figs. 7.2 and 7.3. In stagnant water (origin of the plot in Fig. 7.1), the corrosion rate is low and decreases parabolically with time due to the formation and growth of a corrosion protective film at the surface (curve *a* in Fig. 7.2). At low flow velocities for which laminar and turbulent flow conditions coexist (parts A and B of Fig. 7.1), corrosion stems from a flow-accelerated process. The protective film that forms on the surface by corrosion is dissolved by the flowing water. It is generally accepted that the phenomenon is a steady-state process having linear corrosion kinetics (curve *b* in Fig. 7.2), that is, the dissolved layer at the oxide-water interface is replaced by a new layer of the same thickness.

Erosion–corrosion is encountered most frequently in pumps, valves, centrifuges, elbows, impellers, inlet ends of heat-exchanger tubes, and agitated tanks. Locations in flowing systems where there are sudden changes in direction or flow cross-section, as in heat exchangers where water flows from the water boxes into the tubes, are likely places for erosion–corrosion.

7.2.3 Impurities in the Environment

Some impurities present in the environment even in minute amounts may have more influence on the corrosion behavior of materials than major constituents of the same environments. Sometimes impurities

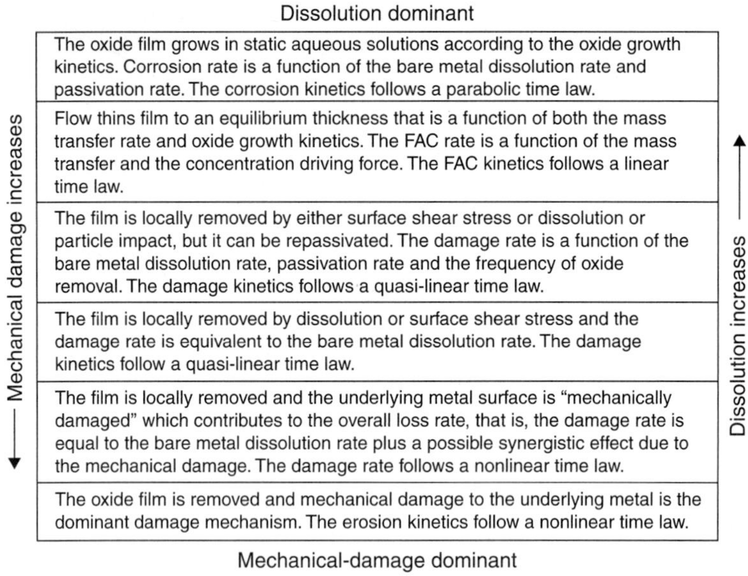

FIGURE 7.3 Summary of damage mechanisms experienced with FAC [5].

in trace quantities may accelerate the corrosion attack while at other times they may act as inhibitors. The introduction of small amounts of ions of metals such as copper, lead, or mercury can cause severe corrosion of aluminum equipment, for example, corrosion of upstream copper alloy equipment can result in contamination of flowing water in a cooling circuit. In this example, copper can plate out on aluminum surfaces downstream as small nodules or deposits setting up local galvanic cells, which can result in severe pitting and perhaps perforation.

There are many examples of the catastrophic role played by impurities in the process industries. Conventional corrosion data on sulfuric acid media, for example, are often based on tests in chemically pure acid or on field exposures of indeterminate chemistry and the effects of contaminants are often overlooked. Serious problems can therefore arise in seemingly straightforward applications [7].

In the manufacturing of concentrated acid by air combustion of various feedstocks, small amounts of nitric oxides may be introduced, depending on burner design, temperature, and so forth. Also, nitric acid is sometimes added to sulfuric acid as antifreeze. The presence of nitric acid can promote the oxidation of the ferrous sulfate film which otherwise protects steel in static exposures. In glacial acetic acid, corrosion of type 316L can be markedly increased by as little as a few parts per million of acetic anhydride. Apparently, the anhydride scavenges the water required to maintain a passive film.

Contaminants may also drastically change the corrosion characteristics of nitric acid. Although the conventional type 300 series austenitic stainless steels, such as annealed type 304L and as-welded types 304L, 321, and 347, are resistant to nitric acid in many instances, there is a general accretion of hexavalent chromium ions (Cr VI) if the acid is recirculated with no provision for dilution elevated temperatures, as with nitric acid condensers. In other instances, contamination of nitric acid with chloride ions results in their being oxidized to nascent chlorine. This is the basis for aqua regia* used to dissolve gold. In 42 percent acid, excessive corrosion of stainless steels results when chloride ion contamination exceeds about 3000 ppm. High initial rates are observed to diminish with time, probably due to the loss of volatile chlorine formed by reaction with the chlorides.

7.2.4 Presence of Microbes

Microbes are present almost everywhere in soils, freshwater, seawater, and air. However, the mere detection of microorganisms in an environment does not necessarily indicate a corrosion problem. Nonetheless, it is well-established that numerous buried steel pipes have suffered severe corrosion as the result of bacterial action.

In unaerated or anaerobic soils, this attack is attributed to the influence of the sulfate-reducing bacteria (SRB). The mechanism is believed to involve both direct attack of the steel by hydrogen sulfide and cathodic depolarization aided by the presence of bacteria. Even in aerated or aerobic soils, there are sufficiently large variations in aeration that the action of SRB cannot be neglected. For example, within active corrosion pits, the oxygen content becomes exceedingly low.

Bacteria, fungi, and other microorganisms can play a major part in soil corrosion. Spectacularly rapid corrosion failures have been observed in soil due to microbial action and it is becoming increasingly apparent that most metallic alloys are susceptible to some form of microbiologically influenced corrosion (MIC).

The presence of aggressive microbes may also be quite severe in industrial water-handling systems, for example, cooling water and injection water systems, heat exchangers, wastewater treatment facilities, storage tanks, piping systems, and all manner of power plants, including those based on fossil fuels, hydroelectric, and nuclear [8]. Table 7.2 lists potential problem areas by industry [9].

7.2.5 Presence of Stray Currents

The corrosion resulting from stray currents coming from external sources is similar to that from galvanic cells that generate their own current. However, the amplitude of stray currents may be much

---

*Aqua regia refers to a mixture in a 3:1 ratio of hydrochloric (HCl) and nitric (HNO_3) acids.

Industry/Application	Potential Problem Sites for MIC	Organisms Responsible
Pipelines—oil, gas, water, wastewater	Internal corrosion primarily at the bottom (6:00) position Dead ends and stagnant areas Low points in long-distance pipes Waste pipes—internal corrosion at the liquid/air interface Buried pipelines—on the exterior of the pipe, especially in wet clay environments under disbonded coating	Aerobic and anaerobic acid producers, SRB, manganese and iron-oxidizing bacteria, sulfur-oxidizing bacteria
Chemical process industry	Heat exchangers, condensers, and storage tanks—especially at the bottom where there is sludge build-up Water distribution systems (See also "Cooling water systems," "Fire protection systems," and "Pipelines" in this table)	Aerobic and anaerobic acid producers, SRB, manganese, and iron-oxidizing bacteria In oil storage tanks also methanogens, oil-hydrolyzing bacteria
Cooling water systems	Cooling towers Heat exchangers—in tubes and welded areas—on shell where water is on shell side Storage tanks—especially at the bottom where there is sludge build-up	Algae, fungi, and other microorganisms in cooling towers Slime-forming bacteria, aerobic and anaerobic bacteria, metal-oxidizing bacteria, and other microorganisms and invertebrates
Fire protection systems	Dead ends and stagnant areas	Anaerobic bacteria, including SRB
Docks, piers, oil platforms, and other aquatic structures	Just below the low-tide line splash zone	SRB beneath barnacles, mussels, and other areas sequestered from oxygen
Pulp and paper	Rotating cylinder machines Whitewater clarifiers	Slime-forming bacteria and fungi on paper-making machines Iron-oxidizing bacteria SRB in waste water
Power-generation plants	Heat exchangers and condensers Firewater distribution systems Intakes	As above for heat exchangers and fire protection systems Under mussels and other fouling organisms on intakes
Desalination	Biofilm development on reverse osmosis membranes	Slime-forming bacteria

Table 7.2 Where MIC Problems are Most Likely to Occur [9]

higher than the current produced by ordinary galvanic cells and as a consequence corrosion may be much more rapid. Another difference between galvanic-type currents and stray currents is that the latter are more likely to operate over long distances since the anode and cathode are more likely to be remotely separated from one another. Seeking the path of least resistance, stray currents from a foreign installation may travel along a pipeline or any other buried metallic structure causing severe corrosion where it leaves the line or current carrier.

Stray currents causing corrosion may originate from direct-current distribution lines, substations, or street railway systems to damage a metallic structure somewhere in its vicinity as described in more details later in the present chapter. Alternating stray currents very rarely cause corrosion.

Investigating the presence of stray currents to prevent or explain corrosion problems is not a new field in corrosion engineering. In fact, as mentioned in App. A, such activities were carried out by probably the first corrosion engineers in North America when the American Committee on Electrolysis was established at the turn of the twentieth century to combat the serious effects of railcar stray currents to underground metal structures (Figs. 7.4 and 7.5).

FIGURE 7.4 An electrolysis surveyor using a McCollum earth current meter from the early 1920s. (Courtesy of East Bay Municipal Utility District).

FIGURE 7.5 Pits in 16-inch sheet iron water main in San Francisco's East Bay that failed in one month due to leakage of current from electric streetcars. (Photo dated August 10, 1926, Courtesy of East Bay Municipal Utility District).

As written in a book published in 1906 [10], "Electrolysis is a disease most largely peculiar to America. The Europeans have experienced little trouble from the corrosive action of return currents, but the reports of the 'horrible' cases in this country aroused their fears and they at once set to work to determine the state of affairs in their own countries."

7.3 Identifying the Corrosion Factors

The factors listed earlier have been organized in a framework of six categories with a number of subfactors as shown in Table 7.3. According to Staehle's materials degradation model, all engineering materials are reactive and their strength is quantifiable, provided that all the variables involved in a given situation are properly diagnosed and their interactions understood [11]. The corrosion-based design analysis (CBDA) approach was further developed from the initial framework as a series of knowledge elicitation steps to guide maintenance and inspection decisions on the basis of first principles [12].

Factor	Subfactors and Contributing Elements
Material	Chemical composition of alloy Crystal structure Grain boundary (GB) composition Surface condition
Environment *Chemical definition* *Circumstance*	Type, chemistry, concentration, phase, conductivity Velocity, thin layer in equilibrium with relative humidity, wetting and drying, heat transfer boiling, wear and fretting, deposits
Stress *Stress definition* *Sources of stress*	Mean stress, maximum stress, minimum stress, constant load/constant strain, strain rate, plane stress/plane strain, modes I, II, III, biaxial, cyclic frequency, wave shape Intentional, residual, produced by reacted products, thermal cycling
Geometry	Discontinuities as stress intensifiers Creation of galvanic potentials Chemical crevices Gravitational settling of solids Restricted geometry with heat transfer leading to concentration effects Orientation vs. environment
Temperature	At metal surface exposed to environment Change with time
Time	Change in GB chemistry Change in structure Change in surface deposits, chemistry, or heat-transfer resistance Development of surface defects, pitting, or erosion Development of occluded geometry Relaxation of stress

Table 7.3 Factors and Contributing Elements Controlling the Incidence of a Corrosion Situation [11]

The two most important of these steps are described in Figs. 7.6 and 7.7 for respectively, the environment and the material definitions. Each of the numbers in brackets in Fig. 7.6 identifies an explicit action that needs to be considered for the environmental definition.

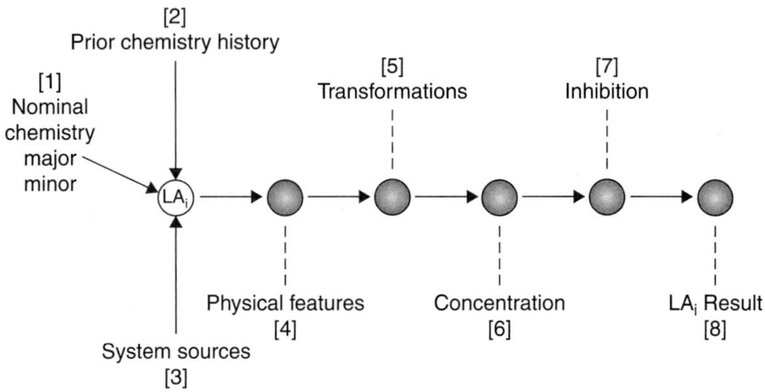

FIGURE 7.6 Analysis sequence for determining environment at a location for analysis.

The end point of the process is an input to a location for analysis (LA) matrix that is illustrated in Fig. 7.8 for the locations in a steam generator. A brief explanation of the individual elements in Fig. 7.6 follows:

[1] Nominal chemistry refers to the bulk chemistry. For components exposed to ordinary air atmospheres, major elements mean humid air. Minor elements refer to industrial contaminants such as SO_2 and NO_2.

[2] Prior chemistry history refers to exposures to environmental species that might still reside on the surfaces or inside crevices.

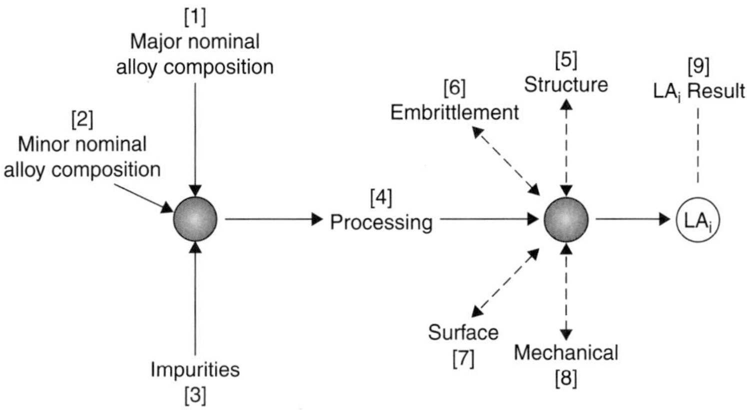

FIGURE 7.7 Analysis sequence for determining materials at a location for analysis matrix.

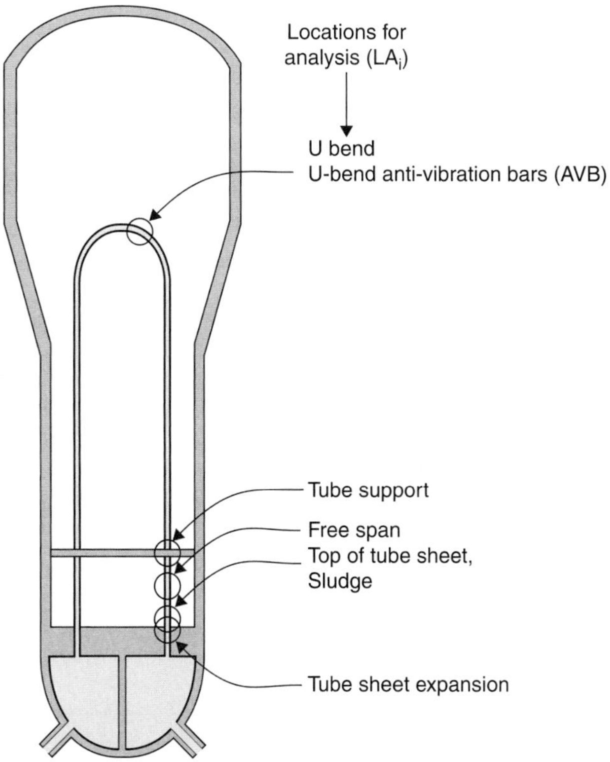

FIGURE 7.8 Schematic view of steam generator with different locations for analysis.

[3] System sources refer to those environments that do not come directly from a component but from an outside source.

[4] Physical features include occluded geometries, flow, and long-range electrochemical cells.

[5] Transformations refer, for example, to microbial actions that can change relatively innocuous chemicals such as sulfates into very corrosive sulfide species that may accelerate hydrogen entry and increase corrosion rates.

[6] Concentration refers to accumulations much greater than that in the bulk environment due to various actions of wetting and drying, evaporation, potential gradients, and crevices actions that prevent dilution.

[7] Inhibition refers to actions taken to minimize corrosive actions. This usually involves addition of oxygen scavengers or other chemicals that interfere directly with the anodic or cathodic corrosion reactions.

The LA template of the locations that correspond to most likely failure sites along tubes in a steam generator of a pressurized water nuclear power plant (Fig. 7.8) is detailed in Table 7.4 for the main failure modes and submodes considered in such analysis. Maintenance and inspection actions can be decided upon by following developing trends monitored in each LA matrix thus produced.

The framework summarized in Table 7.3, which was initially developed to predict the occurrence of stress corrosion cracking (SCC), was extended to other corrosion modes/forms. Additionally, an empirical correlation was established between the factors listed in Table 7.3 and the forms of corrosion described earlier in Chap. 6. Recognized corrosion experts were invited to complete an opinion poll listing the main subfactors and the common forms of corrosion as illustrated in the example shown in Fig. 7.9. Background information on the factors and forms of corrosion was attached to the survey. A total of sixteen opinion polls were completed in this survey and subsequently analyzed.

Factor	**Forms I**			
	Uniform	Pitting	Crevice	Galvanic
Material				
Composition				
Crystal structure				
GB composition				
Surface condition				
Environment				
Nominal				
Circumstantial				
Stress				
Applied				
Residual				
Product built-up				
Cyclic				
Geometry				
Galvanic potentials				
Restricted geometries				
Settling of solids				
Temperature				
Changing T				
T of surface				
Time				
Changes over time				

FIGURE 7.9 Opinion poll-sheet for the most recognizable forms of corrosion problems.

			Modes (MD_j) and Submodes (SD_j) to be Considered												
Locations for Analysis			Submodes of SCC						Submodes of IGC						
LA_i	ID	OD	LPSCC ($j = 1$)	HPSCC ($j = 2$)	AcSCC ($j = 3$)	MRSCC ($j = 4$)	AkSCC ($j = 5$)	PbSCC ($j = 6$)	HPIGC ($j = 7$)	AcIGC ($j = 8$)	AkIGC ($j = 9$)	Wastage ($j = 10$)	Pitting ($j = 11$)	Fatigue ($j = 12$)	Wear ($j = 13$)
Tubular expansion ($i = 1$)	x														
Tubular expansion ($i = 2$)		x													
Top of tube sheet ($i = 3$)	x														
Top of tube sheet ($i = 4$)		x													
Sludge ($i = 5$)		x													
Free span ($i = 6$)	x														
Free span ($i = 7$)		x													
Tube support (hot leg) ($i = 8$)		x													
Tube support (cold leg) ($i = 9$)		x													
U-Bend ($i = 10$)	x														
U-Bend AVB ($i = 11$)		x													

Abbreviations used are low potential (LP), high potential (HP), mid-range pH (MR), alkaline (AK), lead (Pb), stress corrosion cracking (SCC), are intergranular corrosion (IGC).

TABLE 7.4 Matrix for Organizing Mode-Location Cells

Figure 7.10 presents the box-and-whisker plots of the results obtained with pitting corrosion. When presented in this fashion, such results can provide a useful spectrum of factor and subfactor confidence levels for each possible corrosion form. Linking the corrosion factors with possible forms of corrosion in this fashion may provide guidance to inexperienced corrosion failure investigators who have typically limited knowledge of corrosion processes.

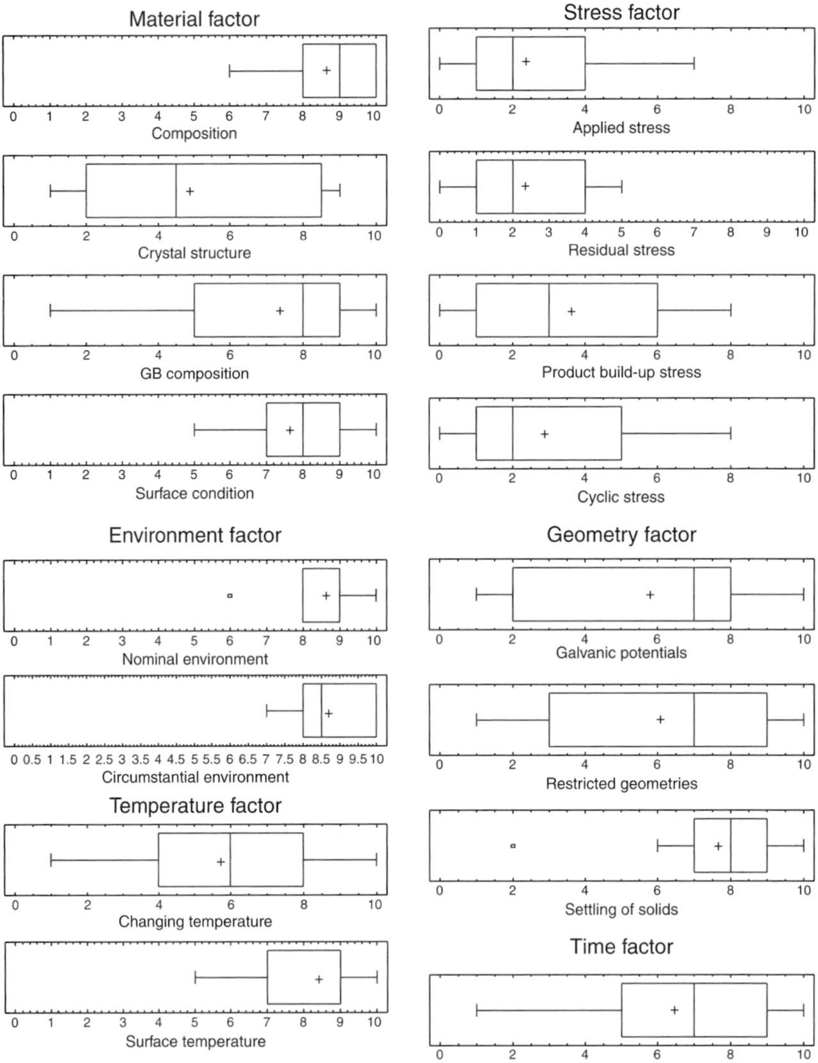

FIGURE 7.10 Box and whisker plots of the survey results obtained for the factors and subfactors underlying the appearance of pitting corrosion.

	Uniform	Pitting	Crevice	Galvanic	Erosion	Cavitation	Fretting	Intergranular	Dealloying	SCC	Fatigue	Scaling	Internal attack
Material													
Composition	■	■		■		■	■	■	■	■	■	■	■
Crystal structure								■					
GB composition								■		■			
Surface condition		■								■			
Environmental													
Nominal	■	■		■	■			■	■	■	■	■	
Circumstantial		■	■							■			
Stress													
Applied										■	■		
Residual										■			
Product build-up										■			
Cyclic											■		
Geometry													
Galvanic potentials				■									
Restricted geometries			■										
Settling of solids		■											
Temperature													
Changing temperature												■	
Temperature of surface		■										■	■
Time													
Changes over time	■											■	■

Based on response where $\overline{x}>7$ and $\sigma<3$

FIGURE 7.11 Results of compiled survey of corrosion experts highlighting the most important correlations between corrosion forms and factors.

A listing of the most important factors should therefore improve the level of awareness on the complexity and interaction of the variables behind most corrosion failures. Some knowledge of how the experts have reduced such complexity to a reduced set of variables, as the compiled results of the survey indicate in Fig. 7.11, may also help to focus an investigation on the most important variables behind a specific corrosion case.

An application of the compiled framework could be to test one's skills against the experts as illustrated in Fig. 7.12. Another application of this practical correlation would be to use the framework of factors versus forms for archiving data in an orderly manner. Analysis of numerous corrosion failure analysis reports has revealed that information on important variables is often lacking [13]. The omission of important information from corrosion reports is obviously not always an oversight by the professional author. In many cases, the desirable information will simply not be (readily) available and require a special investigation to be completed.

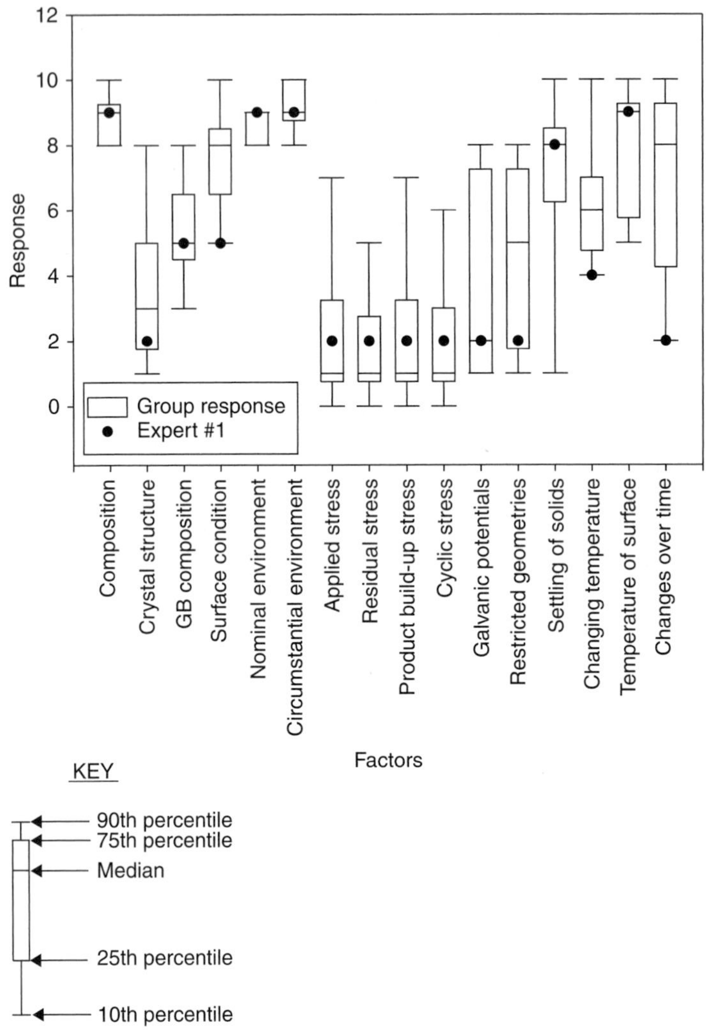

FIGURE 7.12 Comparison of the answers of one expert with some of the compiled expert survey results.

7.4 Examples of Corrosion Cells

A difference in potential will be observed if electrically connected specimens of the same metal are immersed in solutions having different concentrations of ions of the metals. This is one common type of concentration cell. Another type is the oxygen concentration or differential aeration cell.

Local cells may be produced by differences between small nearby areas on the metal surface. They may also be the result from differences in the metal or the environment, from impressed currents, or from local differences in the many factors described earlier that may cause corrosion. Metal variations may be the result of composition differences. A second phase constituent will have a different corrosion potential compared to that of an adjacent solid solution*. The difference may be in the thickness of a surface film at adjacent sites which in turn may reflect metal differences in the substrate.

Any two or more of the cells described in the following sections can operate at the same time. Sometimes they will work in synergy, sometimes they will be opposed. For example, in a well casing, there is usually an oxygen cell and a temperature cell, both tending to make the deep pipe act as an anode. At the same time, there may be two or more ordinary concentration cells, as the casing passes through different strata of soil.

The corrosion of guyed tower anchors such as shown in Fig. 7.13 and Fig. 7.14 and which is becoming an issue of critical importance will be used in the next sections to illustrate some of these corrosion cells. In many cases, the loss of one anchor to corrosion (Fig. 7.15) has resulted in the catastrophic loss of the supported communication tower (Fig. 7.16).

FIGURE 7.13 Basic anchor support design. (Courtesy of Anchor Guard)

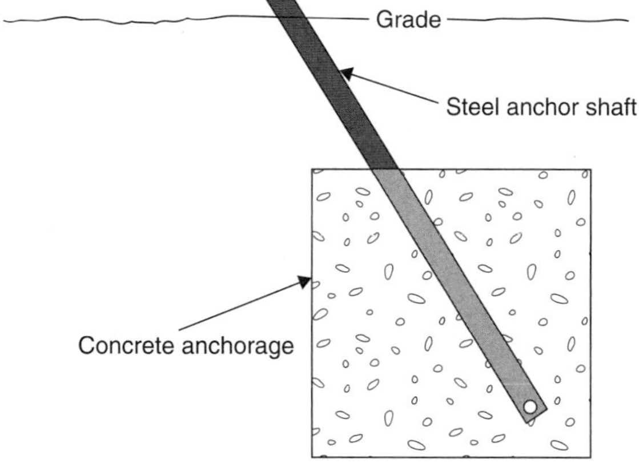

*In metallurgy, a solid solution refers to a homogeneous crystalline structure in which one or more types of atoms or molecules may be partly substituted for the original atoms and molecules without changing the structure.

Figure 7.14 Typical anchor support with a copper rod ground connection. (Courtesy of Anchor Guard)

Figure 7.15 Anchor assembly cut by corrosion at the concrete anchorage. (Courtesy of Anchor Guard)

Figure 7.16 Mangled communication tower fallen due to loss of the anchor shown in Fig. 7.15. (Courtesy of Anchor Guard)

7.4.1 Galvanic Cells

It is important to realize that galvanic corrosion effects can be manifested not only on the macroscopic level but also within the microstructure of a material. Certain phases or precipitates will undergo anodic dissolution under microgalvanic effects. Because the principle of galvanic corrosion is widely known, it is remarkable that it still features prominently in numerous corrosion failures. Figure 7.17 illustrates the main factors affecting the formation of a galvanic cell [14].

Any two metals can be used to make a galvanic cell. Whether a metal will behave as an anode or a cathode in combination with another metal in the same environment can usually be determined by its relative position on a galvanic series. Figure 7.18 shows the galvanic series of many metals and alloys in slow-moving seawater and Table 7.5 presents the galvanic series which was constructed for metals exposed to neutral soils and water [15].

Are these corrosion cells common? The answer is yes. Whenever a copper pipe service line is directly connected to a cast iron gas or water main, a galvanic cell is formed (Fig. 7.19) the soil is the electrolyte, the copper service line is the cathode, the iron (or steel) main is the anode, and the connecting circuit is completed by attaching the line to the main. Such cells may be relatively harmless when the anode or corroding metal occupies a much larger surface than the cathode so that the attack is spread out over a large area.

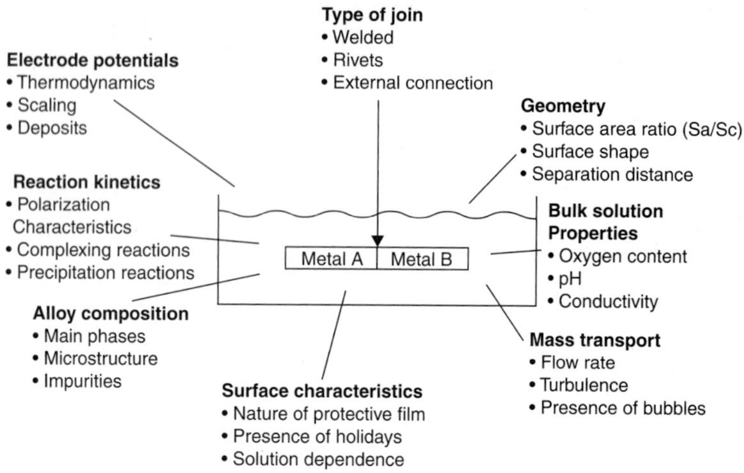

FIGURE 7.17 Factors affecting galvanic corrosion [14].

Metal	Potential V (CCSRE)
Commercially pure magnesium	–1.75
Magnesium alloy (6% Al, 3% Zn, 0.15% Mn)	–1.6
Zinc	–1.1
Aluminum alloy (5% zinc)	–1.05
Commercially pure aluminum	–0.8
Mild Steel (clean and shiny)	–0.5 to –0.8
Lead	–0.5
Cast iron (not graphitized)	–0.5
Mild steel (rusted)	–0.2 to –0.5
Mill scale on steel	–0.2
High silicon cast iron	–0.2
Copper, brass, bronze	–0.2
Mild steel in concrete	–0.2
Platinum	0 to –0.1
Carbon, graphite, coke c	+0.3

Typical potentials normally observed in neutral soils and water, measured in relation to a copper/copper sulfate reference electrode.

TABLE 7.5 Practical Galvanic Series for Metals in Neutral Soils and Water [15]

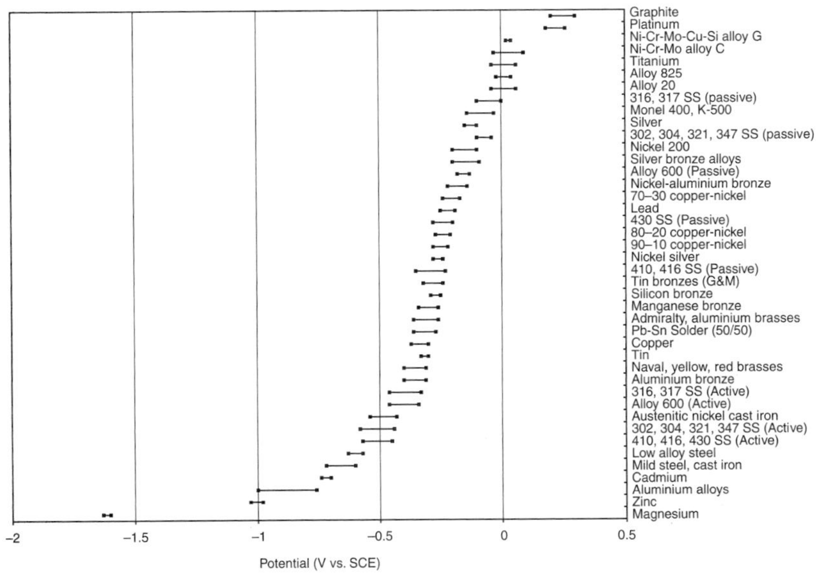

Figure 7.18 Galvanic series for slow-moving seawater.

However, the situation can lead to the corrosion of steel when the surface area of the cathode (copper) approaches that of the anode (steel) such as shown in Fig. 7.20 or when the surrounding environment is more conductive or corrosive in the region where the galvanic coupling exists (Fig. 7.21). In the example shown in Fig. 7.20, the anodic corrosion of the anchor is partly due to the galvanic corrosion cell formed by the

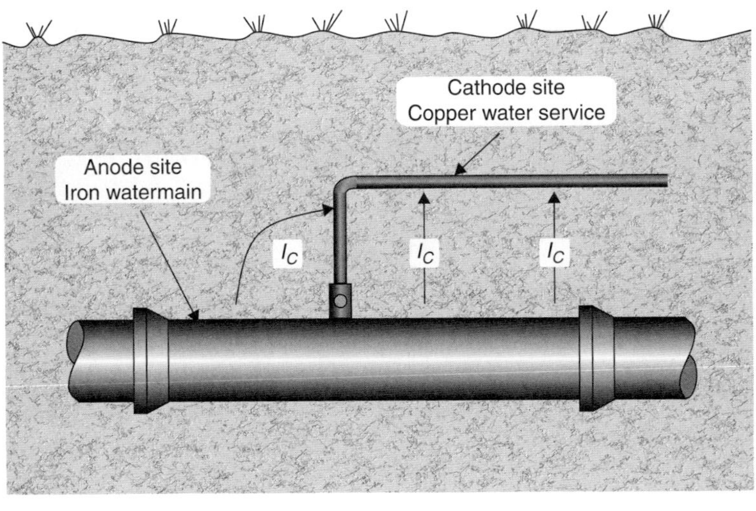

Figure 7.19 Underground corrosion cell involving connection of dissimilar metals.

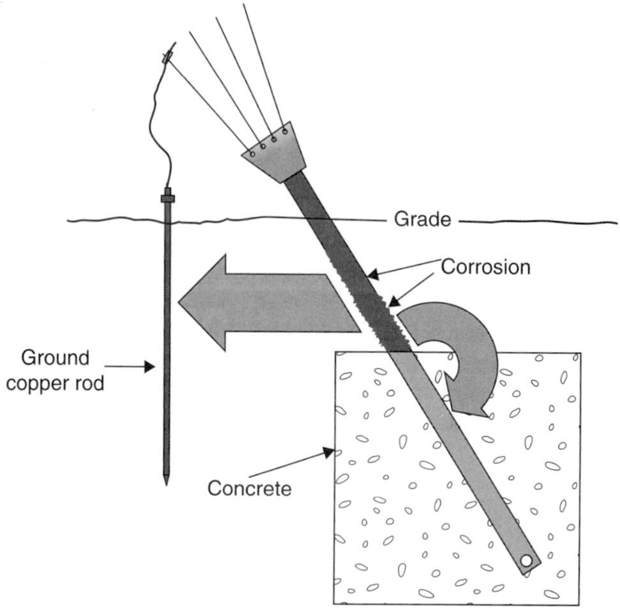

FIGURE 7.20 Schematic of the corrosion cells acting on an anchor support with a ground connection. (Courtesy of Anchor Guard)

FIGURE 7.21 Corrosion of a water main ductile iron adjacent to a copper fitting. (Courtesy of Drinking Water Services, City of Ottawa)

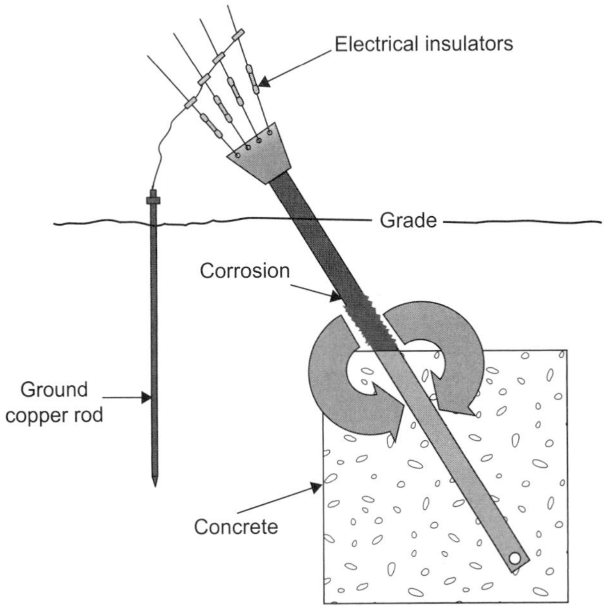

FIGURE 7.22 The addition of insulators between the ground rod and the steel anchor to stop the galvanic action. (Courtesy of Anchor Guard)

connection of ground copper rods. One way to minimize the dissimilar metal interaction that causes the corrosion of the structural steel anchor is to break the electrical path between these two components by adding insulators as shown in Fig. 7.22. However, this measure alone has proven to be only a partial solution to this type of problem since another corrosion cell is still active as will be described in the section "Concentration Cells."

7.4.2 Concentration Cells

Concentration corrosion cells may occur when a metal is exposed to an environment containing varying levels of electrolytes either of different substances or of the same substance in different amounts. For example, if one electrolyte is a dilute salt solution and the other a concentrated salt solution, a concentration cell may be formed. This takes place because one of the factors that determine an electrode potential is the electrolyte concentration.

Dissolved salts in a soil environment can be complex mixtures and include compounds of aluminum, calcium, magnesium, and other metals combined with sulfates, chlorides, hydroxides, or any one of quite a variety of anionic species. The role played by these salts can also vary greatly. Chloride ions, for example, can be quite aggressive toward steel and sulfate ions can serve as nutrients to sulfate reducing bacteria (SRB) which themselves can be extremely damaging to most buried metals.

In the example shown in Fig. 7.22, the main difference in the environment surrounding the steel anchor is due to the variation of pH between the soil typically acidic and the basic pH in the concrete anchorage (pH > 10). Because this higher pH is protective to steel, it naturally follows that the steel in concrete becomes the cathode drawing an anodic current from the adjacent steel in the soil.

A concentration cell common during the corrosion of reinforced concrete exposed to salts containing chloride ions (marine environments or road deicing salts) is illustrated in Fig. 7.23. In such cells the reinforcing steel in contact with the mobile chloride ions may become vulnerable to corrosion due to the depassivation of the steel. The process is further accelerated by the presence of bars embedded deeper in the concrete. These protected bars can then serve as cathodes because of the attaching strings serving as electrical connection. Such situations may result in the formation of stress cells due to the accumulation of corrosion products. Eventually the pressure created* may be high enough to cause cracks and the degradation of reinforced concrete (Fig. 7.24).

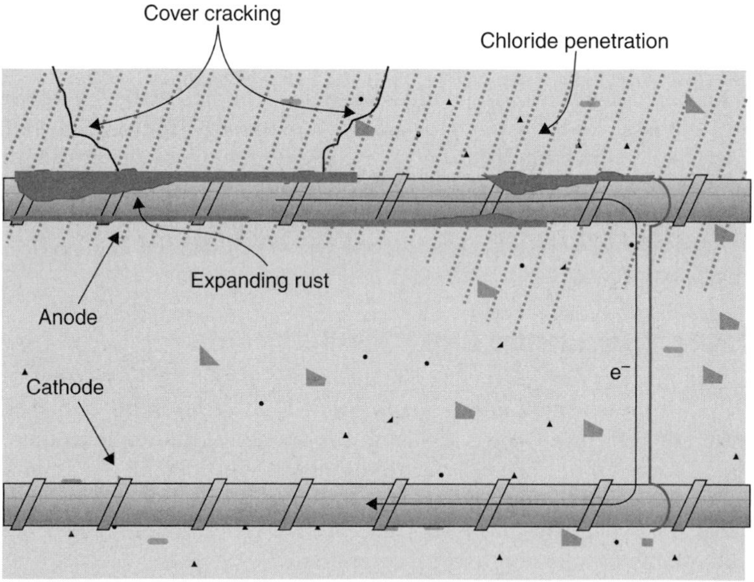

Figure 7.23 Corrosion Cell in Concrete. (Courtesy of Vector Corrosion Technologies)

*An equivalent volume expansion ratio of 3.0 to 3.2 has been measured due to the formation of corrosion products on steel bars embedded in concrete [16].

FIGURE 7.24 Cracking of concrete cover in marine environment.

7.4.3 Differential Aeration: Oxygen Concentration Cells

The oxygen content of any solution ranks high on the list of factors influencing the corrosion of iron and numerous other metals. Elimination of oxygen by deaeration is a practical means of reducing corrosion, as in the case of steam boilers which are operated with deaerated feed water.

Differential aeration cells can be caused by crevices, lap joints, dirt and debris, and moist insulation. Under these conditions, the oxygen starved areas are anodic, while the areas with free access to oxygen are cathodic. Common terms for this type of corrosion include crevice corrosion, oxygen screening, and poultice action.

Oxygen not only enables a corrosion reaction by maintaining a cathodic reaction, but it can promote one. This occurs where there is a difference in the concentration of dissolved oxygen (DO) between two points of the same metal surface. Since the direction of the reaction is toward equilibrium, the only way that equilibrium can be approached by corrosion will be to reduce the concentration of oxygen where it is highest. Such reduction can be done by consuming the oxygen as shown in Eq. (7.1).

$$O_2 + 2H_2O + 4e^- \rightarrow 4OH^- \tag{7.1}$$

The end result is that whenever there is a difference in the concentration of DO at two points on a metal surface, the surfaces in contact with the solution containing the higher concentration of DO will become cathodic to surfaces in contact with a lower concentration of DO causing these surfaces to suffer accelerated corrosion as anodes in an oxygen concentration cell.

It is easy to demonstrate an oxygen concentration cell with an experimental setup using two containers (Fig. 7.25). In this experiment, pieces of steel electrically connected are immersed in a sodium-chloride solution in the two containers. The solution in one container is saturated with oxygen and the solution in the other container is saturated with nitrogen. This establishes a difference in the concentration of DO in contact with the two pieces of steel. The high concentration of DO in one container makes the steel surface cathodic to the steel in the other container.

In the following example the concentration cell is caused by differences in the electrolyte and in oxygen content around a buried anchor rod (Fig. 7.26). In this illustration, the upper soil layer is a loose, somewhat gravelly soil, below which there is a dense clay-type soil. The portion of the shaft in contact with the clay-type soil acts as an anode to the portion of the shaft in contact with the looser gravelly soil, which is consequently the cathode.

Such differential aeration cells are also very common on buried pipes. For example, a pipe usually rests on undisturbed soil at the bottom of a ditch. Around the sides and on top of the pipe is relatively loose backfill that has been replaced in the ditch. A corrosion cell is

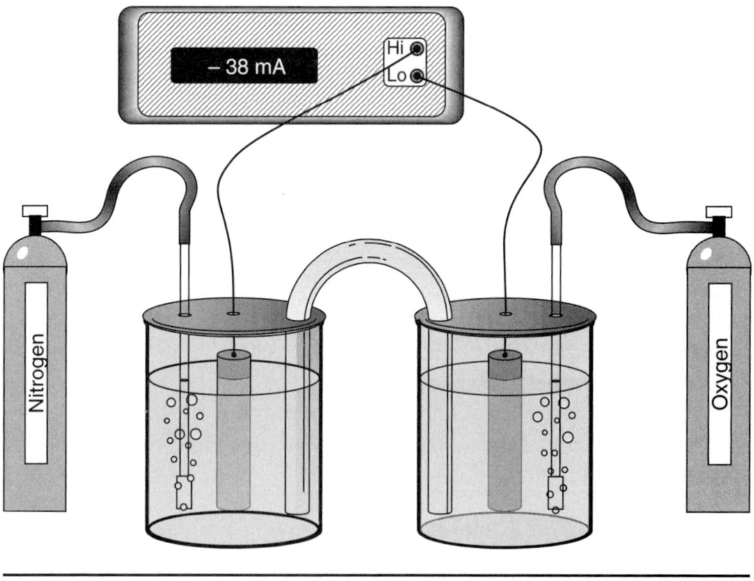

FIGURE 7.25 Experiment to demonstrate generation of corrosion current by an oxygen concentration cell.

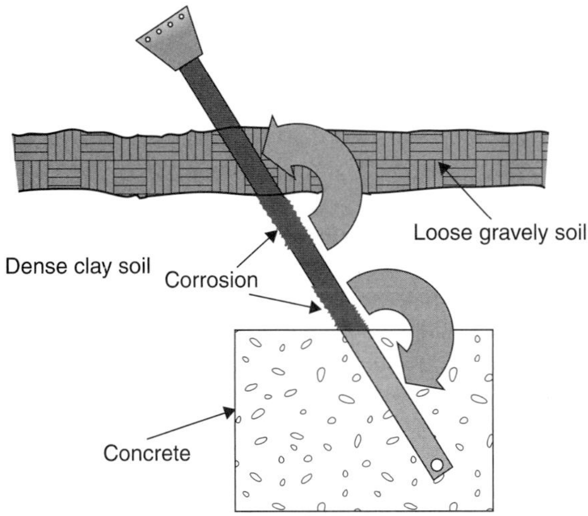

Figure 7.26 Differences in the porosity of the soil that can lead to an oxygen concentration corrosion cell. (Courtesy of Anchor Guard)

formed because the backfill is more permeable to oxygen diffusing down from the surface. In this case, the anode is the bottom surface of the pipe and the cathode is the rest of the surface. The electrolyte is the soil, and the connecting circuit is the metallic pipe itself.

When a pipe or cable crosses a paved road, as in Fig. 7.27 for example, the portion under the paving has less access to oxygen than does the area lying under unpaved soil. Thus, a cell is formed:

- The anode is the pipe under the paving
- The cathode is the pipe outside the paving
- The electrolyte is the soil
- The connecting circuit is the pipe or cable

In this particular example, although the entire length of pipe under the paving is anodic, most of the attack will take place close to the edge due to the normally low resistivity of the soil environment.

7.4.4 Temperature Cells

While concentration and oxygen cells are responsible for perhaps 90 percent of the corrosion in soils and natural waters, other cells, such as the temperature cell shown in Fig. 7.28, may be still quite damaging when they get established. In such cells, the two electrodes are of the same metal, but one is maintained at a higher temperature than the other by some external means. In most cases, the electrode at the higher temperature becomes the anode to cause what has been called thermogalvanic corrosion.

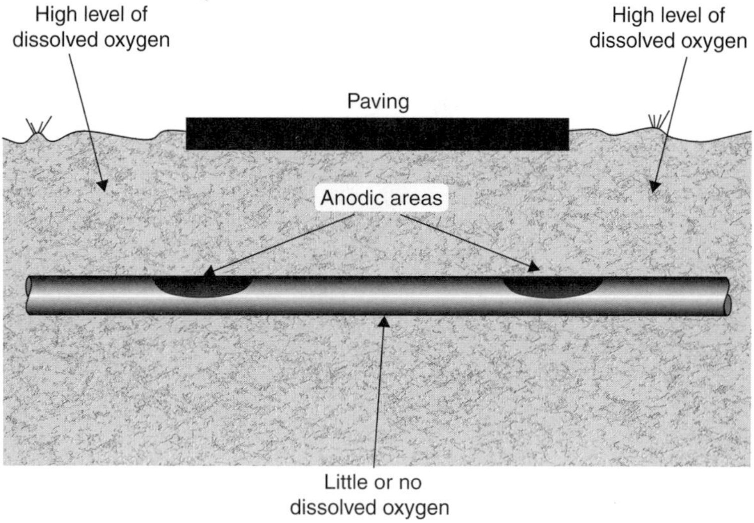

FIGURE 7.27 Oxygen differential cell resulting from burial under paving producing an oxygen concentration cell.

In the gas transmission line example shown in Fig. 7.28, the compressed gas coming out of the station is hot, and, as it travels down the line, it loses heat by transfer to the surrounding soil. The result creates a cell:

- The hot pipe near the compressor is the anode
- The cooler pipe down the line is the cathode

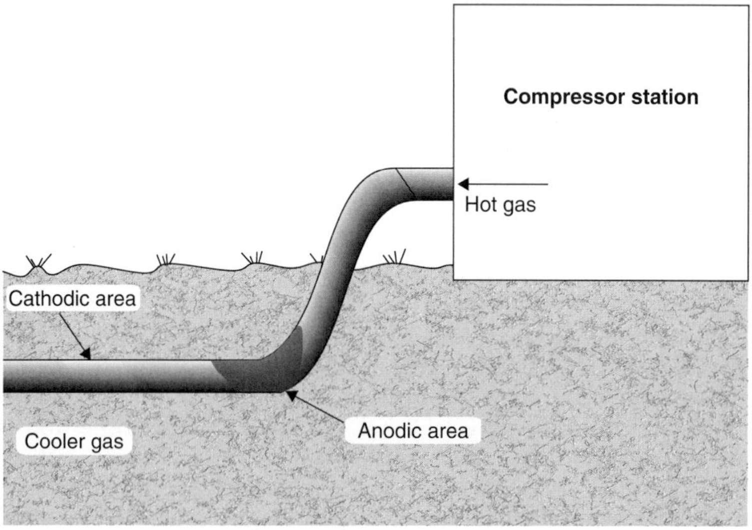

FIGURE 7.28 Concentration cell resulting from heat differential.

- The soil is the electrolyte
- The pipe itself is the connecting circuit

This is a particularly unfortunate type of cell because the high temperature just outside the compressor tends to damage the coating, so that the part of the line which is the anode is also likely to have the poorest coating.

External corrosion of copper hot water tubing underslab has resulted in a great deal of construction defect litigations due to this type of corrosion cell [17]. Oil and gas well casings also experience similar cell attack. The pipe at some depth below the surface is at a higher temperature since temperature increases with depth and thus becomes the anode. The cooler pipe near the surface and the surface piping make up the cathode, the soil is the electrolyte, and the pipe is the electrical connection.

7.4.5 Stray Current Cells

In all the cells described so far, the source of the energy which makes a cell active has been within the cell. However, in stray current cells the energy comes from an electrical current external to the corrosion site per se. The source of energy for such cells may be a distant generator, a direct-current transmission line, a cathodic protection rectifier on some other line, a street car system, or an electric railway (Fig. 7.29).

The corrosion effects of stray current can be easily demonstrated with a simple laboratory setup such as shown in Fig. 7.30(*a*). After only a few minutes of passing a small DC current in the cell containing a dilute saline solution, the formation of hydrogen bubbles is readily visible on the steel nail connected to the negative post of the DC power supply [Fig. 7.30(*b*)], while the nail connected to the positive post shows signs of rapid corrosion a few minutes later [Fig. 7.30(*c*)].

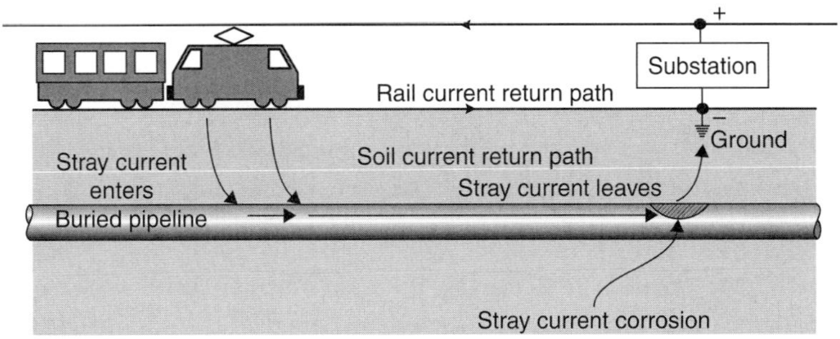

FIGURE 7.29 Stray current corrosion due to the return path of an electric railway.

(a)

(b)

FIGURE 7.30 Experiments to illustrate the effects of stray current showing (*a*) the experimental setup containing a DC power supply plus two steel nails partially immersed in a dilute saline solution and connected to the positive (left) and negative (right) posts of the DC supply; the effects of imposing a DC current on the nails (*b*) after five and (*c*) ten minutes.

(c)

Figure 7.30 *(continued)*

Impressed current cells in which the soil is the electrolyte are of two kinds: accidental and deliberate. An example of a deliberate cell is an impressed current system which supplies cathodic protection to increase the life of a structure by forcing an anodic current to anodes in a remote location. Accidental systems, however, may exist under a variety of circumstances. Any direct current flowing in the soil from any source whatsoever can, if it finds a pipeline or other metal object in its path, collect on the pipe in one area and discharge from it in another. The area where it collects becomes a cathode while the area where it discharges is an anode, and thus corrosion occurs.

7.4.6 Stress Cells

If an ordinary nail is dropped into a vessel of salt water, it will be attacked, and, after some time, rust will be observed discoloring the water. The initial points of attack will be, in almost every case, the point and the head. This is an example of a type of cell known as a stress cell. Figure 7.31 shows how dramatic this effect can be on the heads and nuts used to hold fire hydrants on their base in the soil. In these cases the two electrodes are of the same metal and the electrolyte is uniform. The difference lies in that one electrode is more mechanically stressed than the other. The area of high stress is always the anode of the cell, made that way by the extra energy supplied by the stress itself, and the corrosion reaction is the favored stress relieving path.

FIGURE 7.31 Severe corrosion of fire hydrant bolts and nuts due to stress cells. (Courtesy of Drinking Water Services, City of Ottawa)

Stress cells can take on three basic forms. One, like the nail just described, has its anode established by residual internal stress often produced by cold work or stress left because of something which has happened to the metal. In the case of the nail, the stress was caused by cold-forming of the head and point. If these stresses were relieved by heating the nail at a moderate temperature* and letting it cool slowly, residual stresses would disappear and this type of stress cell eliminated.

In the second type of stress cell, the metal is part of a structure which is under stress. The most highly stressed part of the metal becomes the anode, with the less-stressed or unstressed metal acting as the cathode. Metal atoms are at their lowest strain-energy state when situated in a regular crystal array. Applied stresses may be static or cyclic.

A third type of stress has already been mentioned in the discussion on crevice corrosion in Chap. 6 and in the reinforced concrete example discussed in the previous section on concentration cells.

In these cases, the stresses produced by an accumulation of corrosion products are typically formed in restricted geometries where the specific volume of the corrosion product is greater than

*For carbon steels this temperature would be 75 to 80°C below the A_1 transformation temperature, which is about 727°C. Therefore stress-relieving is done at approximately 650°C for about one hour or until the whole part reaches the temperature.

that of the metal that is corroded. Such stresses can cause cracks to initiate and grow. Stresses from expanding corrosion products can readily cause adjacent metals to flow plastically, as occurs in nuclear steam generators in a process called *denting*. Denting results from the corrosion of the carbon steel support plates and the buildup of corrosion product in the crevices between tubes and the tube support plates. This process is called denting since, when seen from the inside of the tubes, these deformations seem to produce dents at the tubesheet locations.

*Pack rust** is another example of the tremendous forces created by expanding steel corrosion products where forces as high as 70 MPa have been reported [18]. Figure 7.32 illustrates the effect of pack rust that has developed on steel bridge under repair. The force of expansion in this particular example was sufficient to break three of the bridge rivets shown in Fig. 7.32. Such corrosion damage may be a serious derating factor when the load-bearing capacity of a bridge or of any other infrastructure component is evaluated during inspection.

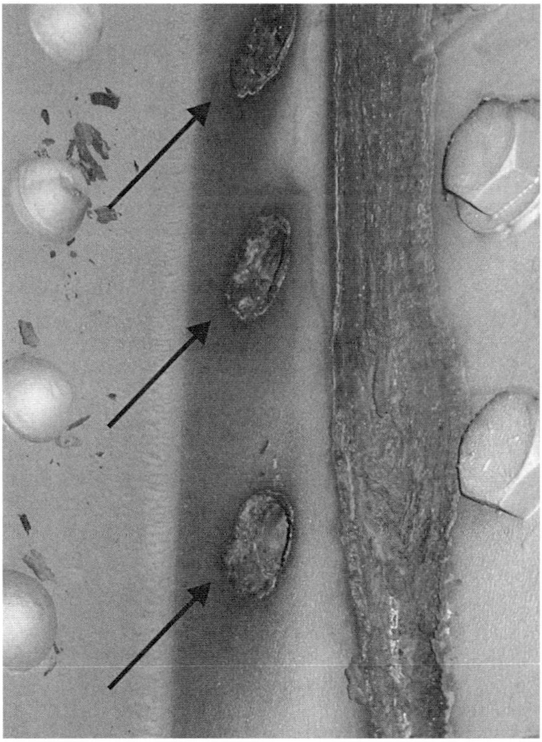

FIGURE 7.32 The effect of *pack rust* that has developed on an important steel bridge under repair. (Courtesy Wayne Senick, Termarust Technologies www.termarust.com).

*The expression "pack rust" is often used in the context of bridge inspection to describe built-up members of steel bridges that are showing signs of rust packing between steel plates.

Figure 7.33 Photograph of the Boeing 737 operated by Aloha Airlines that lost a major portion of the upper fuselage in 1988.

The following example has baffled the airline industry through a few investigations. In this case, the deformation due to the corrosion of aluminum in lap joints of commercial airlines was accompanied by a bulging (pillowing) between rivets, due to the increased volume of the corrosion products over the original material. This problem was said to be the primary cause of the Aloha incident (Fig. 7.33) in which a 19-year-old Boeing 737, operated by Aloha airlines, lost a major portion of the upper fuselage in full flight at 7300 m [19]. The "pillowing" process in which the faying surfaces are forced apart is schematically illustrated in Fig. 7.34.

The prevalent corrosion product in corroded fuselage joints was found to be hydrated alumina [$Al(OH)_3$], with a particularly high volume expansion relative to aluminum as shown in Fig. 7.35 [20].

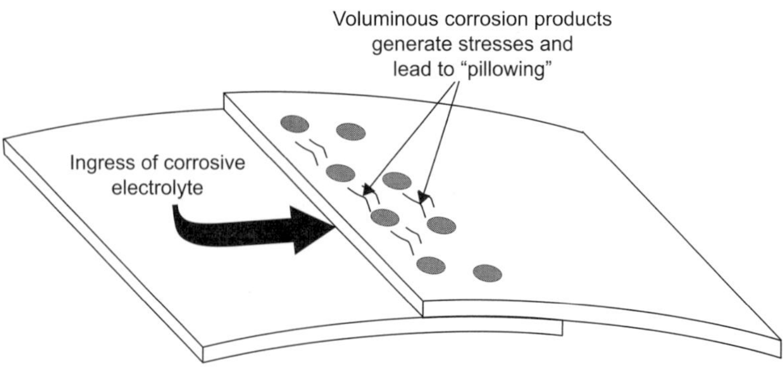

Figure 7.34 Schematic description of the "pillowing" of lap splices.

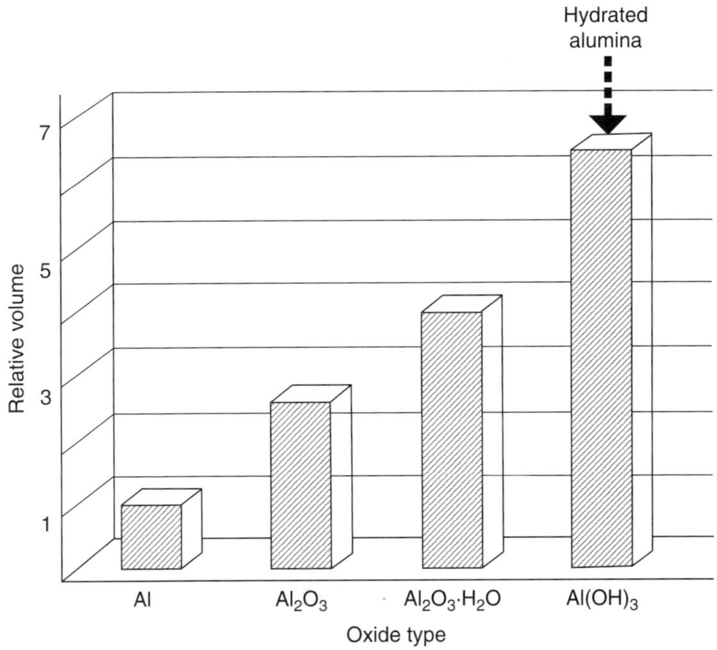

FIGURE 7.35 Relative volume of aluminum corrosion products.

The buildup of voluminous corrosion products when hydrated alumina is formed can produce an undesirable increase in stress levels near critical fastener holes (Fig. 7.36) and subsequent fracture due to the high tensile stresses resulting from the "pillowing."

7.4.7 Surface Film Cells

The surface of a metallic object can be easily reacted in normal atmospheres making it behave quite differently from a piece of unreacted metal, often resulting in a situation resembling dissimilar metal corrosion. A film may be formed which is invisible, actually only a few molecules in thickness, but which may have a potential as much as 0.3 V different from the unfilmed metal. Naturally, such a potential difference is enough to create an active corrosion cell. Steels in soil have a tendency to film with time. "Old" steel, that is, steel which has been in the ground for several years may then become cathodic with respect to "new" steel, even when the two are identical in bulk composition (Fig. 7.37). It is thus strictly a surface phenomenon [21].

A common occurrence of surface film cell is found in older distribution piping systems where a section of pipe has been replaced because of corrosion damage. The new piece of pipe, exposed to the

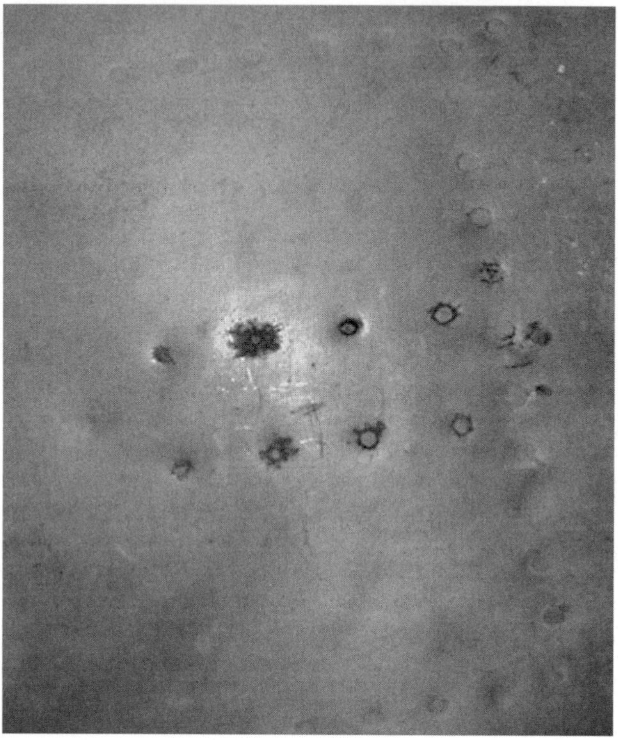

Figure 7.36 Advanced stage on the belly of a Boeing 737, where the corrosion products have expanded to the point where a number of rivets have actually popped their heads (Courtesy Mike Dahlager, Pacific Corrosion Control Corp.).

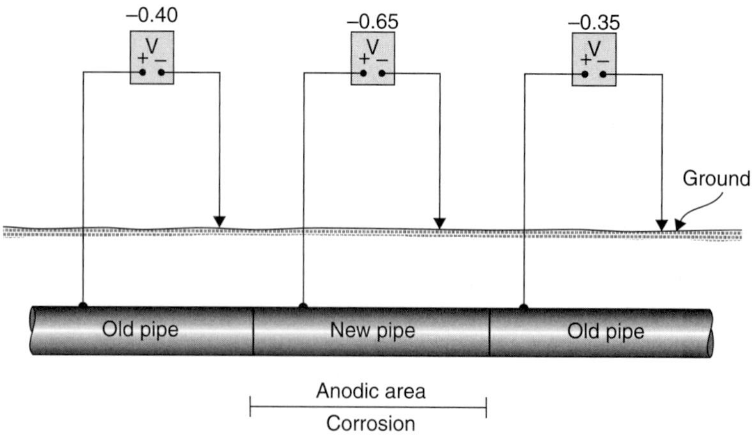

Figure 7.37 Schematic showing a differential corrosion cell created by replacement of a section of pipe.

same corrosion conditions, logically would be expected to last as long as the original section. However, the new section will usually fail sooner than expected unless it is electrically insulated from the remainder of the system.

7.4.8 Microbial Corrosion Cells

The anaerobic corrosion of iron was noted in the nineteenth century and many theories were proposed about its mechanism. Decades of scientific research projects and investigations on the complex influence of microbes on increasing or decreasing corrosion rates have provided a much deeper insight in the role microorganisms play on the life of systems exposed to waters and grounds where they proliferate (see Chap. 10 for more details).

Microorganisms tend to attach themselves to solid surfaces, colonize, proliferate, and form biofilms which may in turn produce an environment at the biofilm/metal interface radically different from the bulk environment in terms of pH, dissolved oxygen, organic and inorganic species. Since the biofilm tends to create nonuniform surface conditions, localized attack might start at some points on the surface leading to localized corrosion, usually in the form of pitting [22].

When microorganisms are involved in the corrosion of metals, the situation is more complicated than for an abiotic environment, because microorganisms not only modify the near-surface environmental chemistry via microbial metabolism but also may interfere with the electrochemical processes occurring at the metal–environment interface. Many industrial systems are likely to contain various structures where MIC and biofouling may cause serious problems: open or closed cooling systems, water injection lines, storage tanks, residual water treatment systems, filtration systems, different types of pipes, reverse osmosis membranes, potable water distribution systems and most areas where water can stagnate.

One noteworthy example that has taken an alarming proportion in recent years is plugging and corrosion of emergency fire sprinkler systems that are typically assumed to be functional until one attempts to use them, often during an emergency situation. Numerous reports in the past decade have described the rapid development of pinhole sized leaks and highly obstructive interior growth developments in sprinkler system piping, fittings, and supply tanks. Some occurrences have been reported after less than one year of system service. In many of these cases, the cause has been attributed to MIC [23].

Such failures can take two forms. First is the failure of a system to hold water. This is most often seen in the development of the pinhole-sized leaks, often considered to be a sure sign of MIC infection. This is also typically the only concern in many treatment investigations.

Second, and more concerning, is the failure of a system to achieve its designed purpose: that of fire control. Several systems with MIC

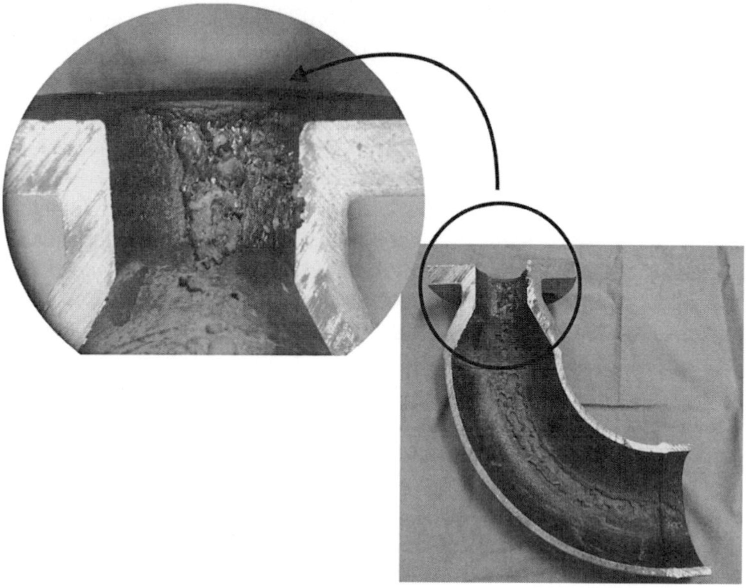

FIGURE 7.38 Cavitation downstream side of the service water system downstream of a valve. (Courtesy of Russ Green, TMI)

have been found with sprinkler drops completely plugged with the biofilm or biosludge debris generated by the MIC process. Sprinkler system feed mains have also been found with up to 60 percent obstruction from biological growth.

The second example relates to service water systems that supply river water flow through emergency cooling coils used by the nuclear industry. Water typically passes through a set of coils before being returned to the river. Such systems are normally kept in standby mode except for a quarterly test during which water is passed through a minimum recirculation valve back to the river. In one case that was fully investigated, cavitation was experienced immediately downstream of the valve disc, cutting many deep grooves in the valve body and the outlet piping. Since the leak was discovered as a drip and not a spray it was postulated that microbiological action caused the final wall breakthrough (Fig. 7.38).

7.5 Corrosion Avoidance

The forms of corrosion discussed in Chap. 6 are generally triggered by one or many of the factors described earlier in the present chapter. Some of these factors are conveniently described as corrosion cells and corrosion avoidance starts by considering how to reduce the effects of these factors and cells.

7.5.1 Pitting Mitigation

Since pitting is an electrochemical process, it can be stopped by cathodic protection. It can also be prevented by the use of inhibitors to alter the electrode reactions of the local cell and remove their driving force. In some cases, agitation of the environment will prevent environmental differences from developing and will prevent pitting that otherwise would occur.

Penetration by pitting is often prevented by coating the surface of a metal with a sacrificial layer of another alloy, such as zinc on steel or Alclad* on aluminum, or by applying protective coatings. A zinc-rich paint is sacrificially active and will prevent the pitting of either steel or aluminum.

Another way to prevent pitting corrosion is simply by choosing a more corrosion resistant material. An empirical ranking scheme that describes the corrosion pitting susceptibility of austenitic and duplex stainless steels can provide some useful guidance. The pitting resistance equivalent number (PREN) predicts the pitting resistance of these alloys as a function of their content in chromium, molybdenum, tungsten, and nitrogen [Eq. (7.2)]:

$$\text{PREN} = \text{Cr} + 3.3\,(\text{Mo} + 0.5\,\text{W}) + 16\text{N} \tag{7.2}$$

7.5.2 Crevice Corrosion Mitigation

Crevice corrosion is ideally prevented at the design stage by avoiding them in the first place or during construction by filling uncorroded dry crevices with a durable jointing compound that will exclude moisture and remain resilient.

Crevices are present in many types of equipment. Examples are gasketed flanges, rolled joints between tubes and tube sheets in heat exchangers, faying surfaces between tanks and supporting structures, and so forth. Moisture and chemical solutions may be trapped within crevices and held stagnant. The use of angle iron construction, skip welding, high outlets in tank bottoms, "dead" flow areas in piping stubs, heat exchangers, and vessels should be avoided. A number of procedures may be used to avoid crevice corrosion at the design stage (see Chap. 11 for more details on this topic).

1. Use butt joints in preference to lap joints
2. Eliminate or carefully seal lap joints so that they will not "open up"

---

*A composite wrought product comprised of an aluminum alloy core having on one or both surfaces a metallurgically bonded aluminum or aluminum alloy coating that is anodic to the core and thus electrochemically protects the core against corrosion.

3. Provide complete drainage
4. Inspect regularly; clean thoroughly
5. Be careful when using packing materials or thermal insulation which can hold moisture in contact with metals

7.5.3 Galvanic Corrosion Mitigation

There are a number of ways that galvanic corrosion may be prevented. These can be used singly or in combination. All of these preventive measures follow directly from the basic mechanism of galvanic corrosion.

1. Avoid the use of dissimilar metals wherever possible. If this is not practical, try to use metals which are close together in the galvanic series (Fig. 6.31 in Chap. 6).
2. Avoid an unfavorable area ratio whenever possible, particularly in the presence of an electrolytically conductive environment.
3. If dissimilar metals are used, insulate these electrically from one another.
4. If it is necessary to use dissimilar metals, and these cannot be insulated, then the more anodic part should be designed for easy replacement or should be constructed of thick materials to longer absorb the effects of corrosion.
5. Coat the cathode (or both anode and cathode) near the junction to reduce the effective cathodic area. Never coat the anode alone since any pinhole or holiday would be the site of rapid anodic attack due to the large S_c/S_a surface area ratio.

7.5.4 Fretting Corrosion Mitigation

Fretting corrosion can be prevented by eliminating any slipping movement between two surfaces in contact. Thus, it is sometimes possible to overcome fretting by increasing the load on the surfaces, if it prevents relative motion. Alternatively, decreasing the load can minimize the effect of vibratory motion. In other cases, roughening the surfaces can increase the friction between the surfaces and stop the movement. Fretting corrosion can also be greatly retarded by lubricating the contacting surfaces with an oil or grease with sufficient load-bearing characteristics that it will separate the surface from the environment.

7.5.5 Mitigation of Stress Corrosion Cracking

In reference to Fig. 7.39, one could be tempted to say that it is easy to prevent mitigation of SCC. Simply eliminate the stress, isolate the

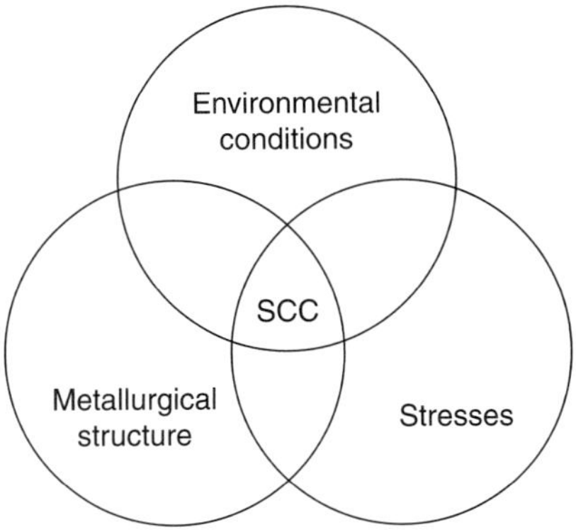

Figure 7.39 The three main factors contributing to the occurrence of SCC.

metal from the environment, or change the environment! This is of course much easier said than done. Nevertheless, the following basic methods of corrosion control are attempts to do exactly that.

Change of material: A total or even partial change of material (e.g., "safe-ending" of heat-exchanger tubes) is a common approach. For complete reliability, use of a superior crack-resistant material is often the most cost-economical approach.

Change of environment: The removal of chlorides, caustic, or other major cracking-type species is an effective solution where possible. However, much less drastic changes are often effective as well, such as

- Removal of oxygen or oxidizing agents.
- Changing the pH.
- Additions of inhibitors may be effective. Nitrates, for example, have been used as an inhibitor against caustic embrittlement of steel.

Barrier coatings: Temporary corrosion prevention compounds have been found helpful against atmospheric cracking of high-strength steels, although good commercial organic coatings are much more reliable. Certain silicone-base paints are routinely used to prevent external chloride induced SCC of insulated stainless steel vessels and piping.

Electrochemical techniques: Cathodic protection has been found effective against anodic SCC. Lead-tin solders and nickel plating have been used to protect stainless steel tube ends against SCC by water. This technique must be carefully controlled to be effective. Austenitic stainless steel has been known to suffer chloride SCC while simultaneously causing galvanic corrosion of steel components.

The following are general recommendations to avoid the initiation of SCC due to the formation of corrosion cells in service.

- Residual stresses may be minimized by careful thermal stress relief appropriate to the alloy
- Compressive stresses may be introduced by controlled shot peening
- Stress raisers should be eliminated or at least minimized
- Open crevices should be eliminated, as well as areas where deposits can accumulate and build up tremendous internal pressure
- Conditions conducive to evaporation and concentration of corrosive species should also be minimized

7.6 Visualizing Corrosion Cells

The existence and location of anodes and cathodes in a corrosion cell can be demonstrated by the changes in color of certain reagents. Such color changes have been very helpful in the early days of corrosion science to study the local interplay of local anodes and cathodes on apparently homogeneous steel surfaces exposed to a corrosive environment. As noted by Cushman and Gardner in their 1910 textbook, it is a matter of common observation that iron usually corrodes rapidly at certain weak points in an effect known as pitting [24].

The interest in using color changes to reveal subtle corrosion mechanisms is still quite modern as attested by two papers recently published in a reputable corrosion journal. In one paper, the corrosion of aluminum and aluminum alloys in chloride-containing agar gels was studied by using a broad-range pH indicator [25]. Distinct changes in pH were observed at low-pH anodic sites and at high-pH cathodes. There was a definite edge effect that occurred when gels were placed directly on the metal, which dominated the pattern of corrosion. Wet-abraded surfaces initially showed a general type of corrosion, whereas dry-abraded surfaces showed localized corrosion under the gel.

In a second paper published recently, paint systems containing color-change or fluorescing compounds were found to be sensitive to underlying corrosion processes by reacting to the pH increase associated with the local cathodic reaction [26]. The sensitivity of acrylic-based coating systems for detection of cathodic reactions associated with corrosion was determined by applying constant cathodic current and measuring the charge at which color change or fluorescence was detected.

As explained in much detail in Chaps. 2 and 3, the cathodic reaction in a corrosion process generally produces an increase in the concentration of hydroxyl ions as a result of removal of hydrogen ions or the reduction of oxygen. Phenolphthalein is a well-known indicator which develops a red color when the concentration of hydroxyl ions is increased. Therefore, if phenolphthalein turns red as corrosion proceeds, it will detect the existence of a cathode and reveal where it is [27].

Similarly, potassium ferricyanide is a reagent which produces, as described in Eq. (7.3), a blue color by reaction with ferrous ions (Fe II) as they form at the anodic areas when iron corrodes. The appearance of this blue color, therefore, demonstrates the existence and location of anodes on iron.

$$3Fe^{2+} + 4[Fe(CN)_6]^{3-}_{(yellow)} \rightarrow Fe_4[Fe(CN)_6]_3 \cdot 4H_2O_{(Prussian\ blue)} + 6CN^- \tag{7.3}$$

$$\qquad Fe^{3+} \qquad\qquad\qquad Fe^{3+}\ Fe^{2+}$$

The combination of these two reagents in a gelling agent to which sodium chloride is added is known as a ferroxyl solution*. Cushman and Gardner made ample use of this medium to reveal local corrosion cells and supported their theories with many pictures in their 1910 textbook.

One classic example of a local cell revealed by the ferroxyl agent is the development and location of the anode and cathode in a cell established on a steel surface within a drop of ferroxyl gel. Oxygen from the air is more accessible to the periphery of the drop and sets up a cathode that becomes visible as a pink color. Simultaneously an anode that develops near the center of the drop which is less accessible to oxygen is revealed by the gel turning blue (Fig. 7.40).

These color-changing reagents have also been used to show the existence and location of anodes and cathodes created by joining dissimilar metals. For example, upon placing a steel nail with copper

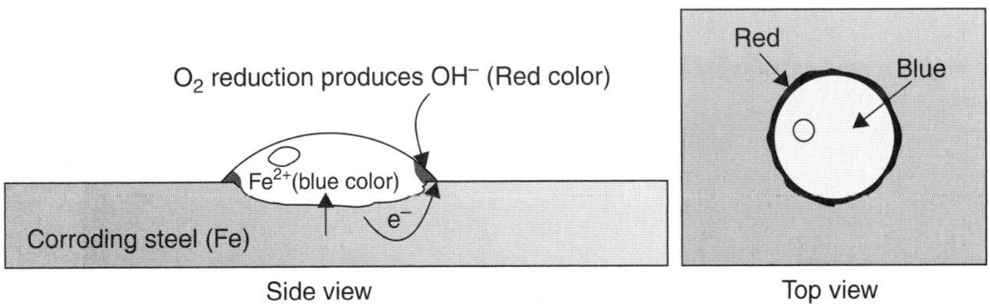

FIGURE 7.40 Ferroxyl drop experiment on a steel plate.

*How to prepare 250 mL of ferroxyl gel containing 3 percent sodium chloride: Dissolve 7.5 g sodium chloride in 250 mL distilled water. Add 5 g powdered agar agar and boil the mixture until the agar is dispersed. Then add 0.5 g potassium ferricyanide and 5 mL of a 0.1 percent phenolphthalein solution.

plate on the pointed half of it in a ferroxyl gel it will develop a pink area around the copper-plated portion and a blue area around the bare steel half (Fig. 7.41) [27]. In this experiment, the colors become more intense with time. This shows that iron continues to corrode as an anode, and, at the same time, there is an increasing concentration of alkali at the copper cathode.

When a steel nail with zinc plated on the pointed end is placed in the same gel, the iron now acts as a cathode and the zinc as an anode (Fig. 7.42). Accordingly, the red color identifying the cathode develops at the iron half. No iron goes into solution on this area, and, therefore, no blue color develops. The zinc corrodes as an anode at the point and is consumed in providing protection to the iron. A white area develops around the zinc because the zinc corrosion products form a white substance in contact with the potassium ferricyanide reagent.

The local action cell on a single metal surface may be revealed by placing an ordinary steel nail in the ferroxyl gel (Fig. 7.43). After a short time, a pink area develops around the shank of the nail, indicating the formation of a cathode. Blue areas develop around the cold worked head and point of the nail where the iron goes into solution, indicating that these areas are the anodes.

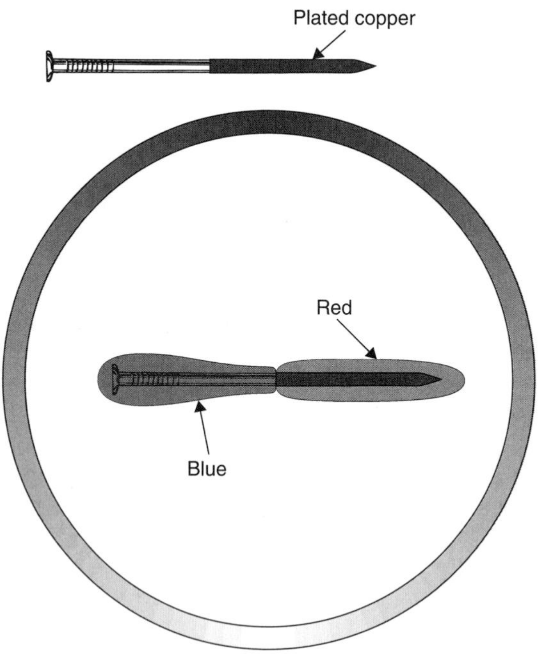

FIGURE 7.41 Steel nail partially plated with copper with anode and cathode revealed by colors developed in a Petri dish containing a ferroxyl gel.

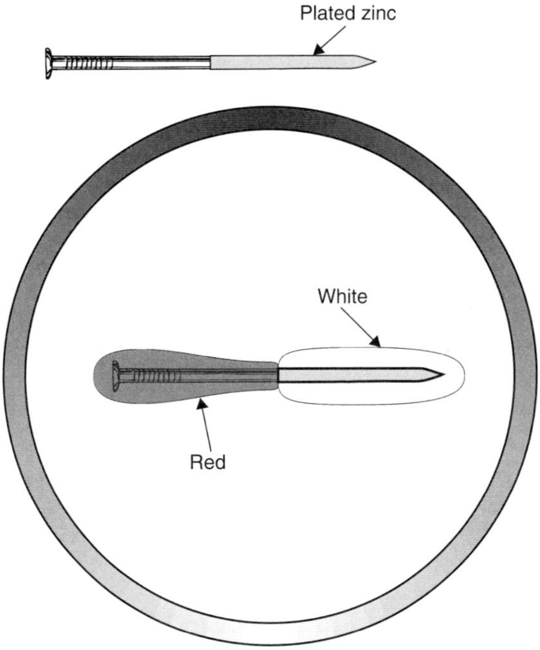

Figure 7.42 Location of anode and cathode areas on a steel nail partially plated with zinc.

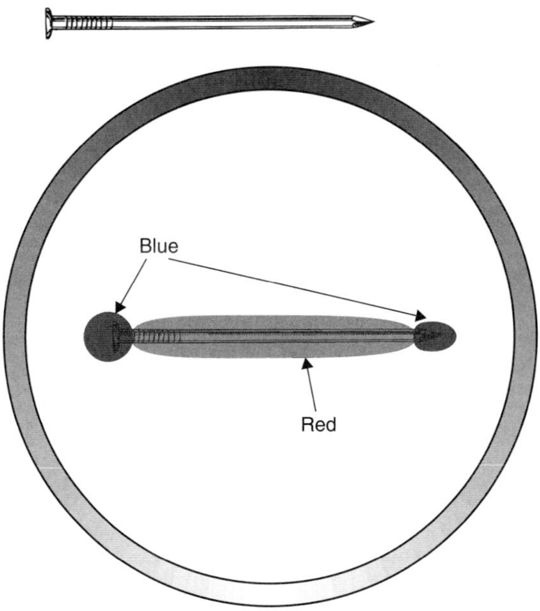

Figure 7.43 Location of local anodes and cathode on a single steel nail.

References

1. Mears RB, Brown RH. Designing to prevent corrosion. *Corrosion*, 1947; 3: 97–120.
2. Hoar TP. Report of the Committee on Corrosion and Protection. London, U.K.: Her Majesty's Stationery Office, 1971.
3. Standard guide for applying statistics to analysis of corrosion data. In: *Annual Book of ASTM Standards*. Philadelphia, Pa.: American Society for Testing of Materials, 1999.
4. Postlethwaite J, Dobbin MH, Bergevin K. The role of oxygen mass transfer in the erosion–corrosion of slurry pipelines. *Corrosion*, 1986; 42: 514–21.
5. Chexal B, Horowitz J, Dooley B, Millett P, Wood C, Jones R. Flow-Accelerated Corrosion in Power Plants. Revision 1. EPRI TR-106611-R1. Palo Alto, Calif.: Electric Power Research Institute, 1998.
6. Roberge PR. *Corrosion Testing Made Easy: Erosion–Corrosion Testing*. Houston, Tex.: NACE International, 2004.
7. Dillon CP. *Unusual Corrosion Problems in the Chemical Industry*. St. Louis, Mo.: Materials Technology Institute (MTI), 2000.
8. Dexter SC. Microbiologically influenced corrosion. In: Cramer DS, Covino BS, eds. Vol. 13A: *Corrosion: Fundamentals, Testing, and Protection*. Metals Park, Ohio: ASM International, 2003; 398–416.
9. Scott PJB. Expert consensus on MIC: Prevention and monitoring. *Materials Performance* 2004; 43: 50–4.
10. Michalke C, Otis AK. *Stray Currents from Electric Railways*. New York, N.Y.: McGraw-Hill, 1906.
11. Cowan RL, Staehle RW. The thermodynamics and electrode kinetic behavior of nickel in acid solution in the temperature range 25° to 300°C. *Journal of the Electrochemical Society*, 1971; 118: 557–68.
12. Staehle RW. Lifetime prediction of materials in environments. In: Revie RW, ed. *Uhlig's Corrosion Handbook*. New York, N.Y.: Wiley-Interscience, 2000; 27–84.
13. Staehle RW. Environmental definition. In: Revie RW, Sastri VS, Ghali E, Piron DL, Roberge PR, eds. *Materials Performance Maintenance*. New York, N.Y.: Pergamon Press, 1991; 3–43.
14. Oldfield JW. Electrochemical theory of galvanic corrosion. In: Hack HP, ed. *Galvanic Corrosion*. Philadelphia, Pa.: American Society for Testing of Materials, 1988; 5–22.
15. Peabody AW, Bianchetti RL. *Peabody's Control of Pipeline Corrosion. 2nd edn*. Houston, Tex.: NACE International, 2001.
16. *Dissimilar Metals*. MIL-STD-889B(3): Army Research Laboratory, Aberdeen, MD:1993.
17. Bell GEC, Schiff MJ, Wilson DF. Field observations and laboratory investigations of thermogalvanic corrosion of copper tubing. *CORROSION 97*, Paper # 568. Houston, Tex.: NACE International, 1997.
18. *External Corrosion—Introduction to Chemistry and Control*, 2nd. Report M27. Denver, Colo.: American Water Works Association, 2004.
19. Miller D. Corrosion control on aging aircraft: What is being done? *Materials Performance* 1990; 29: 10–1.
20. Komorowski JP, Krishnakumar S, Gould RW, Bellinger NC, Karpala F, Hageniers OL. Double pass retroreflection for corrosion detection in aircraft structures. *Materials Evaluation* 1996; 54: 80–6.
21. Beavers JA. Fundamentals of corrosion. In: Peabody AW, Bianchetti RL, eds. *Peabody's Control of Pipeline Corrosion*. Houston, Tex.: NACE International, 2001; 297–317.
22. Al-Darbi MM, Agha K, Islam MR. Modeling and simulation of the pitting microbiologically influenced corrosion in different industrial systems. In: *CORROSION 2005*, Paper # 505. Houston, Tex.: NACE International, 2005.
23. Clarke BH. Microbiologically influenced corrosion in fire sprinkler systems. In: *Fire Protection Engineering*, 2001; 9: 14–22.

24. Cushman AS, Gardner HA. *The Corrosion and Preservation of Iron and Steel.* New York, N.Y.: McGraw-Hill, 1910.
25. Isaacs HS, Adzic G, Jeffcoate CS. Visualizing corrosion. *Corrosion,* 2000; 56: 971–8.
26. Zhang J, Frankel GS. Corrosion-Sensing Behavior of an Acrylic-Based Coating System. *Corrosion* 1999; 55: 957–67.
27. LaQue FL, May TP, Uhlig HH. *Corrosion in Action.* Toronto, Canada: International Nickel Company of Canada, 1955.

CHAPTER 8

Corrosion by Water

8.1 Importance of Water

The amazing view of earth from space shows how the blue planet we live on is unique among the other known celestial bodies. It has water, which covers three-fourths of its surface and constitutes 60 to 70 percent of the living world. Actually, only one percent of this water is fresh and directly usable. Approximately 97 percent of the planet water is salty seawater and two percent is frozen in glaciers and polar ice caps. However, abundant supplies of fresh water are essential to industrial development.

Enormous quantities of water are required for cooling systems, process needs, boiler feed and for sanitary and potable water. It was estimated, in 1980, that the water requirements for industry in the United States approximated 525×10^9 L/day. A substantial quantity of this water was reused. The intake of "new" water was estimated to be about 140×10^9 L/day [1]. If this water was pure and contained no impurities, there would be little need for water conditioning or water treatment.

8.2 Corrosion and Water Quality and Availability

North Americans use a lot of water. Annually, approximately 63×10^9 m^3 of drinking water serves approximately 73 million customers in North America, with the average total water-use rate per customer ranging between 475 and 660 L/capita/day. Recent benchmark estimates by the American Water Works Association (AWWA) on indoor water-use rates, indicated an average use-rate of 245 L/capita/day. The average consumer cost for clean water ranges from \$0.12 to \$0.65/m^3.

According to the AWWA industry database, there was in 1995 approximately 1.5×10^6 km of municipal water piping in the United States. The sewer system consisted of more than 16 thousand publicly owned treatment facilities releasing some 155×10^6 m^3 of wastewater per day. The total annual direct cost of corrosion for the nation's drinking water and sewer systems was estimated at

$36 billion, which included the costs of replacing aging infrastructure, unaccounted-for water through leaks, use of corrosion inhibitors, internal mortar linings, and the cost of external coatings and cathodic protection [2].

Based on AWWA data, there was approximately 1.4×10^6 km of municipal water piping in the United States in 1995. Table 8.1 presents an estimated profile of the different materials that make up these water pipes. New pipes are being installed at a rate which extends the system length by 1.5 percent per year, while an additional 0.5 percent is being replaced annually.

8.2.1 Corrosion Impact

Water possesses several unique properties, one being its ability to dissolve to some degree most of the inorganic substances occurring in nature. Because of this essential property, water typically contains a variety of impurities that may cause problems through the formation of deposits in water lines, on boiler tubes, and on any other surface water comes in contact with. Dissolved oxygen, which supports the aquatic life, is also responsible for the corrosivity of water toward metals. Corrosion jeopardizes the reliability of water delivery systems by causing leaks and breaks and by affecting water quality. Corrosion can have a variety of impacts grouped into the following four categories [3].

Health and Regulations

From a health standpoint, the most significant metal that can enter the drinking water via corrosion is lead. The contribution of drinking water to the total daily intake of lead in the United States is estimated to be approximately 20 percent. In drinking water, the primary sources of lead are lead service lines, lead plumbing, brass fixtures, and 50:50 tin-lead solder used to join copper piping. Lead enters the water when the water has been standing motionless in contact with the lead source for extended time periods.

Material	Percentage
Cast iron	48
Ductile iron	19
Concrete and asbestos concrete	17
PVC	9
Steel	4
Other	2

Table 8.1 Profile of Different Materials Used for U.S. Transmission Water Pipes, as Determined from the 1992 AWWA Water Industry Database [2]

Aesthetics and Customer Perception

The metals of most interest from an aesthetic standpoint are copper, iron, and zinc. The corrosion products of copper piping in drinking water may cause a metallic taste, blue-green staining particles, and discoloration of the water. Iron species from the corrosion of unlined cast-iron piping, unlined steel piping, and galvanized pipe may discolor drinking water, form precipitates, cause rusty water and red staining of laundry and fixtures, impart metallic taste to water, and promote the growth of iron bacteria and sulfate-reducing bacteria. Rusty or red water is one of the most common customer complaints received by water utilities.

Zinc can cause bitter or astringent tastes in drinking water at concentrations of 4 to 5 mg/L. High concentrations may give water a milky appearance at room temperature and a greasy appearance when boiling.

Premature Piping Deterioration and Economic Impacts

Internal corrosion can have significant adverse economic impacts on the water transmission and distribution systems and consumer plumbing. Typically, consumer plumbing is most seriously affected by internal corrosion because piping is normally unlined and its diameter is small. Such piping will have a tendency to leak or develop tubercles that may reduce both flow and pressure (Fig. 8.1).

FIGURE 8.1 Tubercles in a small diameter water pipe. (Courtesy of Public Works and Services, City of Ottawa)

In recent years, many owners of large-diameter transmission and distribution piping systems typically have protected their investment by installing pipe with internal linings, by cleaning and lining pipe in place, or by adding a cathodic protection system for external corrosion (see Chap. 13 for some examples). However, there are still many thousands of kilometers of unlined metallic piping in use throughout North America and Europe. Costs to maintain operability over and above the normal accrue when this piping must be replaced prematurely or cleaned and lined in place.

Environmental Concerns

Corrosion of water distribution piping raises environmental concerns mostly due to the presence of lead, cadmium, zinc, and copper in drinking water. These metals enter the wastewater collection system and eventually accumulate in the sludge and end up in a landfill, on croplands, or other locations, depending on the disposal method. Metals corroded from water piping are the largest source of these contaminants in the wastewater of many communities.

8.2.2 Corrosion Management

System reliability is of the utmost importance to water suppliers and their customers. However, corrosion problems can vary greatly within a single system because many variables affect corrosion, for example, pipe material, pipe age, pipe wall thickness, water additives, corrosion inhibitor treatment, soil chemistry, soil moisture content and/or local groundwater level, and stray currents [2]. Table 8.2 summarizes some of the physical, environmental, and operational factors that can affect the deterioration rate of water distribution systems and lead to their failure [4].

Factor	Explanation
Physical	
Pipe material	Pipes made from different materials fail in different ways
Pipe wall thickness	Corrosion will penetrate thinner walled pipe more quickly
Pipe age	Effects of pipe degradation become more apparent over time
Pipe vintage	Pipes made at a particular time and place may be more vulnerable to failure
Pipe diameter	Small diameter pipes are more susceptible to beam failure
Type of joints	Some types of joints have experienced premature failure (e.g. leadite joints)

TABLE 8.2 Factors that Contribute to Water System Deterioration [4]

Factor	Explanation
Thrust restraint	Inadequate restraint can increase longitudinal stresses
Pipe lining and coating	Lined and coated pipes are less susceptible to corrosion
Dissimilar metals	Dissimilar metals are susceptible to galvanic corrosion
Pipe installation	Poor installation practices can damage pipes, making them vulnerable to failure
Pipe manufacture	Defects in pipe walls produced by manufacturing errors can make pipes vulnerable to failure
Environmental	
Pipe bedding	Improper bedding may result in premature pipe failure
Trench backfill	Some backfill materials are corrosive or frost susceptible
Soil type	Some soils are corrosive; some soils experience significant volume changes in response to moisture changes, resulting in changes to pipe loading
Groundwater	Some groundwater is aggressive toward certain pipe materials
Climate	Climate influences frost penetration and soil moisture. Permafrost must be considered in the north
Pipe location	Migration of road salt into soil can increase the rate of corrosion
Disturbances	Underground disturbances in the vicinity of an existing pipe can lead to actual damage or changes in the support and loading structure on the pipe
Stray electrical currents	Stray currents cause electrolytic corrosion
Seismic activity	Seismic activity can increase stresses on pipe and cause pressure surges
Operational	
Internal or transient water pressure	Changes to internal water pressure will change stresses acting on the pipe
Leakage	Leakage erodes pipe bedding and increases soil moisture in the pipe zone
Water quality	Some water is aggressive, promoting corrosion
Flow velocity	Rate of internal corrosion is greater in unlined dead-ended mains
Backflow potential	Cross connections with systems that do not contain potable water can contaminate water distribution system
O&M practices	Poor practices can compromise structural integrity and water quality

Table 8.2 *(continued)*

The AWWA has developed a six-step procedure shown in Fig. 8.2 for the assessment and control of internal corrosion of water distribution systems [3]. Although the application of this method appears straightforward, working out the details for each system in an actual network can be quite complex. The procedure specifically focuses on older systems and does not consider corrosion prevention for new systems. It assumes that corrosion is already present and

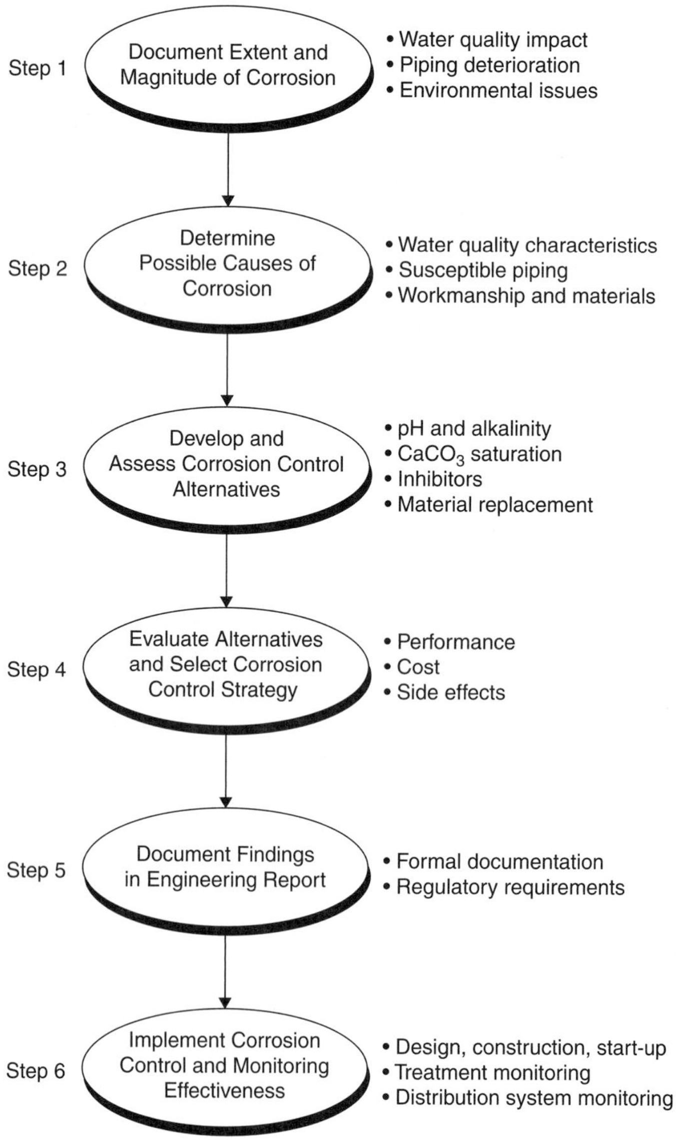

FIGURE 8.2 Corrosion control program implementation flowchart.

that the corrosion occurs only internally. Although the cost per system can be calculated reasonably accurately using this method, interactions with other systems are difficult to evaluate. The system size, location, population served, materials used, water quality, and soil conditions all significantly influence corrosion susceptibility.

Short-Term Corrosion Management

Short-term corrosion problems are often indicated by customer complaints, such as the occurrence of red or yellow "rusty" water or a sudden decrease in water pressure. A reason for rust-colored water is generally the presence of corrosion products that have flaked off of the internal pipe walls, while a water pressure drop may be caused by a leak in the transmission or distribution system.

Finding a leak in an underground pipe system is often difficult because the leak may start small and go undetected for a period of time. Once the leak is so severe that water is literally coming from the ground, it may cause a local flood. In addition to the lost water, the damage can be significant and the repair work is more than what would have been needed to fix a small leak [2].

Long-Term Corrosion Management

Long-term corrosion impact is generally indicated by system integrity studies in which maintenance and inspection teams are tasked to find leaks and failures. Some large utilities have a specialized corrosion team to monitor the water quality, using corrosion loops in which treated water circulates over weight-loss coupons. These coupons are often made from different materials and exposed to various water flow rates. The coupons would be periodically measured to determine average corrosion rates.

Water samples would also be routinely tested to ensure that the water quality is acceptable. The test results are used to make assessments about corrosion as well. For example, the water pH is important both for water consumers and for system integrity. The pH is kept within a predetermined range by adding pH adjusters to the treatment process.

Necessity of Long-Term Corrosion Planning

Because of the long life expectancy of water systems, a long-term vision for corrosion management is required. Unfortunately, some managers give in to short-term cost savings over long-term investments. As an example, the average thickness of cast iron and ductile iron pipe has been continuously decreased over the last 100 years because thinner, higher strength pipe has become available [2]. Unfortunately, corrosion rates are not significantly dependent on the strength of ductile iron or steel as was demonstrated by extensive corrosion studies [5]. As a result, thinner wall pipe will have a smaller corrosion tolerance than thicker wall pipe and will show more frequent failures. The time to corrode through a pipe wall was found to be directly proportional to the square of the wall thickness, for example, for a wall thickness reduction of 50 percent the corrosion life will be reduced to 25 percent of the life of the original pipe thickness.

Framework for Water Pipeline Management

The major tools for managing and preventing pipeline failures until recently have been simple statistical approaches based on numbers of pipe breaks per kilometer and reactive inspection techniques such as leak detection. While these approaches have been useful for managing pipeline failures, new technologies and knowledge about water system piping make it possible to develop more efficient and accurate approaches to maintaining pipeline integrity. The framework illustrated in Fig. 8.3 was recently proposed to help the introduction of these new techniques in service and use them even before the required research and development has been completed [6].

A major component of this framework is the use of nondestructive evaluation techniques to provide information about the condition of the pipeline. All pipes will eventually fail, but the rate of failure will depend on both the pipe material and the actual exposure to environmental and operational conditions. The most important feature of the framework relates to the cyclical nature of pipeline management. Each pipe in the system must be examined periodically, and its condition reassessed in order to determine what action should be taken to maintain or upgrade its condition. The entry point to the cycle is the pipe selection area labeled as "Prioritize Pipes for Analysis" in Fig. 8.3.

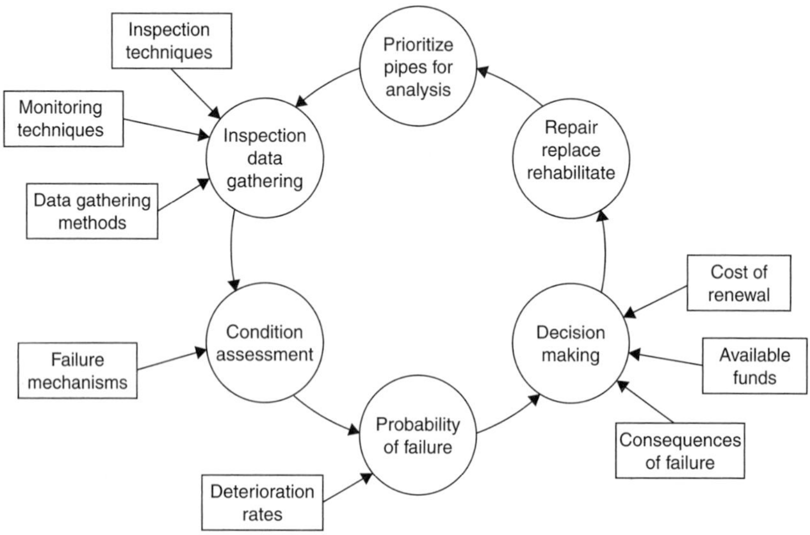

FIGURE 8.3 Water pipeline management cycle [6].

A pipeline must first be selected for analysis before any additional work can be done on it.

A second important point is that the maintenance and inspection strategy should differ between distribution and transmission systems to reflect the much higher failure consequences for transmission pipelines compared to those of a typical distribution line. Approaches that may not be economically viable for a distribution system could be employed advantageously in a transmission system to prevent failures. Essentially, the low consequences associated with a single distribution failure also mean that the emphasis in distribution systems should be on failure management in order to minimize life-cycle costs while a more proactive failure prevention approach would be better suited for transmission systems due to the high consequences of associated failures [6].

8.2.3 Condition Assessment Techniques

The pipes in water distribution and transmission systems are difficult to inspect for damage due to their location below the surface of the ground. This difficulty has led water utilities to rely on techniques such as breakage records, leak detection, and water audits to determine the health of their systems. While these techniques have been shown to be very useful in prioritizing repairs and replacements, they have the disadvantage of being reactive in nature. In each case, problems with the water system only become apparent after the pipes have failed in some manner [7].

Table 8.3 provides a comparative listing of the different techniques for inspecting metallic water pipes. Each technique is named and the relative advantages and disadvantages of the technique are given.

Technique	Advantages	Disadvantages
Zone water audits	Cheap Covers large areas of a city quickly Allows for a comparison of water losses between individual districts Useful as a screening process for other techniques Can be used to evaluate the effectiveness of repair programs	Does not give the precise location of leaks Requires isolation of zones Work must be performed at night Only gives an overview of current problems

TABLE 8.3 Comparison of Diagnostic Techniques for Metallic Water Mains [7]

Technique	Advantages	Disadvantages
Sonic/acoustic leak detection	Widely practiced Known to find leaks accurately Known to find leaks of different sizes Operates from outside the water line	Percentage of leaks missed by the technique is unknown Currently works best in metal water lines Only gives information on the current condition of the line (the tool has little predictive value) Background noise problems
Remote field inspection (hydroscope)	Most advanced technique currently available Detects areas of corrosion pitting, as well as through holes Can be used to give an estimate of the future life of a line	More expensive than leak detection Requires access to the inside of the water line, which may require cleaning Knowledge of the relationship between pit size and residual life of the pipe is not yet complete Limited to pits of less than 3 cm^3 in size.
Magnetic flux leakage	Established technology in oil and gas industry Known to be capable of detecting small defects and through holes in steel pipe	Not yet commercially available for water lines Requires access to and complete cleaning of the inside of the pipe
Ultrasound	Most versatile NDE technique Established technology in oil industry	Technique will not work through tuberculation Not yet commercially available for water lines Requires access to and complete cleaning of the inside of the pipe
Soil corrosivity measurements	Simple to conduct May act as a screening mechanism for more expensive methods	Cities may have similar levels of corrosivity across their region, making use difficult Do not always correlate with corrosion reality
Half-cell potential measurements	Simple to conduct May act as a screening mechanism for more expensive methods Well established as method to detect corrosion activity in buried objects	Factors such as stray currents and soil conditions may affect readings Accuracy of results depends on distance between readings Small areas of localized corrosion can not be detected

Table 8.3 *(continued)*

Table 8.4 shows a similar listing for prestressed concrete pipes. It is apparent from Table 8.3 that most of the techniques available for inspecting metallic water pipes complement, rather than compete with each other. Water audits provide broad, reactive information about the condition of a wide area of a water distribution network.

Technique	Advantages	Disadvantages
Half-cell potential measurements	Standard technique Results easy to interpret Simple to perform Does not require pipe entry	Experience suggests the method is ineffective for use with these lines in an urban setting Indicates the presence of corrosion activity on or nearby the pipe, but not the extent of the damage.
Visual inspection and sounding	Longest record of successfully detecting damaged pipes Examines condition of concrete	Requires man entry into pipes Does not provide direct information on wire breaks Unclear whether all wire breaks will produce noticeable concrete damage Subjective in nature and dependent on skill of inspection team
Acoustic monitoring	Works in an operating pipeline Can detect wire breaks during monitoring period and locate them Works in all types of prestressed concrete pipes	Only detects damage that occurs during monitoring period
Remote field inspection	Can detect single or multiple broken wires Inspection gives complete picture of damage to the pipeline	Currently only available for use in embedded type pipelines Requires man entry into pipes
Impact echo/ spectral analysis of surface waves	Examines condition of concrete Objective measurement system	Does not give direct information on wire breaks Slower inspections than remote field effect Requires an empty pipeline

TABLE 8.4 Comparison of Inspection Methods for Prestressed Concrete Pipes [7]

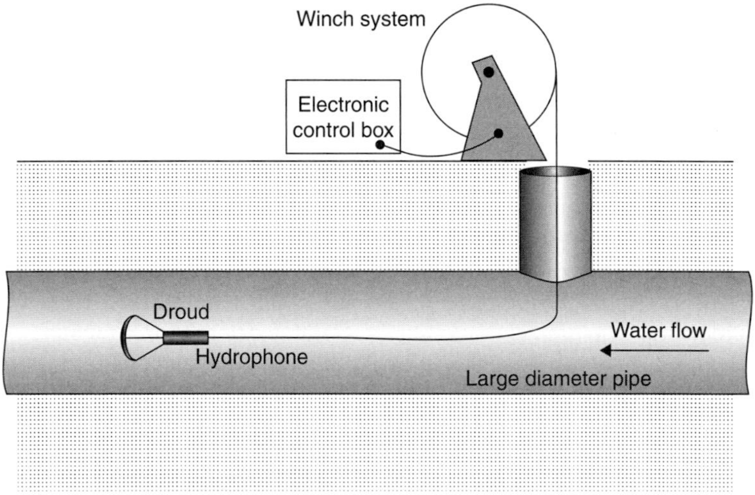

FIGURE 8.4 Schematic description of the Water Research Center's large-diameter pipe leak detection system [7].

Acoustic leak detection methods such as illustrated in Fig. 8.4 may find already broken or damaged pipes [7]. However, corrosion monitoring possibly may identify areas where corrosion activity on pipes is likely while the remote field effect can inspect pipes to find damage before they fail. A complete diagnostic program is likely to use all these inspection methods.

8.3 Types of Water

Water is commonly described either in terms of its nature, usage, or origin. The implications in these descriptions range from being highly specific to so general as to be non-definitive. A more practical description consists in classifying waters according to composition, for example, fresh containing less than 1000 ppm chlorides, brackish having 1000 to 25,000 ppm chlorides, seawater having between 2.5 and 3.5 percent sodium chloride, and brines with still higher chloride concentrations [8].

The corrosive nature of waters varies considerably depending largely on their composition and on the alloy exposed to the aqueous environment. Also very important in many cases is the biological activity of myriad aquatic organisms, whose metabolic products are directly or indirectly corrosive to many metals.

8.3.1 Natural Waters

Natural waters have large seasonal variations in physical, chemical, and biological characteristics. Oxygen, nutrients, pH, and other factors important for fouling and corrosion vary on a more or less predictable schedule during what is known as the "turnover." Any monitoring program that does not consider these changes will have an incomplete picture of the biology and chemistry of the system.

Fresh Water

Fresh water may come from either a surface or ground source, and typically contains less than 0.1 percent sodium chloride (<1000 ppm). It may be either "hard" or "soft" (rich or poor in calcium and magnesium salts) and thus possibly forming insoluble curds with ordinary soap. Actually, there are gradations of hardness, which can be estimated from the Langelier or Ryznar indexes discussed later or accurately determined by titration with standardized chelating agent solutions (e.g., versenates).

The two most important sources of fresh water are surface water and groundwater. A portion of the rain or melting snow and ice at the earth's surface soaks into the ground while part of it collects in ponds and lakes or runs off into creeks and rivers. This latter portion is termed surface water. As the water flows across the land surface, the flowing water dissolves some minerals and carries along finely divided particles and organic matter in suspension. The character of the terrain and the nature of the geological composition of the area influence the nature and quantity of the impurities found in these surface waters.

That portion of water which percolates into the earth's crust and collects in subterranean pools and underground rivers is called groundwater. This is the source of well and spring water. Underground supplies of fresh water differ from surface supplies in three important aspects, two of which are advantageous for industrial use. These are a relatively constant temperature and the general absence of suspended matter. However, groundwater may be higher in mineral content than surface supplies in the same geographic area because of the added solubilizing influence of dissolved carbon dioxide and the long residence time.

The concentrations of various substances in water in dissolved, colloidal, or suspended form are typically low but vary considerably. A hardness value of up to 400 ppm of calcium carbonate, for example, is sometimes tolerated in public supplies, whereas 1 ppm of dissolved iron would be unacceptable. In treated water for high-pressure boilers or where radiation effects are important, as in nuclear reactors, impurities are measured in very small units such as parts per billion (ppb) or 1 μg of contaminant per liter of water. Water analysis for drinking water supplies is concerned mainly with pollution and bacteriological tests. For industrial supplies a mineral analysis is of more interest. The important constituents can be classified as follows [9]:

- Dissolved gases (oxygen, nitrogen, carbon dioxide, ammonia, sulfurous gases)
- Mineral constituents, including hardness salts, sodium salts (chloride, sulfate, nitrate, bicarbonate, and so forth), salts of heavy metals, and silica

- Organic matter, including that of both animal and vegetable origin, oil, trade waste (including agricultural) constituents and synthetic detergents
- Microbiological forms, including various types of algae and slime-forming bacteria

Table 8.5 lists typical analytical determinations made in the examination of most natural waters. The list in Table 8.5 describes the general categories of substances, the difficulties commonly encountered by their presence and water treatment methods used to alleviate the difficulties.

The pH of natural waters errs rarely outside a fairly narrow range between 4.5 and 8.5. High values, at which corrosion of steel may be suppressed, and low values, at which gaseous hydrogen evolution occurs, are not often found in natural waters. Copper exposed to acidic waters might suffer a slight corrosion attack putting small amounts of copper ions in solution that may in turn cause green staining of fabrics and sanitary ware. In addition, redeposition of copper on aluminum, a common radiator material, or on galvanized surfaces might set up a very aggressive corrosion cell resulting in severe pitting of the metal.

Pure water, without dissolved gases (e.g., oxygen, carbon dioxide, and sulfur dioxide) does not cause undue corrosion attack on most metals and alloys at temperatures up to the boiling point of water. Even at temperatures of about 450°C, almost all of the common structural metals, with the exception of magnesium and aluminum, offer adequate corrosion resistance to high-purity water and steam.

From a corrosion standpoint, a significant water component is dissolved oxygen (DO) from ambient air. Oxygen acts both as a cathodic depolarizer and as an oxidizer. As a cathodic depolarizer, DO can remove hydrogen from the cathode during electrochemical corrosion and accelerate the corrosion attack. As an oxidizer, DO can be reduced on the metallic surface and participate directly to the electrochemical processes as described in many examples of Chap. 5.

The effect of DO on the corrosion of carbon steel is illustrated in Fig. 8.5 [10]. It should be noted in Fig. 8.5 that an increasing temperature is accompanied by an increase in corrosion rate of the steel due to faster reaction kinetics. The decreasing solubility of oxygen with temperature and salinity depicted in Table 8.6 only explains the upper limits of each of the three curves in Fig. 8.5.

The effect of oxygen on corrosion with increasing temperature is also shown in Fig. 8.6 that compares the results obtained in a closed vessel with those obtained with an open container that favored deaeration of the water by ebullition [10]. In a closed vessel, the solubility of DO increases with pressure and corrosion continues to increase with

Constituent	Chemical Formula	Difficulties Caused	Means of Treatment
Turbidity	None—expressed in analysis as units.	Imparts unsightly appearance to water. Deposits in water lines, process equipment, and so on. Interferes with most process uses.	Coagulation, settling, and filtration.
Hardness	Calcium and magnesium salts expressed as $CaCO_3$	Chief source of scale in heat exchange equipment, boilers, pipelines, and so on. Forms curds with soap, interferes with dyeing, and so on.	Softening. Demineralization. Internal boiler water treatment. Surface active agents.
Alkalinity	Bicarbonate (HCO_3^-), carbonate (CO_3^{2-}), expressed as $CaCO_3$	Foaming and carry over of solids with steam. Embrittlement of boiler steel. Bicarbonate and carbonate produce CO_2 in steam, a source of corrosion in condensate lines.	Lime and lime soda softening. Acid treatment. Hydrogen zeolite softening. Demineralization. Dealkalization by anion exchange.
Free mineral Acid	H_2SO_4, HCl, expressed as $CaCO_3$	Corrosion.	Neutralization with alkalies.
Carbon dioxide	CO_2	Corrosion in water lines and particularly steam and condensate lines.	Aeration. Deaeration. Neutralization with alkalies.
pH	(H^+)	pH varies according to acidic or alkaline solids in water. Most natural waters have a pH of 6.0–8.0.	pH can be increased by alkalies and decreased by acids.
Sulfate	(SO_4^{2-})	Adds to solids content of water, but in itself is not usually significant. Combines with calcium to form calcium sulfate scale.	Demineralization.
Chloride	Cl^-	Adds to solids content and increases corrosive character of water.	Demineralization.

TABLE 8.5 Difficulties and Means of Treatment for Common Impurities Found in Fresh Water

Constituent	Chemical Formula	Difficulties Caused	Means of Treatment
Nitrate	$(NO_3)^-$	Adds to solids content, but is not usually significant industrially. High concentrations cause methemoglobinemia in infants. Useful for control of boiler metal embrittlement.	Demineralization.
Fluoride	F^-	Cause of mottled enamel in teeth. Also used for control of dental decay. Not usually significant industrially.	Adsorption with magnesium hydroxide, calcium phosphate, or bone black. Alum coagulation.
Sodium	Na^+	Adds to solid content of water. When combined with OH^-, causes corrosion in boilers under certain conditions.	Demineralization.
Silica	SiO_2	Scale in boilers and cooling water systems. Insoluble turbine blade deposits due to silica vaporization.	Hot process removal with magnesium salts. Adsorption by highly basic anion exchange resins, in conjunction with demineralization.
Iron	Fe^{2+} (ferrous) and Fe^{3+} (ferric)	Discolors water on precipitation. Source of deposits in water lines, boilers, and so on. Interferes with dyeing, tanning, papermaking, and so on.	Aeration. Coagulation and filtration. Lime softening. Cation exchange. Contact filtration. Surface-active agents for iron retention.
Manganese	Mn^{2+}	Same as iron.	Same as iron.
Aluminum	Al^{3+}	Usually present as a result of floc carryover from clarifier. Can cause deposits in cooling systems and contribute to complex boiler scales.	Improved clarifier and filter operation.

Table 8.5 *(continued)*

Constituent	Chemical Formula	Difficulties Caused	Means of Treatment
Oxygen	O_2	Corrosion of water lines, heat exchange equipment, boilers, return lines, and so on.	Deaeration. Sodium sulfite. Corrosion inhibitors.
Hydrogen Sulfide	H_2S	Cause of "rotten egg" odor. Corrosion.	Aeration. Chlorination. Highly basic anion exchange.
Ammonia	NH_3	Corrosion of copper and zinc alloys by formation of complex soluble ion.	Cation exchange with hydrogen zeolite. Chlorination. Deaeration.
Dissolved Solids	None	A measure of total amount of dissolved matter, determined by evaporation. High concentrations of dissolved solids are objectionable because of process interference and as a cause of foaming in boilers.	Various softening processes, such as lime softening and cation exchange by hydrogen zeolite, will reduce dissolved solids. Demineralization.
Suspended Solids	None	A measure of undissolved matter, determined gravimetrically. Suspended solids cause deposits in heat exchange equipment, boilers, water lines, and so on.	Subsidence. Filtration, usually preceded by coagulation and settling.
Total Solids	None	The sum of dissolved and suspended solids, determined gravimetrically.	See "Dissolved Solids" and "Suspended Solids."

TABLE 8.5 *(continued)*

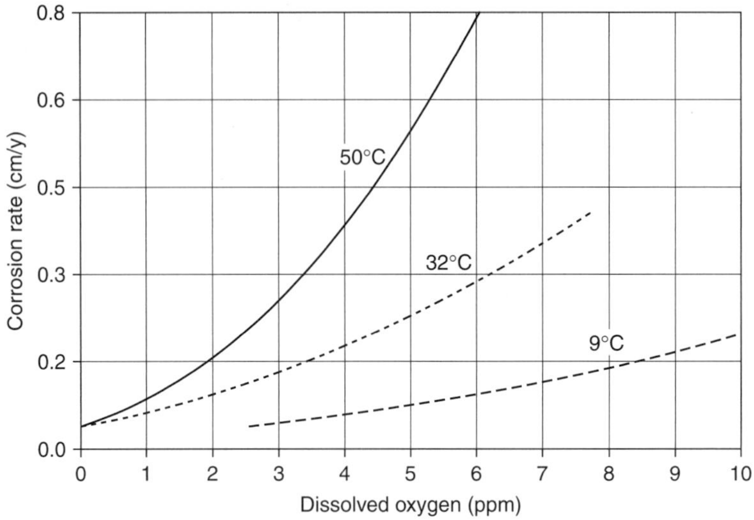

FIGURE 8.5 Effect of oxygen concentration on the corrosion of low-carbon steel in tap water at different temperatures.

temperature, hence the requirement for removing DO from hot water systems and boilers.

Other constituents that contribute to corrosion are chlorides, carbon dioxide and carbonates, and sulfides or ammonia from industrial or

Chlorinity* ($^0/_{00}$)	0	5	10	15	20
Salinity† ($^0/_{00}$)	0	9.06	18.08	27.11	36.11
Temperature(°C)			ppm		
0	14.58	13.70	12.78	11.89	11.00
5	12.79	12.02	11.24	10.49	9.74
10	11.32	10.66	10.01	9.37	8.72
15	10.16	9.67	9.02	8.46	7.92
20	9.19	8.70	8.21	7.77	7.23
25	8.39	7.93	7.48	7.04	6.57
30	7.67	7.25	6.80	6.41	5.37

* chlorinity refers to the total halogen ion content as titrated by the addition of silver nitrate, expressed in parts per thousand, that is $^0/_{00}$.

† salinity refers to the total proportion of salts in sea water, often estimated empirically as chlorinity × 1.80655, also expressed in parts per thousand, that is $^0/_{00}$.

TABLE 8.6 Dissolved Oxygen in Water Exposed to Air at Different Temperatures and Water Salinity

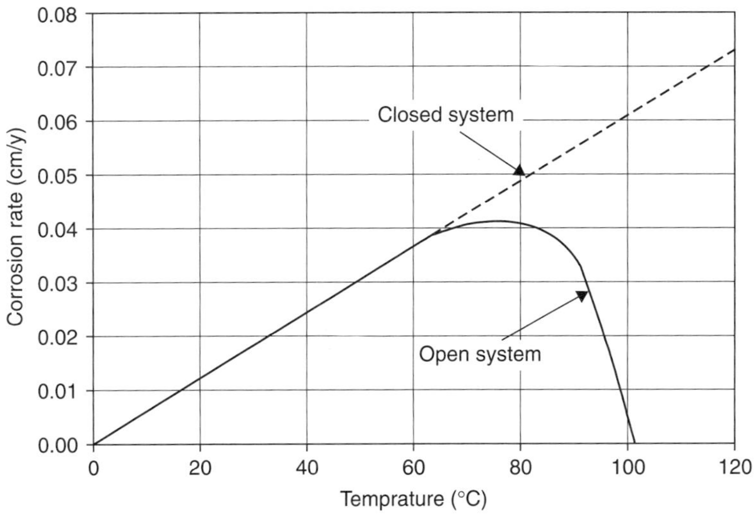

FIGURE 8.6 Effect of oxygen on corrosion of steel.

natural sources. Of course, many other man-made contaminants and pollutants can be found in local areas if industries are permitted to discharge their waste products into water resources. As with other chemical reactions, corrosion increases with elevated temperature, unless stifled by insoluble scales, the removal of corrosive gases, or the addition of corrosion inhibitors.

The formation of scale on a surface plays an ambivalent role. It can be positive by providing a protection of the substrate or negative by forming a poorly adherent deposit accentuating pitting at pores, cracks, or other voids in the scale. If the scale grows to any significant thickness, the loss of heat transfer through the metal and deposited scale can also be a problem in certain applications requiring heat exchange across the metal. Thus, the formation and growth of scales on metal surfaces is an important consideration when using metals in waters.

The effect of oxygen and pH on the corrosion rate of steel at two temperatures is shown in Fig. 8.7 [11]. In a broad range of about pH 5 to 9, the corrosion rate can be expressed simply in terms of the amount of DO present (e.g., micrometer per year per milliliter DO per liter of water). At about pH 4.5, acid corrosion is initiated, overwhelming the corrosion rate by DO. At about pH 9.5 and above, deposition of insoluble ferric hydroxide, $Fe(OH)_3$, or magnetite, Fe_3O_4, tends to slow down the corrosion attack. Amphoteric metals such as aluminum, zinc, and lead, are however additionally sensitive to high pH situations and show a corrosion rate increase in alkaline environments. Figure 8.8 compares the behavior of steel and aluminum as a function of pH.

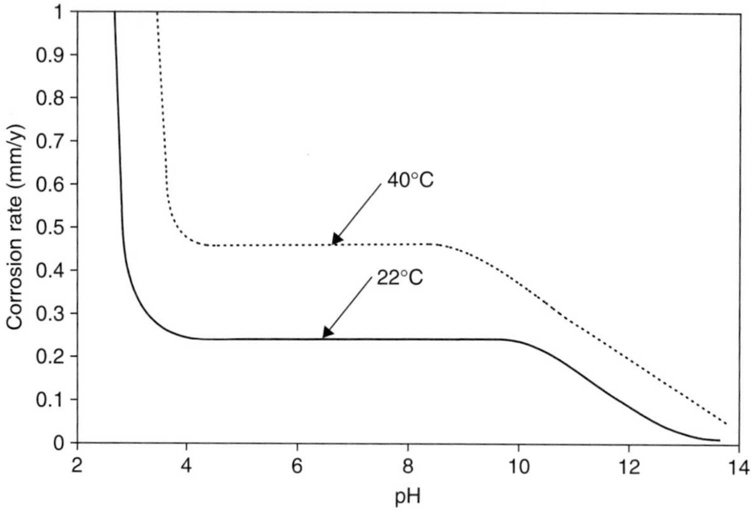

FIGURE 8.7 Corrosion of steel in water containing 5 ppm of dissolved oxygen at two different temperatures as a function of the water pH.

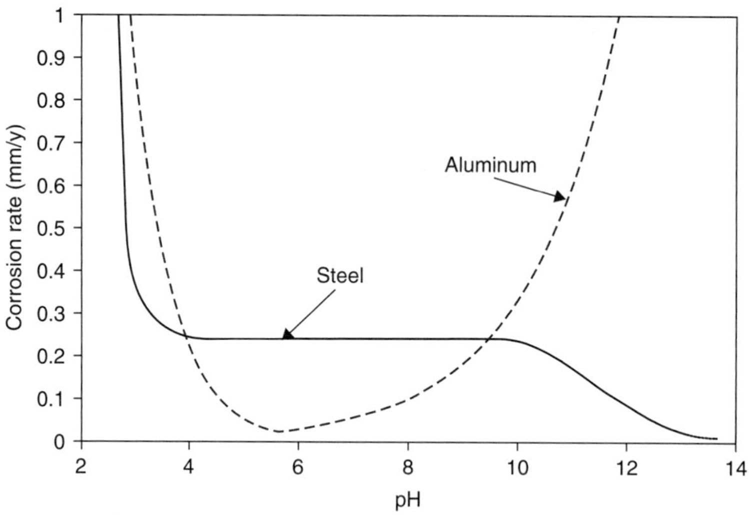

FIGURE 8.8 Corrosion of steel and aluminum as a function of pH at the same temperature (22°C).

Brackish Water

Brackish water contains between 0.1 and 2.5 percent sodium chloride (1000 and 25,000 ppm), either from natural sources (e.g., salt deposits) around otherwise fresh water or by dilution of seawater (e.g., tidal rivers). Brackish water differs from open seawater in other respects. The biological activity, for example, can be significantly modified by higher concentrations of nutrients and fouling is likely to be more severe as a consequence of the greater availability of these nutrients.

Within harbors, bays, and other estuaries, marked differences can exist in the amount and type of fouling agents present in the water. The main environmental factors responsible, singly or in combination, for these differences are the salinity, the degree of pollution, and the prevalence of silt. Moreover, the influence of these factors can be very specific to the type of organism involved. Apart from differences that can develop between different parts of the same estuary, there can also be differences between fouling in enclosed waters and on the open coast. In this respect, the extent of offshore coastal fouling is strongly determined by the accessibility to a natural source of infection. Local currents, average temperature, seasonal effects, depth, and penetration of light are operative factors. The presence of pollutants can also be quite important and highly variable in coastal areas.

Depending on composition brackish waters can be more aggressive than seawater. In tidal estuaries, the highest corrosion rate of carbon steel is just below the tidal zone, while in open seawater the highest corrosion rates are in the splash zone [8].

Seawater

Seawater systems are used by many industries, such as shipping, offshore oil and gas production, power plants, and coastal industrial plants. The main use of seawater is for cooling purposes, but it is also used for firefighting, oilfield water injection, and desalination plants. The corrosion problems in these systems have been well-studied over many decades, but despite published information on materials behavior in seawater, failures still occur.

Most of the elements that can be found on earth are present in seawater, at least in trace amounts. However, 11 of the constituents account for 99.95 percent of the total solutes, as indicated in Table 8.7, with chloride ions being by far the largest constituent. Seawater typically contains about 3.5 percent sodium chloride, although the salinity may be weakened in some areas by dilution with fresh water or concentrated by solar evaporation in others.

A large part of the dissolved components of seawater is present as ion pairs or in complexes, rather than as simple ions. While the major cations are largely uncomplexed, the anions other than chloride are to

Species	Concentration	
	mmol^{-1} kg^{-1}	g kg^{-1}
Na^+	468.5	10.77
K^+	10.21	0.399
Mg^{2+}	53.08	1.290
Ca^{2+}	10.28	0.4121
Sr^{2+}	0.090	0.0079
Cl^-	545.9	19.354
Br^-	0.842	0.0673
F^-	0.068	0.0013
HCO_3^-	2.30	0.140
SO_4^{2-}	28.23	2.712
$B(OH)_3$	0.416	0.0257

TABLE 8.7 Average Concentration of the 11 Most Abundant Ions and Molecules in Clean Seawater (35.00‰ Salinity, Density of 1.023 g mL^{-1} at 25°C)

varying degrees present in the form of complexes. About 13 percent of the magnesium and nine percent of the calcium in ocean waters exist as magnesium sulfate and calcium sulfate, respectively. More than 90 percent of the carbonate, 50 percent of the sulfate, and 30 percent of the bicarbonate exist as complexes. Many minor or trace components occur primarily as complex ions at the pH and the redox potential of seawater. Boron, silicon, vanadium, germanium, and iron form hydroxide complexes. Gold, mercury, and silver, and probably calcium and lead, form chloride complexes. Magnesium produces complexes with fluorides to a limited extent.

Seawater is normally more corrosive than fresh water because of the higher conductivity and the penetrating power of the chloride ion through surface films on a metal. The rate of corrosion is controlled by the chloride content, oxygen availability, and the temperature. The 3.5 percent salt content of seawater produces the possibly most corrosive chloride salt solution (Fig. 8.9) [10]. The combination of high conductivity and oxygen solubility is at a maximum at this point (as shown in Table 8.6, the solubility of oxygen is reduced with increasing salt concentration). The corrosion of numerous metals in a wide range of saline waters is reported in Table 8.8.

Salinity was defined, in 1902, as the total amount of solid material (in grams) contained in 1 kg of seawater when all halides have been replaced by the equivalent of chloride, when all the carbonate is converted to oxide, and when all organic matter is completely oxidized. The definition of 1902 translates in Eq. (8.1),

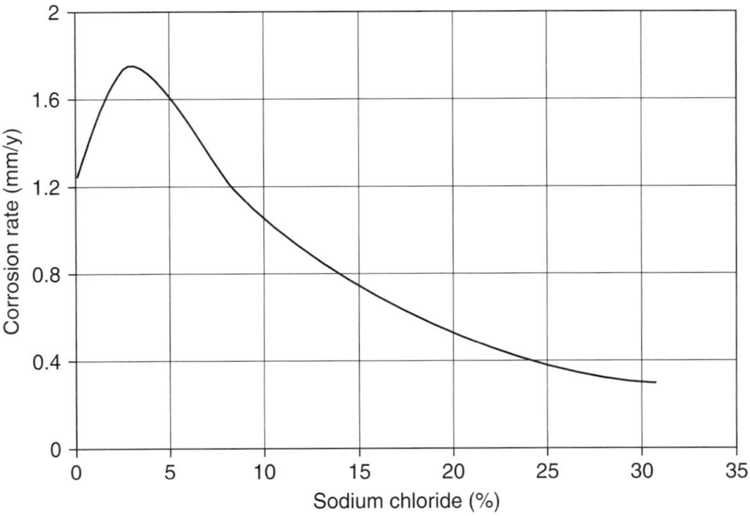

FIGURE 8.9 Corrosion of steel in various sodium chloride solutions. Note the peak in corrosion rate at approximately 3 percent sale.

where the salinity (S) and chlorinity (Cl) are expressed in parts per thousand (‰).

$$S\,(‰) = 0.03 + 1.805\ Cl\,(‰) \tag{8.1}$$

The concern over the fact that the 1902 equation gives a salinity of 0.03‰ for zero chlorinity led the creation of a special United

Alloy	Deepest Pit (mm)	Average Corrosion Rate (mm y^{-1})		
	Quiet Seawater		Flowing Seawater	
			8.2 m s^{-1}	35–42 m s^{-1}
Carbon steel	2.0	0.075	Nil	4.5
Grey cast iron (graphitied)	4.9	0.55	4.4	13.2
Admiralty gunmetal	0.25	0.027	0.9	1.07
85/5/5/5 Cu Zn Pb Zn	0.32	0.017	1.8	1.32
Ni resist cast iron type 1B	Nil	0.02	0.2	0.97
Ni Al bronze	1.12	0.055	0.22	0.97
70/30 Cu Ni + Fe	0.25	< 0.02	0.12	1.47
Type 316 stainless steel	1.8	0.02	< 0.02	< 0.01
6% Mo stainess steel	Nil	0.01	< 0.02	< 0.01
Ni-Cu alloy 40	1.3	0.02	< 0.01	0.01

TABLE 8.8 Effect of Velocity on the Corrosion of Metals in Seawater

Nations Scientific, Education, and Cultural Organization (UNESCO) committee to determine a more precise relation between chlorinity and salinity. The definition of 1969 produced by UNESCO is written in Eq. (8.2):

$$S\,(‰) = 1.80655\,Cl\,(‰) \tag{8.2}$$

The definitions of 1902 and 1969 give identical results at a salinity of 35‰ and do not differ significantly for most applications. The definition of salinity was reviewed again when techniques to determine salinity from measurements of conductivity, temperature, and pressure were developed. The *Practical Salinity Scale* defined in 1978 is a complex function related to the ratio (K) of the electrical conductivity of a seawater sample to that of a potassium chloride (KCl) solution with a mass fraction in KCl of 0.0324356, at the same temperature and pressure.

$$\begin{aligned} S = {} & 0.0080 - 0.1692K^{0.5} + 25.3853\,K \\ & + 14.0941\,K^{1.5} - 7.0261\,K^{2} + 2.7081\,K^{2.5} \end{aligned} \tag{8.3}$$

Note that ‰ is no longer used in this definition. In fact, a value of 35‰ would simply correspond to a value of 35 on the *Practical Salinity Scale*.

Precipitation of Calcareous Deposits. The natural presence of calcium and magnesium in seawater has been advantageously used to coat internal walls of vessels such as ballast tanks with a protective film of calcareous deposits. These films, once they are formed by the cathodic polarization of metal surfaces in seawater, greatly reduce the current density needed to maintain a prescribed cathodic potential. For most cathodic surfaces in aerated waters, the main reduction reaction is described by Eq. (8.4):

$$O_2 + 2\,H_2O + 4e^- \rightarrow 4OH^- \tag{8.4}$$

In cases where the potential is more negative than the reversible hydrogen electrode potential, the production of hydrogen as described in Eq. (8.5) becomes possible:

$$2H_2O + 2e^- \rightarrow H_2 + 2OH^- \tag{8.5}$$

In either case, the production of hydroxyl ions results in an increase in pH for the electrolyte adjacent to the metal surface. This situation causes the production of a pH profile in the diffuse layer where the equilibrium reactions can be quite different from the bulk seawater conditions. Temperature, relative electrolyte velocity and electrolyte composition will all influence this pH profile. In seawater, pH is controlled by the carbon dioxide system described in the following Eqs. (8.6) through (8.8).

$$CO_2 + H_2O \rightarrow H_2CO_3 \quad (8.6)$$

$$H_2CO_3 \rightarrow H^+ + HCO_3^- \quad (8.7)$$

$$HCO_3^- \rightarrow H^+ + CO_3^{2-} \quad (8.8)$$

If OH^- is added to the system as a consequence of one of the above cathodic processes, then the reactions described in the following Eqs. (8.9) and (8.10) become possible, with Eq. (8.11) describing the precipitation of a calcareous deposit.

$$CO_2 + OH^- \rightarrow HCO_3^- \quad (8.9)$$

$$OH^- + HCO_3^- \rightarrow H_2O + CO_3^{2-} \quad (8.10)$$

$$CO_3^{2-} + Ca^{2+} \rightarrow CaCO_3(s) \quad (8.11)$$

The reactions represented by Eqs. (8.6) through (8.11) further indicate that as hydroxyl ions (OH^-) are produced, then reactions in Eqs. (8.7) and (8.8) are displaced to the right, resulting in proton production. This opposition to the rise in pH accounts for the buffering capacity of seawater. Irrespective of this, however, these reactions indicate that the buffering action is accompanied by the formation of calcareous deposits on cathodic surfaces exposed to seawater as shown in Eq. (8.11). It is the main reason why the behavior of $CaCO_3$ in seawater has been so extensively studied. Calcium carbonate sediments are prevalent and widespread in the oceans in two crystalline forms, that is, calcite and aragonite[12].

Magnesium compounds, $Mg(OH)_2$ in particular, can also contribute to the protective character of calcareous deposits. However, calcium carbonate is thermodynamically stable in surface seawater, where it is supersaturated, while magnesium hydroxide is unsaturated and less stable. In fact, $Mg(OH)_2$ would precipitate only if the pH of seawater was to exceed a pH threshold of 9.5.

Since calcite and magnesium carbonate have similar structures, these compounds tend to form solid solutions, the Ca:Mg ratio of these compounds depend on the ratio of these ions in seawater. Theoretical calculations suggest that calcite in equilibrium with seawater should contain between 2 and 7 mol percent $MgCO_3$. But although low magnesium calcite is the most stable carbonate phase in seawater, its precipitation and crystal growth are strongly inhibited by dissolved magnesium. Consequently aragonite is the phase that actually precipitates when seawater is made more basic. The degree of saturation for aragonite is described in Eq. (8.12):

$$K_{sp,\ aragonite} = (Ca^{2+})(CO_3^{2-}) \quad (8.12)$$

Where (Ca^{2+}) and (CO_3^{2-}) are the molalities of the Ca^{2+} and CO_3^{2-} ions, respectively, and $K_{sp,\,aragonite}$ is the solubility product of aragonite (at 25°C $K_{sp,\,aragonite} = 6.7 \times 10^{-7}$)

In order to understand the buildup of carbonate ions at a metallic surface under cathodic protection (CP), one can consider the simplified electrochemical production of carbonate ions described in Eq. (8.13) that summarizes the effect of CP in the presence of dissolved carbon dioxide.

$$H_2O + CO_2 + 2e^- \rightarrow H_2 + CO_3^{2-} \tag{8.13}$$

An expression for the limiting current corresponding to this reaction is described in Eq. (8.14).

$$i_L = nFD_{CO_3^{2-}} \frac{C_{CO_3^{2-}}^{surface} - C_{CO_3^{2-}}^{bulk}}{\delta} \tag{8.14}$$

where $D_{CO_3^{2-}}$ is the diffusion coefficient of carbonate ions and δ the thickness of the diffuse layer as illustrated in Fig. 8.10.

At neutral bulk pH, the concentration of carbonate ions in seawater is basically zero, and the expression of i_L simplified as described in Eq. (8.15).

$$i_L = nFD_{CO_3^{2-}} \frac{C_{CO_3^{2-}}^{surface}}{\delta} \tag{8.15}$$

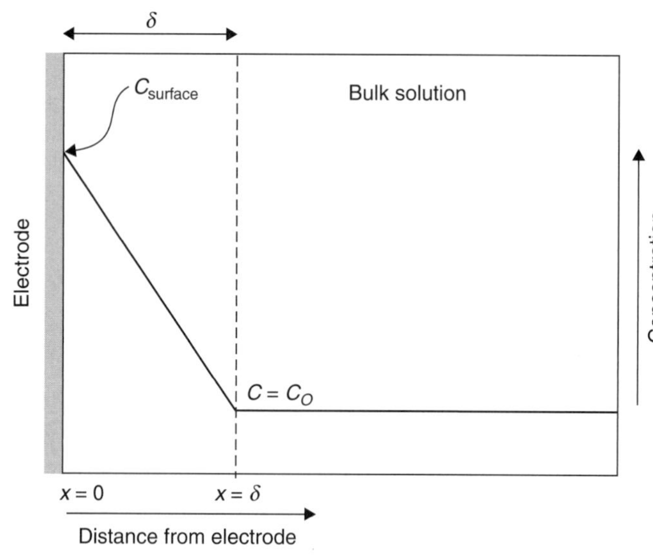

FIGURE 8.10 Schematic of a diffuse layer concentration buildup when a chemical species is produced at an electrode surface.

Calculation Example. For this problem, imagine a situation where you would have to design a sacrificial CP system for the protection of the four main pillars of an oil-drilling platform. These pillars are basically sealed steel cylinders, 1 m in diameter and 25 m long in their immerged section. The sacrificial anodes are zinc bars 100 cm long, 12 cm wide and 12 cm thick, with their back tightly screwed onto the steel surface. This problem is a particular application of the blocking or passivation of an electrode surface by forcing the accumulation of some ions that can eventually precipitate and block the surface.

Answer the following questions knowing that the temperature of the seawater decreases exponentially as a function of depth, changing from 25°C at the surface to 10°C at a depth of 25 m and that the diffusion coefficient of CO_3^{2-} can be described by the following expression:

$$D_t = D_{25°C}(1 - 0.043 \times (25 - t)) \qquad (8.16)$$

where D_t is the diffusion coefficient of CO_3^{2-} at temperature t
t is the temperature in °C
$D_{25°C}$ is the diffusion coefficient of CO_3^{2-} at 25°C (5×10^{-5} cm_2 s^{-1})

Question 1: Given that the corrosion rate of unprotected steel is 1.1 mm y^{-1}, estimate the total corrosion current for one pillar.

Answer
Surface area of one pillar = π × diameter × length = 3.1416 × 1 × 25 m^2
Surface area of one pillar = 78.54 m^2
Corrosion current = surface area × corrosion rate in current density units
Since 1 mm y^{-1} = 0.0863 mA cm^{-2} or 0.863 A m^{-2}
Corrosion current = 1.1 × 0.863 × 78.54 = 74.56 A

Question 2: Given that the zinc anodes can provide a sacrificial current corresponding to a corrosion rate of zinc of 7 mm y^{-1}, evaluate the number of anodes that would be required to reduce the corrosion of steel by a factor of ten. Assume that the total corrosion current calculated in Question 1 remains the same to balance a constant cathodic process, the reduction of oxygen.

Answer
Surface of each anode = face + sides + ends
Surface of each anode = (12 × 100) + 2 × (12 × 100) + 2 × (12 × 12) = 3888 cm^2
Conversion of corrosion rate into current units: 1 mm y^{-1} = n × density × 0.306/M
For zinc, n = 2, density = 7.133 g cm^{-3}, M = 65.38 g mol^{-1}
Corrosion current density of anode = 7 × 2 × 7.133 × 0.306/65.38 = 0.4674 mA cm^{-2}
Corrosion current/anode = 0.4574 × 3888 = 1.817 A
The current required is 90 percent of the current calculated in Question 1 or 0.9 × 74.56 A, that is, 67.1 A
Number of anodes required = 67.1/1.817 = 36.9 anodes

Question 3: In order to reduce the consumption of anodes over time you would like to force the precipitation of calcareous deposits onto your steel surface because you know that by doing so you can cut down the corrosion of steel a hundredfold. Knowing that the level of Ca^{2+} ions in seawater is 0.01 molal and that the diffuse layer is approximately 5 μm thick, calculate:

a) The current density that would be required to force the precipitation of insoluble aragonite onto the steel close to the water line.

Answer

One has to use Eq. in which $n = 2$, $F = 96485$, $D = 5 \times 10^{-5}$ cm^2 s^{-1}, and $\delta = 5 \times 10^{-4}$ cm.

The concentration of CO_3^{2-} ions C can be obtained with the solubility product of aragonite K_{sp} described in Eq. (8.12).

$$C_{CO_3^{2-}} = \frac{K_{sp}}{C_{Ca^{2+}}} = \frac{6.7 \times 10^{-7}}{0.01} = 6.7 \times 10^{-5} \text{mol kg}^{-1} = 6.7 \times 10^{-8} \text{mol cm}^{-3}$$

Hence the limiting current density when the concentration of carbonate ions is high enough to precipitate with the calcium ions is 1.29 10^{-3} A cm^{-2} or 1.29 mA cm^{-2}

b) The number of sacrificial anodes/m^2 required to provide this initial protective current.

Answer

For a 1 m^2 surface the required current would be $10\,000 \times 1.29 \times 10^{-3}$ A = 12.93 A

Each anode provides 1.817 A

The number of anodes required is therefore = 12.93/1.817 = 7.1 anodes

c) What would be the impact on the current density requirements of attempting to deposit aragonite in an agitated sea?

Answer

Agitation would force the limiting current to increase and the number of anodes to do the same.

Question 4: Calculate the current density required to precipitate aragonite at the bottom of the pillar. Assume that the K_{sp} of aragonite has not changed.

Answer

Only the diffusion coefficient for the carbonate ions would change.

Since $D_t/D_{25°C} = 0.355$

The new limiting current density would therefore be $1.29 \times 0.0355 = 0.459$ mA cm^{-2}.

8.3.2 Treated Waters

Potable Water

Potable water is fresh water that is sanitized with oxidizing biocides such as chlorine or ozone to kill bacteria and make it safe for drinking purposes. In general, this is done by public water utilities that are responsible for the treatment and distribution of water to communities. Although in developed countries less than 1 percent of potable water is consumed, all water delivered to homes for food preparation, bathing, washing, watering gardens, heating, and cooling is treated to potable standards.

Water has been treated for thousands of years, but only when the connection between bacteria in sewage and severe epidemics was made were procedures developed for safe water in the nineteenth century [8]. Today the World Health Organization (WHO), European Union (EU), and the United States Environmental Protection Agency (U.S. EPA), along with many other organizations, have a well-defined set of limits and standards for microorganisms and toxic substances in drinking water.

The resistivity of potable water is usually between 1000 and 5000 Ω cm. By definition, certain mineral constituents are also restricted. For example, the chlorinity will be not more than 250 ppm chloride ion in the United States or 400 ppm on an international basis.

Distilled or Demineralized Water

The total mineral content of water can be removed by either distillation or mixed-bed ion exchange. The level of purification may be described qualitatively (e.g., triple-distilled water). However, the most accepted description for both distilled and demineralized water is in terms of its specific conductivity or resistivity. Demineralization actually removes the dissolved minerals and represents the most efficient treatment of hard water. The two most common and cost-effective methods are deionization and reverse osmosis.

Deionization removes minerals by passing water through a mixed resin bed in which two ion exchange resins selectively remove both cations and anions. Cations are replaced by hydronium ions and anions are replaced by hydroxyl ions. Mixed-bed units are extremely effective in reducing hardness levels to almost zero ppm. Such water exhibits a zero ECA* value (neutral ionic charge). Resin beds need to be flushed or backwashed on a regular basis to remove all contaminants and to prevent microbiological interference. Beds also need to be regenerated on a regular basis.

Reverse osmosis forces water through a semipermeable membrane under high pressure and varying flow conditions. This process can remove as much as 95 percent of dissolved minerals. Reverse osmosis units are often used in conjunction with a water softener pretreatment stage to reduce the demand on the osmosis process itself. Filters need to be flushed and replaced as necessary. Microbiological contamination also needs to be regularly monitored.

Steam Condensate

Steam condensate produced from industrial steam approaches distilled water in purity, except for contamination by dissolved gases and the deliberate presence of additives (e.g., neutralizing or filming amines). Corrosion in steam condensate systems is

---

* ECA (electrokinetic charge) is an indication of the charge strength of the water (degree of anionic or cationic charge).

frequently a severe problem mostly caused by dissolved carbon dioxide (CO_2) and aggravated by the presence of DO. Carbon dioxide is produced in the boiler as a result of thermal breakdown of the natural alkalinity, principally bicarbonate ions, which enter with the feedwater. The following reactions are produced at boiler operation temperatures:

$$2NaHCO_3 \rightarrow Na_2CO_3 + CO_2(g) + H_2O \tag{8.17}$$

$$Na_2CO_3 + H_2O \rightarrow 2NaOH + CO_2(g) \tag{8.18}$$

The conversion of bicarbonate alkalinity to gaseous carbon dioxide in Eq. (8.19) is a function of boiler temperature, pressure, and residence time. Because it is extremely volatile, the carbon dioxide exits the boiler with the steam produced. At points of condensation, some fraction of the carbon dioxide present in the steam dissolves in the condensate, forming carbonic acid that hydrolyzes into hydrogen ions:

$$CO_2(g) + H_2O \rightarrow H_2CO_3 \rightarrow H^+ + HCO_3^- \tag{8.19}$$

The hydrogen ions cause acidic corrosion of both iron and copper alloy surfaces in the steam condensate system. The simplified corrosion reaction for iron is

$$Fe(s) + 2H_2CO_3 \rightarrow Fe\,(HCO_3)_2 + H_2(g) \tag{8.20}$$

The $Fe(HCO_3)_2$ formed in this redox reaction is relatively soluble, dissociated and its formation is competing with the formation of insoluble iron(II) oxides such as $Fe(OH)_2$ or colloidal associations such as $Fe\,(OH)_n^{n-2}$ [13].

Dissolved oxygen may be another major cause of condensate system corrosion. Oxygen contamination of steam condensate can occur due to inefficient or improper feedwater deaeration, air leakage at pump seals, receivers and flanges, leaking heat exchangers, and ingress into systems that are under vacuum. In the presence of oxygen, in addition to providing another possible cathodic reaction to pair with the iron oxidation, one more oxidative step is possible as shown in Eq. (8.21). This reaction releases carbon dioxide which makes the process self-perpetuating.

$$4Fe(HCO_3)_2 + O_{2(\text{dissolved})} \rightarrow 2Fe_2O_3(s) + 4H_2O + 8CO_2(g) \tag{8.21}$$

Excessive corrosion of the condensate system can lead not only to costly equipment failure and increased maintenance costs, but can also cause deposition of metal oxide corrosion products on boiler heat transfer surfaces if the condensate is recovered as feedwater. Metal oxide deposition on boiler heat transfer surfaces will result in lower fuel to steam efficiency and higher fuel costs. The deposition may also lead to tube failure due to long-term overheating.

8.4 Cooling Water Systems

Most of the water used industrially removes heat from production processes. It is one of the major applications for water and is a major factor in siting plants and processes. Cooling systems suffer many forms of corrosion and failure. The diversity of attack stems from wide differences in cooling water system design, temperature, flow, water chemistry, alloy composition, and operation. An almost endless variation of process stream chemistries may be involved in cooling water systems. Refinery and chemical process industries can employ hundreds of heat exchangers at a single plant, each with a different process stream chemistry [14].

The three basic types of cooling water systems are once-through, closed recirculating (nonevaporative), and open recirculating (evaporative). True closed systems neither lose nor gain water during service. Open systems, however, must have water added to make up for losses. Open recirculating systems employing cooling towers and spray ponds allow the dissipation of enormous heat loads while limiting the amount of water consumed.

In closed systems (nonevaporative) where water loss is low, the total waterborne material entering the system is limited. Thus, deposited minerals accumulate at a much slower rate than in systems in which large amounts of makeup water are added. Open recirculating (evaporative) and once-through systems are exposed to large quantities of solutes, suspended solids, and biological materials. As a consequence, fouling and associated corrosion are generally more significant in open systems than in true closed systems.

8.4.1 Once-Through Systems

Traditionally, industry has tended to develop in areas with an adequate supply of cooling water. Originally, it was sufficient to pipe water through the plant and discharge it back to its natural source. Only nominal attention was paid to control of water chemistry, and it is in fact economically ridiculous to attempt the chemical treatment of large volumes of once-through water.

However, even such minor additions as the "threshold" treatment with 1 to 2 ppm sodium hexametaphosphate, for example, would be unacceptable to modern environmental standards. Many countries, states, or provinces now forbid the return to source even when the concentrations of natural constituents were higher in the intake water than in the return water. In some areas, thermal pollution is forbidden since the discharge of the same water at a higher temperature than the inlet temperature may be harmful to certain marine species (e.g., oyster beds).

The consensus today for once-through cooling systems seems to be that, where permitted, the materials of construction must be chosen to be resistant to the water, be it fresh water or seawater. All natural waters must be presumed to be corrosive to iron and steel by virtue of aeration. River waters which were once noncorrosive have become corrosive as they are sufficiently cleaned to provide a suitable environment for game fish. Biocidal control is likewise inimical. Treatments against bacteria are incompatible with environmental considerations, unless the effluent is treated for their removal, a somewhat very expensive proposition in most cases.

8.4.2 Recirculated Systems

Closed

Closed recirculated systems are characterized by an essentially permanent charge of water, greatly facilitating the selection of chemical control and permitting the use of relatively large amounts of chemical additives. Typical closed recirculated systems are automobile radiator cooling systems, which are air cooled, and engine jacket cooling systems, which may either be air cooled or have the sensible heat removed by exchange with another type of cooling water system. In some locations, plants may be cooled with a closed loop of treated fresh water, and the fresh water in turn cooled in large exchangers carrying a once-through seawater coolant.

All-weather ethylene glycol-type solutions that automobile radiators require complex mixtures of additives, including corrosion inhibitors, stabilizers, and buffering agents. Without proper commercial inhibition, a 40 percent glycol solution at 70°C would corrode iron and steel at 250 to 500 μm/y, while also attacking the copper, brass, solder, and/or aluminum components at 25 to 50 μm/y.

Open

Open recirculated cooling water systems remove the heat picked up in plant by evaporative cooling. This may be done by a spray pond, for example, combining air conditioning needs with aesthetic considerations in industrial parks. The most common type of evaporative cooling, however, is effective in cooling towers of one type or another (Fig. 8.11).

Cooling towers may operate on natural draft, as in the case of wind-cooled towers for small home air-conditioning systems or the large concrete hyperbolic towers used in power generating stations (Fig. 8.12). In process plants, the towers are more often aided by fans, either forced or induced draft operations, to improve the cooling capacity (Fig. 8.13).

There are certain fundamental considerations which should be understood in relation to open recirculated systems. First is the concept of cycles of concentration. If three cups of boiling water in a tea kettle were allowed to boil away to one cup, the residual cup would contain a threefold concentration of soluble water salts, assuming that only pure steam was driven off. The water would be said to be at three cycles of concentration.

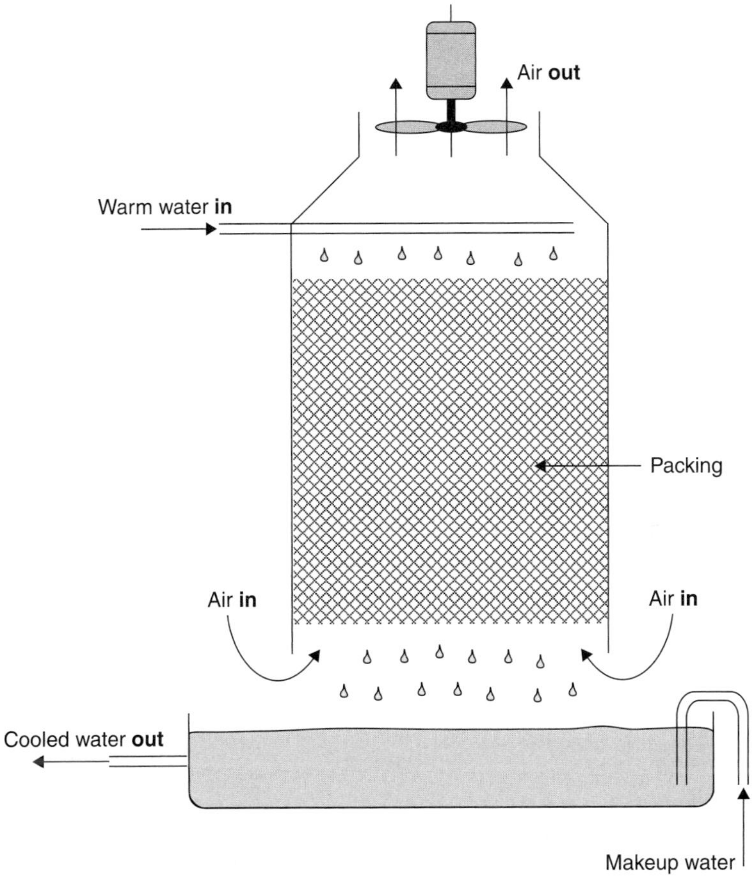

FIGURE 8.11 Schematic of a forced draught cooling tower.

To prevent this accumulation from becoming unacceptable from the standpoint of scale and corrosion, a small amount of blowdown (bleeding of the system) is maintained to control the number of cycles of concentration from evaporation. This means that makeup water must be added to equal the evaporation and blowdown losses, but this is a minor amount compared to the volume of the total system.

For example, if 19,000 Lpm* of cooling water was needed in a system, the cost for treatment in a once-through design would be excessive. However, in a recirculating system, the makeup may only be 380 Lpm, of which only 95 Lpm may need to be treated with inhibitors. This brings chemical treatment into the range of economic feasibility, as compared with a once-through system.

The limits imposed by water chlorinity and hardness on how far the water can be concentrated are such that the savings affected by a

* Liters per minute.

Figure 8.12 Natural draft cooling towers with their characteristic height and distinctive hourglass shape. (Courtesy of Russ Green, TMI)

Figure 8.13 Three cell induced draft counterflow cooling tower made of seamless double wall polyethylene construction. (Courtesy of Delta Cooling Towers)

recirculating system versus a once-through system are maximized at about 4 to 6 cycles of concentration. Below this range, treatment costs become prohibitive. At high cycles (e.g., 8 to 10), the additional water savings are not commensurate with the increased difficulty of effective treatment. If the blowdown is shut off entirely, there is still an effective upper limit of concentration dictated by water losses from drift or windage. The normal upper limits might be about 20 to 22 cycles of concentration for a mechanical-draft tower.

The advantages of water savings affected by the cooling tower also impose certain inherent disadvantages. The water becomes air saturated, ensuring its full corrosion potential, its natural alkalinity tends to increase and aggravate scaling tendencies. The air scrubbing action can contaminate the water with airborne materials, notably dust fines, which form silt in the tower basin, and spores of slime, algae, and fungi that can reproduce in the warm nutrient water of the system.

8.4.3 Heat Exchangers

Heat is removed from exothermic processes, hot gases, and liquids, and to control operating temperatures through heat exchangers cooled with water. Shell and tube heat exchangers consist of a bundle of tubes connected to tube sheets which are then installed into a shell (Fig. 8.14). Tube bundles can be parallel with once-through flow from

(a)

Figure 8.14 Shell and tube heat exchangers being dismantled (*a*) and close-up of heat exchangers head (*b*). (Courtesy of Kingston Technical Software).

(b)

FIGURE 8.14 (*continued*)

one end to the other, or with U-tube in which the hot and cooled fluid enter and exit from the same end of the exchanger.

The selection of materials for these applications is often a compromise between the requirements of the process flow and the type of water. Associated with such heat exchangers are pumps, pipes, and valves to distribute the water and return it to source. The various metals commonly used in heat exchangers have quite different thermal conductivities (Table 8.9). However, the thermal conductivity of the metal wall is only one component of the resistance to heat transfer in a heat exchanger tube. In a condenser (i.e., where steam is condensing on cold tubes), for example, the resistance to heat transfer through a tube wall is made up of five main components as illustrated in Fig. 8.15 [8]. The tube wall resistance is comparatively small so that changes in thermal conductivity from the use of different metals in not necessarily very significant.

Most metals used in heat exchangers perform well in clean water, that is, free of sediment, debris, fouling organisms, and pollutants.

Alloy	Thermal Conductivity (W m^{-1} K^{-1})
Admiralty metal	111
Aluminum brass	100
Aluminum bronze	79.5
90/10 Cu/Ni	44.9
70/30 Cu/Ni	29.4
S30400	15.0
N08367	11.0
S44735	17.0
S44660	17.0
Commercially pure titanium	21.6

Table 8.9 Thermal Conductivity of Various Alloys Used in Heat Exchangers

Sediment or debris can cause underdeposit corrosion or turbulence that can damage or remove the protective film, particularly on the less resistant copper-based alloys. Effective screening or filtering can limit this problem. Copper alloys are, in general, better at resisting the attachment of organisms than stainless steels or nickel alloys.

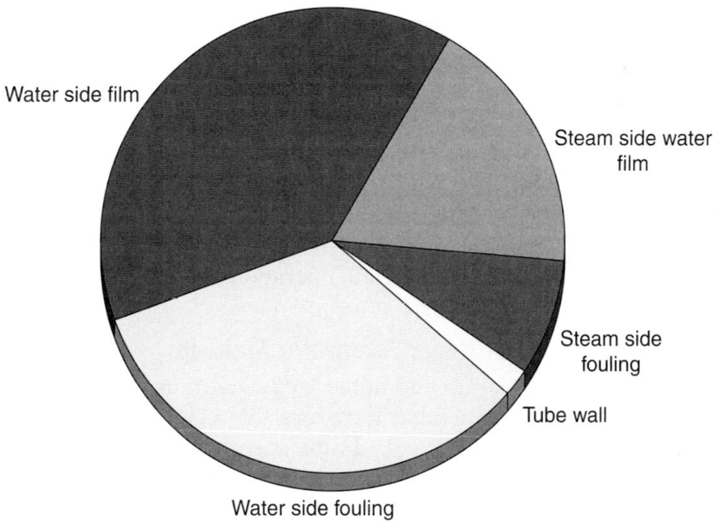

Figure 8.15 Breakdown of the components making up the resistance to heat transfer in a heat exchanger tube.

Chlorides in the water can cause corrosion, particularly in the case of stainless steels. Typically 304 is satisfactory up to about 200 ppm chlorides, while 316 can withstand around 1000 ppm and 4.5 percent molybdenum austenitic stainless steels and duplex stainless steels are known to have suffered from crevice attack at 2000 to 3000 ppm beneath fouling. Titanium and the six percent molybdenum stainless steels have been shown to resist crevice attack in seawater (1900 ppm chlorides) under deposits.

Operational problems can cause corrosion of stainless steel and copper alloy heat exchangers. Stagnant water left from hydrotesting or from inadequate draining at shutdown can cause fouling and/or microbiologically influenced corrosion (MIC). The design and production of the tube to tubesheet joint are critical factors in the successful use of shell and tube heat exchangers. Roll-leaks can be caused by tubes being pulled out of the tubesheet under the action of thermal stresses. This is particularly so where different metals with different expansion coefficients are used for the tubes and the tubesheet. This effect can be prevented by using an expansion joint in the shell. Properly designed and executed seal or strength welds may prevent leakage at this joint. Seamless, as-welded, bead-worked (locally cold-worked) or cold-drawn joints are equally likely to perform well in most services. In critical applications or where localized corrosion may be initiated by surface defects, adequate inspection and testing are recommended regardless of method of tube manufacture.

8.5 Steam Generating Systems

The greatest use of high-temperature water and steam is in electrical power generation. Historically, fossil fuels (i.e., wood, coal, gas, and oil) were used almost exclusively to heat water and make steam until the introduction of nuclear-power steam generators in the second part of the twentieth century. The two types of power plants have much in common, but are sufficiently different to be discussed separately. Both, however, presuppose technically advanced water treatment and control for successful operation.

8.5.1 Treatment of Boiler Feedwater Makeup

Boiler feedwater make-up and boiler feedwater must be softened to prevent scaling and deaerated to reduce the water corrosivity. The extent of the treatment depends on the specific requirements based on a boiler operating temperature and pressure ranges.

A number of lime-softening treatments were used in the past, but these have given way to more sophisticated treatments. Probably the most common for boilers up to 2.8 to 4.0 MPa is Zeolite softening. In this treatment, a sodium salt of a long-chain polymeric organic molecule comprises the ion exchange bed. As the feedwater passes through the bed, sodium ions are exchanged for the calcium and magnesium ions which comprise the water hardness. Other cations are also exchanged. The water then exits from the Zeolite bed softened and at a higher pH than it entered. The beds are regenerated intermittently with sodium chloride solution to backwash the hardness salts and reform the sodium salt.

For purer water quality, the water may be totally demineralized by mixed beds of polymeric resins which exchange in turn hydrogen ions for all cations and hydroxyl ions for all anions, effectively producing pure H_2O from a raw water stream. Such highly purified waters are required for boilers operating from about 6 MPa and for all nuclear boilers (to avoid radioactive half-life of water-borne salts).

The softened water is now in its most corrosive state, being still saturated with DO and having no hardness to form a protective scale. For reasons of economy, the removal of DO is usually first effected in part by thermomechanical deaeration. The boiler feedwater is preheated, then flashed in a deaerator to remove any free carbon dioxide and most of the dissolved oxygen. Then the last traces of DO are chemically scavenged with sodium sulfite (Na_2SO_3) that reacts with oxygen to form sodium sulfate (Na_2SO_4) as described in Eq. (8.22):

$$2\,Na_2SO_3 + O_2 \rightarrow 2\,Na_2SO_4 \tag{8.22}$$

or, if a complete absence of solid precipitates is desired, with hydrazine (N_2H_2) that forms water and gaseous nitrogen in the presence of oxygen:

$$2\,N_2H_2 + O_2 \rightarrow 2\,H_2O + 2\,N_2(g) \tag{8.23}$$

Both sulfite and hydrazine are available in catalyzed form to promote more rapid reaction rates. A boiler designed to be operated with a catalyzed oxygen scavenger must never be operated on the uncatalyzed grades, or severe corrosion will be encountered in the economizers or even the steam drum.

A final step in boiler feedwater treatment consists of pH adjustment as a further aid to corrosion control. Usually the pH is adjusted to a range of 10 to 11 with trisodium phosphate (or combinations of caustic with sufficient mono- or disodium phosphate to form trisodium phosphate upon inadvertent evaporation of the water). This "coordinated phosphate" treatment is intended to preclude the environmental cracking of steel by free sodium hydroxide (caustic embrittlement), a catastrophic form of corrosion described in Chap. 6. Caustic carryover with the steam can present severe corrosion problems (Fig. 8.16).

Nuclear requirements are such that zero solids treatment is required, precluding the addition of sodium salts and necessitating the use of ammonium hydroxide for pH adjustment.

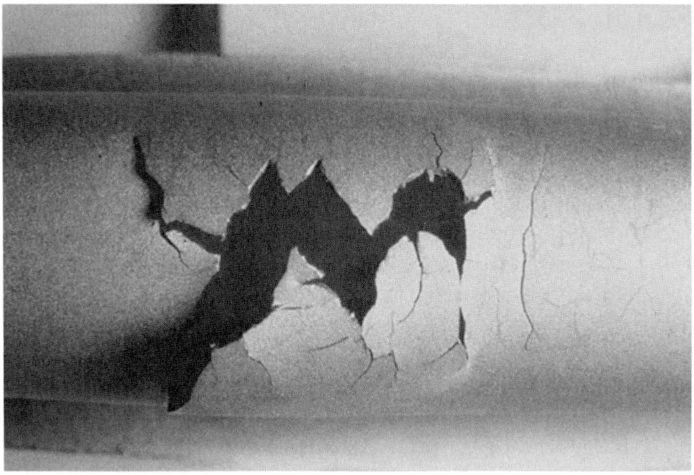

FIGURE 8.16 Type 321 stainless steel expansion joint in 2.8 MPa (400 psi) steam service cracked from caustic carryover in the steam. (*Corrosion Basics: An Introduction*, 2nd edn., NACE International, by permission)

8.5.2 Fossil Fuel Steam Plants

Conventional steam generation plants or more modern cogeneration steam plants (Fig. 8.17) consist of many sections operating at different temperatures and pressures:

- Feedwater heaters with water inside and steam outside the tubes
- A boiler (water inside the tubes, hot combustion product gases outside), where water is heated to high temperatures under pressure and is sometimes flashed to steam
- A steam drum, wherein steam is formed from water and water is separated from the steam (note: the steam drum is omitted in certain once-through systems)
- A superheater, where the steam is further heated to even higher temperatures
- A turbine, where the steam expands against the vanes of a wheel to drive the turbine which generates electricity
- A condenser, where the low-pressure steam is condensed to water and returned to the feedwater heaters

In such plants, there are special corrosion problems in each of these sections. There are also special problems associated with exposure to hot combustion gases. The materials used in high-temperature steam and water include steels, stainless steels, and

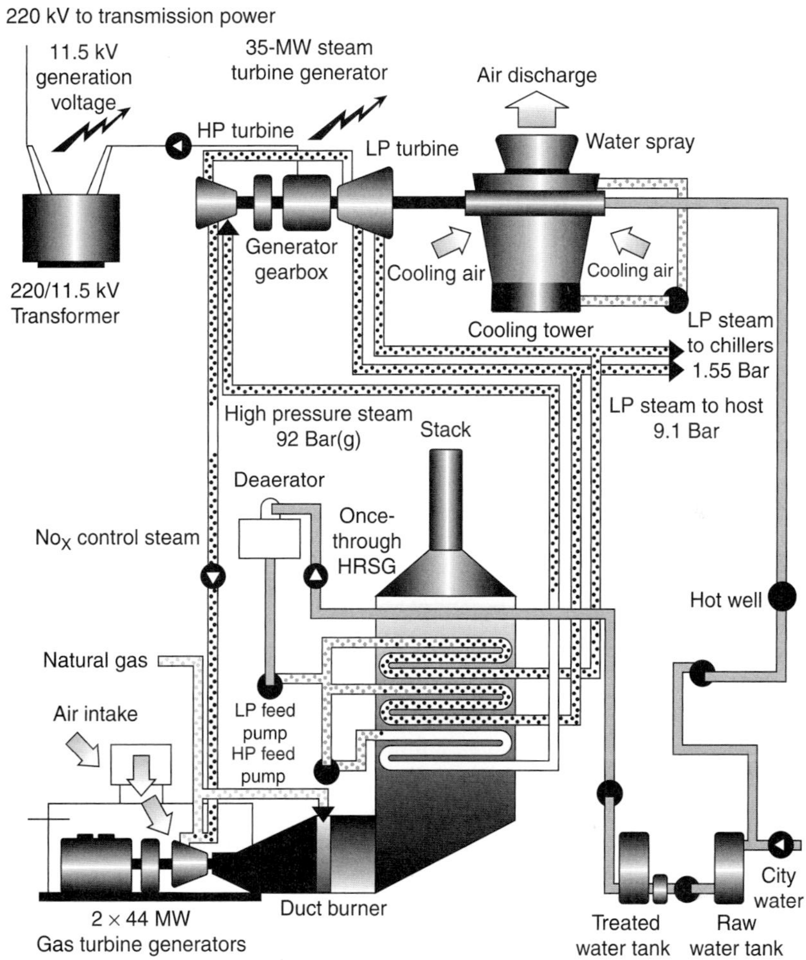

FIGURE 8.17 Schematic of a 123-MW natural gas-fueled cogeneration plant that produces electrical power and industrial use steam. (Courtesy of Mighty River Power, New Zealand)

nickel-base alloys. Copper-base alloys are employed in the intermediate and low temperature ranges. Aluminum alloys generally are not used because of their poor performance above about 200°C.

8.5.3 Supercritical Steam Plants

Supercritical steam plants operating above the water critical temperature (375°C) and pressure (22 MPa) have the same mechanical components as conventional steam plants. Major differences are greater wall thicknesses to withstand the pressure, more corrosion resistant materials, and lower dissolved solids in the water and steam.

Almost all of the components in a supercritical steam plant are made of austenitic stainless steels of the 18-8 variety, for example, S30403 or 31603. These materials are employed to minimize corrosion products and their transport through the system. The steam temperature is no higher than in an ordinary superheater, but the pressures are such that many chemicals and corrosion products may show appreciable solubility in the steam.

Many chemicals exhibit an inverse solubility when the temperature of the steam is above the critical temperature and will therefore deposit out at these higher temperatures. In one supercritical plant, about 140 ppm caustic was accidentally introduced into the plant. Within 30 minutes, caustic stress corrosion cracking (SCC) occurred in that part of the plant where the temperature was about 425°C. This temperature corresponds to the minimum in the caustic solubility-temperature curve. The supercritical steam also undergoes a marked density increase above this temperature range which could accelerate the deposition of chemicals.

For such reasons, the dissolved solids content of the water must be kept as close as possible to zero. A large fraction of the water is continuously cleaned up in a bypass circuit containing ion exchange (demineralizer) beds. At start-up, 100 percent of the water may be passed through the cleanup beds. Chlorides and caustic are the most undesirable salts because they are known to cause SCC of the austenitic stainless steels. Dissolved oxygen may be reduced by the methods described previously to a few parts per billion.

Copper deposition on turbine blades was an early problem with supercritical units. It was found that trace amounts of copper were dissolved from the condenser tubes and recirculated into the boiler section. Because of its excellent solvent properties, the superheated steam carried the dissolved copper into the turbine where, at the lower pressure, copper was deposited upon the turbine blades. This not only affected the efficiency, but threatened to destroy the turbine by mechanical imbalance due to uneven deposition. The problem has been largely eliminated by using stainless steel or titanium in the condenser.

8.5.4 Waste Heat Boilers

In many chemical or petrochemical processes, economy dictates that superfluous exotherms be utilized to generate steam as an energy conservation measure. Examples are the cooling of a butane oxidation reaction or the condensation of hot sulfur vapors, steam pressure being generated on the shell side of specially designed heat exchangers.

Corrosion problems in waste heat boilers usually arise either from unusual materials of construction or from inattention to the required details of water treatment. Austenitic stainless steels may be required from the process-side corrosion aspects, yet be highly susceptible to SCC from boiler feedwater. The most frequent problem, however, is that the operating department personnel, whose primary concern is the manufacture of marketable products, simply do not give adequate technical attention to the details of boiler feedwater chemistry. Many chemical converters, catalytic converters, or other waste heat boilers have failed from steam-generating side corrosion due to such inattention.

8.5.5 Nuclear Boiling Water Reactors

In a nuclear boiling water reactor (BWR), the nuclear fuel boils the water and the steam goes directly to the turbine (Fig. 8.18). Temperatures are approximately 230 to 290°C, and with saturated steam from 2.8 to 7.2 MPa. As with conventional boilers, silica and copper content of the water must be kept low to prevent their transport by steam and deposition on the turbine. Additionally, high-purity neutral water must be used because of the possible deposition of solid chemicals on the fuel elements and the stripping of volatile components from any chemicals that would otherwise be used in water treatment.

Although the water and steam contain appreciable amounts of oxygen, carbon steels and stainless steels exhibit equivalent or superior corrosion resistance to that observed in pressurized water at the same temperatures. However, special problems do exist. Stainless steel fuel cladding has failed by SCC in boiler water reactors.

SCC has also occurred in sensitized stainless steels (i.e., those with chromium carbide precipitation at the grain boundaries) as a result of heating the steel in the 425 to 825°C range by welding or stress relieving. For this reason, Zircaloys are the preferred cladding materials. However, care must be exercised that alkaline components do not enter the water and become concentrated by boiling on the cladding surface because Zircaloys have poor corrosion resistance in high-temperature caustic solutions.

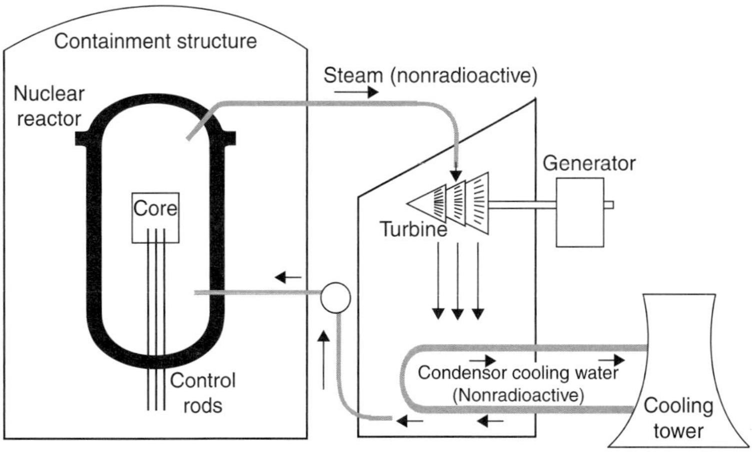

Figure 8.18 Boiling water reactor.

Boiling water reactor designs can be modified to accommodate a superheater to raise the steam temperature for improved efficiency. Experimental nuclear superheaters have been built to superheat the steam from the boiling water reactor. Corrosion rates for stainless steels and nickel alloys under heat transfer in the 600°C temperature range are about 25 μm/y. However, experience has shown that even minute amounts of entrained moisture in the entering steam can carry sufficient chlorides that cracking of the stainless steel cladding of the nuclear fuel elements may ultimately ensue. Alloys with higher nickel contents (e.g., N08800, N06600, or N06690) have resisted SCC under these conditions.

8.5.6 Nuclear Pressurized Water Reactors

In a pressurized water reactor (PWR) (Fig. 8.19), high-purity water is pumped past nuclear fuel elements where it becomes heated. The heated water then goes to the tube side of a heat exchanger, giving up its heat to boil water on the shell side. Steam is generated from the boiler water.

The water adjacent to the core is called primary water. It is usually treated with hydrazine to remove oxygen, with lithium hydroxide or ammonium hydroxide to pH 10 to 11 to minimize corrosion product transport, and with 25 to 50 cc hydrogen per kg of water to suppress radiolytic decomposition of water (thereby minimizing corrosion product transport).

The nuclear fuel must be clad with a corrosion-resistant material to prevent the release of radioactive gases and fission products to the primary water. The fuel itself is usually uranium oxide, which is quite resistant to the primary water. The fuel is clad with austenitic stainless steel or Zircaloy-2 (R60802) or the extra-low nickel Zircaloy-4 (R60804).

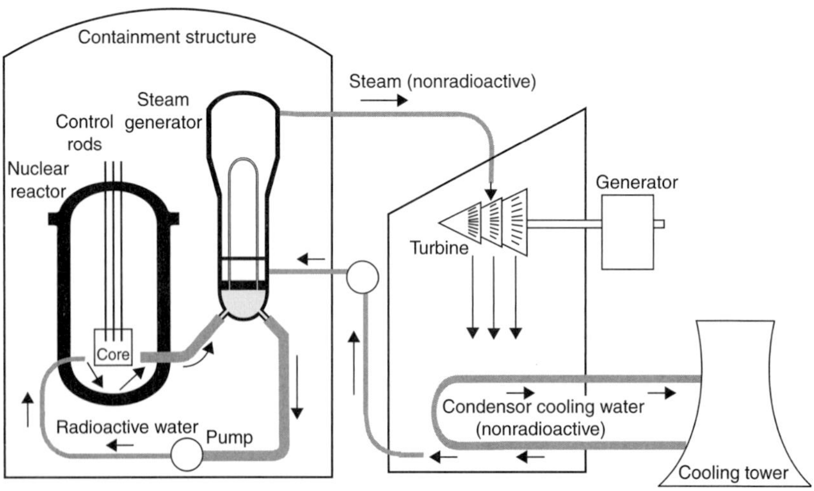

FIGURE 8.19 Pressurized water reactor.

Zircaloys are preferred because they do not "poison" the nuclear reaction as much as do the stainless steels. The Zircaloys perform well in the 285 to 315°C temperature of the primary water for the two to three-year expected life of a nuclear core. They develop a shiny, adherent black oxide film, which is protective and has excellent heat transfer properties. However, after long exposure times (two to four years), or shorter times at higher temperatures, for example, 40 to 120 days at 360 to 400°C, Zircaloys corrosion rate increases and a white, relatively nonadherent insulating film develops.

Stainless steels, on the other hand, develop a relatively thick tarnish film with a thin, powdery surface film. Both films have a nominal composition of M_3O_4 where *M* represents iron, nickel, or chromium. The corrosion rate of stainless steel is not greatly affected by temperature in the range of 260 to 400°C, and it does not exhibit the marked effect of temperature on corrosion rate that is characteristic of Zircaloys.

The piping, steam generator, and pressure vessel comprise much of the total exposed area in the primary section of a pressurized water reactor. The internal surface of the pressure vessel is clad with austenitic stainless steel. The piping and steam-generator components are primarily austenitic stainless steel or Alloy 600 (N06600). Carbon steel has been used in a few systems.

The corrosion rates of austenitic stainless steels and Alloy 600 are about the same, approximately 1.5 μm/y. The corrosion rate of carbon steels is 5 to 10 times higher, to a maximum of about 13 μm/y. These rates are acceptable from a structural standpoint. However, as much as half of this oxide may not remain on the surface, but becomes a radioactive "crud" in its passage through the reactor core. Its subsequent deposition constitutes a personnel hazard.

The isotopes in the corrosion products which contribute most to radioactivity are Co-60, Co-58, Fe-59, Mn-54, and Cr-51. These isotopes are formed from the elements in stainless and nickel-base alloys. Their half-lives* range from 27 days to more than five years. Co-60 has a long half-life, and this is the reason for minimizing cobalt content in nuclear-grade alloys.

It is generally agreed upon that pH in the range of 6 to 10, oxygen up to 5 ppm, or irradiation have little or no effect on the corrosion rate of stainless steels. However, these factors do affect the amount of crud released into the water, the amount increasing with decreasing pH, increasing oxygen content, and irradiation.

Since the primary water is pressurized, it does not boil. Traces of chloride contamination or caustic from the water treatment may therefore concentrate enough to cause SCC. The dissolved chloride content is usually restricted to less than 0.1 ppm. Flow rates are of the order of 7 to 10 m s^{-1}, and the temperature is about 260 to 290°C, with

---

* A half-life of a radioactive element is the period over which one-half of the initial activity decays.

water treatment as described for boiling water reactors. Most structural components are also fabricated from stainless steel with corrosion characteristics as described for the piping.

8.5.7 Corrosion Costs to the Power Industry

A detailed example of cost data for corrosion problems affecting the various broad sectors of the electric power industry has recently been published for the United States [15;16]. The compiled and sorted costs are presented in Table 8.10 with the highest cost listed first. These dollar amounts include corrosion-related costs associated with both operations and maintenance and depreciation. They also include both direct and indirect corrosion-related costs.

Corrosion Problem	Sector	$ million	%
Corrosion product activation and deposition	Nuclear	2205	18.80
Steam generator tube corrosion including IGA and SCC	Nuclear	1765	15.05
Waterside/steamside corrosion of boiler tubes	Fossil	1144	9.76
Heat exchanger corrosion	Fossil & nuclear	855	7.30
Turbine corrosion fatigue (CF) and SCC	Fossil & nuclear	792	6.75
Fuel clad corrosion	Nuclear	567	4.83
Corrosion in electric generators	Fossil & nuclear	459	3.91
Flow-accelerated corrosion	Fossil & nuclear	422	3.60
Corrosion of service water	Fossil & nuclear	411	3.51
Intergranular SCC of piping and internals	Nuclear	363	3.10
Oxide particle erosion of turbines	Fossil	360	3.07
Fireside corrosion of waterwall tubes	Fossil	326	2.78
Primary water SCC of non–steam-generator alloy 600 parts	Nuclear	229	1.95
Corrosion of concentric neutrals	Distribution	178	1.52
Copper deposition in turbines	Fossil	149	1.27
Fireside corrosion of superheater and reheater tubes	Fossil	149	1.27

TABLE 8.10 1998 Costs of Corrosion: Problems from All Power Industry Sectors [16]

Corrosion Problem	Sector	$ million	%
Corrosion of underground vault equipment	Distribution	142	1.21
Corrosion of flue gas desulfurization system	Fossil	131	1.12
Corrosion in valves	Nuclear	120	1.03
Liquid slag corrosion of cyclone boilers	Fossil	120	1.02
Back-end dewpoint corrosion	Fossil	120	1.02
Atmospheric corrosion of enclosures	Distribution	107	0.91
Hot corrosion of combustion turbine (CT) blades and vanes	Combustion turbine	93	0.79
Boric acid (H3BO3) corrosion of CS and low-alloy steel parts	Nuclear	93	0.79
Irradiation-assisted SCC of reactor internals	Nuclear	89	0.76
Corrosion in pumps	Nuclear	72	0.61
Corrosion of tower footings	Transmission	45	0.38
Hot oxidation of CT blades and vanes	Combustion turbine	35	0.30
Corrosion of boiling water reactor control blades	Nuclear	32	0.27
Corrosion of anchor rods	Transmission	27	0.23
Corrosion of tower structures	Transmission	27	0.23
Heat recovery steam generator (HRSG) CF	Combustion turbine	20	0.17
Conductor deterioration	Transmission	18	0.15
HRSG flow-accelerated corrosion	Combustion turbine	10	0.09
HRSG underdeposit corrosion	Combustion turbine	10	0.09
Corrosion of CT compressor section	Combustion turbine	9	0.08
Corrosion of CT exhaust section	Combustion turbine	9	0.08
Corrosion of splices	Transmission	9	0.08
Corrosion of shield wires	Transmission	9	0.08
Corrosion of substation equipment	Transmission	5	0.04

Table 8.10 *(continued)*

The total cost of the listed items is approximately $11.7 billion or 76 percent of the total $15.4 billion cost of corrosion for 1998. The balance of the corrosion cost ($3.7 billion) likely stems from many miscellaneous less costly corrosion problems. As shown in Table 8.10, corrosion costs in the nuclear power and fossil steam power sectors dominate corrosion costs in the electric power industry. The very large cost problems in the nuclear and fossil sectors at the top of the list warrant serious attention.

Some special monitoring techniques are available to supplement normal water chemistry control in these important industrial applications. These tools are designed to help operators carry out specific monitoring tasks, for example, scale and deposits, composition of moisture droplets and liquid film in two-phase regions, in situ corrosion potential, at-temperature pH, and exfoliation in the superheater and reheater. Table 8.11 provides a brief description of the results that can be obtained with such devices and Fig. 8.20 illustrates where, for example, the measurement points would be for monitoring a PWR steam generator [17].

Device	Applications	Monitoring Results
Steam turbine deposit collector/ simulator	High-pressure (HP), intermediate-pressure (IP), and low-pressure (LP) turbines	Deposit composition, morphology, and rate of deposition vs. operation
Converging-diverging nozzle for LP turbines	Fossil and nuclear LP turbines. Simulates HP turbine deposition	Quantity and types of impurities depositing on LP turbine blades and corrosiveness of the environment
Converging nozzle for HP turbines	Simulates moisture drying on hot surfaces in LP turbines	Types of impurities depositing on HP turbine blades
Drying probe for wet steam stages	Boilers/turbines	Deposits of low-volatility impurities in LP turbines are collected
Boiler carryover monitors	LP turbines, boilers, condensers	Mechanical carryover
Early condensate samplers	Piping, turbines; also used to monitor effectiveness of steam blow and foreign object damage	Chemistry of water droplets formed in the final stages of the LP turbine, and so on.
Particle flow monitor for exfoliated oxides	Condensers, cooling towers, piping, heat exchangers, boilers	Number and size distribution of oxide particles in superheated and reheated steam

Table 8.11 Monitoring Related to Water and Steam Chemistry, Scale, and Deposits [17]

Device	Applications	Monitoring Results
Biofouling monitor	Boiler tubes	Detection of biofouling, collection of organic matter
In situ pH and corrosion potential	Boiler tubes	Susceptibility to general and localized corrosion
Heat flux gauge	Deposit buildup in turbines	Value of local heat flux, type of boiling, potential for impurity concentration
Chordal thermocouples	Deposits and erosion in turbines	Boiler tube temperature vs. scale
Rotor position and thrust bearing wear		Damage to thrust bearing caused by deposit accumulation on blades
Turbine first-stage pressure		Degree of deposition or erosion of control stage

TABLE 8.11 *(continued)*

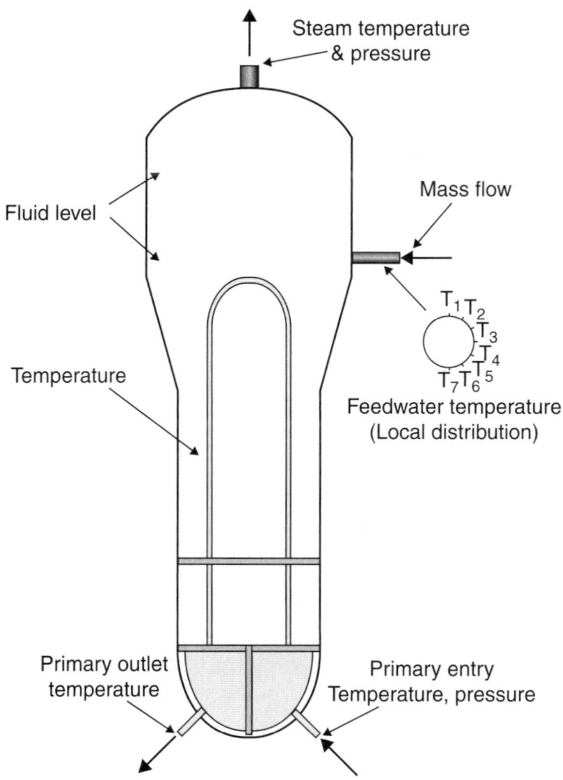

FIGURE 8.20 Measurement points for monitoring of a pressurized water reactor (PWR) steam generator.

Two factors that are not monitored often enough in the power producing industry are thermal stresses in heavy sections, which can lead to low-cycle corrosion fatigue (CF), and vibration in rotating machinery, which can lead to high-cycle CF. Table 8.12 summarizes the applications of inspection and corrosion monitoring tools that can be used for field monitoring these problems.

As also revealed in Table 8.10, the corrosion costs of some distribution, transmission, and combustion turbine sector problems may be substantial and warrant attention. Detailed discussions of the

Device	Applications	Monitoring Results
Corrosion product monitor	Feedwater systems including feedwater heaters	Quantitative determination of corrosion product transport
Erosion–corrosion (flow accelerated corrosion)	Piping components	Thinning rate for materials of concern in the specific suspected areas of piping
U-bend and double U-bend specimens	LP turbines, piping, feedwater heaters, condensers, boilers	Detects general corrosion, pitting, and SCC; double U-bends simulate crevice and galvanic effects
Fracture mechanics specimens	LP turbine disks and rotors, piping, headers, deaerators	Stress corrosion and CF crack growth rate and crack incubation times
Heat exchanger or condenser test tube	Condensers, feedwater heaters, and other heat exchangers; installed within the circuit	Accumulation of scale and corrosion pertinent to plant-specific conditions
Model crevice	Studies of crevice chemistry and corrosion heat exchangers, PWR steam generators	Crevice chemistry and corrosion data for specific conditions
Corrosion hydrogen monitor	Boiler tubes, feedwater system, and PWR steam generator corrosion	Detect general corrosion vs. load and chemistry; hydrogen damage, caustic gouging in boiler tubes, potential for impurity concentration
Vibration signature	Turbines and pumps—periodic monitoring	Detection of cracks and other distress

TABLE 8.12 Summary of Devices for Field Corrosion Monitoring in Power Plants [17]

Device	Applications	Monitoring Results
Boiler tube leak monitor	Fossil boilers	Early detection and location of a tube leak
Acoustic emission leak detector	Feedwater heaters and other heat exchangers	Early detection of tube leaks
Cavitation monitor	Feedwater piping and pumps	Early detection of cavitation noises
Stress and condition monitoring system	All types of steam cycles and major components	Actual on-line stresses, temperatures, and other conditions; used to determine damaging conditions and residual life
Turbine blade telemetry	LP turbines	Resonant frequencies and alternating stresses

Table 8.12 *(continued)*

items listed in Table 8.10 can be found in a special Electric Power Research Institute (EPRI) report [16].

8.6 Water Treatment

Through the natural water cycle, the earth has recycled and reused water for millions of years. Water recycling, though, generally refers to projects that use technology to speed up these natural processes. Water recycling is often characterized as *unplanned* or *planned*. A common example of unplanned water recycling occurs when cities draw their water supplies from rivers, such as the Colorado River and the Mississippi River, that receive wastewater discharges upstream from those cities. Water from these rivers has been reused, treated, and piped into the water supply a number of times before the last downstream user withdraws the water. For example, it is estimated that the water flowing in the Mississippi River has been used approximately seven times by the time it reaches the Gulf of Mexico. Planned projects are those that are developed with the goal of beneficially reusing a recycled water supply.

In many areas, the availability of new intake water is limited. Thus, in those industries requiring large amounts of cooling water, it is necessary to conserve available supplies by recirculating water. Industries producing primary metals, petrochemicals and paper typically require large volumes of water in the manufacturing processes. Recycled water can satisfy most water demands, as long as it is adequately treated to ensure water quality appropriate for the use. In uses where there is a greater chance of human exposure to the water, more treatment is required. As for any water source that is not properly treated, health problems could arise from drinking or being exposed to recycled water if it contains disease-causing organisms or other contaminants.

There is no single valid solution with regard to water treatments. The specific conditions of water supplies can be vastly different, even in systems separated by only a few meters. The evaluation of the water quality is typically determined by a chemical water analysis [9]. As shown in Table 8.13, there are basically two general categories of water treatment methods:

a) Chemical procedures based on material modifications due to chemical reactions. These can be monitored by analyzing the water before and after the treatment (softening, respective demineralization).

b) Physical treatments to alter the crystal structure of the deposits.

Chemical procedures	
Pretreatment	**Methods for clarifying:**
	• Coagulation
	• Flocculation
	• Sedimentation to clear floating and grey particles
In operation	**Softening methods:**
	• Lime milk/soda principle
	• Cations exchange (full softening)
	• Acid dosage (partly softening)
	Demineralization method:
	• Cation and anion exchanges (presently the most effective and economical method).
	Hardness stabilization:
	• Inhibitor dosage, also as dispersion and corrosion protection agents.
Post-treatment	Acid and caustic solution for cleaning of polluted thermal systems inclusively the neutralization of applied chemical detergents.
Physical procedures	
Pretreatment	Filtration of the subsoil water predominantly sand as filtering medium, in pressure and gravity filters.

TABLE 8.13 Methods of Water Treatment

Physical procedures	
In operation	Reverse osmosis for demineralization by use of diaphragms.
	Transformation of the crystal structures of the hardening causing substances:
	• Magnetic field method by means of electrical alternating or permanent magnet
	• Electrostatic method by applied active anodes
Post-treatment	Automatic cleaning of tubes heat exchangers by sponge rubber balls or brushes without operating interrupt of the plant.

TABLE 8.13 *(continued)*

8.6.1 Corrosion Inhibitors

By definition, a corrosion inhibitor is a chemical substance that, when added in small concentration to an environment, effectively decreases the corrosion rate. The efficiency of an inhibitor can be expressed by a measure of this improvement as indicated in Eq. (8.24):

$$\text{Inhibitor efficiency (\%)} = 100\,\frac{(\text{CR}_{\text{uninhibited}} - \text{CR}_{\text{inhibited}})}{\text{CR}_{\text{uninhibited}}} \tag{8.24}$$

where $\text{CR}_{\text{uninhibited}}$ is the corrosion rate of the uninhibited system and $\text{CR}_{\text{inhibited}}$ the corrosion rate of the inhibited system

In general, the efficiency of an inhibitor increases with an increase in inhibitor concentration (e.g., a typically good inhibitor would give 95 percent inhibition at a concentration of 0.008 percent and 90 percent at a concentration of 0.004 percent). There is often some positive synergism between different inhibitors and therefore mixtures are commonly chosen in commercial formulations. The scientific and technical corrosion literature has descriptions and lists of numerous chemical compounds that exhibit inhibitive properties. Of these, only very few are actually used in practice. This is partly because the desirable properties of an inhibitor usually extend beyond those simply related to metal protection. Considerations of cost, toxicity, availability, and environmental friendliness are of considerable importance.

The use of chemical inhibitors to decrease the rate of corrosion processes is quite varied. In the oil extraction and processing industries, for example, corrosion inhibitors have always been considered to be the first line of defense against corrosion. Table 8.14 presents some inhibitors that have been used with success in typical water environments to protect the metallic elements of industrial systems.

Commercial formulations generally consist of one or more inhibitor compounds with other additives such as surfactants, film enhancers, de-emulsifiers, oxygen scavengers, and so forth. The inhibitor solvent package used can be critical in respect to the solubility/dispersibility characteristics and hence the application and performance of the products.

Type of Water	Inhibitor	Metals	Concentration
Potable	$Ca(HCO_3)_2$	Steel, cast iron	10 ppm
	Polyphosphate	Fe, Zn, Cu, Al	5–10 ppm
	$Ca(OH)_2$	Fe, Zn, Cu	10 ppm
	Na_2SiO_3		10–20 ppm
Cooling	$Ca(HCO_3)_2$	Steel, cast iron	10 ppm
	Na_2CrO_4	Fe, Zn, Cu	0.1%
	$NaNO_2$	Fe	0.05%
	NaH_2PO_4		1%
	Morpholine		0.2%
Boilers	NaH_2PO_4	Fe, Zn, Cu	10 ppm
	Polyphosphate		10 ppm
	Morpholine	Fe	variable
	Hydrazine		O_2 scavenger
	Ammonia		neutralizer
	Octadecylamine		variable
Engine coolants	Na_2CrO_4	Fe, Pb, Cu, Zn	0.1–1%
	$NaNO_2$	Fe	0.1–1%
	Borax		1%
Glycol/water	Borax + MBT	All	1% + 0.1%
Oil field brines	Na_2SiO_3	Fe	0.01%
	Quaternaries		10–25 ppm
	Imidazoline		10–25ppm
Seawater	Na_2SiO_3	Zn	10 ppm
	$NaNO_2$	Fe	0.5%
	$Ca(HCO_3)_2$	All	pH dependent
	$NaH_2PO_4 + NaNO_2$	Fe	10 ppm + 0.5%

MBT = mercaptobenzotriazole

TABLE 8.14 Modified Typical Water Systems and the Inhibitors Used to Protect Them

Corrosion inhibition in recirculated cooling water systems historically depended on the oxidizing inhibitors such as chromates and nitrites. These days are gone since nitrites have been found to be highly conducive to organic growths and chromates have been phased out for environmental toxicity reasons.

The higher the chlorinity of the water, the more it requires a good dosage of corrosion inhibitors. Traditional glassy metaphosphates lose much of their effectiveness because they revert in time to orthophosphates. There is also an attendant possible danger of precipitation of tricalcium phosphate. This is a function of pH, calcium, orthophosphate, and solids concentration of the water. Silicates have also been used in water systems, but tend to have an adverse effect on heat transfer because of the deposition of relatively heavy films.

8.6.2 Scale Control

Encrustation of tubing, boilers, coils, jets, sprinklers, cooling towers, and heat exchangers arise wherever hard water is used. Scale formation can greatly affect heat transfer performance. One mm thick scale, for example, can add 7.5 percent to energy costs, while 1.5 mm adds 15 percent and 7 mm can increase cost by over 70 percent. Many factors can affect scaling. *Scaling*, which is basically the deposition of mineral solids on the interior surfaces of water lines and containers, most often occurs when water containing the carbonates or bicarbonates of calcium and magnesium is heated. There are basically three alternatives to prevent scale deposition:

- pH control
- the addition of scale inhibitors
- removal of scaling species

One of the most effective methods for controlling crystallization fouling is by adding chemical inhibitors to a scaling water. Commonly used antiscalants are derived from chemicals such as condensed polyphosphates, organophospates, and polyelectrolytes.

Scale control in open recirculated systems is affected primarily by limiting the concentration of the scale-forming species and related parameters, especially by controlled acid additions to bring the water to a suitable pH range. Many of the commercial inhibitor formulations include scale-controlling additives (e.g., polyphosphates and chelating agents).

8.6.3 Microorganisms

Algae and slimes are naturally occurring where sunlight and airborne contamination are to be expected. Slimes normally contain fungi, yeasts, bacteria, and entrapped quantities of inorganic and/or organic material. Some of these are described in more details in Chap. 10.

Conditions at the base of even thin slime films (biofilms) can be ideal for the growth of anaerobic bacteria, with high organic nutrient status, no oxygen, low redox potential, and protection from biocides. Figure 8.21 illustrates the steps in biofilm formation [18]. Sulfate reducing bactreria (SRB), for example, can produce active sulfide corrosion even in systems where the bulk liquid phase has a low nutrient status, a high oxygen concentration, and will not support growth of anaerobic aerobic bacteria [19]. Slime deposits can cause oxygen concentration cell-type corrosion and pitting. If sloughed off, they can contribute toward fouling of filters and other equipment (Fig. 8.22).

Chlorination to residual chlorine content of a few tenths of a ppm is the usual treatment for control of biological growths in open recirculated systems. In order to discourage the development of chlorine-resistant strains, it is advisable to surprise the microbes at irregular intervals with other biocides (e.g., quaternary amines). Other oxidizing biocides (e.g., chlorine dioxide and ozone) used in municipal waters are normally too expensive for cooling water systems.

Historically, copper sulfate had been used to control algae in some specific applications. It is effective at concentrations as low as 1 ppm, but must be used at a sufficiently low pH to prevent precipitation of insoluble copper hydroxide. The use of copper sulfate as a biocide is incompatible with aluminum equipment because of deposition and subsequent galvanic corrosion. Copper plating may also occur on zinc, on galvanized steel, and even steel.

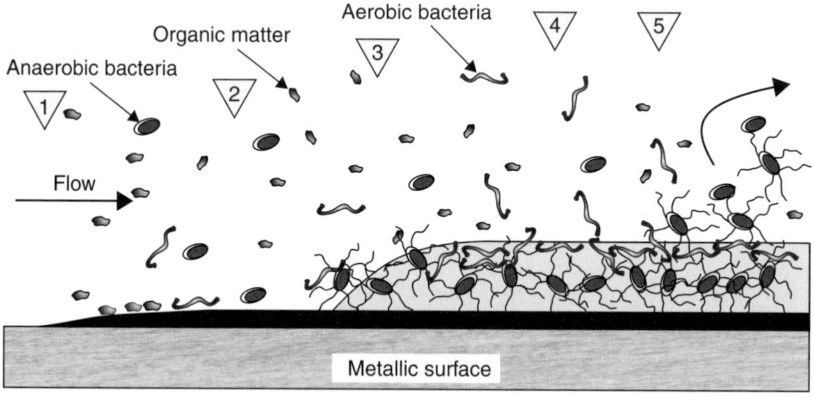

FIGURE 8.21 Steps in biofilm formation. Formation is initiated when small organic molecules become attached to an inert surface (1) and microbiological cells are adsorbed onto the resulting layer (2). The cells send out hair like exopolymers to feed on organic matter (3), adding to the coating (4). Flowing water detaches some of the formation (5), producing an equilibrium layer.

Figure 8.22 Fouled chiller unit. (Courtesy of Defence R&D, Canada-Atlantic)

8.7 Scaling Indices

The saturation level (SL) of water in a mineral phase is a good indicator of the potential for scaling due to that specific scalant. *Saturation level* is a ratio between the ion activity product (IAP) and the thermodynamic solubility product (K_{sp}) of a specific compound in that water. For example, when calcium carbonate ($CaCO_3$) is the scalant SL is defined as

$$SL = \frac{a_{Ca^{2+}} \cdot a_{CO_3^{2-}}}{K_{sp}} \tag{8.25}$$

where $a_{Ca^+} \cdot a_{CO_3^{2-}}$ is the IAP of the two ions involved in the formation of $CaCO_3$, that is, Ca^{2+} and CO_3^{2-}.

K_{sp} is a measure of ionic concentration when dissolved ions and undissolved ions are in equilibrium. When a saturated solution of sparingly or slightly soluble salt is in contact with undissolved salt, equilibrium is established between the dissolved ions and the undissolved salt. In theory, this equilibrium condition is based upon an undisturbed water maintained at constant temperature and allowed to remain undisturbed for an infinite period of time.

Water is therefore said to be undersaturated (SL value less than 1.0) if it can dissolve the scalant, that is, calcium carbonate in the present example. When water is at equilibrium, SL will be 1.0 by definition. Supersaturated water (SL value greater than 1.0) will precipitate calcium carbonate from water if allowed to rest. As the saturation level increases beyond 1.0, the driving force for the precipitation of calcium carbonate increases.

The following sections describe some indices that are used to indicate the tendency of given waters to deposit scales on metal substrates and by extension to predict the corrosivity of specific waters. Generally speaking, scales precipitated onto metal surfaces can provide protection of the substrate from general corrosion. If on the other hand, the scales are defective and contain voids and/or cracks, they could lead to localized corrosion. However, the assumption that water below saturation with respect to calcium carbonate is corrosive, while occasionally correct, is not always reliable.

8.7.1 Langelier Saturation Index

The Langelier saturation index (LSI) is probably the most widely used indicator of a water scale potential. This index indicates the driving force for scale formation and growth in terms of pH as a master variable. In order to calculate the LSI, it is necessary to know the alkalinity (mg L^{-1} as $CaCO_3$ or calcite), the calcium hardness (mg L^{-1} Ca^{2+} as $CaCO_3$), the total dissolved solids (mg L^{-1}, TDS), the actual pH, and the temperature of the water (°C). If TDS is unknown, but conductivity is, one can estimate mg L^{-1} TDS using a conversion table (Table 8.15). LSI is defined as

$$\text{LSI} = \text{pH} - \text{pH}_s \tag{8.26}$$

where pH is the measured water pH

pH_s is the pH at saturation in calcite or calcium carbonate and is defined as:

$$\text{pH}_s = (9.3 + A + B) - (C + D) \tag{8.27}$$

$$A = \frac{[\text{Log}_{10}(\text{TDS}) - 1]}{10} \tag{8.28}$$

$$B = -13.12 \times \text{Log}_{10}(^\circ\text{C} + 273) + 34.55 \tag{8.29}$$

$$C = \text{Log}_{10}(\text{Ca}^{2+}\text{as CaCO}_3) - 0.4 \tag{8.30}$$

$$D = \text{Log}_{10}(\text{alkalinity as CaCO}_3) \tag{8.31}$$

As for the SL reasoning described earlier, the LSI indicates three situations:

- If LSI is negative: No potential to scale, the water will dissolve $CaCO_3$.
- If LSI is positive: Scale can form and $CaCO_3$ precipitation may occur.
- If LSI is close to zero: Borderline scale potential. Water quality or changes in temperature, or evaporation could change the index.

Conductivity (μS/cm)	TDS (mg/L as $CaCO_3$)
1	0.42
10.6	4.2
21.2	8.5
42.4	17.0
63.7	25.5
84.8	34.0
106.0	42.5
127.3	51.0
148.5	59.5
169.6	68.0
190.8	76.5
212.0	85.0
410.0	170.0
610.0	255.0
812.0	340.0
1008.0	425.0

TABLE 8.15 Conversion Table between the Conductivity of Natural Water and the TDS It Contains

Calculation example. Suppose the drinking water supplied to animals has the following analysis:

$$\text{pH} = 7.5$$

$$\text{TDS} = 320 \text{ mg/L}$$

$$\text{Calcium} = 150 \text{ mg L}^{-1} \text{ (or ppm) as } CaCO_3$$

$$\text{Alkalinity} = 34 \text{ mg L}^{-1} \text{ (or ppm) as } CaCO_3$$

The LSI index is calculated at two temperatures, that is, 25°C (room temperature) and 82°C (cage wash cycle). The colder incoming water will warm to room temperature in the manifolds. Residual water in the rack manifold can be heated to 82°C when the rack is in the cage washer.

LSI formula:

$$\text{LSI} = \text{pH} - \text{pH}_s$$

$$\text{pH}_s = (9.3 + A + B) - (C + D)$$

where $A = [\text{Log}_{10}(\text{TDS}) - 1]/10 = 0.15$

$B = -13.12 \times \text{Log}_{10}(°\text{C} + 273) + 34.55$

$= 2.09$ at 25°C and 1.09 at 82°C

$C = \text{Log}_{10}(\text{Ca}^{2+} \text{ as CaCO}_3) - 0.4 = 1.78$

$D = \text{Log}_{10}(\text{alkalinity as CaCO}_3) = 1.53$

Calculation at 25°C:

$$\text{pH}_s = (9.3 + 0.15 + 2.09) - (1.78 + 1.53) = 8.2$$

$$\text{LSI} = 7.5 - 8.2 = -0.7$$

Hence *no tendency to scale*

Calculation at 82°C:

$$\text{pH}_s = (9.3 + 0.15 + 1.09) - (1.78 + 1.53) = 7.2$$

$$\text{LSI} = 7.5 - 7.2 = +0.3$$

Hence *slight tendency to scale*

8.7.2 Other Indices

The *Ryznar stability index (RSI)* uses a correlation established between an empirical database of scale thickness observed in municipal water systems and associated water chemistry data. As with the LSI, the RSI has its basis in the concept of saturation level. The RSI takes the following form:

$$\text{RSI} = 2(\text{pH}_s) - \text{pH}$$

The empirical correlation of the RSI can be summarized as follows:

- RSI < 6 the scale tendency increases as the index decreases
- RSI > 7 the calcium carbonate formation probably does not lead to a protective corrosion inhibitor film
- RSI > 8 mild steel corrosion becomes an increasing problem

The *Puckorius scaling index (PSI)* is based on the buffering capacity of the water, and the maximum quantity of precipitate that may form in bringing water to equilibrium. Water high in calcium, but low in alkalinity and buffering capacity can have a high calcite saturation level that increases the ion activity product. Such water might have a high tendency to form scale, but the scale itself might be of such a small quantity as to be unobservable. The water has the driving force but not the capacity and ability to maintain pH as precipitate matter forms.

The PSI index is calculated in a manner similar to the RSI. Puckorius uses an equilibrium pH rather than the actual system pH to account for the buffering effects:

$$PSI = 2\,(pH_s) - pH_{eq}$$

where pH_s is still the pH at saturation in calcite or calcium carbonate

$pH_{eq} = 1.465 \times \log_{10}[\text{Alkalinity}] + 4.54$

$[\text{Alkalinity}] = [HCO_3^-] + 2\,[CO_3^{2-}] + [OH^-]$

The *Larson-Skold index* is based upon evaluation of in situ corrosion of mild steel lines transporting Great Lakes waters. Extrapolation to other waters than the Great Lakes, such as those of low alkalinity or extreme alkalinity, goes beyond the range of the original data. The index is the ratio of equivalents per million (epm) of sulfate (SO_4^{2-}) and chloride (Cl^-) to the epm of alkalinity in the form bicarbonate plus carbonate:

$$\text{Larson-Skold index} = (\text{epm } Cl^- + \text{epm } SO_4^{2-}) / (\text{epm } HCO_3^- + \text{epm } CO_3^{2-}) \tag{8.32}$$

The index has proven a useful tool in predicting the aggressiveness of once-through cooling waters. The Larson-Skold index might be interpreted by the following guidelines:

- Index < 0.8 chlorides and sulfate probably will not interfere with natural film formation.
- 0.8 < index < 1.2 chlorides and sulfates may interfere with natural film formation. Higher than desired corrosion rates might be anticipated.
- Index > 1.2 the tendency toward high corrosion rates of a local type should be expected as the index increases.

The *Stiff-Davis index* attempts to overcome the shortcomings of the LSI with respect to waters with high total dissolved solids and the impact of "common ion" effects on the scale formation driving force. Like the LSI, the Stiff-Davis index has its basis in the concept of saturation level. The solubility product used to predict the pH at saturation (pHs) for a water is empirically modified in the Stiff-Davis index. The Stiff-Davis index will predict that water is less scale forming than the LSI calculated for the same water chemistry and conditions. The deviation between the indices increases with ionic strength. Interpretation of the index is by the same scale as for the LSI.

The *Oddo-Tomson index* accounts for the impact of absolute pressure and partial pressure of carbon dioxide on the pH of water, and on the solubility of calcium carbonate [20]. This empirical model also incorporates corrections for the presence of two or three phases (water, gas, and oil). Interpretation of the index is by the same scale as for the LSI and Stiff-Davis indices.

8.8 Ion-Association Model

The saturation indices discussed previously can be calculated based upon total analytical values for all possible reactants. Ions in water, however, do not tend to exist totally as free ions [21]. Calcium, for example, may be paired with sulfate, bicarbonate, carbonate, phosphate, and other species. Bound ions are not readily available for scale formation and such binding, or reduced availability of the reactants, decreases the effective ion-activity product. Saturation indices such as the LSI are based upon total analytical values rather than free species primarily because of the intense calculation requirements for determining the distribution of species in water. Speciation breakdown of all species in a given water requires numerous computer iterations to achieve the following [22]:

- The verification of electroneutrality via a cation-anion balance, and balancing with an appropriate ion (e.g., sodium or potassium for cation-deficient waters; sulfate, chloride, or nitrate for anion-deficient waters).
- Estimating ionic strength; calculating and correcting activity coefficients and dissociation constants for temperature; correcting alkalinity for noncarbonate alkalinity.
- Iterative calculation of the distribution of species in the water from dissociation constants. A partial listing of possible ion pairs is given in Table 8.16.
- Verification of mass balance and adjustment of ion concentrations to agree with analytical values.
- Repeating the process until the desired resolution is achieved.
- Calculation of saturation levels based upon free concentrations of ions estimated using the ion-association model (ion pairing).

The ion-association model has been used by major water treatment companies since the early 1970s. When indices are used to establish operating limits such as maximum concentration ratio or maximum pH, the differences between indices calculated using ion pairing can have some serious economic significance. For example, experience on a system with high-TDS water may be translated to a system operating with a lower-TDS water. The high indices found acceptable in the high-TDS water may be unrealistic when translated to a water where ion pairing is less significant in reducing the apparent driving force for scale formation. Table 8.17 summarizes the impact of TDS upon LSI when it is calculated using total analytical values for calcium and alkalinity, and when it is calculated using the free calcium and carbonate concentrations determined with an ion-association model.

Indices based upon ion-association models provide a common denominator for comparing results between systems. For example, calcite saturation level calculated using free calcium and carbonate

ALUMINUM
[Aluminum] = $[Al^{3+}] + [Al(OH)^{2+}] + [Al(OH)_2^+] + [Al(OH)_4^-] + [AlF^{2+}] + [AlF_2^+] + [AlF_3] + [AlF_4^-] + [AlSO_4^+] + [Al(SO_4)_2^-]$
BARIUM
[Barium] = $[Ba^{2+}] + [BaSO_4] + [BaHCO_3^+] + [BaCO_3] + [Ba(OH)^+]$
CALCIUM
[Calcium] = $[Ca^{2+}] + [CaSO_4] + [CaHCO_3^+] + [CaCO_3] + [Ca(OH)^+] + [CaHPO_4] + [CaPO_4^-] + [CaH_2PO_4^+]$
IRON
[Iron] = $[Fe^{2+}] + [Fe^{3+}] + [Fe(OH)^+] + [Fe(OH)^{2+}] + [Fe(OH)_3^-] + [FeHPO_4^+] + [FeHPO_4] + [FeCl^{2+}] + [FeCl_2^+] + [FeCl_3] + [FeSO_4] + [FeSO_4^+] + [FeH_2PO_4^+] + [Fe(OH)_2^+] + [Fe(OH)_3] + [Fe(OH)_4^-] + [Fe(OH)_2] + [FeH_2PO_4^{2+}]$
MAGNESIUM
[Magnesium] = $[Mg^{2+}] + [MgSO_4] + [MgHCO_3^+] + [MgCO_3] + [Mg(OH)^+] + [MgHPO_4] + [MgPO_4^-] + [MgH_2PO_4^+] + [MgF^+]$
POTASSIUM
[Potassium] = $[K^+] + [KSO_4^-] + [KHPO_4^-] + [KCl]$
SODIUM
[Sodium] = $[Na^+] + [NaSO_4^-] + [Na_2SO_4] + [NaHCO_3] + [NaCO_3^-] + [Na_2CO_3] + [NaCl] + [NaHPO_4^-]$
STRONTIUM
[Strontium] = $[Sr^{2+}] + [SrSO_4] + [SrHCO_3^+] + [SrCO_3] + [Sr(OH)^+]$

TABLE 8.16 Example of Ion Pairs Used to Estimate Free Ion Concentrations

Water	LSI		TDS Impact on LSI
	Low TDS	High TDS	
High chloride			
• No pairing	2.25	1.89	–0.36
• With pairing	1.98	1.58	–0.40
High sulfate			
• No pairing	2.24	1.81	–0.43
• With pairing	1.93	1.07	–0.86

FIGURE 8.17 Impact of Ion Pairing on the Langelier Scaling Index (LSI).

concentrations has been used successfully as the basis for developing models which describe the minimum effective scale inhibitor dosage that will maintain clean heat-transfer surfaces [23]. The following cases illustrate some practical usage of the ion-association model.

8.8.1 Limiting Halite Deposition in a Wet High-Temperature Gas Well

There are several gas-well fields that produce hydrocarbon gas associated with very high TDS connate waters. Classical oilfield scale problems (e.g., calcium carbonate, barium sulfate, and calcium sulfate) are minimal in these fields. Halite (NaCl), however, can be precipitated to such an extent that production is lost in hours. As a result, a bottom-hole fluid sample is retrieved from all new wells. Unstable components are "fixed" immediately after sampling, and pH is determined under pressure. A full ionic and physical analysis is also carried out in the laboratory.

Some of the analyses were run through an ion-association model computer program to determine the susceptibility of the brine to halite precipitation. If a halite precipitation problem was predicted, the ion-association model was run in a "mixing" mode to determine if mixing the connate water with boiler feedwater would prevent the problem. This approach has been used successfully to control salt deposition in the well with the composition outlined in Table 8.18. The ion-association model evaluation of the bottom-hole chemistry indicated that the water was slightly supersaturated with sodium chloride under the bottom-hole conditions of pressure and temperature. As the fluids cooled in the well bore, the production of copious amounts of halite was predicted.

	Bottom Hole Connate	Boiler Feed Water
Temperature (°C)	121	70
Pressure (bars)	350	1
pH (site)	4.26	9.10
Density (kg/m^3)	1.300	1.000
TDS (mgl^{-1})	369,960	<20
Dissolved CO_2 (mgl^{-1})	223	<1
H_2S (gas phase) (mgl^{-1})	50	0
H_2S (aqueous phase) (mgl^{-1})	<0.5	0
Bicarbonate (mgl^{-1})	16	5.0
Chloride(mgl^{-1})	228,485	0

TABLE 8.18 Hot Gas-Well Water Analysis

	Bottom Hole Connate	Boiler Feed Water
Sulfate (mgl^{-1})	320	0
Phosphate (mgl^{-1})	<1	0
Borate (mgl^{-1})	175	0
Organic acids <C_6 (mgl^{-1})	12	<5
Sodium (mgl^{-1})	104,780	<1
Potassium (mgl^{-1})	1600	<1
Calcium (mgl^{-1})	30,853	<1
Magnesium (mgl^{-1})	2910	<1
Barium (mgl^{-1})	120	<1
Strontium (mgl^{-1})	1164	<1
Total iron (mgl^{-1})	38.0	<0.01
Lead (mgl^{-1})	5.1	<0.01
Zinc (mgl^{-1})	3.6	<0.01

Table 8.18 *(continued)*

In this case, the ion-association model predicted that the connate water would require a minimum dilution with boiler feedwater of 15 percent to prevent halite precipitation (Fig. 8.23). The model also predicted that over-injection of dilution water would promote barite (barium sulfate) formation (Fig. 8.24). Although the well produced H_2S at a concentration of 50 mg/L, the program did not predict the formation of iron sulfide because of the combination of low pH and high temperature. Boiler feedwater was injected into the bottom of the well using the downhole injection valve normally used for corrosion inhibitor injection. Injection of dilution water at a rate of 25 to 30 percent has allowed the well to produce successfully since start-up. Barite and iron sulfide precipitation have not been observed, and plugging with salt has not occurred.

8.8.2 Identifying Acceptable Operating Range for Ozonated Cooling Systems

It has been well-established that ozone is an efficient microbiological control agent in open recirculating cooling-water systems (cooling towers). It has also been reported that commonly encountered scales have not been observed in ozonated cooling systems under conditions where scale would otherwise be expected. The water chemistry of 13 ozonated

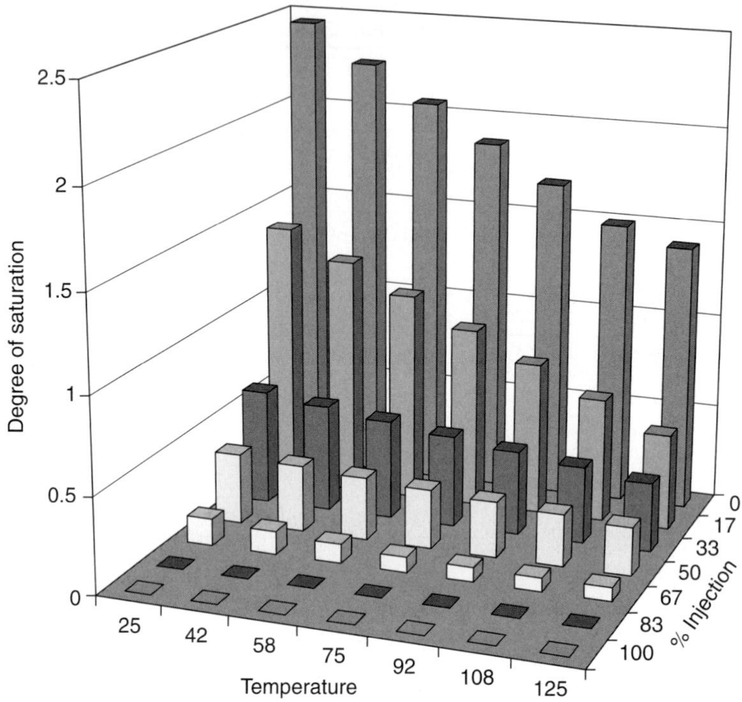

FIGURE 8.23 Hot gas-well halite degree of saturation as a function of temperature and reinjected boiler water (DownHole SAT).

cooling systems was evaluated using an ion-association model. Each system was treated solely with ozone on a continuous basis at the rate of 0.05 to 0.2 mg/L based upon recirculating water flow rates [24].

The saturation levels for common cooling-water scales were calculated, including calcium carbonate, calcium sulfate, amorphous silica, and magnesium hydroxide. Brucite saturation levels were included because of the potential for magnesium silicate formation as a result of the adsorption of silica upon precipitating magnesium hydroxide. Three categories of systems were encountered [24]:

- *Category 1:* The theoretical chemistry of the concentrated water was not scale-forming (i.e., undersaturated).
- *Category 2:* The concentrated recirculating water would have a moderate to high calcium carbonate scale-forming tendency. Water chemistry observed in these systems was similar to that in systems run successfully using traditional scale inhibitors such as phosphonates.
- *Category 3:* These systems demonstrated an extraordinarily high-scale potential for at least calcium carbonate and brucite. These systems operated with a recirculating water chemistry

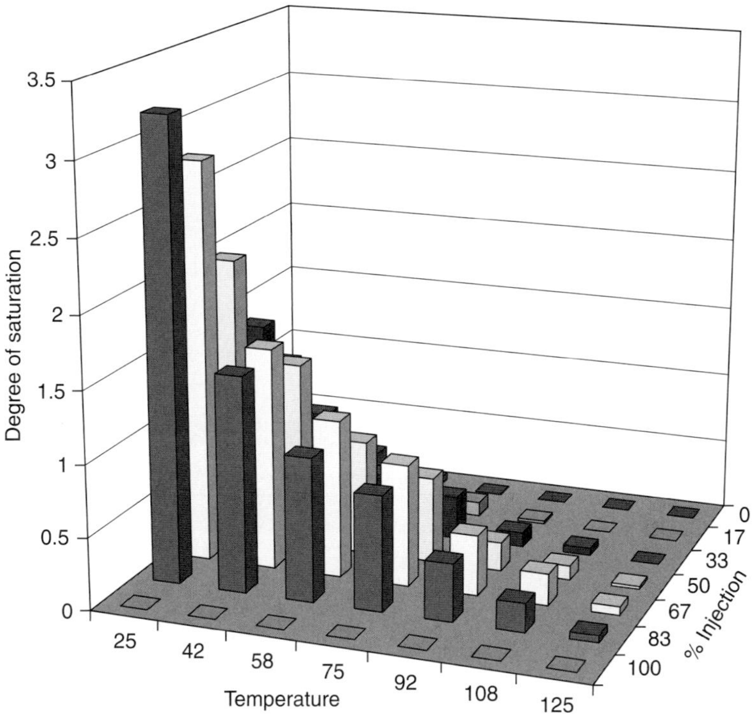

FIGURE 8.24 Hot gas-well barite degree of saturation as a function of temperature and reinjected boiler water.

similar to that of a softener rather than of a cooling system. The Category 3 water chemistry was above the maximum saturation level for calcium carbonate for which traditional inhibitors such as phosphonates are able to inhibit scale formation.

Table 8.19 outlines the theoretical versus actual water chemistry for the 13 systems evaluated. Saturation levels for the theoretical and actual recirculating water chemistries are presented in Table 8.20. A comparison of the predicted chemistries to observed system cleanliness revealed the following:

- Category 1 (recirculating water chemistry undersaturated). The systems did not show any scale formation.
- Category 2 (conventional alkaline cooling system control range). Scale formation was observed in eight of the nine Category 2 systems evaluated.
- Category 3 (cooling tower as a softener). Deposit formation on heat-transfer surfaces was not observed in most of these systems.

System	Calcium			Magnesium			Silica			
(Category)	T^a	A^b	Δ^c	T^a	A^b	Δ^c	T^a	A^b	Δ^c	System Cleanliness
1 (1)	56	43	13	28	36	–8	40	52	–12	No scale observed
2 (2)	80	60	20	88	38	50	24	20	4	Basin buildup
3 (2)	238	288	–50	483	168	315	38	31	7	Heavy scale
4 (2)	288	180	108	216	223	–7	66	48	18	Valve scale
5 (3)	392	245	147	238	320	–82	112	101	11	Condenser tube scale
6 (3)	803	163	640	495	607	–112	162	143	19	No scale observed
7 (3)	1464	200	1264	549	135	414	112	101	11	No scale observed
8 (3)	800	168	632	480	78	402	280	78	202	No scale observed
9 (3)	775	95	680	496	78	418	186	60	126	No scale observed
10 (3)	3904	270	3634	3172	508	2664	3050	95	2995	Slight valve scale
11 (3)	4170	188	3982	308	303	5	126	126	0	No scale observed
12 (3)	3,660	800	2860	2623	2972	–349	6100	138	5962	No scale observed
13 (3)	7930	68	7862	610	20	590	1952	85	1867	No scale observed

T^a = theoretical (ppm)
A^b = actual (ppm)
Δ^c = differencel (ppm)

TABLE 8.19 Theoretical vs. Actual Recirculating Water Chemistry

System	Calcite		Brucite		Silica		
(Category)	T^a	A^b	T^a	A^b	T^a	A^b	Observation
1 (1)	0.03	0.02	<0.001	<0.001	0.20	0.25	No scale observed
2 (2)	49	5.4	0.82	0.02	0.06	0.09	Basin buildup
3 (2)	89	611	2.4	0.12	0.10	0.12	Heavy scale
4 (2)	106	50	1.3	0.55	0.13	0.16	Valve scale
5 (3)	240	72	3.0	0.46	0.21	0.35	Condenser tube scale
6 (3)	540	51	5.3	0.73	0.35	0.49	No scale observed
7 (3)	598	28	10	0.17	0.40	0.52	No scale observed
8 (3)	794	26	53	0.06	0.10	0.33	No scale observed
9 (3)	809	6.5	10	<0.01	0.22	0.27	No scale observed
10 (3)	1198	62	7.4	0.36	0.31	0.35	Slight valve scale
11 (3)	1670	74	4.6	0.36	0.22	0.44	No scale observed
12 (3)	3420	37	254	0.59	1.31	0.55	No scale observed
13 (3)	7634	65	7.6	0.14	1.74	0.10	No scale observed

T^a = theoretical (ppm)
A^b = actual (ppm)

TABLE 8.20 Theoretical vs. Actual Recirculating Water Saturation Level

The study revealed that calcium carbonate (calcite) scale formed most readily on heat-transfer surfaces in systems operating in a calcite saturation level range of 20 to 150 mg/L, the typical range for chemically treated cooling water. At much higher saturation levels, in excess of 1000, calcite precipitated in the bulk water. Because of the overwhelming high surface area of the precipitating crystals relative to the metal surface in the system, continuing precipitation would lead to growth on crystals in the bulk water rather than on heat-transfer surfaces. The presence of ozone in cooling systems did not appear to influence calcite precipitation and/or scale formation.

8.8.3 Optimizing Calcium Phosphate Scale Inhibitor Dosage in a High-TDS Cooling System

A major manufacturer of polymers for calcium phosphate scale control in cooling systems has developed laboratory data on the minimum effective scale inhibitor (copolymer) dosage required to prevent calcium phosphate deposition over a broad range of calcium and phosphate concentrations, and a range of pH and temperatures. The data were developed using static tests, but have been observed to correlate well with the dosage requirements for the copolymer in operating cooling systems. The data were developed using test waters with relatively low levels of dissolved solids. Recommendations from the data were typically made as a function of calcium concentration, phosphate concentration, and pH. This database was used to project the treatment requirements for a utility cooling system that used geothermal brine for makeup water. An extremely high dosage (30 to 35 mg/L) was recommended based upon the laboratory data [22].

It was believed that much lower dosages would be required in the actual cooling system because of the reduced availability of calcium anticipated in the high-TDS recirculating water. As a result, it was believed that a model based upon dosage as a function of the ion-association model saturation level for tricalcium phosphate would be more appropriate, and accurate, than a simple lookup table of dosage versus pH and analytical values for calcium and phosphate. Tricalcium phosphate saturation levels were calculated for each of the laboratory data points. Regression analysis was used to develop a model for dosage as a function of saturation level and temperature.

The model was used to predict the minimum effective dosage for the system with the makeup and recirculating water chemistry found in Table 8.21. A dosage in the range of 10 to 11 mg/L was predicted, rather than the 30 mg/L derived from the lookup tables. A dosage minimization study was conducted to determine the minimum effective dosage. The system was initially treated with the copolymer at a dosage of 30 mg/L in the recirculating water. The dosage was decreased until deposition was observed. Failure was noted when the recirculating water concentration dropped below 10 mg/L, validating the ion–association-based dosage model.

Water Analysis at 6.2 Cycles		Deposition Potential Indicators	
• Cations		• Saturation Level	
Calcium (as $CaCO_3$)	1339	Calcite	38.8
Magnesium (as $CaCO_3$)	496	Aragonite	32.9
Sodium (as Na)	1240	Silica	0.4
• Anions		Tricalcium phosphate	1074
Chloride (as Cl)	620	Anhydrite	1.3
Sulfate (as SO_4)	3384	Gypsum	1.7
Bicarbonate (as HCO_3)	294	Fluorite	0.0
Carbonate (as CO_3)	36	Brucite	<0.1
Silica (as SiO_2)	62	• Simple Indices	
Parameters		Langelier	1.99
PH	8.40	Ryznar	4.41
Temperature (°C)	36.7	Practical	4.20
1/2 Life (hours)	72	Larson-Skold	0.39
Recommended Treatment			
100% Active Copolymer (mg/L)	10.53		

Table 8.21 Calcium Phosphate Inhibitor Dosage Optimization Example

References

1. *Petroleum and Natural Gas industries—Materials for Use in H_2S-Containing Environments in Oil and Gas Production*. NACE MR0175/ISO 15156. Houston, Tex.: NACE International, 2001.
2. Koch GH, Brongers MPH, Thompson NG, Virmani YP, Payer JH. *Corrosion Costs and Preventive Strategies in the United States*. FHWA-RD-01-156. Springfield, Va.: National Technical Information Service, 2001.
3. Kirmeyer GJ, Wagner I, Leroy P. Organizing corrosion control studies and implementing corrosion control strategies. In: *Internal Corrosion of Water Distribution Systems*. Denver, CO: American Water Works Association, 1996; 487–540.
4. Deterioration and inspection of water distribution systems: a best practice by the national guide to sustainable municipal infrastructure. *Deterioration and Inspection of Water Distribution Systems*. Issue No. 1.1. Ottawa, Canada: National Research Council of Canada, 2003.
5. Romanoff M. *Underground Corrosion*. Houston, Tex.: NACE International, 1989.
6. Makar JM, Kleiner Y. *Maintaining Water Pipeline Integrity*. NRCC-43986. Ottawa, Canada, National Research Council, 2001.
7. Makar JM, Chagnon N. *Inspecting Systems for Leaks, Pits, and Corrosion*. American Water Works Association. 1999; 91: 36–46.
8. Davies M, Scott PJB. *Guide to the Use of Materials in Waters*. Houston, Tex.: NACE International, 2003.

9. Roberge PR. *Handbook of Corrosion Engineering*. New York, N.Y.: McGraw-Hill, 2000.
10. Berry WE. Water corrosion. In: Van Delinder LS, Brasunas Ad, eds. *Corrosion Basics*. Houston, Tex.: NACE International, 1984; 149–76.
11. Uhlig HH. Iron and steel. In: Uhlig HH, ed. *The Corrosion Handbook*. New York, N.Y.: John Wiley and Sons, 1948; 125–43.
12. Hartt WH, Culberson CH, Smith SW. Calcareous deposits on metal surfaces in seawater—A critical review. *Corrosion*. 1984; 40: 609–18.
13. Crovetto R, Murtagh E. Novel boiler condensate corrosion inhibitor with FDA approval. *CORROSION 2007*, Paper #073. Houston, Tex.: NACE International, 2007.
14. Herro HM, Port RD. *The NALCO Guide to Cooling Water Systems Failure Analysis*. New York, N.Y.: McGraw-Hill, 1993.
15. Syrett BC, Gorman JA. Cost of corrosion in the electric power industry—An update. *Materials Performance* 2003; 42: 32–8.
16. *Cost of Corrosion in the Electric Power Industry*. EPRI 1004662. Palo Alto, Calif.: Electric Power Research Institute, 2001.
17. Jonas O. Monitoring of steam plants. *Materials Performance* 2003; 42: 38–42.
18. Garey JF, Jorden RM. *Condenser Biofouling Control*. Ann Arbor, MI: Ann Arbor Science Publishers,1980.
19. Stott JFD. Evaluating microbiologically influenced corrosion. In: Cramer DSaCBS, ed. Vol. 13A: *Corrosion: Fundamentals, Testing, and Protection*. Metals Park, Ohio: ASM International, 2003; 644-9.
20. Oddo JE, Tomson MB. Scale control, prediction and treatment or how companies evaluate a scaling problem and what they do wrong. *CORROSION 92*, Paper #34. Houston, Tex.: NACE International, 1992.
21. Truesdell AH, Jones BF. A computer program for calculating chemical equilibria of natural waters. *Journal of Research, U.S.Geological Survey* 1974; 2: 233–48.
22. Ferguson RJ, Freedman AJ, Fowler G, Kulik AJ, Robson J, Weintritt DJ. The practical application of ion association model saturation level indices to commercial water treatmant problem solving. In: Amjad Z, ed. *Mineral Scale Formation and Inhibition*. New York, N.Y.: Plenum Press, 1995; 323–40.
23. Ferguson RJ, Codina O, Rule W, Baebel R. Real time control of scale inhibitor feed rate. *International Water Conference*, 49th Annual Meeting. IWC-88–57. Pittsburg, Pa. 1988.
24. Ferguson RJ, Freedman AJ. A comparison of scale potential indices with treatment program results in ozonated systems. *CORROSION 93*, Paper #279. Houston, Tex.: NACE International, 1993.

CHAPTER 9
Atmospheric Corrosion

9.1 Introduction

Atmospheric corrosion is surely the most visible of all corrosion processes, for example, rusty bridges, flag poles, buildings, and outdoor monuments. The large segment of the paint industry committed to the manufacture and application of products for the protection of metals, as well as the large-scale operations of the galvanizing industry attest to the importance of controlling atmospheric corrosion.

Economic losses caused by atmospheric corrosion are tremendous and therefore account for the disappearance of a significant portion of metal produced. Consider, for instance, agricultural machinery, steel structures, fences, exposed metals on buildings, automobile mufflers or bodies, and the myriad of other metal items that are sent to the scrap yard when they become unusable as a result of corrosion. These constitute direct losses from corrosion.

Atmospheric corrosion has been reported to account for more failures in terms of cost and tonnage than any other type of material degradation processes. This particular type of material degradation has recently received more attention, particularly by the aircraft industry, since the Aloha incident in 1988, when a Boeing 737 lost a major portion of the upper fuselage in full flight at 7300 m [1].

All of the general types of corrosion attack occur in the atmosphere. Since the corroding metal is not bathed in large quantities of electrolyte, most atmospheric corrosion operates in highly localized corrosion cells, sometimes producing patterns difficult to explain as in the example of the rusting galvanized roof shown in Fig. 9.1.

Thus, calculation of the electrode potentials on the basis of ion concentration, the determination of polarization characteristics, and other electrochemical operations are not as simple in atmospheric corrosion as they are in liquid immersion corrosion. However, all of the electrochemical factors which are significant in corrosion processes do operate in the atmosphere.

FIGURE 9.1 Rusting galvanized roof with regular bands of rust.

9.2 Types of Corrosive Atmospheres

While atmospheres have been traditionally classified into four basic types, most environments are in fact mixed and present no clear demarcation. Furthermore, the type of atmosphere may vary with the wind pattern, particularly where corrosive pollutants are present, or with local conditions (Fig. 9.2) [2].

9.2.1 Industrial

An industrial atmosphere is characterized by pollution composed mainly of sulfur compounds such as sulfur dioxide (SO_2), a precursor to acid rain, and nitrogen oxides (NO_x), the backbone of smog in modern cities. Sulfur dioxide from burning coal or other fossil fuels is picked up by moisture on dust particles as sulfurous acid. This is oxidized by some catalytic process on the dust particles to sulfuric acid, which settles in microscopic droplets and fall as acid rain on exposed surfaces. The result is that contaminants in an industrial atmosphere, plus dew or fog, produce a highly corrosive, wet, acid film on exposed surfaces.

In addition to the normal industrial atmosphere in or near chemical plants, other corrosive pollutants may be present. These are usually various forms of chloride which may be much more corrosive than the acid sulfates. The reactivity of acid chlorides with most metals is more pronounced than the reactivity of other pollutants such as phosphates and nitrates.

FIGURE 9.2 Bird damage to a 1931 bronze statue commemorating the 21st Battalion battles in WW I. (Courtesy of Kingston Technical Software)

9.2.2 Marine

A marine atmosphere is laden with fine particles of sea mist carried by the wind to settle on exposed surfaces as salt crystals. The quantity of salt deposited may vary greatly with wind velocity and it may, in extreme weather conditions, even form a very corrosive salt crust, similar to what is experienced on a regular basis by sea patrolling aircraft or helicopters [Figs. 9.3(*a*) and (*b*)].

The quantity of salt contamination decreases with distance from the ocean, and is greatly affected by wind currents. The marine atmosphere also includes the space above the sea surfaces where splashing and heavy sea spray are encountered. The equipment exposed to these splash zones are indeed subjected to the worst conditions of intermittent immersion with wet and dry cycling of the corrosive agent (Fig. 9.4).

9.2.3 Rural

Rural atmospheres are typically the most benign and do not contain strong chemical contaminants, that is unless one is close to a farm

(a)

(b)

Figure 9.3 Sea salt deposited on the external surface (*a*) of a Cormorant sea and rescue helicopter radar antenna, and (*b*) salt causing corrosion to the internal components of the antenna due to a broken seal. (Courtesy Major S.J.R. Giguère)

FIGURE 9.4 Aircraft carrier's top deck. (Courtesy of Mike Dahlager, Pacific Corrosion Control Corporation)

operation where byproducts made of various waste materials can be extremely corrosive to most metals.

Arid or tropical atmospheres are special variations of the rural atmosphere. In arid climates there is little or no rainfall, but there may be a high relative humidity and occasional condensation. This situation is encountered along the desert coast of northern Africa. In the Tropics, in addition to the high average temperature, the daily cycle includes a high relative humidity, intense sunlight, and long periods of condensation during the night. In sheltered areas, the wetness from condensation may persist long after sunrise. Such conditions may produce a highly corrosive environment.

9.2.4 Indoor

Normal indoor atmospheres are generally considered to be quite mild when ambient humidity and other corrosive components are under control. However, some combinations of conditions may actually cause relatively severe corrosion problems. While there is no typical contaminant or set of conditions associated with an indoor atmosphere, any enclosed space which is not evacuated or filled with a liquid can be considered an indoor atmosphere. If not ventilated, such an environment may contain fumes, which in the presence of condensation or high humidity could prove to be highly corrosive.

Even in the absence of any other corrosive agent, the constant condensation on a cold metallic surface may cause an environment similar

FIGURE 9.5 Electric junction box badly corroded only four years after a new residence was completed. (Courtesy of Kingston Technical Software)

to constant immersion for which a component may not have been chosen or prepared for. Such systems are commonly encountered in confined areas close to ground level or, worse, below ground where high humidity may prevail. Figure 9.5 shows the advanced corrosion of the frame and contacts in an electric junction box only four years after a building was completed. While the junction box in this example was only at the ground level, the wires coming to the box were buried without additional insulation and in constant contact with much cooler ground than ambient air in the room. The repeated condensation of ambient humidity on the electrical box support and on many of its connections caused enough corrosion in such a short period of time to require the complete replacement of the system to avoid unscheduled power interruptions.

9.3 Factors Affecting Atmospheric Corrosion

The most important factor in atmospheric corrosion, overriding the presence of any other surface contamination, is moisture, either in the form of rain, dew, condensation, or high relative humidity (RH).

In the absence of moisture, most contaminants would have little or no corrosive effect.

Rain is not always corrosive. It may even have a beneficial effect by washing away atmospheric pollutants that have settled on exposed surfaces. This effect is particularly noticeable in marine environments. On the other hand, if the rain collects in pockets or crevices, it may accelerate corrosion by supplying continued wetness in such areas as shown in Fig. 9.6.

Dew and condensation are undesirable from a corrosion standpoint if they are not accompanied by frequent rain washing to dilute or eliminate surface contamination. A film of dew, saturated with sea salt or acid sulfates (acid rain), and acid chlorides coming from an industrial atmosphere provide an aggressive electrolyte for the promotion of corrosion.

Temperature plays an important role in atmospheric corrosion in two ways. First, there is a normal increase in corrosion activity which can theoretically double for each 10° increase in temperature. Secondly, a little-recognized effect is the temperature lag of metallic objects, due to their heat capacity, behind changes in the ambient temperature. The period of wetness is often much longer than the time the ambient air is at or below the dew point and varies with the section thickness of the metal structure, air currents, RH, and direct radiation from the sun.

FIGURE 9.6 Galvanized bolting assembly after 10 years of exposure to a deicing salt environment.

Cycling temperature may accelerate these processes and produce severe corrosion on metal objects in unheated warehouses, and on metal tools or other objects stored in humid environments. Since the dew point of an atmosphere indicates the equilibrium condition of condensation and evaporation from a surface, it is advisable to maintain the temperature some 10 to 15°C above the dew point to ensure that no corrosion will occur by condensation on a surface that could be colder than the ambient environment.

As the ambient temperature drops during the evening, metallic surfaces tend to remain warmer than the humid air surrounding them and do not begin to collect condensation until some time after the dew point has been reached. As the temperature begins to rise in the surrounding air, the lagging temperature of the metal structures will tend to make them act as condensers, maintaining a film of moisture on their surfaces.

In the following example, a complete truckload of expensive vehicle parts were lost due to this effect. In the "Just in Time" manufacturing era, system parts can be produced at different plants within a few driving hours of the final assembly line. These parts will therefore only be in transit for a few hours and usually do not require much protection against corrosive elements.

On a spring Friday, two trailer trucks were loaded at the part-manufacturing facility, with engine parts packed in their habitual dunnage* for delivery at the assembly plant seven driving hours away. The first truck left early that morning and the parts were received in good condition at the assembly plant. The second truck was loaded later during the day. However, a long weekend was coming and the decision was made by the transport company to park the trailer truck near the manufacturing area for delivery to the assembly plant the following week. The second truck left the manufacturing area on the following Monday afternoon to be delivered the next day to the assembly plant where it was discovered that all the parts were surface rusted and deemed unusable for production.

Figure 9.7 shows a plot of the ambient and dew-point temperatures extracted from the weather data file for the three days the loaded truck spent in the manufacturing area while parked outside without any particular care or attention. An obvious feature in Fig. 9.7 is the very small difference between the ambient and dew-point temperatures during the three nights the truck was parked idle in the manufacturing area. However, this factor alone would not be sufficient to explain the serious corrosion experienced by all the parts stored in the second truck.

A more dramatic situation is revealed by considering that a certain time lag existed between the ambient temperature and the temperature of the steel parts enclosed in their donnage. The five hours lag between these two temperatures, shown as a dotted line in Fig. 9.8, is

* Dunnage is the name for the materials used in holds and containers to protect goods and their packaging from moisture, contamination, and mechanical damage.

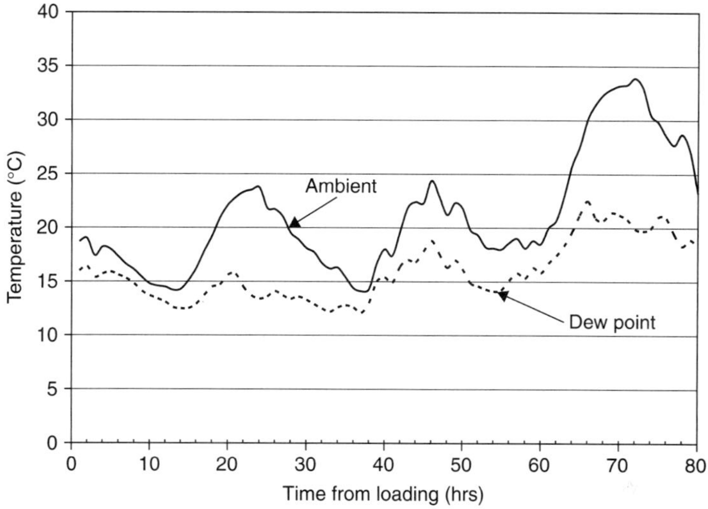

Figure 9.7 Ambient and dew point temperatures as a function of the time of loading of steel parts stored in a truck parked outdoor for three days.

realistic considering the mass of the steel parts and their partial insulation from the ambient environment. Figure 9.8 clearly shows that surface condensation on the vehicle parts was highly probable during the three nights spent by the truck in the parking lot, with the second night being the most dramatic.

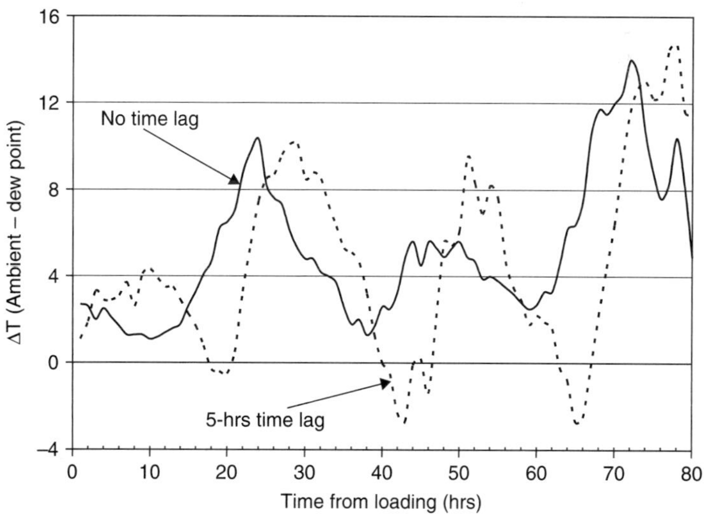

Figure 9.8 Difference between ambient, with and without a time lag, and dew point temperature as a function of the time of loading of steel parts stored in a truck parked outdoor for three days.

9.3.1 Relative Humidity and Dew Point

Relative humidity is defined as the ratio of the quantity of water vapor present in the atmosphere to the saturation quantity at a given temperature, and it is expressed as percent. A fundamental requirement for atmospheric corrosion processes is the presence of a thin film electrolyte that can form on metallic surfaces when exposed to a critical level of humidity. While this film is almost invisible, the corrosive contaminants it contains are known to reach relatively high concentrations, especially under conditions of alternate wetting and drying.

The critical humidity level itself is a variable that depends on the nature of the corroding material, the tendency of corrosion products and surface deposits to absorb moisture, and the presence of atmospheric pollutants [3]. It has been shown, for example, that this critical humidity level is 60 percent for steel when the environment is free of pollutants.

In the presence of thin film electrolytes, atmospheric corrosion proceeds by balanced anodic and cathodic reactions described, respectively, in Eqs. (9.1) and (9.2). The anodic oxidation reaction involves the corrosion attack of the metal, while the cathodic reaction is naturally the oxygen reduction reaction (Fig. 9.9).

$$\text{Anode reaction: } 2Fe \rightarrow 2Fe^{2+} + 4e^- \tag{9.1}$$

$$\text{Cathode reaction: } O_2 + 2H_2O + 4e^- \rightarrow 4OH^- \tag{9.2}$$

Marine environments typically have high RH, as well as salt rich aerosols. Studies have shown that the thickness of the adsorbed layer of water on a zinc surface increases with percent RH and that corrosion rates increase with the thickness of the adsorbed layer. There also seems to be a finite thickness to the water layer that, when exceeded, can limit the corrosion reaction due to limited oxygen diffusion [4]. However, when metallic surfaces become contaminated with hygroscopic salts their surface can be wetted at a lower RH. The presence of magnesium chloride ($MgCl_2$) on a metallic surface can make a surface apparently wet at 34 percent RH while sodium chloride (NaCl) on the same surface requires 77 percent RH to create the same effect [5].

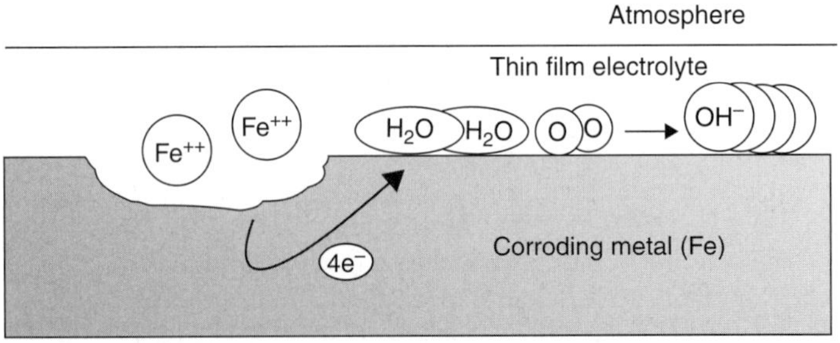

FIGURE 9.9 Schematic description of the atmospheric corrosion of iron.

In absence of any particular surface effects, the dew point corresponds to the temperature at which condensation occurs. A high RH would therefore be associated with a dew point close to the current air temperature. If the RH was 100 percent, for example, the dew point would be equal to the current temperature. Given a constant dew point, an increase in temperature will lead to a decrease in relative humidity. Equation (9.3) provides a convenient way to calculate the dew point as a function of temperature to within ±0.4°C [6]:

$$t_d = \frac{B\left(\ln RH + \frac{At}{B+t}\right)}{A - \ln RH - \frac{At}{B+t}} \tag{9.3}$$

where $A = 17.625$
$B = 243.04°C$
RH is the relative humidity as a fraction (not percent)
t is the surface temperature (°C)
t_d is the dew point temperature (°C)

Equation (9.3) is valid for $0°C < t < 100°C$, $0.01 < RH < 1.0$, and $0°C < T_d < 50°C$. Figure 9.10 illustrates the relationship between the dew point temperature and relative humidity for selected surface temperatures.

9.3.2 Pollutants

Sulfur dioxide (SO_2), a gaseous product of the combustion of fuels containing sulfur such as coal, diesel, gasoline, and natural gas, has

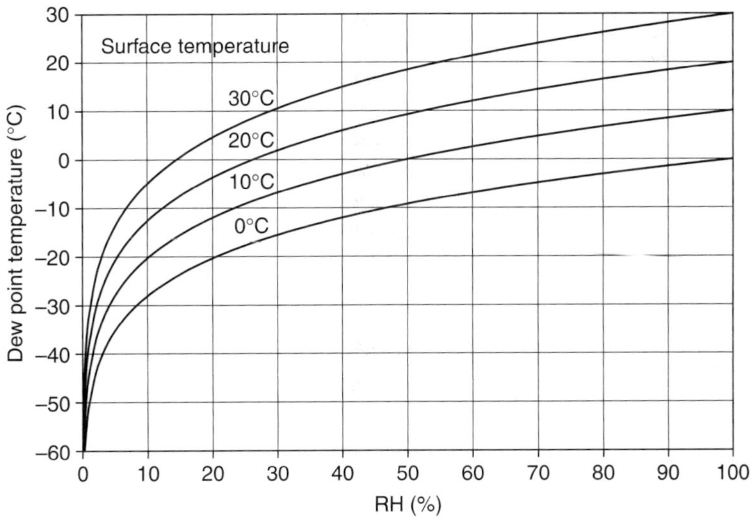

FIGURE 9.10 Relationship between dew point temperature and relative humidity for selected surface temperatures.

been identified as one of the most important air pollutants that contribute to the corrosion of metals.

Less recognized as corrosion promoters, are the nitrogen oxides (NO_x), which are also products of combustion. A major source of NO_x in urban areas is the exhaust fumes from vehicles. Sulfur dioxide, NO_x and airborne aerosol particles can react with moisture and UV light to form new chemicals that can be transported as aerosols. A good example of this is the summertime haze over many large cities. Up to 50 percent of this haze is a combination of sulfuric and nitric acids.

9.3.3 Deposition of Aerosol Particles

The behavior of aerosol particles in outdoor atmospheres is explained by laws that govern their formation, movement, and capture. These particles are present throughout the planetary boundary layer and their concentrations depend on a multitude of factors including location, time of day or year, atmospheric conditions, presence of local sources, altitude, and wind velocity.

The highest concentrations are usually found in urban areas, reaching up to 10^8 and 10^9 particles per cm^3, with particle size ranging from around 100 μm to a few nanometer. Size is normally used to classify aerosol because it is the most readily measured property and other properties can be inferred from size information [7]. The highest mass fraction of particles in an aerosol is characterized by particles having a diameter in the range of 8 to 80 μm [8].

Some studies have indicated that there is a strong correlation between wind speed and the deposition and capture of aerosols. In such a study of saline winds in Spain a very good correlation was found between chloride deposition rates and wind speeds above a threshold of 3 $m\ s^{-1}$ or 11 $km\ h^{-1}$ [9].

Aerosols can either be produced by ejection into the atmosphere, or by physical and chemical processes within the atmosphere (called primary and secondary aerosol production, respectively). Examples of primary aerosols are sea spray and wind-blown dust. Secondary aerosols are produced by atmospheric gases reacting and condensing, or by cooling vapor condensation. Once an aerosol is suspended in the atmosphere, it can be altered, removed, or destroyed.

Aerosol particles do not stay in the atmosphere indefinitely, and average lifetimes are of the order of a few days to a week, depending on their size and location. Aerosol particles have a finite mass and are subject to the influence of gravity, wind resistance, droplet dry-out, and possibilities of impingement on a solid surface. Studies of the migration of aerosols inland of a sea coast have shown that typically the majority of the aerosol particles are deposited close to the shoreline (typically 400 to 600 m) and consist of large particles (>10 μm diameter), which have a short residence time and are controlled primarily by gravitational forces [8;9].

9.3.4 Deicing Salts

Snow fighting has a long history. The first use of salt for deicing roads can be traced back to the 1930s. However it was not until the 1960s that the use of salt in conjunction with plowing became widespread after winter maintenance personnel learned its effectiveness. Figure 9.11 presents the usage of deicing salt or rock salt (mostly sodium chloride) in the United States between 1940 and 2005 (Data from the Salt Institute www.saltinstitute.org). Initially confined to the "snowbelt," ice fighting has become a priority activity in the "sunbelt" too.

The Benefits of Salting

There is a vast international experience in effective snow fighting and the efficient use of deicing salts. A 1972 study by Paul J. Claffey presented to the Highway Research Board (now Transportation Research Board) concluded that the roughness of road ice and slippage of wheels can result in an increase in fuel consumption around 35 percent in averages and as much as 50 percent when there is 5 cm of snow on the road. A 1976 report by the Institute for Safety Analysis (TISA) listed the following cost benefits (reported here using 1976 prices and rates) of using salt in the United States to deice highways:

- Reduces wages lost due to lateness to work by $7.6 billion
- Saves $3 billion in wage loss because of absenteeism
- Reduces production losses by $7 billion

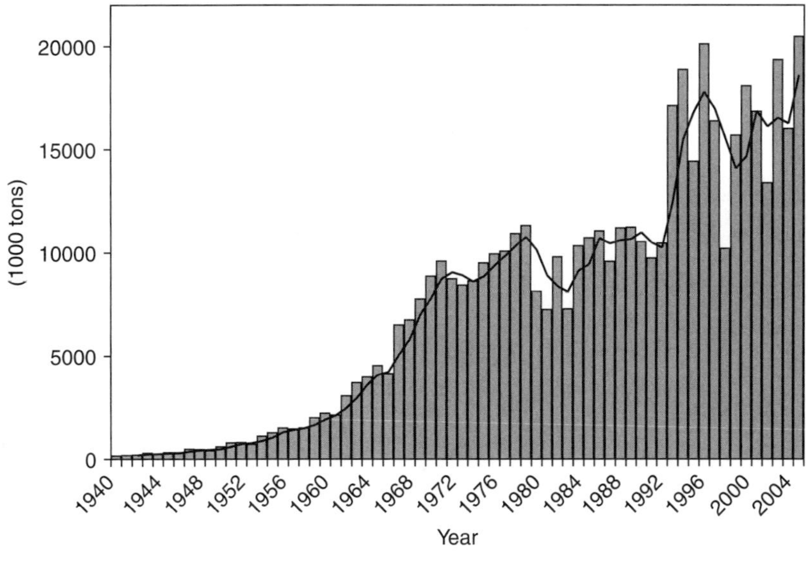

FIGURE 9.11 Usage of rock salt for deicing in the United States in thousands of tons.

- Reduces losses in goods shipment by $600 million
- Saves 1.4 to 4.5 billion L of fuel
- Has an 18:1 benefit/cost ratio

A study published in 1993 [10] concluded that "As a winter maintenance service, deicing pays for itself within the first 25 minutes after the first hour that salt is spread on two-lane highways. Then, during the first four hours after the hour of application of salt, the direct road user benefits were $6.50 for every $1.00 spent on direct maintenance costs for the operation." The study found that costs related to accidents, including medical expenses, emergency services, workplace costs, travel delay, property damage, and administration and legal expenses decrease by 88 percent after the application of deicing salt. Figure 9.12 illustrates the results of two studies of traffic accident rates carried out in two different countries (Germany and the United States) before and after salt spreading at the onset of a snow storm.

Use of salt, in conjunction with a good plowing program, is the fastest and most efficient means of snow and ice removal. The use of abrasives requires at least seven times more material to treat a given distance of roadway. Therefore, it takes seven loads and seven round trips to the loading point, compared to just one for salt, resulting in a greater use of fuel, increased manpower and more time to treat roads during a storm. Studies by the Salt Institute have determined that a

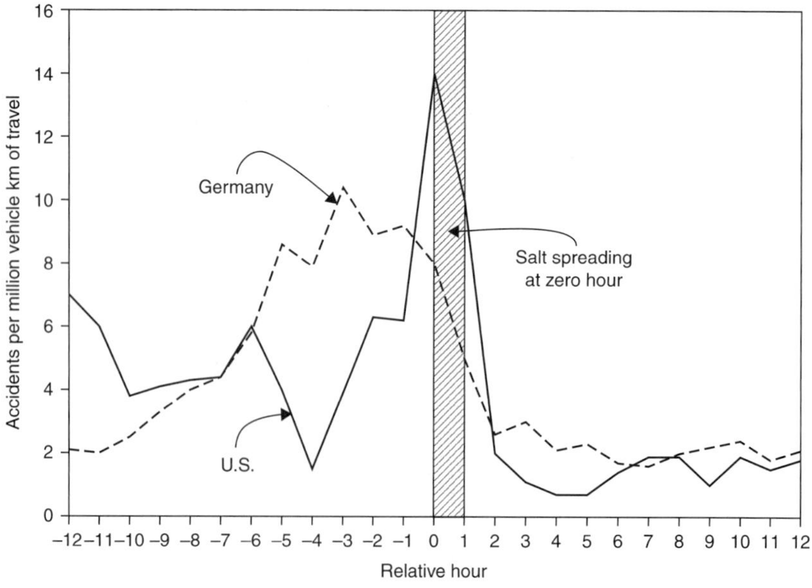

FIGURE 9.12 Statistics revealing the number of accidents before and after applying deicing salts.

loaded salt truck, spreading at the generally accepted rate of 140 kg per two-lane km for general storm conditions, can treat a 36 km stretch of roadway, traveling a total of 72 km. A sand truck requires seven loads, must travel a total distance of 300 km to treat the same section of road and that truck requires four times more fuel. In more ways than one, salt used in snow and ice control contributes to energy savings.

The Bad and the Ugly Corrosion

The massive spreading of salt on roads and highways unfortunately also has some serious negative effects. Besides environmental concerns, one of the major criticisms of salt use for deicing is its contribution to corrosion of metal in steel bridges (Fig. 9.13), road vehicles (Fig. 9.14), reinforced concrete (Fig. 9.15) (bridge decks, parking garages), and any other metallic objects in close proximity to roads and highways (lampposts, statues, buildings, and so forth).

Although another effective but less corrosive deicing agent is commercially available (calcium magnesium acetate or CMA), its price seems to be too high for wide usage. One ton of CMA costs $300 to $600 as opposed to $20 to $70 for rock salt. The use of CMA has

FIGURE 9.13 Pack rust under a steel bridge.

Figure 9.14 A truck all corroded on the driver side due to salt on the road.

Figure 9.15 Reinforced concrete spalling off due to corrosion of steel rebar.

thus been confined to areas where corrosion could cause critically important damage such as for airport winter maintenance. It can therefore be expected that the road environment will likely remain corrosive well into the future.

The impact of salts on protective coatings is widely recognized. Any breach or holiday in the coating will let salts reach the metallic substrate and initiate a very aggressive environment that in turn will force the coating to blister and peel off (Figs. 9.16 and 9.17), sometimes until complete perforation of the metal (Fig. 9.18).

The effect of deicing salts extends much beyond the immediate vicinity where the salts are spread because these salts can travel as aerosol particles generated by the traffic circulation. Figure 9.19 summarizes the results of a study using standard corrosion coupons [11] deployed on a pedestrian walkway across a well-traveled road during the winter months of a moderately cold Canadian city where rock salt is the deicing agent of choice. These results clearly indicate that the corrosion rates (percent mass loss) while being highest closer to the ground (at the bottom of the pillars) are still appreciable many meters above the traffic level. For comparison, similar measurements made in non-trafficked areas of the same city typically showed corrosion rates fifty times smaller during the same exposure period, that is, 0.2 percent mass loss.

In another study, it was demonstrated that high corrosion rates could be measured more than 100 m downwind of a major highway in similar winter conditions Fig. 9.20 [12]. The corrosion trends measured

FIGURE 9.16 Blistering and peeling of protective coating subjected to regular deicing salt application.

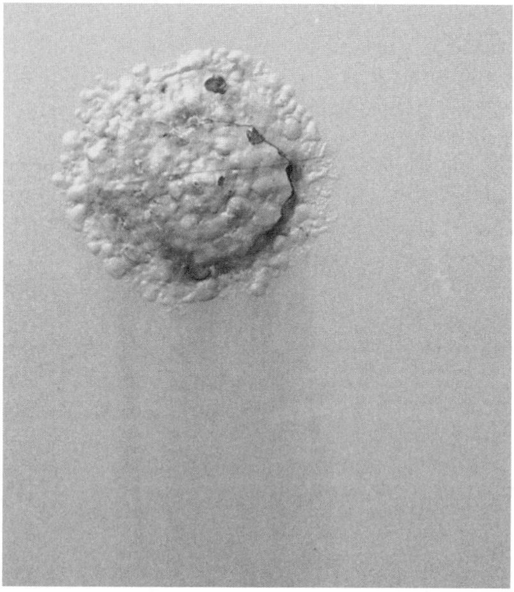

Figure 9.17 Filiform corrosion and paint blistering accelerated by deicing salt.

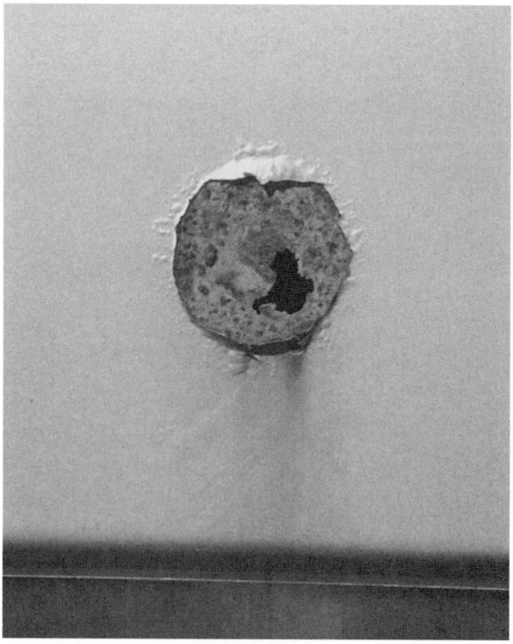

Figure 9.18 Fallen blister showing complete perforation.

FIGURE 9.19 Results of a study investigating the transport and effects of deicing salt-laden aerosols.

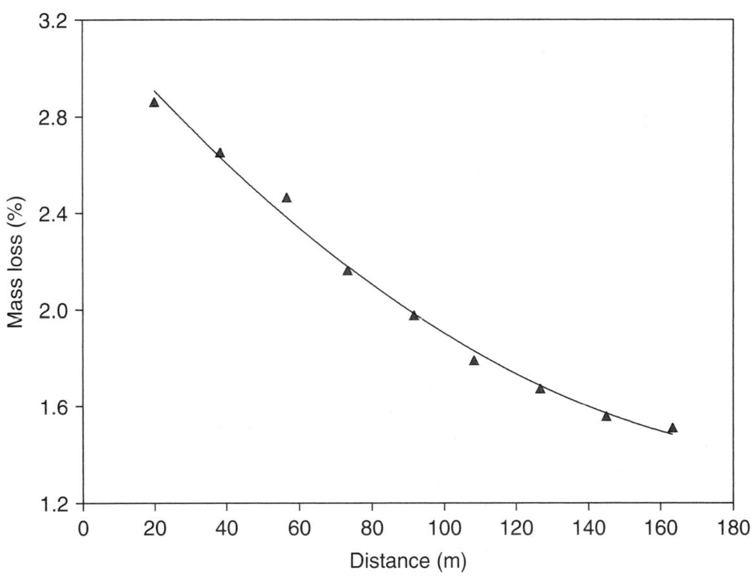

FIGURE 9.20 The corrosion impact of deicing salts downwind of a major highway.

Figure 9.21 Decreasing corrosion of a sign post as a function of height.

in these studies also leave a visible imprint on many structures adjacent to roads and highways. The shop sign post shown in Fig. 9.21, for example, has a decreasing level of coating blistering and peeling as a function of height while the overpass bridge in New York (Fig. 9.22) shows a clear imprint of the corrosion due to chloride rich aerosols.

Figure 9.22 Corroded overpass steel bridge above the most traveled lane of a two-lane highway, a telltale symptom of deicing salt corrosion.

9.4 Measurement of Atmospheric Corrosivity Factors

Various methods have been developed for measuring many of the factors that influence atmospheric corrosion. The quantity and composition of pollutants in the atmosphere, the amount collected on surfaces under a variety of conditions, and the variation of these with time have been determined. Temperature, RH, wind direction and velocity, solar radiation, and amount of rainfall are easily recorded. Not so easily determined are dwelling time of wetness (TOW), and the surface contamination by corrosive agents such as sulfur dioxide and chlorides. However, methods for these determinations have been developed and are in use at various test stations. By monitoring these factors and relating them to corrosion rates, a better understanding of atmospheric corrosion can be obtained.

9.4.1 Time of Wetness

Time of wetness is an estimated parameter based on the length of time when the relative humidity is greater than 80 percent at a temperature greater than 0°C. It can be expressed as the hours or days per year or the annual percentage of time.

A method of measuring the TOW has been developed by Sereda and correlated with the corrosion rates encountered in the atmosphere [13]. The moisture sensing elements in this sensor are manufactured by plating and selective etching of thin films of appropriate anode (copper) and cathode (gold) materials in an interlaced pattern on a thin nonconductive substrate (Fig. 9.23). When moisture condenses on the sensor it activates the cell, producing a small voltage (0 to 100 mV) across a 10^7 Ω resistor.

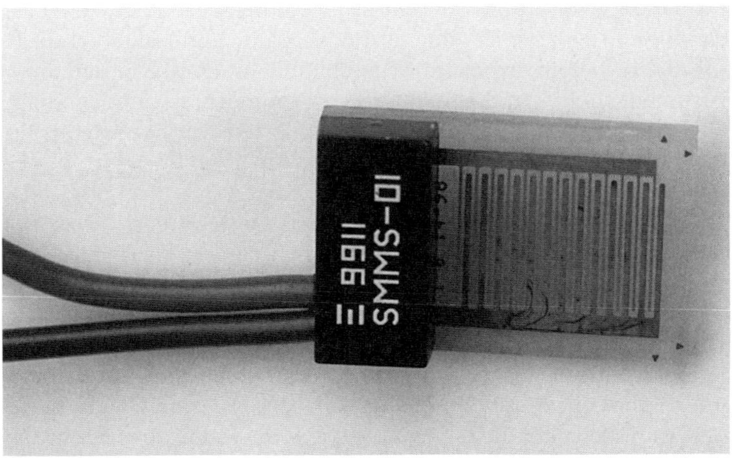

FIGURE 9.23 Interlocking combs of gold and copper electrodes in a Sereda humidity sensor.

Thin sensing elements are preferred in order to preclude influencing the surface temperature to any extent. Although a sensor constructed using a 1.5-mm thick glass-reinforced polyester base has been found to be satisfactory on plastic surfaces, this will not be the case with the same sensing element on a metal surface with a high thermal conductivity [14]. For metal surfaces, the sensing element should be appreciably thinner. Commercial epoxy sensor backing products of a thickness of 1.5 mm, or less, are suitable for this purpose.

9.4.2 Sulfur Dioxide

Sulfur dioxide is usually measured in terms of its concentration in air in units of $\mu g/m^3$. Precise methods are available to monitor continuously the amount of sulfur dioxide in a given volume of air. The pollution levels may also be measured in terms of the concentration of dissolved sulfate (SO_4^{2-}) in rain water. However, this is only indirectly related to the effect of sulfur dioxide on corrosion since only the actual amount of hydrated sulfur dioxide or sulfur trioxide deposited on metal surfaces is important.

There are two widely used methods for determining the sulfur dioxide (SO_2) concentration in the atmosphere of interest. Both employ the affinity of lead oxide to react with gaseous SO_2 to form lead sulfate. The most common method used in corrosion work is the sulfation. These devices can be either purchased or prepared in the laboratory. They consist of small disks of lead oxide that are exposed facing the ground under a small shelter to prevent the reactive paste from being removed by the elements [15].

The disk surface is thus exposed only to gaseous SO_2, not particulates. The American Society for Testing of Materials (ASTM) procedure suggests a 30-day exposure, followed by a standard sulfate analysis [16]. The other method sometimes used is the peroxide candle, similar in its function to the chloride candle, but again using lead peroxide to capture SO_2. In this procedure, a lead peroxide paste is applied to a paper thimble in the laboratory, and allowed to dry thoroughly before exposure. The thimble is then exposed in an instrument shelter to the test yard environment. In both cases, the SO_2 deposited results are appropriately reported in terms of deposition rate on the surface in units of $mg/m^2/day$.

9.4.3 Airborne Chlorides

Airborne salinity refers to the content of gaseous and suspended salt in the atmosphere. It is measured by the concentration in the air in units of $\mu g/m^3$. Since it is the salt that is deposited on the metal surface that affects the corrosion, it is usually reported in terms of deposition rate in units of $mg/m^2/day$. Chloride levels can also be measured in terms of the concentration of the dissolved salt in rain water.

A number of methods have been employed for determining the contamination of the atmosphere by aerosol transported chlorides, for example, sea salt and road deicing salts. The *wet candle method*, for example, is relatively simple, but has the disadvantage that it also collects particles of dry salt that might not deposit otherwise [17]. This technique uses a wet wick of a known diameter and surface area to measure aerosol deposition (Fig. 9.24). The wick is maintained wet using a reservoir of water or 40 percent glycol/water solution. Particles of salt or spray are trapped by the wet wick and retained. At intervals, a quantitative determination of the chloride collected by the wick is made and a new wick is exposed.

In reality, the wet candle method gives an indication of the salinity of the atmosphere rather than the contamination of exposed metal surfaces. The technique is considered to measure the total amount of chloride arriving to a vertical surface and its results may not be truly significant for corrosivity estimates.

In order to understand the chloride deposition rates in a confined spaces, such as ventilated subfloors in a sea coastal area, a special collecting box shown in Fig. 9.25 was used in which the airborne chlorides were collected on horizontal and vertical filter papers positioned at different locations from the box openings (Fig. 9.26).

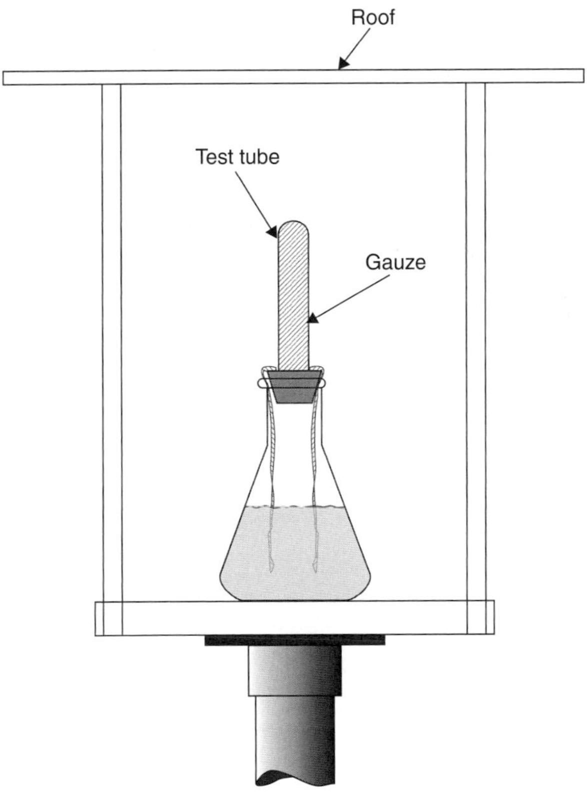

FIGURE 9.24 Schematic of a wet candle chloride apparatus.

FIGURE 9.25 Subfloor simulation box surmounted by a wet candle chloride measuring device that has been installed at several sites in Australia and New Zealand. (Courtesy of Branz, New Zealand)

FIGURE 9.26 Horizontal and vertical filter papers installed inside a subfloor simulation box. (Courtesy of Branz, New Zealand)

The roof protected the surface from direct rain but the filter papers were still exposed to deposition of the airborne chlorides and possibly some rain. Filter papers were removed every month for chemical analysis. A wet candle with its own roof cover was installed on each box (Fig. 9.25) to provide a measure of the level of chlorides outside the boxes.

9.4.4 Atmospheric Corrosivity

The simplest form of direct atmospheric corrosion measurement is by coupon exposures. Subsequent to their exposure, coupons can be subjected to weight-loss measurements, pit density and depth measurements and to other types of examination. The main drawback associated with conventional coupon measurements is that extremely long exposure times are usually required to obtain meaningful data, even on a relative scale. It is not uncommon for such programs to run for 20 years or longer.

Some variations of the basic coupon specimens can provide rapid material corrosivity evaluations. The helical coil adopted in the ISO 9226 methodology is a high surface area/weight ratio coupon that gives a higher sensitivity than panel coupons of the same material. The use of bimetallic specimens in which a helical A91100 aluminum wire is wrapped around a coarsely threaded bolt may provide additional sensitivity and forms the basis of the Classify Industrial and Marine Atmospheres (CLIMAT) coupon [11;18].

The mass loss of the aluminum wire of a CLIMAT coupon after 90 days of exposure is considered to be a relative measure of atmospheric corrosivity. However, the results vary greatly between the various combinations of materials suggested in the ASTM standard [11]. The aluminum wire on copper bolts has been found by many to be the most sensitive of the three proposed arrangements in the ASTM standard. The use of triplicate coupons on a single holder additionally provides an indication of the reproducibility of the measurements and the use of vertical rods can reveal directional information on the corrosive agents as will be illustrated in the following examples.

A CLIMAT coupon with three copper rods installed at the NASA Kennedy Space Center (KSC) beach corrosion test site (Fig. 9.27) is shown immediately after it had been installed [Fig. 9.28(*a*)], after 30 days [Fig. 9.28(*b*)], and after 60 days [Fig. 9.28(*c*)]. KSC having the highest corrosivity of any test site in the continental United States [19], the mass loss recorded even after a shorter exposure than usual can be very high. In the present example it was already 16 percent of the original aluminum wire after 60 days. The base support of these CLIMATs having been purposefully installed parallel to the sea coast the directional effect of the marine salts may be illustrated by comparing the front and back of the exposed CLIMATs [Fig. 9.29(*a*) and (*b*)].

In another study focused on the shielding effects of buildings in a marine environment, the directional impact of marine aerosols was revealed by comparing the level of patina on CLIMAT's copper rod

FIGURE 9.27 Aerial view of the NASA Kennedy Space Center beach corrosion test site where atmospheric corrosivity is the highest corrosivity of any test site in the continental United States.

(a)

FIGURE 9.28 A CLIMAT coupon with three copper rods immediately after it was installed at the Kennedy Space Center beach corrosion test site (*a*), after 30 days (*b*), and after 60 days (*c*).

(b)

(c)

Figure 9.28 *(continued)*

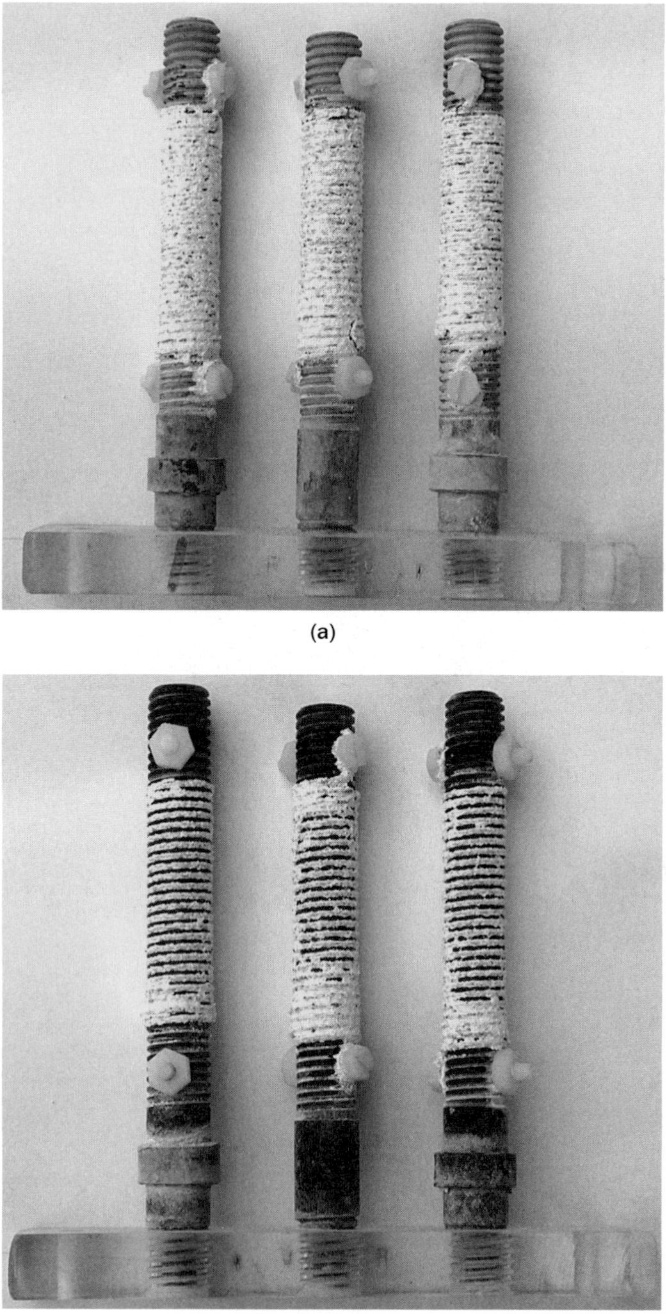

(a)

(b)

FIGURE 9.29 Close-up pictures of a CLIMAT coupon exposed at KSC for two months: (*a*) seaside and (*b*) backside.

exposed for three winter months at facilities on the Pacific coast [20]. What became evident in that study was that the pattern of the bluish-green patina attributed to the corrosion product $CuCl_2 \cdot 2H_2O$ was not uniformly distributed around the circumference of each copper rod.

A template with 16 points of the compass was placed onto the outside of each copper rod that was on the most boldly exposed CLIMATs of that study in order to visually quantify the intensity of the bluish-green color. The relative degree of corrosion for each compass point was assessed by assigning a number between 0 and 10 with 0 corresponding to 0 bluish-green patina and 10 corresponding to 100 percent coverage of the colored corrosion product. The average corrosion index for each of the 16 points of the compass for the copper rods is shown in Fig. 9.30.

One attempt to correlate the directional corrosivity observed on the copper rods with weather data was to plot the fraction of time

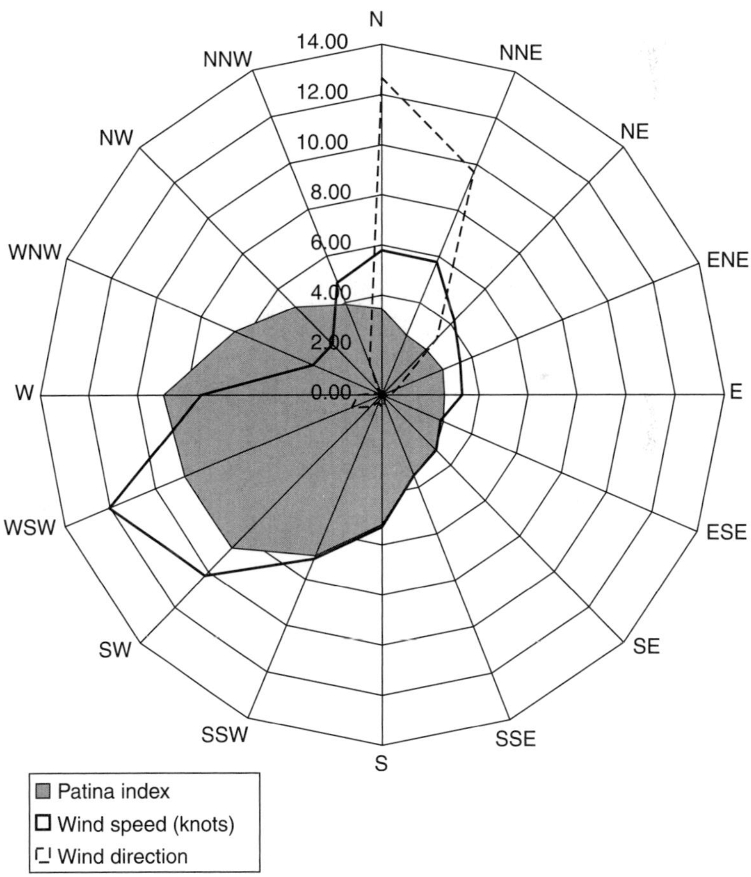

FIGURE 9.30 Average corrosion index for the copper rods exposed on the rooftop and the average wind speed recorded at the local weather station as a function of the 16 points of the compass.

that winds came from the 16 points of the compass during the three-month exposure period. The dominant direction was the north to north-east. However, the pattern of corrosion product did not correspond to the most frequent wind direction but did correspond to the direction with the highest wind speeds, which are in the west to south quadrant at this particular site (Fig. 9.30).

9.5 Atmospheric Corrosivity Classification Schemes

9.5.1 Environmental Severity Index

An environmental corrosivity scale, based on atmospheric parameters, has been developed over the years for the maintenance management of structural aircraft systems [21]. A corrosion damage algorithm (CDA) was proposed as a guide for anticipating the extent of corrosion damage and planning maintenance operations. This classification scheme was developed primarily for uncoated aluminum, steel, titanium, and magnesium alloys exposed to the external atmosphere at ground level.

The section of the CDA algorithm presented in Fig. 9.31 considers the distance to salt water, leading either to the very severe AA rating for close distance to seashore or a consideration of moisture factors. Following the moisture factors, pollutant concentrations are compared with values of Working Environmental Corrosion Standards (WECS). The WECS values were adopted from the 50th percentile median of a study aimed at determining ranges of environmental parameters in the United States and represent "average of averages." For example, a severe A rating would be given if any of the three pollutants considered in this scheme, that is, sulfur dioxide, total suspended particles, and ozone levels, would exceed the WECS values in combination with a high moisture factor.

The environmental corrosivity, predicted from the CDA algorithm, of six marine air bases has been compared to the actual corrosion maintenance effort expended at each base. Considering the simplicity of the algorithms and simplifying assumptions in obtaining relevant environmental and maintenance data the correlation obtained was considered to be reasonable. However, subsequent attempts to enhance the PACER LIME algorithm by using the results obtained from broad-based corrosion testing programs have failed to provide enough differentiation between moderately corrosive environments [22].

In order to remedy the deficiencies in the CDA scheme, Battelle Memorial Institute has been tasked to monitor the atmospheric corrosivity of air force and other sites worldwide [23]. The database describing the relative corrosive severity levels of different locations and actual corrosion rates of a variety of metals has now grown to more than 100 sites worldwide. The metals included in that study are three aluminum alloys (A92024, A96061, and A97075), copper, silver, and steel.

Data have been collected for metals directly exposed to the outdoor environment in a standard sample mounting configuration

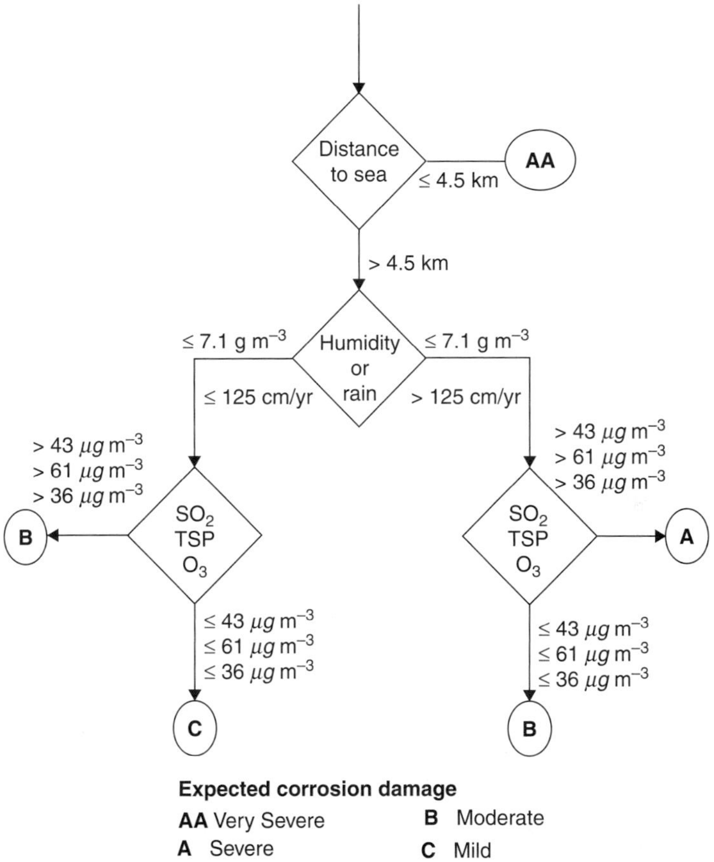

FIGURE 9.31 The PACER LIME algorithm for determining the corrosion severity for a given location.

and test package. A typical plastic test rack with its metallic coupons is shown in Fig. 9.32 besides a CLIMAT coupon exposed at the KSC beach corrosion test site. Figure 9.33(*a*) shows a close-up view of the coupons before exposure. Once exposed to the environment for a given period of time, the corroded metal strips [(Fig. 9.33(*b*)] are sent back to the laboratory for mass loss measurements following standard methods [24] and further analysis.

These efforts established a correlation between the corrosivity of various air force base environments to aluminum and the costs associated with the maintenance of various aircraft. Figure 9.34 summarizes three years of corrosion data compared to maintenance records.

Figure 9.32 Metal coupons in a plastic test rack exposed at the Kennedy Space Center beach corrosion test site besides a CLIMAT coupon. (Courtesy of Battelle)

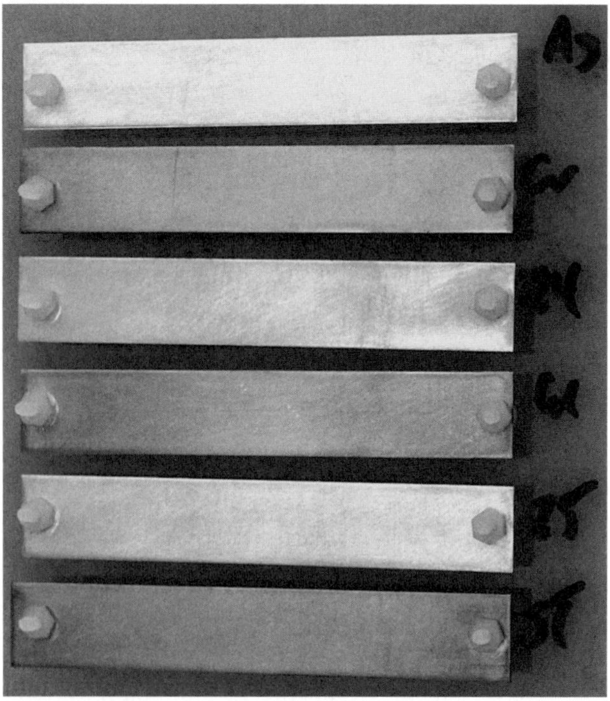

(a)

Figure 9.33 Metal coupons before exposure to the environment (*a*) and after a three month exposure in a rural environment (*b*). (Courtesy of Battelle)

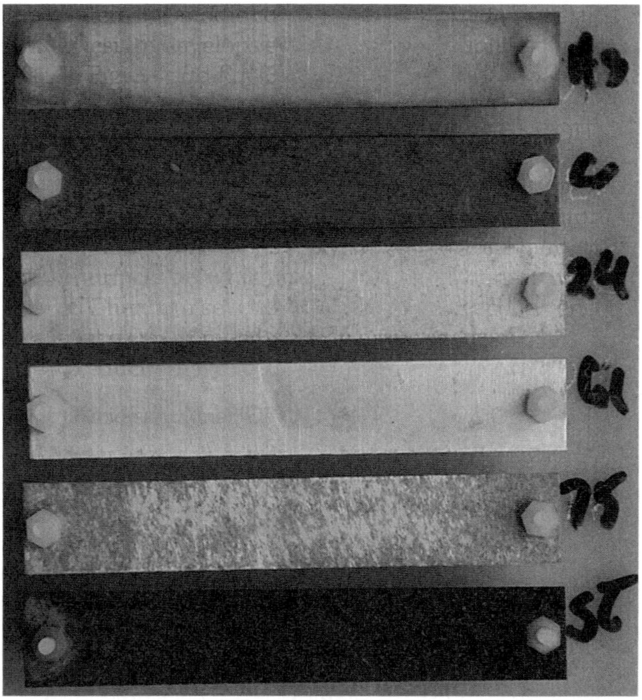

(b)

Figure 9.33 *(continued)*

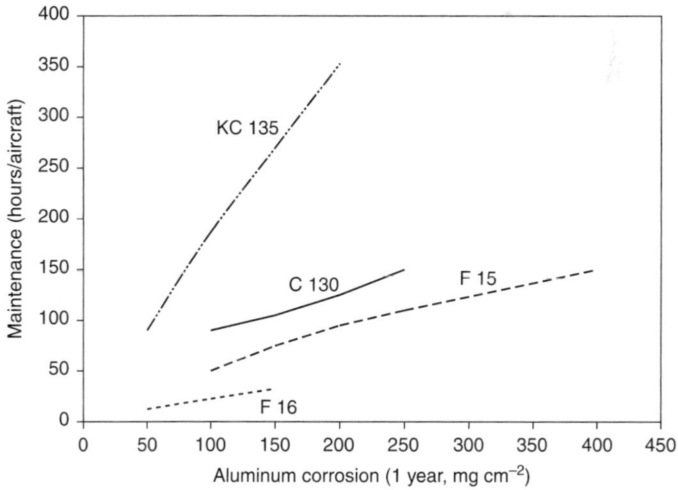

Figure 9.34 Maintenance costs for various aircraft as a function of the corrosivity of air force base environments toward the corrosion of aluminum. (KC 135 is the refueling support aircraft; CC130 is the Hercules airlift mission aircraft; F15 and F16 are fighter aircraft).

9.5.2 ISO Classification of Corrosivity of Atmospheres

The International Organization for Standardization (ISO) has implemented a worldwide atmospheric exposure program known as ISO CORRAG. This program included exposure of steel, copper, zinc, and aluminum for one, two, four, and eight years at 51 sites located in 14 nations in order to generate the necessary data to produce a classification system for predicting atmospheric corrosivity from commonly available weather data [25].

Triplicate specimens were used for each exposure. The metals selected were a low carbon steel from a single supplier and commercially pure zinc, copper, and aluminum. These nonferrous metals were obtained from local sources in each of the participating nations. The program was initiated in 1986 and closed in 1998. After a planned exposure, each specimen was sent to the laboratory that had done the initial weighing for cleaning and evaluation.

Mass loss values were obtained and converted to corrosion thickness loss values [25]. Based on these data, a simple classification scheme of five corrosivity classes was established for each metal (Table 9.1). The environmental and weather data gathered in this program are based on SO_2, and Cl^- deposition rates (Table 9.2) combined with TOW measurements (Table 9.3) at each site to provide an estimated corrosivity class (Table 9.4).

Experience from applying the ISO classification system has shown, however, that certain observations need further clarification. Substantial corrosion has, for example, been measured on specimens exposed at temperatures well below 0°C, which is in direct contradiction with the ISO criterion which assumes that no corrosion occurs when the temperature falls below 0°C [26]. It has been proposed on the basis of such results, that the TOW in these environments be estimated as the length of time that relative humidity exceeds 50 percent and the ambient temperature exceeds –10°C.

9.5.3 Maps of Atmospheric Corrosivity

Maps are powerful tools for communicating information related to geographical landscapes and corrosivity maps of various countries have been drawn to illustrate the corrosion severity of regions of these

Corrosion Category	Steel (g/m² y)	Copper (g/m² y)	Aluminum (g/m² y)	Zinc (g/m² y)
C_1	≤10	≤0.9	negligible	≤0.7
C_2	11–200	0.9–5	≤0.6	0.7–5
C_3	201–400	5–12	0.6–2	5–15
C_4	401–650	12–25	2–5	15–30
C_5	651–1500	25–50	5–10	30–60

Table 9.1 ISO 9223 Corrosion Rates after One Year of Exposure Predicted for Different Corrosivity Classes

Sulfur Dioxide Category	Sulfur Dioxide Deposition Rate (mg/m² day)	Chloride Category	Chloride Deposition Rate (mg/m² day)
P_0	≤10	S_0	≤3
P_1	11–35	S_1	4–60
P_2	36–80	S_2	61–300
P_3	81–200	S_3	301–1500

Table 9.2 ISO 9223 Classification of Sulfur Dioxide and Chloride Pollution Levels

Wetness Category	Time of Wetness (percent)	Time of Wetness (hours per year)	Examples of Environments
T_1	<0.1	<10	Indoor with climatic control
T_2	0.1–3	10–250	Indoor without climatic control
T_3	3–30	250–2500	Outdoor in dry, cold climates
T_4	30–60	2500–5500	Outdoor in other climates
T_5	>60	>5500	Damp climates

Table 9.3 ISO 9223 Classification of Time of Wetness

TOW	Cl^-	SO_2	Steel	Cu & Zn	Al
T_1	S_0 or S_1	P_1	1	1	1
		P_2	1	1	1
		P_3	1–2	1	1
	S_2	P_1	1	1	2
		P_2	1	1	2
		P_3	1–2	1–2	2–3
	S_3	P_1	1–2	1	2
		P_2	1–2	1–2	2–3
		P_3	2	2	3
T_2	S_0 or S_1	P_1	1	1	1
		P_2	1–2	1–2	1–2
		P_3	2	2	3–4

Table 9.4 ISO 9223 Corrosivity Categories of Atmosphere

TOW	Cl^-	SO_2	Steel	Cu & Zn	Al
	S_2	P_1	2	1–2	2–3
		P_2	2–3	2	3–4
		P_3	3	3	4
	S_3	P_1	3–4	3	4
		P_2	3–4	3	4
		P_3	4	3–4	4
T_3	S_0 or S_1	P_1	2–3	3	3
		P_2	3–4	3	3
		P_3	4	3	3–4
	S_2	P_1	3–4	3	3–4
		P_2	3–4	3–4	4
		P_3	4–5	3–4	4–5
	S_3	P_1	4	3–4	4
		P_2	4–5	4	4–5
		P_3	5	4	5
T_4	S_0 or S_1	P_1	3	3	3
		P_2	4	3–4	3–4
		P_3	5	4–5	4–5
	S_2	P_1	4	4	3–4
		P_2	4	4	4
		P_3	5	5	5
	S_3	P_1	5	5	5
		P_2	5	5	5
		P_3	5	5	5
T_5	S_0 or S_1	P_1	3–4	3–4	4
		P_2	4–5	4–5	4–5
		P_3	5	5	5
	S_2	P_1	5	5	5
		P_2	5	5	5
		P_3	5	5	5
	S_3	P_1	5	5	5
		P_2	5	5	5
		P_3	5	5	5

Table 9.4 *(continued)*

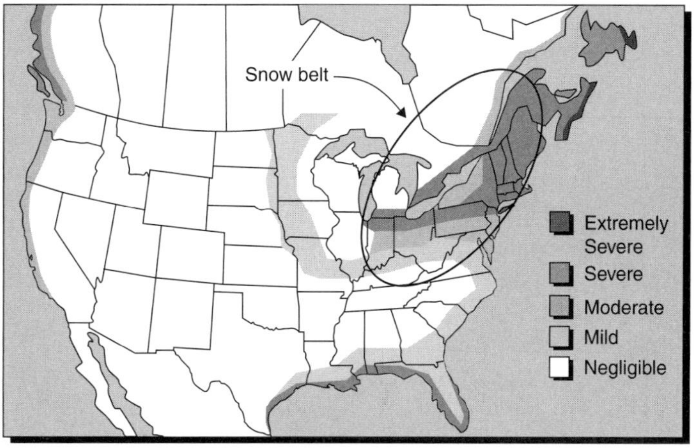

Figure 9.35 Corrosivity map of North America showing the particular aggressiveness of the snowbelt region.

countries [27]. One of the very first atmospheric corrosivity maps was produced to summarize many years of results obtained by exposing bare steel coupons attached to different vehicles in the northeastern United States and Canada (Fig. 9.35) [28]. The results presented in this figure reveal the corrosive effects of using deicing salts since this is the main factor that can explain the higher level of vehicle corrosion in the snowbelt region when compared to adjacent non-marine regions.

Similar corrosivity maps have been created from data available in the literature, such as the corrosivity map of Mexico shown in Fig. 9.36 [29], the map of Japan (Fig. 9.37) that shows an increasing corrosivity from north to south regions [30], and the map of India (Fig. 9.38) [31].

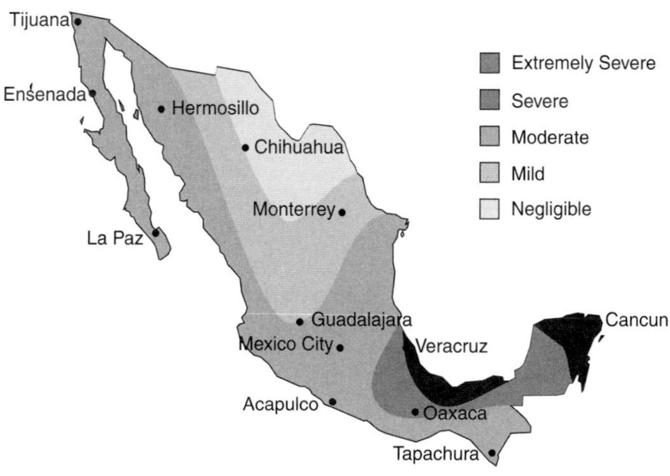

Figure 9.36 Corrosivity map of Mexico.

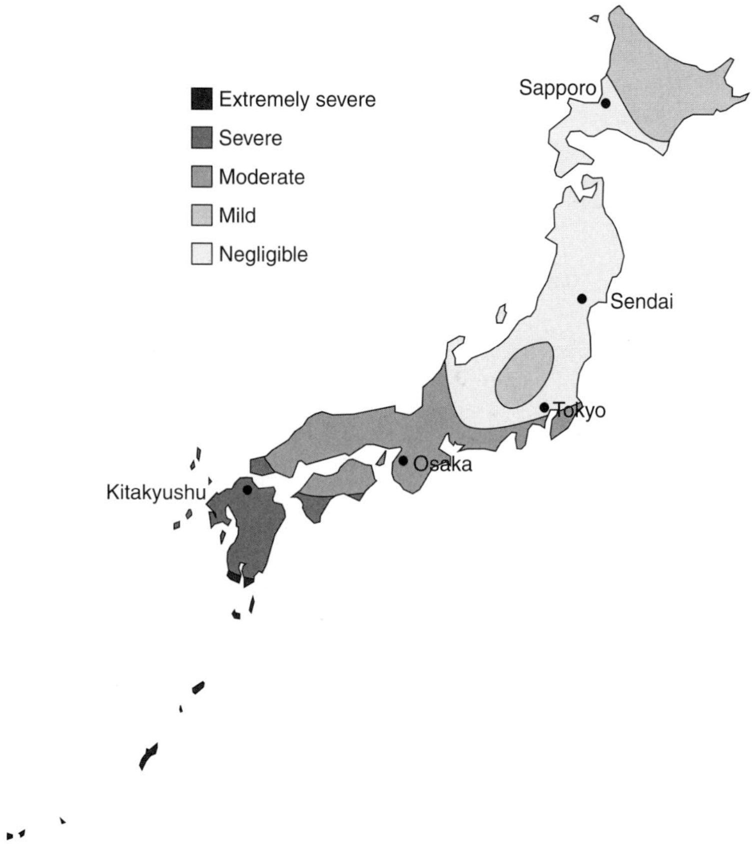

FIGURE 9.37 Corrosivity map of Japan.

9.6 Atmospheric Corrosion Tests

For many years, different individuals and organizations have been putting specimens out in the atmosphere in all kinds of tests and shapes. There has been an attempt to standardize some of these tests, as well as the method of reporting the results. When initiating corrosion tests in the atmosphere, these standardized procedures should be consulted and used whenever possible.

Flat panels exposed boldly to the environment on special exposure racks are a common method for testing materials and protective coatings (Fig. 9.39). Various other specimen configurations have been used, including stressed U-bend or C-ring specimens for stress corrosion cracking (SCC) studies, and crevice assemblies to verify the susceptibility of specific configurations (Fig. 9.40).

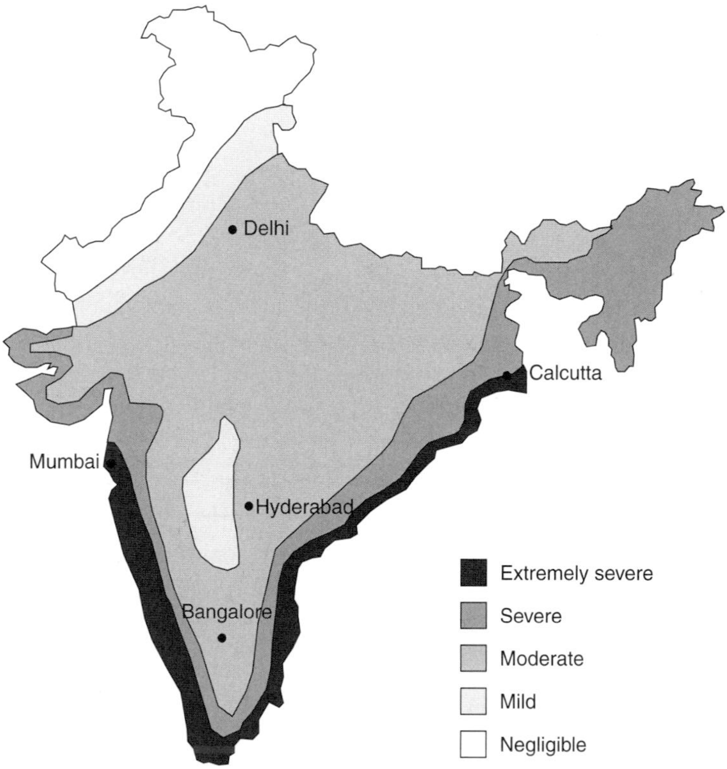

FIGURE 9.38 Corrosivity map of India.

FIGURE 9.39 Atmospheric test rack exposed to marine West Coast environments. (Courtesy Defence R&D Canada-Atlantic)

FIGURE 9.40 Aluminum and magnesium test specimens equipped with crevice spacers and mounted on a rack to be exposed at the previous test station. (Courtesy of DSTO Australia)

In addition to the atmospheric corrosivity conditions at the test site the following factors are important for the design and interpretation of atmospheric corrosion tests:

- Shape of the specimen
- Direction it faces
- Amount of shelter, drip, or runoff from other specimens
- Elevation
- Shading
- Unusual contamination

Panel specimens are typically placed in racks at a 30-degree angle to the horizontal, facing the source of corrosive elements. They are electrically insulated from the racks on which they are mounted and are arranged so that drip from neighboring panels does not contaminate them (Fig. 9.39). Cylindrical specimens are mounted horizontally, facing the same direction as the panels. They may be exposed fully to the weather or be partially sheltered, depending upon the requirements of the test. There is a great variety of test fixtures and shape of specimens used for environmental cracking tests. Figures 9.41 and 9.42 show respectively a full exposure and a semi-sheltered test stations equipped with TOW galvanic sensors, temperature, and SO_2 level recorded with a data logger.

In most exposure tests, enough specimens are used so that removals may be made after predetermined periods of 1 to 20 years. Very short-term tests usually can be misleading in that the condition

Figure 9.41 Outdoor exposure test station with specimen rack on top and data logger attached below at an army base in Northern Australia. (Courtesy of DSTO Australia)

Figure 9.42 Semisheltered test station with test coupons, wet candle apparatus, and data logger attached at an army base in Northern Australia. (Courtesy of DSTO Australia)

FIGURE 9.43 Coated panels exposed on an atmospheric test rack in a marine environment with CLIMAT coupons for corrosivity calibration. (Courtesy of Defence R&D Canada-Atlantic)

of the metal surfaces during the first few days of exposure may affect the initial corrosion rate, or average weather conditions may not be encountered during the initial exposure period.

Some tests are continued until failure, as in SCC testing. However, in tests of protective coatings, as well as in others, periodic measurements of typical properties such as degree of undercutting are more adequate (Fig. 9.43).

9.7 Corrosion Behavior and Resistance

The following sections contain a brief description of the atmospheric corrosion performance of some metallic materials commonly used in atmospheric conditions.

9.7.1 Iron, Steel, and Stainless Steel

Iron, in its various forms, is exposed to all kinds of environments. It tends to be highly reactive with most of them because of its natural tendency to form iron oxide. When it does resist corrosion it is due to the formation of a thin film of protective iron oxide on its surface by reaction with oxygen in the air. This film can prevent rusting in air at 99 percent RH, but a contaminant such as acid rain may destroy the effectiveness of the film and permit continued corrosion. Thicker films of iron oxide may act as protective coatings, and after the first year or so, could reduce the corrosion rate, as shown in Fig. 9.44.

While the corrosion rate of bare steel tends to decrease with time, the difference in corrosivity of different atmospheres toward unalloyed cast irons or steels can be quite dramatic. The relative corrosivity for open-hearth steel in atmospheres ranging from a desert to the spray zone on an ocean beach is shown in Table 9.5. Similar ranges in corrosivity were determined by the ISO 9223 corrosion rates for steel (Table 9.1). In a few cases, the corrosion rates of ferrous metals have been reported as increasing with time, and careful analysis of the exposure conditions generally reveals that an accumulation of contaminating corrosive agents has occurred, thus changing the severity of the exposure.

It is generally conceded that steels containing only very low amounts of copper are particularly susceptible to severe atmospheric corrosion. In one test over a 3½-year period in both a marine and an industrial atmosphere, a steel containing 0.01 percent copper corroded at a rate of 80 μm/y, whereas increasing the copper content by a factor of five reduced the corrosion rate to only 35 μm/y. Further additions of small amounts of nickel and chromium reduced the corrosion rate to 10 μm/y.

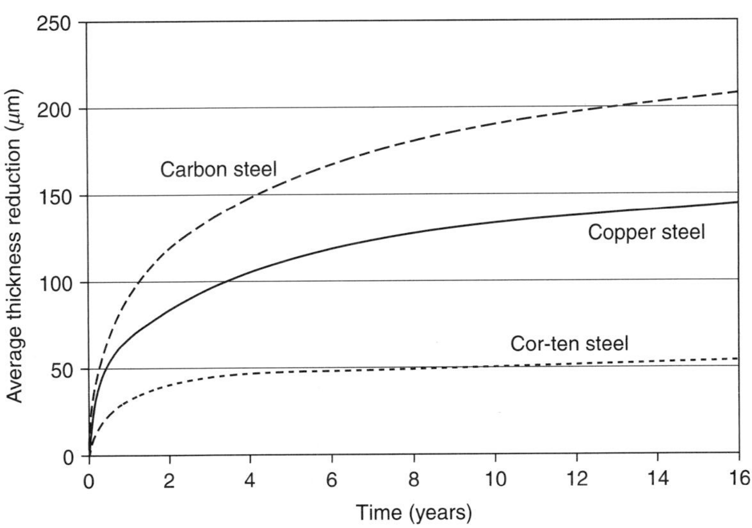

FIGURE 9.44 Time–corrosion curves of three steels in industrial atmosphere, Kearny, N.J.

Khartoum, Egypt	1
Abisco, North Sweden	3
Singapore, Malaysia	9
Daytona Beach, Fla. (Inland)	11
State College, Pa.	25
South Bend, Pa.	29
Miraflores, Canal Zone, Panama	31
Kure Beach, N.C., 250 m. from ocean	38
Sandy Hook, N.J.	50
Kearny, N.J.	52
Vandegrift, Pa.	56
Pittsburgh, Pa.	65
Frodingham, U.K.	100
Daytona Beach, Fla., near ocean	138
Kure Beach, N.C., 25 m. from ocean	475

TABLE 9.5 Relative Atmospheric Corrosion Susceptibility of Open-Hearth Steel

As shown in Table 9.5, it is impossible to give a corrosion rate for steel in the atmosphere without specifying the location, composition, and certain other factors. If one can relate exposure conditions to those described in the literature, a fairly good estimate can be made of the probable corrosion behavior of a selected material. However, all aspects of the exposure of the metal surface must be considered.

A high-strength low-alloy (HSLA) steel, also called *weathering steel*, may show an advantage in corrosion resistance of 12:1 over carbon steel when freely exposed in a mild environment. This is why weathering steel has been the choice of many designers for the construction of boldly exposed surfaces, from buildings to utility poles (Fig. 9.45). While weathering steel does not require any particular care once installed, it may suffer surprising corrosion attack in crevice areas. As the severity or the physical conditions of exposure change, the HSLA steel will show less superiority. In crevices or on the backside of structural forms in a corrosive atmosphere, HSLA steel will in fact not perform better than plain carbon steel (Fig. 9.46).

Very little needs to be said about the behavior of stainless steels (Types 200 and 300), which contain high percentages of nickel and chromium, except that they can keep their shiny aspect without tarnishing for many decades, as illustrated in "The Triad," a tall sculpture erected in the busy part of Toronto, Canada in 1984 (Fig. 9.47). Stainless steel has also been used to great advantage in more notorious buildings and monuments.

Figure 9.45 A 35-m-high highway lamppost made of weathering steel.

The Chrysler Building completed in New York, in 1930 was the first high-profile stainless steel application in the world. Type 302 stainless steel was used for the production of six rows of arches topped with a stainless steel spire. Stainless steel gargoyles were installed on the 31st and 61st floors. The present day condition is estimated to be very good. Only small dents from the cleaning equipment have been observed. Minor pitting above the 61st floor balcony has been removed with polishing.

The exterior of the Empire State Building completed one year later (1931), only a few blocks from the Chrysler Building, was made of stainless steel, gray limestone, and dark gray aluminum. Over 300 t of 1.3-mm-thick Type 302 stainless steel have been used for its construction. This historic landmark with a height of 282 m was the world's tallest building for 41 years. The stainless steel was estimated to be in excellent condition 70 years after the construction of this historical landmark.

804 metric tons of Type 304 stainless steel plates, 6.3 mm thick, were used more recently for the construction of the Gateway Arch erected in 1965 in St. Louis, Mo. on the bank of the Mississippi River. The 190-m-high

FIGURE 9.46 Pack rust developing at the base of the weathering steel highway lamppost shown in Fig. 9.45, only 15 years after its installation.

monument is presently the tallest memorial in the United States and surely St. Louis most famous icon. The arch is exactly as wide as it is tall.

The excellent performance of austenitic stainless steel in normal atmospheres combined with the possibility to spot-weld this structural material, tempted a major manufacturer of stainless steel trains in the United States to bid for a contract to build two hundred all-stainless steel air freighters to carry troops, tanks, and guns during the World War II efforts. In fact, the first entry of the Budd Company in the world of aeronautics was made in 1930 through the contract manufacture of aircraft wheels and stainless steel wing ribs, following which a complete aircraft was designed and built by the company in 1931, the BB-1 Pioneer amphibian (Fig. 9.48).

The first stainless steel air freighter, named RB-1 Conestoga, first flew on October 31, 1943. During testing the RB-1 crashed but the test pilot reported that the stainless steel construction of the plane contributed to saving his life. However, the RB-1 had poor flying characteristics and was plagued with construction delays due to cost overruns and problems with stainless steel fabrication. Two other prototypes were built followed

Figure 9.47 *Triad* by Ted Bieler completed in 1984 using stainless steel.

by a short production of 17 planes for the U.S. Navy. Unfortunately, the Conestoga was not a success, due to undesirable handling qualities, and a further contract for 180 planes was canceled.

9.7.2 Copper and Copper Alloys

Copper and its alloys are not exposed to the atmosphere in great quantities when compared with steel. However, this material brings aesthetic value to building construction, in addition to excellent corrosion resistance. The black and then green patina formed on the surface provides an attractive decorative finish, while sealing the metal from further corrosion. As a consequence, some copper has been used for roofs, gutters, and as flashings on wood or composition shingled roofs.

Extensive tests have been made of the corrosion resistance of copper and its alloys. Various alloys were exposed in rural, industrial, and marine

FIGURE 9.48 The Budd BB-1 Pioneer built in 1931 and mounted on a pylon outside the Franklin Institute Science Museum in Philadelphia. The lower wing and empennage fabric has been lost to the weather.

atmospheres for periods of up to 20 years. From data accumulated in these tests and the calibrations of relative corrosivity of the test sites, a fairly clear picture can be obtained of the corrosion behavior of copper.

In addition to the corrosion penetration rates, one must be mindful of dezincification of brasses and selective attack on some bronzes as well as SCC of yellow brass or bronze (Fig. 6.46 in Chap. 6). These types of corrosion contribute to the failure of the material in mechanical respects without significant weight changes or losses in thickness.

Where copper is used as flashing on roofs, corrosion has been encountered at the edge of the shingles as a continuous groove. This effect is more pronounced when the atmosphere contains both chlorides and sulfides, and with wood shingles as compared with roofs of other composition. Tests indicate that all-copper roofs 0.5 mm or more in thickness would last several centuries in urban atmospheres.

If the green patina on copper alloys is desired for aesthetic reasons, pretreatment of the surface with appropriate passivating solutions is recommended. If oxidation by sulfur compounds precedes the desired reaction, the surface will present only a dark brown color for many years. The behavior of copper and copper alloys in three typical atmospheres is summarized in Table 9.6.

9.7.3 Nickel and Nickel Alloys

Electrodeposited nickel and electroless nickel are widely used as a protective coating for atmospheric exposure, and some nickel alloys, while, selected for other reasons, are also exposed to atmospheric corrosion.

Material	Urban industrial[†]	Rural[‡]	Marine[§]
T.P. copper	1.42	0.58	1.45
Phosphor bronze	1.91	0.56	2.51
Red brass	1.80	0.51	0.58
Yellow brass	2.16	0.58	0.46
90–10 Cupro Ni	3.78	2.08	2.90
Tin brass	1.73	0.76	0.61
Admiralty metal	2.51	0.51	0.33
Manganese bronze	8.64	0.48	2.02
Si Al bronze	1.27	0.33	0.33
10% Ni silver	1.96	0.64	0.58
18% Ni silver	2.01	0.61	0.53
Advance	1.55	0.25	0.46
Be copper	1.75	0.76	1.02
12 different coppers	0.81	0.46	1.35

[†] Newark, N.J.
[‡] State College, Pa.
[§] Kure Beach, N.J.; La Jolla, Calif.

TABLE 9.6 Corrosion of Copper and Copper Alloys in μm/y

The results shown in Table 9.7 were obtained for several representative alloys. As can be concluded from these data, nickel tends to be passive in a marine atmosphere. The ratio between the corrosion rate for nickel exposed to the industrial atmosphere and that exposed to rural or marine atmospheres was 28:1.

9.7.4 Aluminum and Aluminum Alloys

Aluminum, in its many forms is exceeded only by steel in tonnage directly exposed to the elements. It is produced in the form of wrought products, extrusions, and castings with a large variety of alloying elements to impart desired mechanical properties. Before anodizing*, the atmospheric corrosion behavior of aluminum products fits into some fairly well-defined patterns that are related to composition.

While pure aluminum has excellent atmospheric corrosion resistance and is used extensively as a cladding materials for this very

* See Chap. 5 for a detailed description of aluminum anodizing and Chap. 4 for a description of the thermodynamic corrosion stability of pure aluminum.

Material	Urban industrial*	Rural†	Marine‡
Nickel	0.15	3.66	0.22
	0.24	1.65	0.23
Monel	0.16	1.57	0.17
	3.56	0.81	0.21
Incoloy	0.06	0.02	0.02
Inconel	0.05	0.03	0.02
Nionel	0.03	0.02	0.02

* Newark, N.J.; New York City
† State College, Pa.
‡ Kure Beach, N.C.; La Jolla, Calif.

TABLE 9.7 Corrosion in μm/y of Nickel and Its Alloys in Various Atmospheres

reason, alloys containing copper and silicon as the principal alloying constituents are susceptible and should be used with care.

In a rural atmosphere, the corrosion rate for most alloys is approximately 0.06 μm/y, while those containing large amounts of copper about double this low rate. In a marine environment, the differences between alloys appear as a tenfold increase, from about 0.6 μm/y for the less corrosion-resistant materials, to about 0.7 μm/y for the better materials. Pitting also is about 10 times greater in marine atmospheres and corrosion can be much greater in a severe industrial atmosphere than in the marine atmosphere.

Some aluminum alloys develop severe pitting and a voluminous white corrosion product under some exposure conditions such as marine atmospheres. Aluminum roofs have been known to corrode severely at the overlaps. Some aluminum alloys may also be attacked in their intergranular regions when exposed after certain manufacturing processes involving cold work or precipitation hardening. General intergranular attack or exfoliation can then occur. In extreme cases, the edges of the affected area are leaf-like and resemble the separated pages of a wetted book that has become swollen and begun to open up (Fig. 9.49). The attack tends to start at sheared edges or punched holes, but is not restricted to these areas. Aircraft manufacturers, in particular, must guard against this type of corrosion.

In designing aluminum equipment, care must be exercised to avoid dissimilar metal couples and the attendant galvanic corrosion. Copper and rusty steel are particularly detrimental to aluminum. Where it is impractical to avoid dissimilar metals, the aluminum should be electrically insulated from the more noble metal by means of washers, sleeves, and so forth. In some instances, covering the

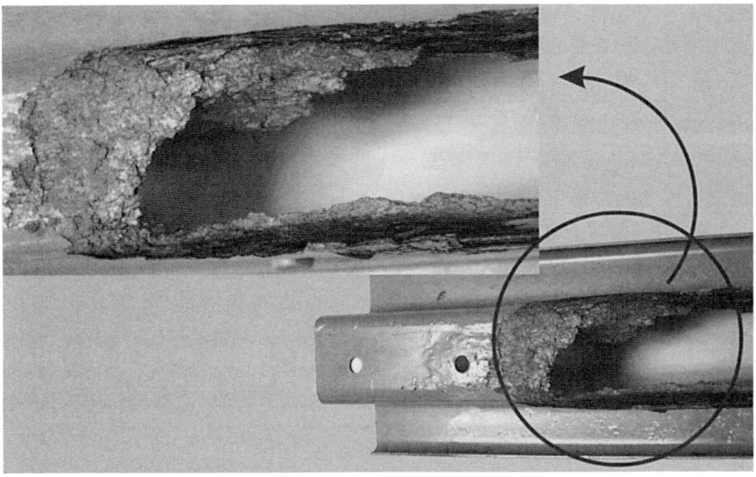

Figure 9.49 Leaf-like exfoliation corrosion of corrugated A92024 aluminum alloy. (Courtesy of Kingston Technical Software)

noble metal with an organic finish is sufficient to greatly reduce galvanic couple corrosion. Due to the passive film on stainless steel, it has been used in contact with aluminum with little expectation of accelerated corrosion, despite the difference in potential.

As would be expected, constant exposure to moisture with a limited supply of oxygen to the aluminum surface leads to the rapid corrosion of any aluminum apparatus or equipment component. This is due to the highly reactive nature of aluminum that leads to formation of oxides or hydroxides. In the presence of oxygen, a protective aluminum oxide film which is substantially unreactive, develops on any aluminum surface. If the film is removed by mechanical or chemical means and the aluminum exposed to water, a rapid reaction sets in and the fresh aluminum surface is converted to the hydroxide and subsequently to the oxide.

9.7.5 Zinc and Zinc Alloys

Zinc is exposed to the atmosphere in the form of galvanized sheet, as in flashings on roofs; as die castings, and as coatings on steel, either hot-dipped or electroplated. The general behavior of zinc metal and zinc coatings is described in the ISO tables presented earlier. Note the particularly low rates of attack on zinc as compared with steel in marine exposures where chloride deposition is important. Such excellent resistance is acquired by the hard, dense, protective products of corrosion in a chloride atmosphere. Similar results cannot be obtained in a sulfurous atmosphere where the corrosion products are soft, voluminous, and non-protective.

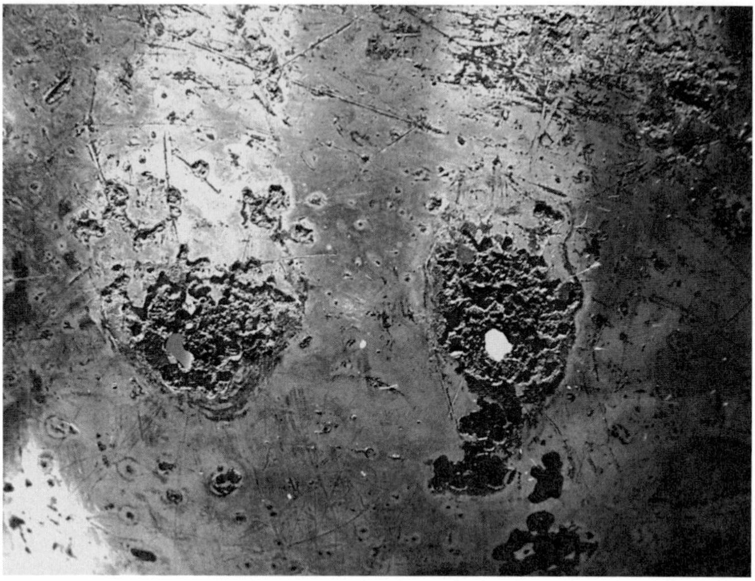

Figure 9.50 Close-up of corrosion pits in a water jug (Fig. 6.8) made of a silver-plated zinc casting.

Zinc-base die castings are not usually exposed boldly to the outside atmosphere without a protective coating. When breaks or pits occur in plated coatings such as nickel, chromium, or even silver (Fig. 9.50), the corrosion of the die casting may be accelerated due to galvanic corrosion and thus give a false impression of the corrodibility of zinc. Many small parts of machinery, household appliances, and hardware are made of zinc-base die castings or "white metal"* and are exposed to indoor atmospheres where their corrosion behavior is usually very good. In these cases where severe corrosion is encountered in this relatively noncorrosive environment, the cause may be improper alloy selection or the use of material containing too high a percentage of impurities.

Galvanized steel is the most important zinc application. Galvanizing produces a zinc coating on the steel surface and is one of the most effective methods for corrosion protection of steel. Worldwide, the use of zinc for galvanizing is estimated at more than 3×10^6 ton/y, constituting nearly one-half of the world production [32]. Most of these coatings are hot-dipped galvanized coatings containing a small amount of aluminum. The thickness of electroplated coatings is considerably lower than those applied by the hot-dip process.

Corrosion rates for zinc may vary by as much as two orders of magnitude depending on the specific environmental conditions. It is

* White metal is typically 93 to 96 percent Zn, 4 percent Al, 0.05 percent Mg, and sometimes 1 to 3 percent Cu.

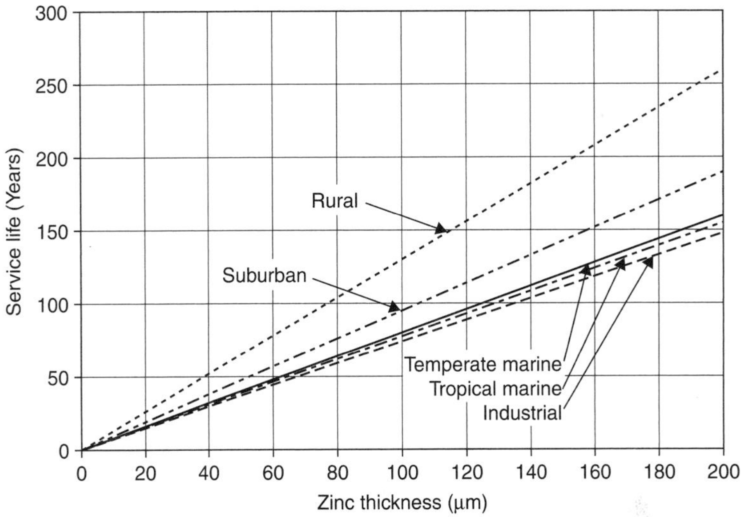

FIGURE 9.51 Service life of hot-dipped galvanized coatings as a function of zinc thickness and specific environments.

therefore important to know the specific corrosion rate in a given application environment in order to effectively use zinc-coated steels in outdoor structures. A widely used method for corrosion life estimation of the various types of galvanized steels provides generalized corrosion rate values for five predetermined atmospheric environments as a function of zinc coating thickness (Fig. 9.51) [33]. Service life in Fig. 9.51 is defined as the time to 5 percent rusting of the steel surface. This method may be used to estimate the service life of a given coating thickness or to specify a coating for a given environment.

This graphical method is applicable to zinc-coated steel produced by batch or continuous galvanizing, including hot-dip, electrogalvanized, and thermal sprayed coatings. However, it does not apply to coatings containing more than one percent alloying elements. The method also assumes that the galvanized product is free of significant defects that could accelerate corrosion. Additionally, the service life prediction does not consider issues of water entrapment that can create severe crevice chemistry as shown earlier in Fig. 9.6.

9.7.6 Polymeric Materials

Essentially all polymers freely exposed to the elements will change in some manner. The active rays of the sun become potent agents of change in the organic materials. Further polymerization of the resin can occur to produce embrittlement. Other types of new bonding can be triggered to make polymers more crystalline. Any volatile component of the material, such as a plasticizer, can be evaporated. The polymer chains may be simply oxidized and broken up to destroy the product.

The external evidence of attack may be blushing (loss of gloss), chalking, change in color of the product, or extreme brittleness as in Fig. 9.52. These effects are often readily observed on epoxy and polyester polymers when they have been boldly exposed to the environment. However, mechanical tests will usually be required to reveal the extent of degradation of either thermoplastic or thermosetting resins unless they have been exposed for decades to direct sunlight, as in Fig. 9.52.

The effect of high atmospheric temperatures or heating from direct exposure to the sun can be particularly severe on thermoplastic polymers. Creep or distension of the polyvinyl chloride and polyethylene plastics will occur readily unless provision is made to prevent overheating or stressing of the materials.

Polymeric materials should be thoroughly tested if they are to be exposed freely in the atmosphere. ASTM Recommended Practice D1435 describes the appropriate conditions for such test exposures and suggests tests that might be used to evaluate changes in the materials.

FIGURE 9.52 Polycarbonate (Nalgene) bottle and polypropylene funnel degraded over a few years of exposure to sun rays on a window sill.

Changes in mechanical and physical properties of the polymers are determined for definitive results. Weight gains or losses may be of interest, but typically do not provide substantive results.

Accelerated cabinet testing of plastic materials can be performed with essentially the same validity as when testing metallic materials. Poor materials can be eliminated, but extrapolation of data to forecast the life of plastic parts in the atmosphere should not be attempted. ASTM D1499, D2565, and G26 may be used for guidance on the control of the test cabinet when testing plastics and ASTM D750 should be used for the evaluation of elastomers.

The preventive measures used to ensure the integrity of polymeric materials are similar to those for metals, that is, the external surface should be either painted or metal plated to provide a barrier between the material and the atmosphere. Alloying of the basic material with small amounts of other resins can often upgrade the stability of these materials.

References

1. Miller D. Corrosion control on aging aircraft: What is being done? *Materials Performance* 1990; 29: 10–1.
2. Selwyn LS, Roberge PR. Corrosion of outdoor monuments. In: Cramer DS, Covino BS, eds. Vol. 13C: *Corrosion: Environments and Industries*. Metals Park, Ohio: ASM International, 2006; 289–305.
3. Roberge PR. *Handbook of Corrosion Engineering*. New York, N.Y.: McGraw-Hill, 2000.
4. Chung SC, Lin AS, Chang JR, Shih HC. EXAFS Study of atmospheric corrosion products on zinc at the initial stage. *Corrosion Science* 2000; 42: 1599–610.
5. Duncan JR, Ballance JA. Marine salts contribution to atmospheric corrosion. In: Dean SW, Lee TS, eds. *Degradation of Metals in the Atmosphere*. Philadelphia, Pa.: ASTM, 1988; 316–26.
6. Lawrence MG. *The Relationship between Relative Humidity and the Dewpoint Temperature in Moist* Air—A Simple Conversion and Applications. Bulletin. American Meteorological Society 2005; 86: 225–33.
7. Hidy GM. *Aerosols: An Industrial and Environmental Science*. Orlando, Fla.: Academic Press, 1984.
8. Feliu S, Morcillo M, Chico B. Effect of distance from sea on atmospheric corrosion rate. In: *Corrosion* 1999; 55: 883–91.
9. Morcillo M, Chico B, Mariaca L, Otero E. Salinity in marine atmospheric corrosion: Its dependence on the wind regime existing in the site. In: *Corrosion Science* 2000; 42: 91–104.
10. Kuemmel DA, Hanbali RM. Accident analysis of ice control operations. *Public Works* 1993; 124: 48-51.
11. *Standard Practice for Conducting Wire-on-Bolt Test for Atmospheric Galvanic Corrosion*. ASTM G116-99 ed. West Conshohocken, Pa.: American Society for Testing of Materials, 1999.
12. Klassen RD, Roberge PR. Aerosol transport modeling as an aid to understanding atmospheric corrosivity patterns. *Materials and Design* 1999; 20: 159–68.
13. Sereda PJ, Croll SG, Slade HF. Measurement of the time-of-wetness by moisture sensors and their calibration. In: Dean SW, Rhea EC, eds. *Atmospheric Corrosion of Metals*. Philadelphia, Pa.: ASTM, 1982; 48.
14. *Standard Practice for Measurement of Time-of-Wetness on Surfaces Exposed to Wetting Conditions as in Atmospheric Corrosion Testing*. Annual Book of ASTM Standards, West Conshohocken, Pa.: American Society for Testing of Materials. 1999; 03(02): G84–89.

15. Lawson HH. *Atmospheric Corrosion Test Methods*. Houston, Tex.: NACE International, 1995.
16. *Standard Practice for Monitoring Atmospheric* SO_2 *Using the Sulfation Plate Technique*. Annual Book of ASTM Standards. West Conshohocken, Pa.: American Society for Testing of Materials. 1997; 03(02): G91–97,
17. *Standard Test Method for Determining Atmospheric Chloride Deposition Rate by Wet Candle Method*. Annual Book of ASTM Standards. Philadelphia, Pa.: American Society for Testing of Materials. 2002; 03(02): G140–02.
18. Doyle DP, Wright TE. Rapid method for determining atmospheric corrosivity and corrosion resistance. In: Ailor WH, ed. *Atmospheric Corrosion*. New York, N.Y.: John Wiley and Sons, 1982; 227–43.
19. Coburn S. Atmospheric corrosion. In: Korb, LJ. *Metals Handbook*, 9th ed, Vol. 1, *Properties and Selection, Carbon Steels*. Metals Park, Ohio:: American Society for Metals, 1978; 720.
20. Klassen RD, Roberge PR, Lenard DR, Blenkinsop GN. Corrosivity patterns near sources of salt aerosols. In: Townsend HE. *Outdoor and Indoor Atmospheric Corrosion*, ASTM STP 1421. West Conshohocken, Pa.: American Society for Testing and Materials, 2002; 19–33.
21. Summitt R, Fink FT. *PACER LIME: An Environmental Corrosion Severity Classification System*. AFWAL-TR-80-4102, Air Force Wright Aeronautical Laboratories, Wright-Patterson Air Force Base, OH,1980.
22. Dean SW. Corrosion testing of metals under natural atmospheric conditions. In: Baboian R, Dean SW, eds. *Corrosion Testing and Evaluation: Silver Anniversary Volume*. Philadelphia, Pa.: American Society for Testing of Materials, 1990; 163–76.
23. Abbott WH, Kinzie R. Corrosion Monitoring in Air Force Operating Environments. Tri-Service Corrosion Conference. Las Vegas, Nev.: U.S. Department of Defense, 2003.
24. *Standard Practice for Preparing, Cleaning, and Evaluating Corrosion Test Specimens*. Annual Book of ASTM Standards. West Conshohocken, Pa.: American Society for Testing of Materials, 2003; 03(02): G1–03.
25. Leygraf C, Graedel TE. *Atmospheric Corrosion*. New York, N.Y.: John Wiley & Sons, 2000.
26. King G, Ganther W, Hughes J, Grigioni P, Pellegrini A. Studies in Antarctica help to better define the temperature criterion for atmospheric corrosion. Northern Area Region Conference, Feb. 26–28 2001. Anchorage, Alaska: NACE International.
27. Roberge PR. *Corrosion Basics—An Introduction. 2nd edn*. Houston, Tex.: NACE International, 2006.
28. Steinmayer RF. Land vehicle management. In: *AGARD Lecture Series No. 141*. Neuilly-sur-Seine, France: NATO, 1985.
29. Cook DC, Van Orden AC, Reyes J. Atmospheric corrosion in marine environments along the Gulf of México. In: Dean SW, Hernandez-Duque Delgadillo G, Bushman JB, eds. *Marine Corrosion in Tropical Environments*. Philadelphia, Pa.: ASTM, 2000; 75–97.
30. Fushoku, Boushoku, Gijutsu, Kyoukai. *Q & A on Corrosion and Protection for Metals*, Japan: Japanese Society of Corrosion and Protection, 1988.
31. Rao KNP, Lahiri AK. *Corrosion Map of India*. Jamshedpur, India: Council of Scientific and Industrial, 1971.
32. *Zinc and Lead Statistics*. London, U.K.: International Lead Zinc Study Group, 1996.
33. *Galvanizing for Corrosion Protection: A Special Guide*. Centennial, Colo.: American Galvanizers Association, 1990.

CHAPTER 10

Corrosion in Soils and Microbiologically Influenced Corrosion

10.1 Introduction

Microbiologically influenced corrosion (MIC) is not restricted to soils and as illustrated in Fig. 10.1 [1] may occur in most environments where deposits can form. The muddy bottom of any relatively stagnant body of water with a high biological oxygen demand often supports massive growth of sulfate reducing bacteria (SRB), as may waterlog soils. Any metallic installations buried or immersed in such environments can be expected to suffer badly from microbiological corrosion. The most serious economic problem is the corrosion of pipelines, although sheet piles and piers are frequently attacked. In some instances, cast iron pipes of 6.3 mm thickness have perforated within the first year of operation under such conditions, while perforation in three years is common. Even if MIC problems are not restricted to soils, MIC is such an important factor in understanding the corrosivity of most soils that it was decided to combine these two topics in a single chapter.

10.2 Corrosion in Soils

It has long been observed that a piece of iron buried in a dry soil suffers much less corrosion than when it is buried in a wet soil. However, soils are commonly wet due to rain, natural springs, and rivers soils. Corrosion in soils is a major concern, especially as much of the buried infrastructure is aging. Increasingly stringent environmental protection requirements are also bringing corrosion issues to the forefront for many systems either buried or aboveground and possibly in contact

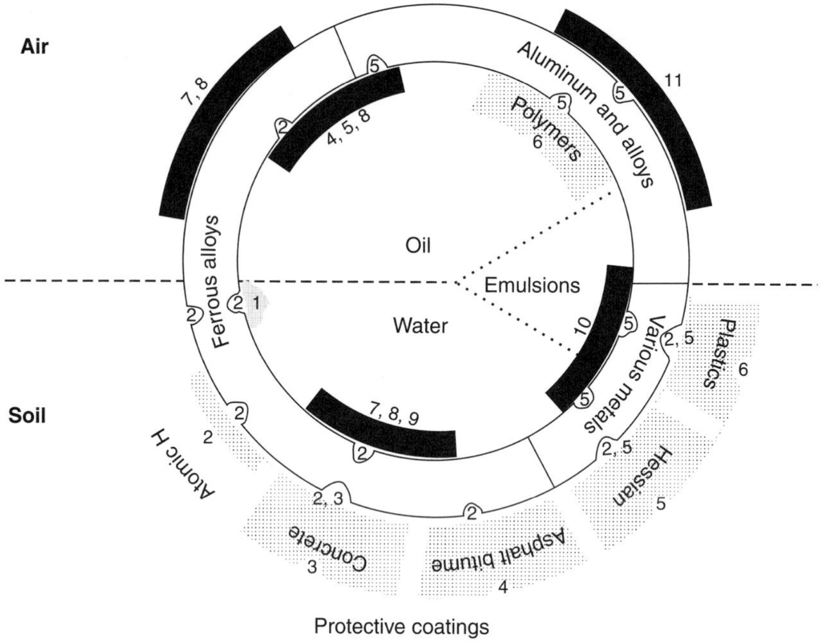

FIGURE 10.1 Schematic illustration of the principal methods of microbial degradation of metallic alloys and protective coatings. 1: Tubercle leading to differential aeration corrosion cell and providing environment for "2"; 2: Anaerobic SRB; 3: Sulfur oxidizing bacteria, producing sulfates and sulfuric acid; 4: Hydrocarbon utilizers, breaking down aliphatic and bitumen coatings and allowing access of "2" to underlying metallic structure; 5: Various microbes producing organic acids as end-products of growth, attacking mainly nonferrous metals/alloys and coatings; 6: Bacteria and molds breaking down polymers; 7: Algae forming slimes on above-ground damp surfaces; 8: Slime forming molds and bacteria, which may produce organic acids or utilize hydrocarbons, providing differential aeration cells and growth conditions for "2"; 9: Mud on river bottoms providing matrix for heavy growth of microbes (including anaerobic conditions for "2"); 10: Sludge (inorganic debris, scale, corrosion products, and so on) providing matrix for heavy growth and differential aeration cells, and organic debris providing nutrients for growth;11: Debris (mainly organic) on metal above ground providing growth conditions for organic acid-producing microbes.

with soils. The following applications are typical examples where corrosion is a concern:

- Oil, gas, and water pipelines
- Buried storage tanks (a vast number are used by gas stations)
- Electrical communication cables and conduits
- Anchoring systems for communication and power transmission (see Chap. 7 for more details and examples)

- Road culverts
- Steel piling for the foundations of bridges and buildings
- Well and shaft casings.

The equipment in these systems is typically expected to function reliably and continuously over several decades.

10.2.1 Soil Classification

Soil is an aggregate of minerals, organic matter, water, and gases (mostly air). It is formed by the combined weathering action of wind and water, and also organic decay. The proportions of the basic constituents vary greatly in different soil types. For example, humus has very high organic matter content, whereas the organic content of beach sand is practically nil. The properties and characteristics of soil obviously vary as a function of depth. A vertical cross-section taken through the soil is known as a soil profile, and the different layers of soil are known as soil horizons. The following soil horizons have been classified:

A: Surface soil (usually dark in color due to organic matter)

O: Organic horizon (decaying plant residues)

E: Eluviation horizon (light color, leached)

B: Accumulation horizon (rich in certain metal oxides)

C: Parent material (largely nonweathered bedrock)

The distribution and size of mineral particles in a soil is described by its texture. Sand (rated from coarse to very fine), silt, and clay refer to textures of decreasing particle coarseness (Table 10.1). Soils with a high proportion of sand have very limited storage capacity for water, whereas clays are excellent in retaining water. A popular soil

Category	Diameter (mm)
Sand (very coarse)	1.00–2.00
Sand (coarse)	0.50–1.00
Sand (medium)	0.25–0.50
Sand (fine)	0.10–0.25
Sand (very fine)	0.05–0.10
Silt	0.002–0.05
Clay	< 0.002

Table 10.1 Particle Sizes in Soil Texture

identification scheme defines eleven soil types on the basis of their respective proportions of clay, silt, and sand, that is, sand, loamy sand, sandy loam, sandy clay loam, clay loam, loam, silty loam, silt, silty clay loam, silt clay, and clay.

A newer soil classification system has evolved in the United States to classify soils globally, at any location. In this universal classification system, soils are considered as individual three-dimensional entities that may be grouped according to their physical, chemical, and mineralogical properties. The system uses a hierarchical approach, with the amount of information about a soil increasing down the classification ladder. From top to bottom, the hierarchy is structured in the following categories: order, suborder, great groups, subgroups, families, and series. Further details are provided in Table 10.2.

Category	Basis for Classification	Example(s)	Comments
Order	Differences in measurable and visible characteristics of soil horizons.	Entisol, Vertisol, Inceptisol, Aridisol, Mollisol, Spodosol, Alfisol, Ultisol, Oxisol, Histosol	Nine orders for mineral soils and one order for all organic soils.
Suborder	Differences in development characteristics.	Aquod, Udult	Grouping according to accumulation of soluble materials, presence or absence of B horizons, mineralogy, and chemistry.
Great group	Presence or absence of certain horizons	Kandihumult	Relative thickness of horizons is important.
Subgroup	Typical or dominant concept of the great group.	Typic Kandihumult	Coded as either the great group name with the "typic" prefix, or a combination of great group names.

TABLE 10.2 Soil Classification System Using Hierarchical Approach

Category	Basis for Classification	Example(s)	Comments
Family	Differences in textural classes, mineralogy, acidity, and temperature.	Clayey oxidic isothermic Typic Kandihumult	Plants generally react in a similar manner to the same soil family.
Series	Differences in texture.	Paaola	Usually named after the location where the soil was first described.

TABLE 10.2 (continued)

10.2.2 Soil Parameters Affecting Corrosivity

Several variables are known to have an influence on corrosion rates in soil; these include water, degree of aeration, pH, redox potential, resistivity, soluble ionic species (salts), and microbiological activity. The complex nature of selected variables is presented graphically in Fig. 10.2 [2].

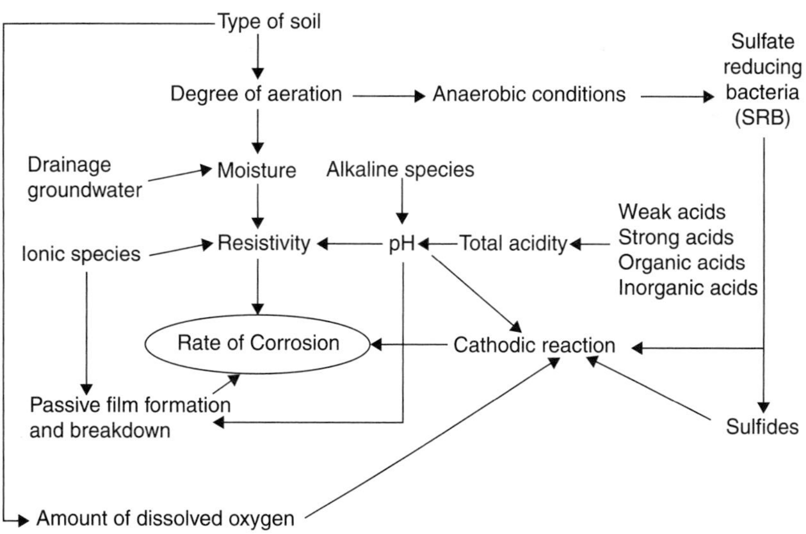

FIGURE 10.2 Relationship of variables affecting the rate of corrosion in soil. For simplicity, only the MIC effects of sulfate reducing bacteria are shown.

Water

Water constitutes the essential electrolyte that supports electrochemical corrosion reactions in water saturated or unsaturated soils. The groundwater level is important in this respect. It fluctuates from area to area, with water moving from the water table to higher soil, against the direction of gravity. Saturated water flow is dependent on pore size and distribution, texture, structure, and organic matter. Water movement in soil can occur by gravity, capillary action, osmotic pressure (from dissolved species), and/or electrostatic interaction with soil particles. The water-holding capacity of a soil is strongly dependent on its texture. Coarse sands retain very little water, while fine clay soils store water to a high degree.

Degree of aeration

Oxygen concentration typically decreases with increasing soil depth. In neutral or alkaline soils, oxygen concentration has an important effect on corrosion rate as a result of its participation in the cathodic reaction. However, in the presence of certain microbes (such as sulfate-reducing bacteria), corrosion rates can be very high, even under anaerobic conditions. Oxygen transport is more rapid in coarse-textured, dry soils than in fine, waterlogged textures. Excavation can obviously increase the degree of aeration in soils. It is generally accepted that corrosion rates in disturbed soil with greater oxygen availability are significantly higher than in undisturbed soil.

pH

Soil pH typically varies between 5 and 8. In this range, pH is generally not considered to be the dominant variable affecting corrosion rates. More acidic soils produced by mineral leaching, decomposition of acidic plants (e.g., coniferous tree needles), industrial wastes, acid rain, and certain forms of microbiological activity represent a serious corrosion risk to common construction materials such as steel, cast iron, and zinc coatings. On the other hand, alkaline soils tend to have high sodium, potassium, magnesium, and calcium contents, with the latter two elements forming possibly protective calcareous deposits on buried structures.

Soil resistivity

Resistivity has historically been used as an indicator of soil corrosivity. Since ionic current flow is associated with soil corrosion reactions, high soil resistivity will usually slow down corrosion reactions. Soil resistivity generally decreases with increasing water content and the concentration of ionic species. Soil resistivity is by no means the only parameter affecting the risk of corrosion damage and a high soil resistivity alone will not guarantee the absence of corrosion. Soil resistivity variations along the length of a pipeline, for example, may lead to the formation of macro corrosion cells. Therefore, for structures such as pipelines, the merit of a corrosion risk classification based on an absolute value of soil resistivity is limited. Soil resistivity can be measured by a few traditional techniques such as the Wenner four-pin method described in Chap. 5 or, more recently, by electromagnetic measurements. The latter allows measurements in a convenient manner and at different soil depths. Another option for soil resistivity measurements is the soil box method, also described in Chap. 5, in which a sample is taken during excavation, preferably in the immediate vicinity of the buried structure being investigated.

Redox potential

The redox potential is essentially a measure of the degree of aeration in a soil. A high redox potential indicates a high oxygen level. Low redox values may provide an indication that conditions are conducive to anaerobic microbiological activity. Sampling of soil will obviously lead to oxygen exposure, and unstable redox potentials are thus likely to be measured in disturbed soil.

Chlorides

Chloride ions generally participate in the dissolution reactions of many metals. Furthermore, their presence tends to decrease the soil resistivity. Chlorides may be found naturally in soils as a result of brackish groundwater and historical geological seabeds or come from external sources such as deicing salts applied to roadways.

Sulfates

Sulfate ions are generally considered to be more benign in their direct corrosive action toward metallic materials than chlorides. However, concrete may be attacked as a result of high sulfate levels. The presence of sulfates also poses a major risk for metallic materials since these ions are nutrients to SRBs that convert these benign ions into highly corrosive sulfides.

10.2.3 Soil Corrosivity Classifications

For design and corrosion risk assessment purposes, it is desirable to estimate the corrosivity of soils without necessarily conducting exhaustive corrosion testing. Corrosion testing in soils may be complicated by the need of long exposure periods since buried structures are usually expected to last for several decades during which many soil conditions may be encountered. Considering the complexity of the parameters affecting soil corrosion, it is obvious that the use of relatively simple soil corrosivity models is bound to be inaccurate.

One of the simplest classifications is based on a single parameter, soil resistivity. Table 10.3 shows the generally adopted corrosion severity ratings. Sandy soils are high on the resistivity scale and therefore are considered to be the least corrosive. Clay soils, especially

Soil Resistivity (Ω cm)	Corrosivity Rating
> 20,000	Essentially noncorrosive
10,000 to 20,000	Mildly corrosive
5,000 to 10,000	Moderately corrosive
3,000 to 5,000	Corrosive
1,000 to 3,000	Highly corrosive
<1,000	Extremely corrosive

Table 10.3 Corrosivity Ratings Based on Soil Resistivity

those contaminated with saline water, are on the opposite end of the corrosivity scale. The soil resistivity parameter is very widely used in practice and is generally considered to be the dominant variable in absence of microbial activity.

The American Water Works Association (AWWA) has developed a more complex numerical soil corrosivity scale that is applicable to cast iron alloys. A severity ranking is generated by assigning points for different variables presented in Table 10.4 [3]. When the total

Soil Parameter	Assigned Points
Resistivity (Ω cm)	
< 700	10
700–1000	8
1000–1200	5
1200–1500	2
1500–2000	1
> 2000	0
PH	
0–2	5
2–4	3
4–6.5	0
6.5–7.5	0
7.5–8.5	0
> 8.5	3

Table 10.4 Point System for Predicting Soil Corrosivity According to the AWWA C-105 Standard

Soil Parameter	Assigned Points
Redox Potential (mV)	
> 100	0
50–100	3.5
0–50	4
< 0	5
Sulfides	
Positive	3.5
Trace	2
Negative	0
Moisture	
Poor drainage, continuously wet	2
Fair drainage, generally moist	1
Good drainage, generally dry	0

Table 10.4 *(continued)*

points of a soil in the AWWA scale are 10 or higher, corrosion protective measures such as cathodic protection (CP) would be recommended. It should be appreciated that this rating scheme remains a relatively simplistic and subjective procedure. It should therefore be used as a broad indicator of local soil corrosivity.

A 25-point scoring risk assessment method was also proposed that considers, in addition to soil properties, pipe factors such as: pipe location and leak repair difficulty, pipe minimum design life, pipe maximum design surge pressure, and pipe size. In this method, the soil corrosivity potential is divided into four categories: mild, moderate, appreciable, and severe [4]. The methodology is not intended to be definitive, but it provides a preliminary outline that may be modified to meet individual project or owner needs. This method may be used, for example, as a checklist to prioritize various inspection and maintenance tasks.

An extensive soil evaluation program carried in Europe has resulted in the production of a worksheet method for estimating the probability of corrosion damage to metallic structures in soils [5]. The worksheet consists of 12 individual ratings (R1 to R12), listed in Table 10.5. This methodology is very detailed and comprehensive.

Rating Number	Parameter	Rating
R1	Soil type	See Table 10.6
R2	Resistivity (Ω)	> 50,000 (+4); > 20,000 (+2); > 5000 (0); > 2000 (−2); 1000 to 2000 (−4); < 1000 (−6)
R3	Water content (%)	≤ 20% (0); >20% (−1)
R4	pH	> 9 (+2); > 5.5 (0); 4.0 to 5.5 (−1); < 4 (−3)
R5	Buffering capacity	See Table 10.6
R6	Sulfide content (mg/kg)	< 5 (0); 5 to 10 (−3); > 10 (−6)
R7	Neutral salts (mmol/kg)	< 3 (0); 3 to 10 (−1); > 10 to 30 (−2); > 30 to 100 (−3); > 100 (−4)
R8	Sulfates (mmol/kg)	< 2 (0); 2 to 5 (−1); > 5 to 10 (−2); > 10 (−3);
R9	Groundwater	no groundwater (0); presence of groundwater (−1)
R10	Horizontal homogeneity	See Table 10.6
R11	Vertical homogeneity	See Table 10.6
R12	Redox potential	See Table 10.6

Table 10.5 Variables Considered in Dechema Soil Corrosivity Worksheet [5]

For example, the effects of vertical and horizontal soil homogeneity are included, as outlined in Table 10.6. Even details such as the presence of coal or coke and other pollutants in the soil are considered.

The assessment is directed at ferrous materials (steels, cast irons, and high-alloy stainless steels), hot-dipped galvanized steel, and copper and copper alloys. Summation of the individual ratings produces an overall corrosivity classification into one of four categories with scores less than −10 indicating a highly corrosive soil and positive values (>0), a noncorrosive environment (Table 10.7). It has been pointed out that sea or lake beds cannot be assessed using this worksheet.

10.2.4 Auxiliary Effects of Corrosion Cells

Chapter 7 has provided many examples of corrosion cells that may happen in a soil environment. The following sections illustrate some effects these corrosion cells may have on buried structures with a particular focus on pipelines which are, by far, the greater portion of

R1—Type of Soil (Texture and Structure)	**(%)**	**Rating**
Coherency: elutriable fraction	≤ 10%	+4
	> 10 to 30%	+2
	> 30 to 50%	0
	> 50 to 80%	−2
	> 80%	−4
Soils containing organic carbon, e.g., muddy or swampy soils: peat, fen, mud, marsh, and organic carbon	> 5%	−12
Severely polluted soil: pollution by fuel ash, slag coal, coke, refuse, rubbish or waste water		−12
R5—Buffering Capacity	**mmol/kg**	**Rating**
Acid capacity to pH 4.3: (alkalinity K_a 4.3)	> 1000	+3
	200 to 1000	+1
	< 200	0
Base capacity to pH 7 : (acidity K_b 7.0)	< 2.5	0
	2.5 to 5	−2
	> 5.0 to 10	−4
	> 10 to 20	−6
	> 20 to 30	−8
	>30	−10
R10—Horizontal Soil Homogeneity		**Rating**
Resistivity variation between adjacent domains (all positive R2 values are treated as equal)		
R2 difference < 2		0
R2 difference >= 2 and <= 3		−2
R2 difference > 3		−4
R11—Vertical Soil Homogeneity		
Adjacent soils with same resistivity	Embedded in soils with same structure or in sand	0
	Embedded in soils with different structure or containing foreign matter	−6
Adjacent soils with different resistivity	R2 difference >= 2 and <= 3	−1
	R2 difference > 3	−6

Table 10.6 Ratings 1, 5, 10, 11, and 12 in Dechema Soil Corrosivity Worksheet [5]

R12—Redox Potential*	Rating
mV vs. $Cu/CuSO_4$	
−500 to −40	−3
> −400 to −300	−8
> −300	−10

* When no measurement is possible (e.g., soil survey with no structure present) R12 should be set at −10 when pieces of coal or coke are present (e.g., foreign cathodes).

TABLE 10.6 *(continued)*

the metal threatened by corrosion in soils. For example, when a new line is laid in the earth a vast number of cells come into being:

- The line passes through a variety of soils giving rise to concentration cells
- There are variations in oxygen access setting up oxygen cells
- There are impurities and differences on the surface of the pipe creating galvanic cells
- Cold bends and welding stresses set up stress cells
- Other less known types of cells may also exist given the complexity of most soils

These cells are of various sizes and shapes. In some cases, anodes and cathodes are separated by only a few centimeters apart; in others, they may be kilometers apart. The cells may also greatly vary in strength, from a few millivolts to sometimes 0.5 V. Anodic areas are of all sizes, tiny to large, as are the cathodes.

As soon as these areas become active, they begin changing. On cathodic areas, hydrogen may begin to form, as described in Chap. 2. These changes may affect both cell resistance and potential. Anodes begin corroding, which puts new ions into solution at the surface and as these react with various components in the environment, their respective concentrations change.

Summation of R1 to R12 Ratings	Soil Classification
>= 0	Virtually noncorrosive
−1 to −4	Slightly corrosive
−5 to −10	Corrosive
<= −10	Highly corrosive

TABLE 10.7 Overall Soil Corrosivity Classification in Dechema Soil Corrosivity Worksheet [5]

Cell potentials also change; some increase, while others decrease. Some areas which were initially anodic become cathodic. The reverse also may take place, although not as often. Generally, as time passes, the total anodic area on a line gets smaller, although total activity does not decrease at the same rate. The result is that the rate of attack at the worst locations tends to increase. Finally, among all of the various cells along the entire line, the most active anode loses enough metal by corrosion to penetrate the pipe wall completely.

When an impressed current cathodic protection (ICCP) system is in full operation there is a high possibility for oxygen to be produced at the anode, and in nearly all cells, hydrogen is formed at the cathode. If chloride ions are present, chlorine gas may be formed at the anode. This generation of gas, either oxygen or chlorine, at the anode is not nearly as likely to occur in a natural corroding cell as it is when an ICCP system is used, particularly when inert anodes are used.

Hydrogen

The formation of hydrogen at the cathode, however, can occur during both normal corrosion and under sacrificial CP or ICCP. Initially, it may be formed as nascent hydrogen, that is, single hydrogen atoms are produced. These single atoms may then combine with oxygen to form water, or with some other ion in the environment to form another compound. Nascent hydrogen may also dissolve in the metal (cathode area), or combine to form ordinary gaseous hydrogen.

Atomic hydrogen formed at the cathode may diffuse into the steel and recombine at some depth in the metal to form molecular gaseous hydrogen that unlike atomic hydrogen cannot migrate any further through the metal crystal structure. Consequently, an accumulation of hydrogen gas within the metal may build up to generate enough pressure to blister or split the solid steel. Hydrogen can also react with the metal causing hydrogen induced cracking. Chapter 6 provides more details on this particular type of corrosion damage.

When gaseous hydrogen is formed on the metallic surface, it normally escapes readily in the environment without leaving a trace. After all, hydrogen is the lightest gas known. However, if the production of the gas happens to be under a partially defective coating, the formation of this hydrogen gas may contribute to disbondment and acceleration of the coating deterioration as explained in more details in Chap. 14.

Electroendosmosis

Another side effect of the passage of electrical current through a porous medium like soil is electroendosmosis, during which water is "carried along" with the current. Normally this effect is significant only with higher current densities than natural corrosion cells would produce. Like hydrogen formation, it is more likely to be associated with the use of ICCP in which case water molecules would be carried to the cathode and away from the anode resulting in an unwanted increase in anode resistance.

Cathode Scale

A third auxiliary effect is the deposition of compounds from the soil on the buried system. This effect is extremely pronounced in seawater, where a cathode scale resembling hard glossy enamel can build up over time and grow as thick as 2 cm. When deposited in seawater, this material is a complex mixture of calcium and magnesium oxides, hydroxides, and carbonates. The composition of this hard deposit depends upon CP current density, among other factors (see Chap. 8 for details on calcareous deposits).

In fresh or brackish water cathode scale composition is even more variable than the scale formed in seawater since its formation is based on highly variable ion concentrations. This is also true in soils. Some soils will not form a visible scale, although there is enough calcium present in most soils to favor scale formation. Often, when a system has been polarized by CP, no scale will be visible until the system surface dries out. It then appears as a whitish coloration. On a buried system which has not been under CP, cathode scale is frequently found in an irregular mottled pattern; this makes the actually active cathodic areas visible to the eye.

Pitting

When a buried system that has been in corrosive soil without adequate protection for some time is examined, it is usually found that by far the greater part of the area is unaffected. Where corrosion has taken place, it is in the form of pits, which are relatively small areas where the attack has been deep (Fig. 10.3).

In general, there will always be one spot on any buried system where all of the conditions combine to give the highest rate of penetration. This hot spot is exactly where the first pit through perforation will occur. Since the underground system is invisible to the observer, the resulting leak will be the first place where corrosion will make its presence known.

10.2.5 Examples of Buried Systems

Pipelines

Pipelines carrying oil, gas, and water are surely the most considerable assets buried in soils from the deepest subsea exploitation fields to the most remote tropical regions of the world. The subject of both external and internal gas pipeline corrosion is covered in many details in Chap. 12 and many details of the corrosion management of water lines are discussed in Chap. 8.

Some pipelines deteriorate slowly, and in certain cases pipeline life has been reliably targeted at 70 years or more. Other pipelines have been built which have exhausted their useful life after one year of operation. Apart from the quality of the construction, coatings,

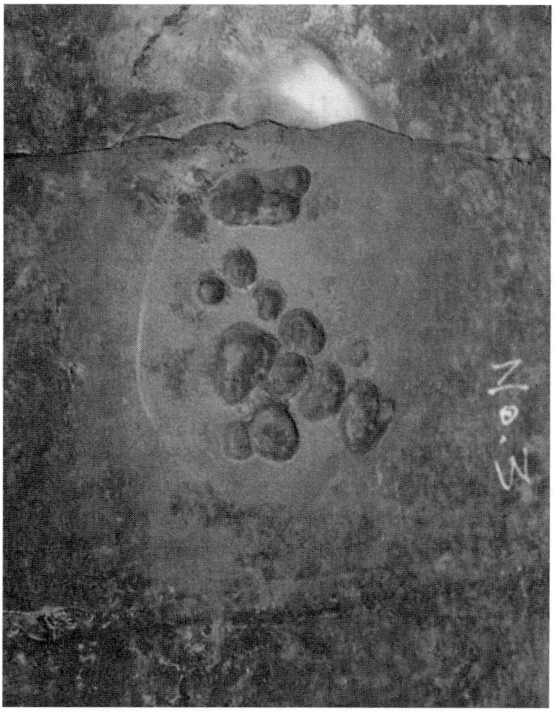

Figure 10.3 Typically deep, sharp-sided pits usually found under disbonded coatings or hard accumulations on a pipe surface. When disturbed the area may smell of hydrogen sulfide, a telling sign of microbial activity. (Courtesy of MACAW's Pipeline Defects, published by Yellow Pencil Marketing Co.)

CP systems, and so forth, the factors which affect pipeline life include nature of the product, nature of the external environment, operating conditions, and quality of maintenance. Regular inspections to assess the rate of change in physical condition provide a more accurate assessment of how much longer a pipeline can be expected to operate safely and productively. These inspections may also provide the essential information to plan for remedial action if the estimated life is below requirement [6].

In addition to corrosion protection, many pipelines require thermal insulation to prevent hydrocarbons to produce waxes or hydrates. These heavier components can clog lines and require immediate attention. There is thus a continuous need of improvements in coating as oil and gas operations extend to unprecedented depths and temperatures [7]. As indicated in Fig. 10.4, pipelines have been coated with a variety of protective coatings with a wide performance range over the past 50 years. The advantages and disadvantages of the main coatings used for pipeline protection are summarized in Table 10.8.

Coating type	Advantages	Disadvantages
Asphalt/coal tar	Easy to apply Minimal surface preparation required Long track record in certain environments without failure Permeable to cathodic protection in event of failure	Subject to oxidation and cracking Soil stress has been an issue Limitations at low application temperatures Environmental and exposure concerns Associated with corrosion and stress crack corrosion failures
Tape wrap (two layer)	Simple application	Poor shear stress resistance Many documented failures related to corrosion and stress crack corrosion Shielding of cathodic protection Adhesives subject to biodegradation
Two-layer extruded polyethylene	Excellent track record Good handling	Limited temperature range Poor shear stress resistance Limited pipe sizes [< 24 in (60 cm) outside diameter]
Fusion-bonded epoxy	Excellent adhesion and corrosion resistance Does not shield cathodic protection	Low impact resistance High moisture absorption and permeation
Three-layer polyotefin	Excellent combination of properties	Best suited for electrical resistance welded pipes High thickness to eliminate weld tenting
Composite coating	Excellent combination of properties Conforms well to external raised weld profiles	Suitable only for large diameter pipes and is not designed for small diameter pipes (< 40 cm OD)

Table 10.8 Comparison of Pipeline Coatings Systems Used in Recent History

Distribution Systems

Distribution systems generally are composed of pipes of many different sizes and ages. The age factor is surely the most important consideration. Distribution systems are much more subject to accidental contacts with other lines, typically water lines. It is also common for distribution systems to be made of a greater variety of materials. For example,

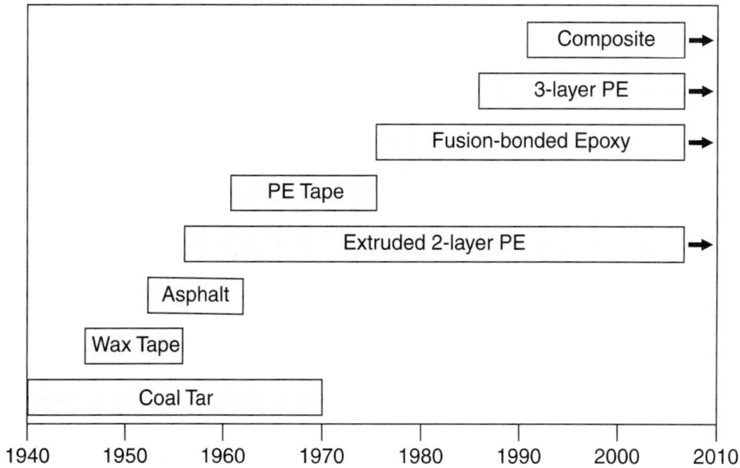

Figure 10.4 External pipeline coating development.

copper service lines may be used. Because of these differences, the approach to the protection of such components is likely to differ in that more emphasis is placed on the use of small ICCP units and galvanic anodes, even when the metal surface is protected by a coating.

Gathering Systems

Gathering systems do not ordinarily differ much from pipelines. They offer more variety and often have more accidental contacts. ICCP systems may encounter more interference problems in both directions. One common problem is that gathering lines, particularly in the early development of a field, may be laid in great haste, with little or no thought to the corrosion control problem. Although this is not, strictly speaking, a difference in corrosion behavior, it does account for some serious problems.

Plant Piping

The corrosion activity in a refinery complex piping network or in any other industrial plant has several unique features:

- First, there are always combinations of aboveground and underground piping
- A considerable variety of sizes, materials, and functions
- Almost invariably a great variety of coatings applied in original construction, creating a range of surfaces from bare to very well-coated

Another factor present in almost every plant is the large amount of bare copper in the ground system, typically a collection of ground rods all possibly interconnected. This can cause much difficulty, both in causing corrosion when no protective measures are taken, and again as a complicating factor when CP is applied. Further problems may arise because of the presence of many paved and unpaved areas that set up troublesome aeration cells. Furthermore, the usual construction activity ultimately results in badly mixed soils. Soil resistivity maps, covering entire plant areas, are useful both in anticipating trouble areas and in planning control measures.

Well Casings

Although a well casing may be considered as merely a vertical pipeline, it also has several unique features. It costs much more per unit length, so that leak repair is enormously expensive, if at all possible. The corrosive exposure is also complicated by the variety of soil strata through which it passes. Normally, these are much more varied than would be found on an equal length of horizontal pipeline and casings are usually bare. Any investigation is severely hampered by the fact that only one end of the pipe is accessible, so that most conventional surveys cannot be conducted.

Underground Tanks

Underground tanks usually are large enough to contact more than one stratum of soil, so they may be subject to concentration cell action. They are almost always subject to oxygen cell attack, even when under pavement. Often, there are fittings of different metals, and seldom is the coating as good as that used on pipelines. One of the major problems in dealing with these structures is that it is difficult to justify an engineering study for each tank, yet they cannot all be treated alike.

Steel Piling

An important difference with steel piling is that a few pits or even holes have little effect on its structural strength. Consequently, much more corrosion can be tolerated than with pipelines or any other vessels for which leaks cannot be tolerated. Piling is almost always bare, vertical, and hence subject to the same kinds of cells that attack oil well casings. Bonding often may be a problem because individual piles may not be interconnected electrically, a condition that makes both investigation and protection a problem.

Transmission and Communication Towers

Electrical lines often have built-in problems because of the common use of extensive copper ground wire. The galvanic cells thus created can be ruinous. While this problem can be avoided by not interconnecting, there is still a problem with the corrosion of a large number of small units, which makes both study and protection difficult (see Chap. 7 for some examples).

10.2.6 Corrosion of Materials Other Than Steel

Cast Iron

The corrosion rates of pure iron, wrought iron, or mild steel are quite similar in soil environments. However, as described under graphitic corrosion in Chap. 6, when grey cast iron corrodes only a network of graphite particles may be left behind by the dealloying process. Some cast iron water mains that are almost one century old are in this condition and continue to adequately serve their purpose. Once the cast iron is graphitized, the exterior becomes an extremely noble electrode in any galvanic couple. Thus, care should be used in attaching uncoated or unprotected cast iron to other metals in the soils. Cast iron may also suffer pitting and depending on soil corrosivity and the presence of bacteria can lead to complete perforation of the pipe wall as in Fig. 10.5.

Coating of the exterior of cast iron pipe with bituminous or lower oil coatings has often been used to provide a measure of protection. More recently tape wrapping with self adherent tapes have been found to be a very cost-economical method of providing an efficient barrier between the metal and a corrosive soil Fig. 10.6. CP can then be applied provided the pipe sections are electrically joined.

FIGURE 10.5 A corrosion pit that led to the perforation and leak of a water main. (Courtesy of Drinking Water Services, City of Ottawa)

Figure 10.6 Wrapping a water main pipe with adhesive tape to provide a barrier against soil corrosion. (Courtesy of Drinking Water Services, City of Ottawa)

Aluminum

Aluminum alloys are used relatively rarely in buried applications, although some pipelines and underground tanks have been constructed from these alloys. Like stainless steels, aluminum alloys tend to undergo localized corrosion damage in chloride-contaminated soils. Protection by coatings is essential to prevent localized corrosion damage. Cathodic protection criteria for aluminum alloys to minimize the risk of generating undesirable alkalinity are available (see Fig. 4.14). Aluminum alloys can undergo accelerated attack under the influence of microbiological effects. Documented mechanisms include attack by organic acid produced by bacteria and fungi and the formation of differential aeration cells [8].

Zinc

Zinc may be used as a reference half-cell in soils. However, the main application of zinc in buried applications is in galvanized steel, for example, in the fabrication of culverts. Performance may be adequate unless soils are poorly aerated, acidic, or highly contaminated with chlorides, sulfides, and other corrosive ions. Well-drained soils with a coarse texture (the sandy type) provide a high degree of aeration. It should also be borne in mind that zinc corrodes rapidly under highly alkaline conditions. Such conditions can arise on the surface of cathodically overprotected structures. The degree of corrosion protection afforded by galvanizing obviously increases with the thickness of the galvanized coating. Additional protection can be afforded by so-called duplex systems, in which additional paint coatings are applied to galvanized steel.

Lead

At one time, lead was used extensively as sheathing for telephone cables, but it has been phased out in favor of plastics in most applications, in great part due to the toxicity of lead ions. Although lead is amphoteric*, a reasonable corrosion resistance was shown in most soils. The corrosion resistance of lead and lead alloys in soils is generally regarded as being in-between those of steel and copper. The corrosion resistance of buried lead sheathing for power and communication cables has usually been satisfactory. Caution needs to be exercised in soils containing nitrates and organic acids such as acetic acid. Excessive corrosion is also found under highly alkaline soil conditions. Silicates, carbonates, and sulfates tend to retard corrosion reactions by their passivating effects on lead.

Stainless Steels

Stainless steels are rarely used in soil applications, as their corrosion performance in soil is generally poor and not better than bare steel. Localized corrosion attack is a particularly serious concern. The presence of chloride ions and concentration cells developed on the surface of these alloys tends to induce localized corrosion damage. Since pitting tends to be initiated at relatively high corrosion potential values, higher redox potentials increase the localized corrosion risk. Common grades of stainless steel and even the very highly alloyed versions are certainly not immune to MIC.

Copper and Alloys

Copper is generally considered to have good resistance to corrosion in soils. Corrosion concerns are mainly related to highly acidic soils and the presence of carbonaceous contaminants such as cinder. Sulfides, often produced by SRBs, also greatly increase the risk of corrosion damage.

In the case of brasses, consideration must be given to the risk of dezincification, especially at high zinc levels. Soils contaminated with detergent solutions and ammonia also pose a higher corrosion risk for copper and copper alloys. Additional corrosion protection for copper and copper alloys is usually considered only in highly corrosive soil conditions. CP, the use of acid-neutralizing backfill such as limestone, and protective coatings can be used in these applications.

Concrete

Concrete is extensively used underground as footings, piers, tanks, piping, and so forth. The material is normally used with reinforcing steel.

---

* Amphoteric signifies that the metal is attacked by either acidic or basic environments.

Concrete is stable in most soils but serious corrosion can occur from a number of sources. When appraising the potential corrosivity of a soil toward concrete, both the soil chemistry and the effect of groundwater must be considered. The soil must be judged aggressive if

1. pH is less than 6
2. Sulfate or sulfide content is high
3. Magnesia content is high

Sulfate and magnesium ions can be corrosive to concrete even in neutral soils. Reactions occur with the calcium salts in the concrete to destroy the cohesion and produce a soft, porous mass. A Type-V Portland will resist the sulfate more capably, but aluminate cements are typically used to prevent magnesium salt attack.

Organic compounds, particularly esters, can rapidly degrade the usual concrete. Detergents can accelerate the rate of degradation. The freezing of absorbed water in the concrete can cause spalling or cracking. Thus, when proposing the use of concrete underground one should be attentive to water levels in the ground, the selection of the proper concrete, the proper cure of the material, the density of the finished product, the cleanliness of the water and sand used, the depth of coverage of the reinforcing metal, the possible need to seal the exterior, and the need to maintain the reinforcing metal as an electrically continuous structure if CP is considered.

The surface can be treated chemically with fluoride treatments or sodium silicate washes to densify and harden the exterior as a method of corrosion protection. Bituminous coatings are often applied to seal the exterior. Epoxy coatings are most compatible with the concrete as a coating, patching compound, or adhesive.

Polymeric Materials

Polymeric materials or "plastics" of construction have revolutionized many of the underground materials applications. Although certainly not immune to failure, the use of plastics negates the pitting, galvanic action, and other forms of localized attack experienced when using steel in buried applications. The smooth interior of tubing material will normally remain clean for the easy conveyance of liquids or gases. In diameters of 15 cm or less, the laying of pipe in the ground can be accomplished at the pace of a slow walk, which is an important economic advantage when combined with the savings from unneeded external protection.

The three major polymeric materials used in soils are thermoplastics: polyvinyl chloride (PVC), acrylonitrile-butadiene-styrene (ABS), and polyethylene (PE) of various densities. These materials are not attacked by the concentration of acids, alkalies, or solvents encountered in soil environments. Mechanical support of the materials by soil at a constant temperature overcomes two of the major problems associated with the use of plastics: the low modulus and strength, and the large thermal expansion property. When installed at sufficient depth to prevent collapse from heavy topsoil loading by trucks or tractors, for example, these materials will provide many years of trouble-free service.

When piping installations are made, care should be used in making up the joints to ensure their integrity. Extra care should be taken to avoid making notches on the outer surface and the pipe should be laid on a uniform base free of rocks or any other hard, irregular pieces.

Other plastics, such as polypropylene, polybutylene, and glass-reinforced polyester or epoxy, are also used for specific services at greater expense. The glass-reinforced thermosetting plastics have been particularly successful as tank materials or the underground storage of a wide range of products from water to gasoline. Long exposures in wet soils show only a superficial attack on the exterior of such thermoset polymers.

10.3 Microbiologically Influenced Corrosion

As mentioned in Chap. 7, microbes are present in almost all environments. These potentially corrosive agents flourish in a wide range of habitats and show a surprising ability to colonize water-rich surfaces wherever nutrients and physical conditions allow. A significant feature of microbial problems is that they often appear when conditions are favorable to an exponential growth of the organisms [9]. Because they are largely invisible, it has taken considerable time to establish a solid scientific basis for defining their role in materials degradation. Many engineers still continue to be surprised that such small organisms can lead to spectacular failures of large engineering systems.

Figure 10.7 shows the pitted area of a 15-cm circulating water line from the supply to the auxiliary vacuum pumps. Note the cluster of hemispherical pits and the long striated grooves due to the MIC attack. Each pit represents a localized cluster of anaerobic bacteria that have become destructive to the pipe wall. The grooves are the results of bacteria attacking along the steel structure probably due to the drawing process in manufacturing the pipe. Figure 10.8 shows a pit and perforation of 6.3-cm internal diameter carbon steel pipe carrying heavy oil. The pit morphology is typical of sulfate reducing MIC attack.

As illustrated in the form of a pipe cross-section in Fig. 10.1, MIC is responsible for the degradation of a wide range of materials [1]. Most metals and their alloys, for example, stainless steels, aluminum and copper alloys, polymers, ceramic materials, and concrete can be attacked by microorganisms. The synergistic effect of different microbes and degradation mechanisms should also

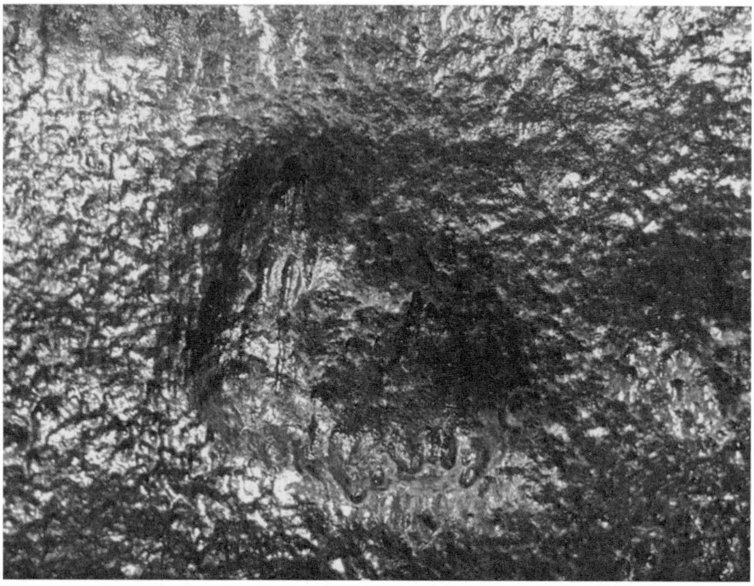

Figure 10.7 Circulating water MIC cell corrosion footprint. (Courtesy of Russ Green, TMI)

Figure 10.8 Pit and perforation of 6.3-cm internal diameter carbon steel pipe carrying heavy oil. The pit morphology is typical of sulfate reducing MIC attack [8]. (Courtesy of Kingston Technical Software)

be noted in Fig. 10.1. The mechanisms potentially involved in MIC are summarized as

- Cathodic depolarization, whereby the cathodic rate limiting step is accelerated by microbiological action
- Formation of occluded surface cells, whereby microorganisms form "patchy" surface colonies. Sticky polymers attract and aggregate biological and nonbiological species to produce crevices and concentration cells, the basis for accelerated attack
- Fixing of anodic reaction sites, whereby microbiological surface colonies lead to the formation of corrosion pits, driven by microbial activity and associated with the location of these colonies
- Underdeposit acid attack, whereby corrosive attack is accelerated by acidic products of the MIC "community metabolism," principally short-chain fatty acids.

In order to influence either the initiation or the rate of corrosion in the field, microorganisms usually must become intimately associated with the corroding surface. In most cases, they become attached to the metal surface in the form of either a thin, distributed film, or a discrete biodeposit. The thin film, or biofilm, is most prevalent in open systems exposed to flowing seawater, although it can also occur in open freshwater systems. Such thin films start to form within the first 2 to 4 hours of immersion, but often take weeks to mature. These films will usually be spotty rather than continuous but will nevertheless cover a large portion of the exposed metal surface [10].

Biodeposits differ from distributed films and may be up to several square centimeters covering typically only a small fraction of the total exposed metal surface, possibly leading to localized corrosion effects. The organisms in these deposits have generally a large effect on the chemistry of the environment at the metal/film or the metal/deposit interface without having any measurable effect on the bulk electrolyte properties. However, organisms will occasionally be concentrated enough in the environment to influence corrosion by changing the bulk chemistry. This is sometimes the case in anaerobic soil environments, where the organisms do not need to form either a film or a deposit in order to influence corrosion [10].

10.3.1 Planktonic or Sessile

Microorganisms that are attached to a surface are termed *sessile organisms* and these are most often present as a consortium or community of organisms, collectively referred to as a biofilm. Complex assemblages of various species may occur within both planktonic and sessile microbial populations. The environmental conditions largely dictate whether these microorganisms exist in a planktonic or sessile state.

Sessile microorganisms do not attach directly to the actual surface, but rather to a thin layer of organic matter adsorbed on the surface (Fig. 10.9, Stages 1 and 2). As microbes attach to and multiply,

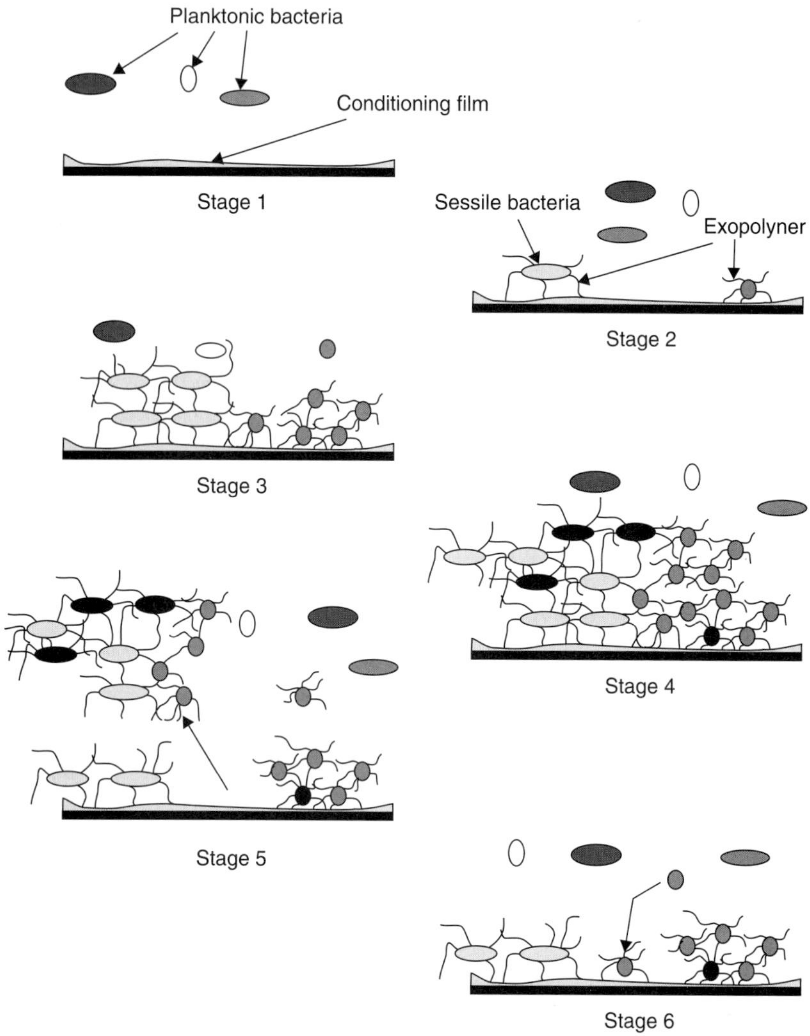

FIGURE 10.9 Different stages of biofilm formation and growth. Stage 1: Conditioning film accumulates on submerged surface; Stage 2: Planktonic bacteria from the bulk water colonize the surface and begin a sessile existence by excreting exopolymer that anchors the cells to the surface; Stage 3: Different species of sessile bacteria replicate on the metal surface; Stage 4: Microcolonies of different species continue to grow and eventually establish close relationships with each other on the surface. The biofilm increases in thickness. Conditions at the base of the biofilm change; Stage 5: Portions of the biofilm slough away from the surface; Stage 6: The exposed areas of surface are recolonized by planktonic bacteria or sessile bacteria adjacent to the exposed areas [11].

a biofilm composed of immobilized cells and their extracellular polymeric substances builds up on the surface.

The growing biofilm increasingly prevents the diffusion of dissolved gases and other nutrients coming from the bulk environment. These changing conditions become inhospitable to some microorganisms at the base of the biofilm and eventually many of these cells die, for example, on the internal wall of a water handling system. As the foundation of the biofilm weakens, shear stress due to adjacent fluid flow may cause sloughing of cell aggregations exposing the bare surface to the bulk fluid in localized areas (Fig. 10.9, Stage 5). The exposed areas are subsequently recolonized and new microorganisms and their exopolymers are woven into the fabric of the existing biofilm (Fig. 10.9, Stage 6). This phenomenon of biofilm instability occurs even when the physical conditions in the bulk liquid remain constant. Thus, biofilms are constantly in a state of flux [11].

Depending on the type of industrial system, planktonic organisms may include, besides bacteria, unattached algae, diatoms, fungi, and other microorganisms present in a system bulk fluids. In most cases, it is planktonic bacteria that are the focus of monitoring for MIC using microbiological detection techniques since system fluids are generally easier to sample than metallic surfaces. Unfortunately, the levels of planktonic bacteria present in the liquids are not necessarily indicative of MIC problems or their severity [12].

Monitoring sessile organisms either requires that the system be regularly opened for sampling or that accommodations be made in the system to allow for regular collection or on-line tracking of attached organisms while the system continues to operate. Because the presence of viable sessile organisms rarely correlates to the corrosivity of an environment, it is a good practice to use additional methods that directly determine the presence of active MIC. At best, the detection of viable planktonic bacteria may serve as an indicator that living microorganisms are present in a particular system, some of these organisms being capable of participating in the microbial attack.

10.3.2 Microbes Classification

One useful microbe classification consists in describing microorganisms according to their oxygen tolerance [10]:

- Strict (or obligate) anaerobes, which will not function in the presence of oxygen
- Aerobes, which require oxygen in their metabolism
- Facultative anaerobes, which can function in either the absence or presence of oxygen
- Microaerophiles, which use oxygen but prefer low levels

Strictly anaerobic environments are quite rare in nature. However, strict anaerobes are commonly found flourishing within anaerobic microenvironments (under deposits and fouling) even in highly aerated systems. The pH conditions and availability of nutrients also play a role in determining what type of microorganisms can thrive in a particular soil environment. Another way of classifying organisms is according to their metabolism:

- The compounds or nutrients from which they obtain their carbon for growth and reproduction
- The chemistry by which they obtain energy or perform respiration
- The elements they accumulate as a result of these processes

All microorganisms are not equally aggressive to metals. What is important, from a corrosion standpoint, is the number of microorganisms of the specifically corrosive types. Microorganisms associated with corrosion damage are classified as

- Fungi that may produce corrosive byproducts in their metabolism, such as organic acids. Apart from metals and alloys these can also degrade organic coatings and wood.
- Slime formers that may produce concentration corrosion cells on surfaces.
- Anaerobic bacteria that produce highly corrosive species as part of their metabolism.
- Aerobic bacteria that produce corrosive mineral acids.

Fungi

Fungi can be separated into yeasts and molds. Certain fungi are capable of producing organic acids and have been blamed for corrosion of steel and aluminum, as in the corrosion failures of aluminum aircraft fuel tanks. In addition, fungi may produce anaerobic sites for SRB and produce metabolic byproducts that help the growth of various bacteria leading to fouling and associated corrosion problems. In general, molds are considered to be of greater importance in corrosion problems than yeasts [13].

Algae

Algae are photosynthetic organisms with relatively simple nutritional requirements of light, water, air and certain inorganic compounds. Algae are found in wide ranges of salinity, from seawater to distilled water, and survive in different degrees of light intensity. Algae have been associated with fouling and related corrosion problems through the photosynthetic production of gaseous oxygen when exposed to light and the production of corrosive species such as organic acids.

Bacteria

Bacteria are generally small, with lengths typically under 10 μm. Collectively, they tend to live and grow under wide ranges of temperature, pH, and oxygen concentration. Carbon molecules represent an important nutrient source for bacteria.

Bacteria can exist in several different metabolic states. Bacteria that are actively respiring, consuming nutrients, and proliferating are said to be in a growth stage while bacteria that are simply existing, but not growing because of unfavorable conditions, are said to be in a resting state. Some strains, when faced with unacceptable surroundings, form spores that can survive extreme temperatures and long periods without moisture or nutrients, yet produce actively growing cells quickly when conditions again become acceptable. Cells that actually die are usually consumed rapidly by other organisms or enzymes. When looking at an environmental sample under a microscope, therefore, it should be assumed that most or all of the cell forms observed were alive or capable of life at the time the sample was taken.

Sulfate Reducing Bacteria SRBs have been implicated in the corrosion of cast iron and steel, ferritic stainless steels, 300 series stainless steels and other highly alloyed stainless steels, copper nickel alloys, and high nickel molybdenum alloys. They are almost always present at corrosion sites because they are in soils, surface water streams and waterside deposits in general. The key symptom that usually indicates their involvement in the corrosion process of ferrous alloys is localized corrosion filled with black sulfide corrosion products.

SRBs are anaerobes that are sustained by organic nutrients. Generally they require a complete absence of oxygen and a highly reduced environment to function efficiently. Nonetheless, they circulate (probably in a resting state) in aerated waters, including those treated with chlorine and other oxidizers, until they find an ideal environment supporting their metabolism and multiplication.

SRBs are usually lumped into two nutrient categories, those that can use lactate and those that cannot. The latter generally use acetate and are difficult to grow in the laboratory on any medium. Lactate, acetate, and other short chain fatty acids usable by SRB do not occur naturally in the environment. Therefore, these organisms depend on other organisms to produce such compounds.

SRBs reduce sulfate to sulfide, which usually shows up as hydrogen sulfide or, if iron is available, as black ferrous sulfide (Fig. 10.10). In the absence of sulfate, some strains can function as fermenters and use organic compounds such as pyruvate to produce acetate, hydrogen, and carbon dioxide. Many SRB strains also contain hydrogenase enzymes, which allow them to consume hydrogen. Most common strains of SRB grow best at temperatures from 25 to 35°C. A few thermophilic strains capable of functioning efficiently at more than 60°C have been reported.

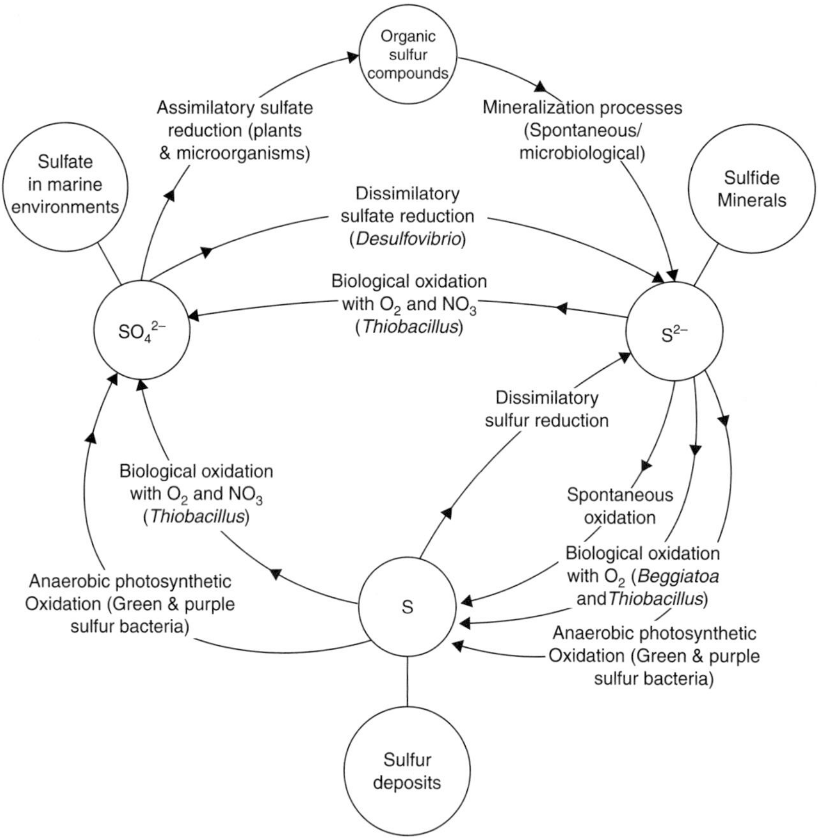

Figure 10.10 The sulfur cycle showing the role of bacteria in oxidizing elemental sulfur to sulfate and in reducing sulfate to sulfide [27].

Tests for the presence of SRB have traditionally involved growing the organisms on laboratory media, quite unlike the natural environment in which they were sampled. These laboratory media will only grow certain strains of SRB, and even then some samples require a long lag time before the organisms will adapt to the new growth conditions. As a result, misleading information has been obtained regarding the presence or absence of SRB in field samples.

Sulfur/Sulfide Oxidizing Bacteria This broad family of aerobic bacteria derives energy from the oxidation of sulfide or elemental sulfur to sulfate (Fig. 10.10). Some types of aerobes can oxidize sulfur to sulfuric acid, with pH values as low as one reported. These *Thiobacillus* strains are most commonly found in mineral deposits, and are largely responsible for acid mine drainage, which has become an environmental concern. They proliferate inside sewer lines and can cause rapid deterioration of concrete mains and the reinforcing steel therein.

They are also found on stone buildings and statues and probably account for much of the accelerated damage commonly attributed to acid rain. Where *Thiobacillus* bacteria are associated with corrosion, they are almost always accompanied by SRB. Thus, both types of organisms are able to draw energy from a synergistic sulfur cycle.

Iron/Manganese Oxidizing Bacteria Bacteria that derive energy from the oxidation of Fe^{2+} to Fe^{3+} are commonly reported in deposits associated with MIC. They are almost always observed in tubercles (discrete hemispherical mounds) over pits on steel surfaces. The most common iron oxidizers are found in the environment in long protein sheaths or filaments. While the cells themselves are rather indistinctive in appearance, these long filaments are readily seen under the microscope and are not likely to be confused with other life forms. The observation that filamentous iron bacteria are omnipresent in tubercles might be, therefore, more a matter of their easy detection than of their relative abundance. An intriguing type of iron oxidizers is the *Gallionella* bacterium, which has been blamed for numerous cases of corrosion of stainless steels.

Besides the iron/manganese oxidizers, there are organisms that simply accumulate iron or manganese. Such organisms are believed to be responsible for the manganese nodules found on the ocean floor. The accumulation of manganese in biofilms is blamed for several cases of corrosion of stainless steels and other ferrous alloys in water systems treated with chlorine or chlorine/bromine compounds.

Methane Producers Methane producing bacteria (methanogens) have only been added in recent years to the list of organisms believed responsible for corrosion. Like many SRB, methanogens consume hydrogen and thus are capable of performing cathodic depolarization. While methane producers normally consume hydrogen and carbon dioxide to produce methane, in low nutrient situations these strict anaerobes will become fermenters and consume acetate.

Organic Acid Producing Bacteria Various anaerobic bacteria such as *Clostridium* are capable of producing organic acids. Unlike SRBs, these bacteria are not usually found in aerated macroenvironments such as open, recirculating water systems. However, they are a problem in gas transmission lines and could be a problem in closed water systems that become anaerobic.

Aerobic Slime Formers Aerobic slime formers are a diverse group of aerobic bacteria. They are important in managing corrosion mainly because they produce extracellular polymers that make up what is commonly referred to as slime, a natural polymer that is actually a sophisticated network of sticky strands that bind cells to the surface and control what permeates through the deposit.

The stickiness of slime formers traps all sorts of particulates that might be floating by, which, in dirty water, can result in the impression that the deposit or mound is an inorganic collection of mud and debris. Slime formers can be efficient "scrubbers" of oxygen, thus preventing oxygen from reaching the underlying surface. This creates an ideal site for SRB growth.

10.3.3 Monitoring Microbiologically Influenced Corrosion

An effective biocorrosion mitigation program needs to include corrosion monitoring as a periodic or continual means of assessing whether program goals are being achieved. Monitoring techniques that detect the presence of microbes, especially on metallic surfaces, can provide an early indication of incipient MIC or the potential for MIC. A number of methods for the detection of microorganisms, including specific types of organisms and estimates of their numbers and activity, have been developed [12].

The first biocorrosion monitoring systems were focused on assessing the number of microbes per unit volume of water sampled from a system. Data obtained with these systems were combined with electrochemical corrosion measurements, using electrical resistance (ER) or linear polarization resistance (LPR) probes in addition to coupon weight loss measurements. The problem with this approach is that the number of free-floating planktonic organisms in the water does not correlate well with the organisms present in biofilms on the metal surface where the corrosion actually takes place. An effective monitoring scheme for controlling both biofouling and biocorrosion should include data gathering using as many of the following techniques as possible [14]:

- Sessile bacterial counts, by either conventional biological techniques or optical microscopy of organisms present in the biofilm, on the metal surface.
- Direct observation of the community structure of the biofilm. Several types of probe systems are commercially available for holding and inserting metallic coupons into the system. Examination of the biofilm has been done by scanning electron microscopy, epifluorescence optical microscopy, or confocal laser scanning microscopy.
- Identification of the microorganisms found in both the process water and on the metal surface.
- Surface analysis to obtain chemical information on corrosion products and biofilms.
- Evaluation of the morphology of the corrosive attack on the metal surface after removal of biological and corrosion product deposits with conventional macrophotography, stereomicroscopy, optical and scanning electron microscopy, or other metallographic techniques.

- Electrochemical corrosion measurements.
- Water quality and redox potential measurements.
- Other types of information specific to each operational system, including duty cycle and downtime information, concentrations, and timing for addition of biocides and other chemical inputs, local sources and nature of pollutants, and so forth.

Sampling

Samples for analysis are usually obtained by scraping accessible surfaces. In open systems or on external surface of pipelines or other underground facilities this can be done directly. For low pressure water systems, bull plugs, coupons or inspection ports can provide a way to expose specimens representative of internal surfaces [15]. However, more sophisticated devices are required for pressurized systems to allow mounting an assembly on a standard pressure fitting [16].

If the biofilm developing is to be representative of the behavior in a system it is important for the sampling coupons to be made of a material similar to the system material and that the coupons be flush-mounted in the wall of the system to have the same flow effects as those of the surrounding surface. While pressure fittings allow inserting coupons directly in process units, these fittings can be expensive. Pressure vessel codes and accessibility can also restrict possible locations. For these reasons sidestream installations are often preferred.

Handling of field samples should be done carefully to avoid contamination with foreign matter including biological materials. Many different types of sterile sampling tools and containers are available and for anaerobic systems, special handling and transport tools are essential to avoid exposure to oxygen from the air. One option is to analyze samples on the spot with special kits. Where transportation to a laboratory is required, Torbal jars or similar anaerobic containers can be used [17]. In many cases, simply placing samples directly in a large volume of the process water in a completely filled screw cap container is adequate.

Processing in the lab should also be done anaerobically using special techniques or in anaerobic chambers designed for this purpose. Because viable organisms are involved, processing should be done quickly to avoid growth or death of cells stimulated or inhibited by changes in temperature, oxygen exposure, or other factors [9].

Biological Assessment

Biological assays are performed on liquid samples or on suspensions of solid deposits to identify and enumerate viable microorganisms, quantify metabolic or specific enzyme activity, or determine the concentration of key metabolites [18]. Table 10.9 summarizes some of the methods that have been used to detect and describe

Assay	Method	Comments
Microorganisms Cell numbers Specific organisms	*Microscopic examination:* Cell numbers are obtained with a Petroff-Hausser counting chamber. Fluorescent dyes can light up specific microbes.	Requires a microscope with high magnification, phase contrast, and ultraviolet fluorescence, as appropriate. Cell counts are straightforward, but particulates and fluorescent materials may interfere.
Viable cell counts SRB APB Other organisms	*Most probable number:* Sample is diluted into a series of tubes of specific growth medium. Cell numbers based on growth at various dilutions. *Dip slides:* A tab coated with growth medium is immersed, then incubated two to five days. Microbial colonies are counted visually.	Commercially available kits can be inoculated in the field, but growth takes days to weeks. Crude numbers of bacteria, yeast, or fungi can be estimated in contaminated crankcase oil or fuels. Commercial kits require minimal cost and expertise.
Identification of microorganisms	*Fatty acid methyl ester analysis:* Methyl esters of fatty acids from field samples are analyzed by gas chromatograph. *Probes based on nucleic-acid base sequences:* Probes bind to DNA or RNA for specific proteins or organisms. *Reverse sample genome probing:* DNA from MIC microbes is spotted on a master filter. Labeled DNA from field samples will bind to that DNA, if the reference organism is present.	Fatty acid composition fingerprints specific organisms. The technique is available commercially. Probes require special expertise and lab facilities. The assay answers the question, "Is this organism present?" It can also be used to track changes in a population after chemical treatment but it requires special expertise and equipment.

Table 10.9 Inspection, Growth, and Activity Assays for Microbial Populations

Assay	Method	Comments
Biomass	*Protein, lipopolysaccharide, nucleic acid analysis:* Established methods, widely available.	Concentration of cell constituents correlates to level of organisms present in field sample.
Metabolic activity	*Adenosine triphosphate (ATP):* Fluorometer and supplies commercially available for field use.	ATP level correlates well with microbial metabolism level. Its concentration reflects the level of activity in a sample.
Sulfate reduction	^{35}*S-sulfate reduction:* Radioactively labeled sulfate is incubated in field sample. H_2S formed is liberated by acid addition, trapped on a zinc acetate wick, and measured by scintillation counter.	A specialized lab technique useful in discovering nutrient sources for MIC and in quickly screening biocide activity in samples taken from the target system.
Enzyme activity	*Hydrogenase assay or sulfate reductase assay:* Measures enzyme activity in a field sample as a rate of reaction.	Commercial kits are available. Hydrogenase activity may be related to MIC, while sulfate reductase assesses the presence of SRB.

Table 10.9 *(continued)*

microorganisms in terms of total cells present, viable cell numbers, and metabolic activity. Table 10.9 also identifies assays that can be used to establish the presence of biomass in a field sample, identify the organisms present, and assess the activity of enzymes, such as hydrogenase, that are thought to accelerate corrosion through cathodic depolarization. A more detailed description of these assays and other techniques is provided in the following sections.

Direct Inspection Direct inspection is best suited for the enumeration of planktonic organisms suspended in relatively clean water. In liquid suspensions, cell densities greater than 107 cells cm^{-3} may cause the sample to appear turbid. Quantitative enumerations using phase contrast microscopy can be done quickly using a counting chamber, which holds a known volume of fluid in a thin layer and visualization of microorganisms can be enhanced by fluorescent dyes that cause cells to light up under ultraviolet radiation [19].

Cells separated by filtration from large aliquots of water can be visualized and counted on a 0.25 μm filter using the epifluorescent technique and a stain such as acridine orange. Stains such as fluorescein diacetate, 5-cyano-2,3-ditolyltetrazolium chloride or *p*-iodonitrotetrazolium violet indicate active metabolism by the formation of fluorescent products [9]. Antibody fluorescence microscopy is similar to general fluorescent microscopy, except that the fluorescent dye used is bound to antibodies specific to SRBs. Only bacteria recognized by the antibodies fluoresce. Results can be analyzed within two hours. The technique detects viable and nonviable bacteria but it is limited to the type of SRB used in the manufacture of these antibodies [19].

Identification of organisms can be accomplished by use of antibodies generated as an immune response to the injection of microbial cells into an animal, typically a rabbit. These antibodies will selectively bind to the target organism in a field sample. A second antibody tagged with a fluorescent dye is then used to light up the rabbit antibody bound to the target cells. In effect the staining procedure can selectively light up target organisms in a mixed population or in difficult soil, coating, or oily emulsion samples [20]. Such techniques can provide insight into the location, growth rate, and activity of specific kinds of organisms in mixed biofilm populations.

Growth Assays A widely used method to assess microbial populations in water samples is through growth tests using commercially available growth media for groups of organisms most commonly associated with industrial problems. These are packaged in a convenient form suitable for use in the field. Serial dilutions of suspended samples are grown on solid agar or liquid media. Based on the growth observed for each dilution, estimates of viable cells, most probable number

(MPN) [21] are made. The test can show some results in a few days, but the usual incubation period for the test is 14 to 28 days [19].

However, despite the common use of these assays, only a small fraction of wild organisms actually grow in commercially available artificial media. For example, estimates for SRB in marine sediments indicate that only one in a thousand of the organisms present actually show up in standard growth tests [22].

Activity Assays

Whole Cell The conversion of a radioisotopically labeled substrate can be used to assess the potential activity of microbial populations in field samples. The technique depends on bacterial growth for detection, but it generates results in about two days and is specific to SRBs. The sample is incubated with a known trace amount of radioactive-labeled sulfate. (SRBs reduce sulfate to sulfide.) After incubation, the reaction is terminated by adding an acid to kill the cells and the radioactive sulfide is fixed with zinc acetate for evaluation. This is a highly specialized technique, involving expensive laboratory equipment and the handling of radioactive substances [19].

The radiorespirometric method, however, can use field samples directly without the need to separate organisms and it is very sensitive. Selection of the radioactively labeled substrate is key to interpretation of the results. The method can provide insights into factors limiting growth by comparing activity in native samples with supplemented test samples under various conditions. Oil degrading organisms, for example, can be assessed through the mineralization of ^{14}C labeled hydrocarbon to carbon dioxide.

Radioactive methods are not routinely used by field personnel. However, they have been particularly useful in a number of applications including biocide screening programs, identification of nutrient sources, and assessment of key metabolic processes in various corrosion situations [9].

Enzyme-Based Assays An increasingly popular approach is the use of commercial kits to assay the presence of enzymes associated with microorganisms suspected to cause problems. For example, kits are available for the sulfate reductase enzyme common to SRBs [23]. This technique takes advantage of the fact that SRBs reduce sulfate to sulfide through the APS-reductase enzyme, common to all SRBs. Measurement of the amount of APS-reductase in a sample gives an estimate of SRBs present. The test does not require bacterial growth and the entire test takes only 15 to 20 minutes [19].

Another example is the hydrogenase enzyme implicated in the acceleration of corrosion through rapid removal of cathodic hydrogen formed on the metal surface [24]. The test analyzes for the hydrogenase enzyme that is produced by bacteria able to use hydrogen as an energy source. The test is usually performed on sessile samples by exposing a sample to an enzyme-extracting solution and monitoring the degree of hydrogen oxidation in an oxygen-free atmosphere after adding a dye [19].

Performance of several of these kits has been assessed by field personnel in round robin tests. Correlation of activity assays and population estimates is variable. In general these kits have a narrower range of application than growth-based assays and it is therefore important to select a kit with a detection range appropriate to the problem under consideration [25].

Metabolites Adenosine triphosphate (ATP) is present in all living cells but it disappears rapidly on death. A measure of ATP may thus quantify the presence of living material. The ATP can be measured using an enzymatic reaction, which generates flashes of light that are detected by a photomultiplier [19]. Commercial instruments are available that measure the release of light by the firefly luciferin/luciferase with ATP. The method is best suited to clean aerobic aqueous samples since suspended solids and chemical quenching can affect the results. Detection of metabolites such as organic acids in deposits or gas compositions including methane or hydrogen sulfide by routine gas chromatography can also indicate biological involvement in industrial problems [9].

Cell Components Biomass may be generally quantified by assays for protein, lipopolysaccharide, or other common cell constituents but the information gained is of limited value. An alternate approach is to use cell components to define the composition of microbial populations with the hope that the insight gained may allow future damaging situations to be recognized and managed. Fatty acid analysis and nucleic acid sequencing provide the basis for the most promising methods in this category.

Fatty Acid Profiles Analyzing fatty acid methyl esters derived from cellular lipids can fingerprint organisms rapidly and, provided pertinent profiles are known, organisms in industrial and environmental samples may be identified with confidence. The immediate impact of events such as changes in operating conditions or the application of biocides can be monitored by such analysis. Problem populations of certain organisms may also be identified in order to implement appropriate management responses in a timely fashion.

Nucleic Acid–Based Methods Specific DNA probes have been developed to detect segments of genetic material coding for known enzymes. A gene probe developed to detect the hydrogenase enzyme which occurs broadly in SRB from the genus *Desulfovibrio* has been tested on samples from an oilfield waterflood plagued with iron-sulfide related corrosion problems. The enzyme was detected with this probe in 12 of 20 samples, suggesting that sulfate reducers which did not have this enzyme were also present in the operation [26]. In principle, probes could be developed to detect potentially all sulfate-reducers. However, the operation of a battery of probes could be a daunting task where a large number of field samples have to be analyzed.

The reverse sample genome probe (RSGP) has been developed to overcome this obstacle. With RSGP, the DNA from organisms previously isolated from samples acquired from known field problems is spotted on a master filter, following which DNA isolated from a new sample is labeled with either a radioactive or fluorescent indicator and placed on the filter. Labeled DNA from the new sample sticks to the corresponding spot on the master filter when complementary strands of DNA are present. Organisms represented by the labeled spots are then known to be in the new field sample [26].

Detailed Coupon Examinations A great deal of information can be learned by careful, in-depth examination of corrosion coupon surfaces using commonly available analytical techniques. A wide variety of samplers for introducing metallic surfaces of interest into a system are available. A popular sampling device is shown in Fig. 10.11.

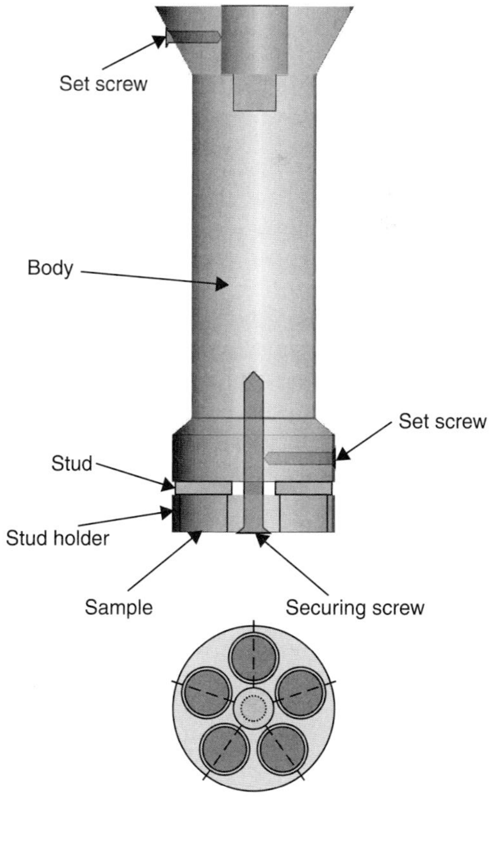

Figure 10.11 Biofilm sampling device with removable "buttons."

Special handling of coupons after removal from the system being monitored is crucial to ensure that subsequent laboratory tests provide representative information. Biofilms in particular are highly sensitive to dehydration, exposure to air, temperature, mechanical damage, and other gross environmental changes that can occur during removal and transport of the coupons [12].

Examination of coupons for microbial populations can be performed either directly or indirectly, using histological embedding techniques to preserve and remove the biofilm. Although fairly involved, the embedding technique offers several advantages over direct observation in that the biofilm and corrosion products are preserved for future analysis. Environmental scanning electron microscopy (ESEM) may also be utilized to examine biofilms on test coupons. However, exopolymers and corrosion products often obscure the presence of cells, making quantification and identification difficult with this method.

Monitoring Microbiologically Influenced Corrosion Effects

The presence of a biofilm on a metallic surface can greatly alter the local corrosion processes. In addition to the electrochemical changes that directly affect corrosion, biofilms can also modify other readily measured characteristics such as pressure drop or heat transfer resistance. Monitoring such microbiological influences can provide a useful indicator that a biofilm is present and that action should be taken to mitigate potential MIC.

Deposition Accumulation Monitors Methods for monitoring deposits can provide an indication of the accumulation of biofilm and other solids on surfaces or in orifices. For example, monitoring the pressure drop across an orifice provides a simple method for continuous monitoring of deposit accumulation and biofilm accumulation. The main disadvantage of the pressure drop technique is that it is not specific to the biofilm buildup since it detects the total scaling and deposition effects in a line [12].

These measurements can be made on actual operating units on line but they may also be done using model heat exchanger units or instrumented pipe loops run in parallel to system flow. Figure 10.12 shows such an instrumented pipe loop test unit with five parallel, instrumented pipe runs. Water flow from the target system is diverted through this unit, so that conditions are representative of the actual operating system [18].

Measurements of friction factors and heat exchange efficiencies can indicate fouling. While all sorts of deposits can affect flow and heat transfer in an operating industrial system, biofilms are especially effective. A 165-pm thick biofilm shows 100 times the relative roughness of a calcite scale and a thermal conductivity close to that of water, that is, almost 100 times less than carbon steel [9].

Figure 10.12 Oil recovery microbial test lines to evaluate biocide programs. (Courtesy of Kingston Technical Software)

The assumption is generally made that a system showing extensive fouling is prone to MIC. Systems with extensive fouling are operating inefficiently and may in any case warrant remedial action.

Electrochemical Methods An electrochemical method for on-line monitoring of biofilm activity has been developed for continuous monitoring of biofilm formation without the need for excessive involvement of plant personnel [Figure 10.13(a) and 10.13(b)]. The series of stainless steel or titanium disks in the monitoring device are exposed to the plant environment with only one set of disks being polarized relative to the other sets for a short period of time each day. The electrodes are typically connected through a shunt the remainder of the time. Biofilm activity, which is also an electrochemical process, is monitored by tracking changes in the applied current when the external potential is on and the generated current when the potential is off [12].

The onset of biofilm formation on the probe is indicated when either of these independent indicators deviates from the baseline level (Fig. 10.14). Such a departure would then trigger the alarm located in a control box (Fig. 10.15). The level of biofilm activity is also measured by the amount of variations from the baseline assuming that the applied and generated currents from a well-controlled system produce typically a flat line devoid of any significant deviations.

Another experimental approach to detect MIC with an electrochemical signals is based on the use of small silver sulfide and silver chloride electrodes capable of detecting sulfides or chlorides by changes in the potential between the Ag/AgCl or Ag/Ag_2S

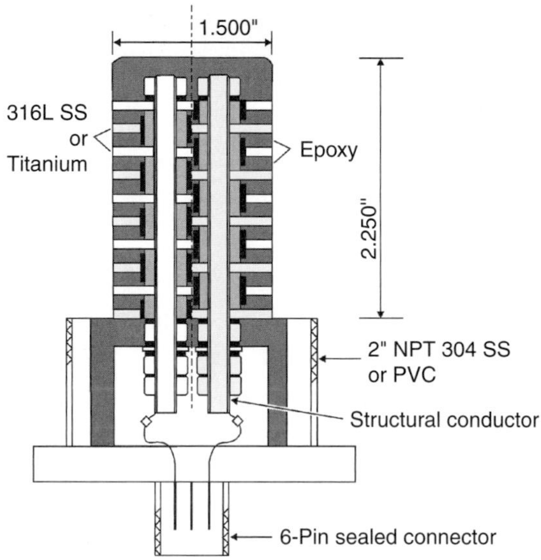

(a)

(b)

Figure 10.13 (*a*) Schematic and (*b*) picture of the BioGEORGE MIC detection probe. (Courtesy of George Licina, Structural Integrity Associates, Inc)

electrode pairs [12]. The production of sulfides or concentration of chlorides by microbial action has been shown to produce a nearly instantaneous change in the potential of the electrode pair, thus providing an indication of specific microbial activity.

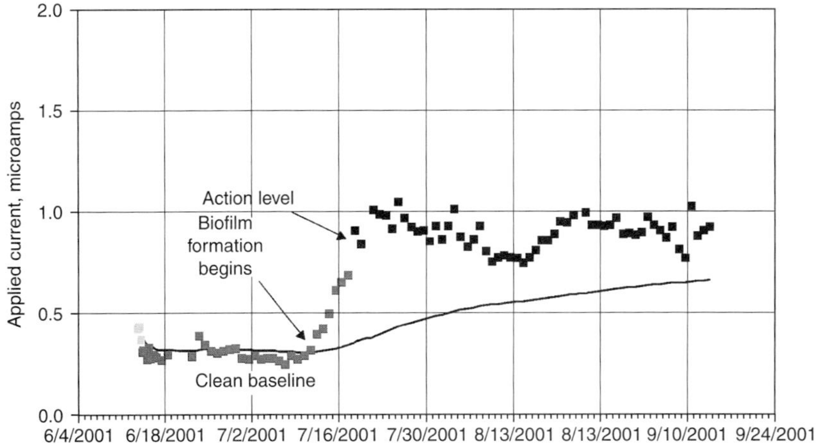

FIGURE 10.14 A plot of signals generated by a BioGEORGE MIC detection probe. (Courtesy of George Licina, Structural Integrity Associates, Inc).

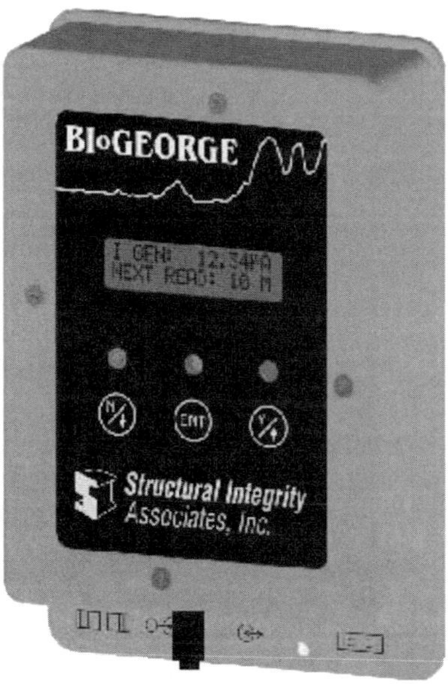

FIGURE 10.15 Control box of a BioGEORGE MIC detection probe. (Courtesy of George Licina, Structural Integrity Associates, Inc)

References

1. Hill EC. *Microbial Aspects of Metallurgy*. New York, N.Y.: American Elsevier, 1970.
2. Robinson WC. Testing soil for corrosiveness. *Materials Performance* 1993; 32: 56–8.
3. Palmer JD. environmental characteristics controlling the soil corrosion of ferrous piping. In: Chaker V, Palmer JD, eds. *Effects of Soil Characteristics on Corrosion*. Philadelphia, Pa.: American Society for Testing and Materials, 1989; 5–17.
4. Spickelmire B. Corrosion considerations for ductile iron pipe. In: *Materials Performance* 2002; 41: 16–23.
5. Heim M, Schwenk W. Corrosion in aqueous solutions and soil. In: von Baeckmann W, Schwenk W, Prinz W, eds. *Handbook of Cathodic Protection*. 1997; 139–52.
6. Roberge PR. *Corrosion Inspection and Monitoring*. New York, N.Y.: John Wiley & Sons, 2007.
7. Wilmott M, Highams J, Ross R, Kopystinski A. Coating and thermal insulation of subsea or buried pipelines. *Journal of Protective Coatings & Linings* 2000; 17: 47–54.
8. Wagner P, Little B. Impact of alloying on microbiologically influenced corrosion—A review. *Materials Performance* 1993; 32: 65–8.
9. Jack TR. Monitoring microbial fouling and corrosion problems in industrial systems. *Corrosion Reviews* 1999; 17: 1–31.
10. Tatnall RE. Introduction Part I. In: Kobrin G, ed. *Microbiologically Influenced Corrosion*. Houston, Tex.: NACE International, 1993.
11. Geesey GG. Introduction Part II—Biofilm formation. In: Kobrin G, ed. *Microbiologically Influenced Corrosion*. Houston, Tex.: NACE International, 1993.
12. Zintel TP, Licina GJ, Jack TR. Techniques for MIC monitoring. In: Stoecker II JG, ed. *A Practical Manual on Microbiologically Influenced Corrosion*. Houston, Tex.: NACE international, 2001.
13. Pope DH, Duquette D, Wayner Jr. PC, Johannes AH. *Microbiologically Influenced Corrosion: A State-of-the-Art Review. 2nd edn*. Columbus, Ohio: Materials Technology Institute, 1989.
14. Dexter SC. Microbiologically influenced corrosion. In: Cramer DS, Covino BS, eds. *Vol. 13A: Corrosion: Fundamentals, Testing, and Protection*. Metals Park, Ohio: ASM International, 2003; 398–416.
15. Sanders PF. Monitoring and control of sessile microbes: Cost effective ways to reduce microbial corrosion. In: Sequeira CAC, Tiller AK, eds. *Microbial Corrosion–1*. New York, N.Y.: Elsevier Applied Science, 1988; 191–223.
16. Gilbert PD, Herbert BN. Monitoring microbial fouling in flowing systems using coupons. In: Hopton JW, Hill EC, eds. *Industrial Microbiological Testing*. London, U.K.: Blackwell Scientific Publications, 1987; 79–98.
17. Gerhardt P, Murray RGE, Costilow RN et al. *Manual of Methods for General Bacteriology*. Washington, D.C.: American Society of Microbiology, 1981.
18. Jack TR. Biological Corrosion Failures. In: Shipley RJ, Becker WT, eds. *ASM Handbook Vol. 11: Failure Analysis and Prevention*. Materials Park, Ohio: ASM International, 2002.
19. *Techniques for Monitoring Corrosion and Related Parameters in Field Applications*. NACE 3T199. Houston, Tex.: NACE International, 1999.
20. Hunik JH, van den Hoogen MP, de Boer W, Smit M, Tramper J. Quantitative determination of the spatial distribution of nitrosomonas europaea and nitrobacter agilis cells immobilized in k-carrageenan gel beads by a specific fluorescent-antibody labelling technique. *Applied and Environmental Microbiology* 1993; 59: 1951–4.
21. Costerton JW, Colwell RR. *Native Aquatic Bacteria: Enumeration, Activity and Ecology*. [STP 695]. Philadelphia, Pa.: American Society for Testing and Materials, 1977.

22. Jorgenson BB. A comparison of methods for the quantification of bacterial sulfate reduction in coastal marine sediments. *Geomicrobiology Journal* 1978; 1: 49–64.
23. Odom JM, Jessie K, Knodel E, Emptage M. Immunological cross-reactivities of adenosine-5′-phosphosulfate reductases from sulfate-reducing and sulfide oxidizing bacteria. *Applied and Environmental Microbiology* 1991; 57: 727–33.
24. Bryant RD, Jansen W, Boivin J, Laishley EJ, Costerton JW. Effect of hydrogenase and mixed sulfate-reducing bacterial populations on the corrosion of steel. *Applied and Environmental Microbiology* 1991; 57: 2804–9.
25. Scott PJB, Davies M. Survey of field kits for sulfate reducing bacteria. *Materials Performance* 1992; 31: 64–8.
26. Voordouw G, Telang AJ, Jack TR, Foght J, Fedorak PM, Westlake DWS. Identification of sulfate-reducing bacteria by hydrogenase gene probes and reverse sample genome probing. In: Minear RA, Ford AM, Needham LL, Karch MJ, eds. *Applications of Molecular Biology in Environmental Chemistry*. Boca Raton, Fla.: Lewis Publishers, 1995.
27. Borenstein SW. *Microbiologically Influenced Corrosion Handbook*. New York, N.Y.: Industrial Press, 1994.

CHAPTER 11

Materials Selection, Testing, and Design Considerations

11.1 Materials Selection

Materials are sometimes chosen by trial and error or simply on the basis of what has been used before. While this approach frequently works, it does not always lead to optimization or innovation. The advent of computers has revolutionized the design process. It is now possible to make design changes and prototype components rapidly with very little effort. On the materials side, it is now possible to generate databases of properties that allows instant retrieval of desired properties which can in turn be integrated into the design process. As a result, materials selection has undergone radical transformation over the years.

An example of such an activity is the development of a materials selection methodology by Michael Ashby of Cambridge University. In this approach, materials selection is well integrated into the design process. The constraints that are identified based on function and design are used to identify the class of potential materials from the broad spectrum of available materials. As the design is optimized and detailed, additional high-precision data from this subset of materials is considered in order to narrow down the choice to a single material. Factors like cost and fabricability can also be introduced into the factor that represents design and functional constraints thus making it a rather powerful and cost-effective method for materials selection.

In Ashby's model, a material has *attributes:* its density, strength, cost, electrical properties, and so forth [1]. A design demands, perhaps, a certain profile of low density, high strength, modest cost, and high electrical conductivity. The problem is that of identifying the desired attribute profile and then comparing it with those of real engineering materials to find the best match.

In the Materials Selector software system developed following Ashby's principles, this is achieved by first *screening and ranking* the candidates to give a shortlist, and then seeking detailed *supporting information* for each short-listed candidate, allowing a final choice. It is important to start with a full menu of materials in mind; failure to do so may mean a missed opportunity. If an innovative choice is to be made, it must be identified early in the design process. Later, too many decisions have been taken and commitments made to allow radical change: it is now or never.

The immensely wide choice is narrowed, first, by applying *property limits* which screen out the materials which cannot meet the design requirements. Further narrowing is achieved by ranking the candidates by their ability to maximize performance. Performance is generally limited not by a single property, but by a combination of them. The results of a search with Materials Selector in its supporting databases can be presented in a convenient selection chart such as shown in Fig. 11.1 for two basic mechanical properties, that is, strength and Young's modulus.

However, there is no obvious provision to include corrosion resistance considerations in Ashby's model. The main reason for this may be that unlike other design considerations such as fatigue, designing with corrosion in mind is far from being a straightforward process since each form of corrosion is essentially a separate failure mode that needs

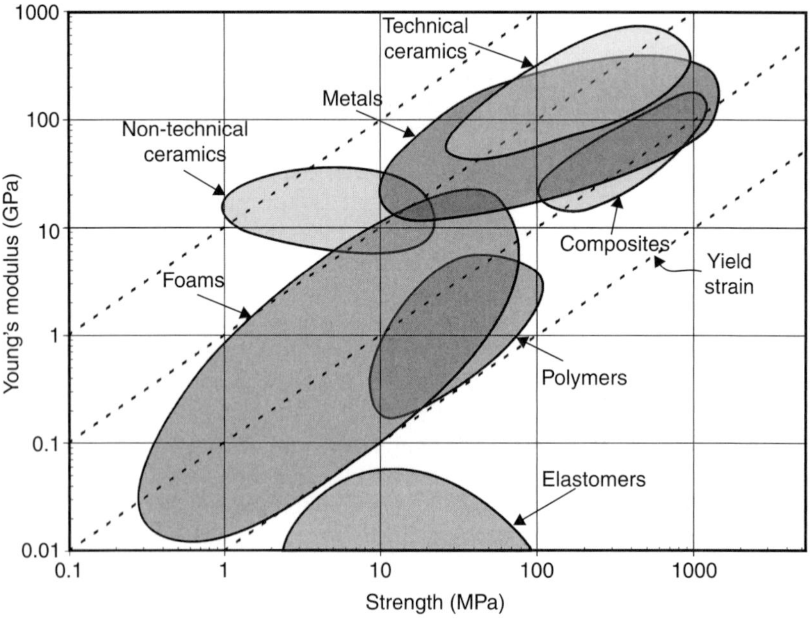

FIGURE 11.1 Ashby materials selection chart based on strength properties.

to be considered individually. The challenge that designers face is balancing the degree of detail put forth in the analysis against their available resources including time and budgetary concerns [2].

11.2 Complexity of Corrosion Conscious Materials Selection

Unanticipated corrosion problems will always occur, and correcting them may require a reactive approach using the service of a trained specialist. However, corrosion should ideally be minimized by proactively considering corrosion resistance early during system design, when material selection has not been "locked in." However, because of its complexity, corrosion analysis is really more a practice or discipline, such as reliability engineering, than a simple process of designing fatigue resistance into a structure. The impact of improper considerations given to corrosion is also similar to improper consideration of reliability: unanticipated problems that result in reduced readiness and increased life-cycle costs [2].

Unfortunately, materials selection is rarely taught to designers with a focus on corrosion prevention and control. One might wonder why a practical, design-based material selection process that considers corrosion prevention and control has not already been developed. The following list identifies some of the factors that contribute to the difficulty in developing such a simplified, readily understood corrosion conscious materials selection process.

11.2.1 Multiple Forms of Corrosion

There are multiple and often competing forms of corrosion to consider for every design. As described previously in Chap. 6, corrosion damage can take many forms that may be triggered, as explained and illustrated in Chap. 7, by often relatively complex invisible cells. One commonly encountered situation is the stress corrosion cracking (SCC) or the corrosion fatigue of a component following the formation of corrosion pits in the stressed area. In such cases the initiation of a corrosion process is governed by a pitting mechanism while the subsequent propagation phase is governed by a cracking mode.

Another common occurrence of multiple forms of corrosion is what is called *pack rust* in civil engineering circles and pillowing in aerospace engineering. Both types of corrosion attack start by the formation of a special crevice corrosion environment that is often exacerbated by successive wetting and drying cycles. The end result is a continuous buildup of internal stresses that were never considered during the initial design. Forces as high as 70 kPa have been measured in the case of pack rust affecting steel components [3]. The buildup of voluminous corrosion products on aluminum parts may also lead to an undesirable increase in stress levels near critical fastener holes in aircraft in which rivets have been known to fracture due to the high tensile stresses resulting from pillowing [4].

11.2.2 Multiple Material/Environment Combinations

There are a vast number of material/operational environment combinations, and each can form the basis for potential corrosion problems to occur. Of the 103 plus elements in a Periodic Table in App. B, approximately 80 are metals, each possessing different mechanical, chemical, and physical properties. Although all these metals can corrode, they do it in many different manners. Furthermore, most of these metals have been alloyed to *make tens of thousands of different alloys*.

From a purely technical standpoint, an obvious answer to corrosion problems would be to use more resistant materials. In many cases, this approach is an economical alternative to other corrosion control methods. Table 11.1 lists the questions that should be answered in order to estimate the corrosion behavior of materials either in service or considered for such usage [5].

A common form of representing the corrosion resistance of materials is what is known as *isocorrosion diagrams*. The use of the prefix "iso" refers to lines (or regions) of constant corrosion behavior across variations in concentration and temperature. The diagram shown in Fig. 11.2 illustrates how some stainless steels compare to

• **Corrosive Agent Variables**
Main constituents (identity and amount)
Impurities (identity and amount)
Temperature
pH
Degree of aeration
Velocity or agitation
Pressure
Estimated range of each variable
• **Type of Application**
What is the function of part or equipment?
What effect will uniform corrosion have on serviceability?
Are size change, appearance or corrosion product a problem?
What effect will localized corrosion have on usefulness?
Will there be stresses present?
Is SCC a possibility?
Is design compatible with the corrosion characteristics of the material?
What is the desired service life?

TABLE 11.1 Information Required for Estimating the Corrosion Performance of a Material

• **Experience**
Has the material been used in identical situation?
With what specific results?
If equipment is still in operation, has it been inspected?
Has the material been used in similar situations?
What are the differences in performance between the old and new situations?
Any pilot-plant experience?
Any plant corrosion-test data?
Have laboratory corrosion tests been run?
Are there any available reports?

TABLE 11.1 (*continued*)

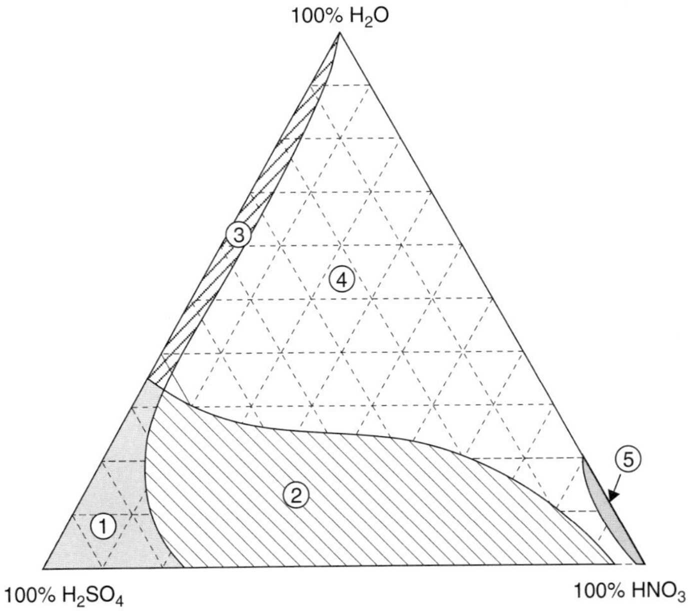

Materials in shaded zones have reported corrosion rates of < 0.50 mm/y

Zone 1	**Zone 2**	**Zone 3**	**Zone 4**	**Zone 5**
20Cr 30Ni	18Cr 8Ni	20Cr 30Ni	18Cr 8Ni	18Cr 8Ni
Gold	20Cr 30Ni	Gold	20Cr 30Ni	20Cr 30Ni
Lead	Cast iron	Platinum	Gold	Aluminum
Platinum	Gold	Silicon iron	Platinum	Gold
Silicon iron	Lead	Tantalum	Silicon iron	Platinum
Steel	Platinum		Tantalum	Silicon iron
Tantalum	Silicon iron			Tantalum
	Tantalum			

FIGURE 11.2 Mixed-acid isocorrosion diagram.

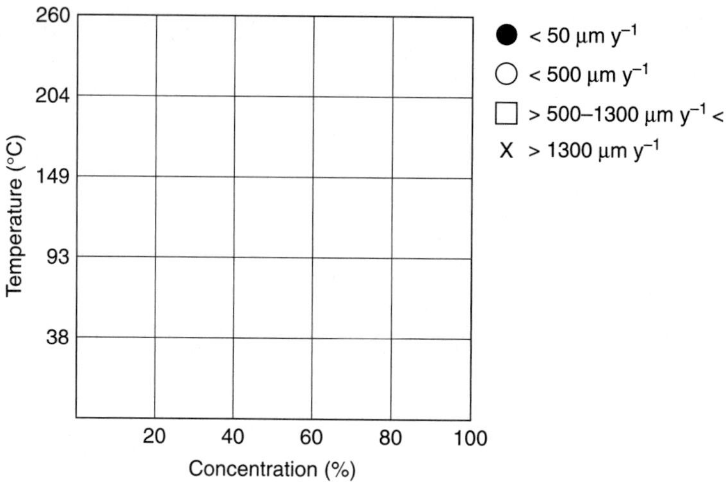

Figure 11.3 Template used to express the corrosion resistance of most metals exposed to industrial environments.

other metallic materials for handling industrially important nitric or sulfuric acids and their mixtures [6].

Corrosion Data Survey of Metals [6] and *Corrosion Data Survey of Nonmetals* [7] are two publications using such a scheme for representing corrosion information. In these publications the corrosion behavior of metals is expressed in units of penetration rates, that is, micrometer per year (μm y^{-1}) or milli-inch per year (mpy), while the corrosion behavior of nonmetals is expressed in qualitative terms such as "recommended," "questionable," and "not recommended." Figure 11.3 explains how the template in *Corrosion Data Survey of Metals* is used to express the corrosion resistance of the most commonly used metals and Fig. 11.4 illustrates how the template serves to summarize the

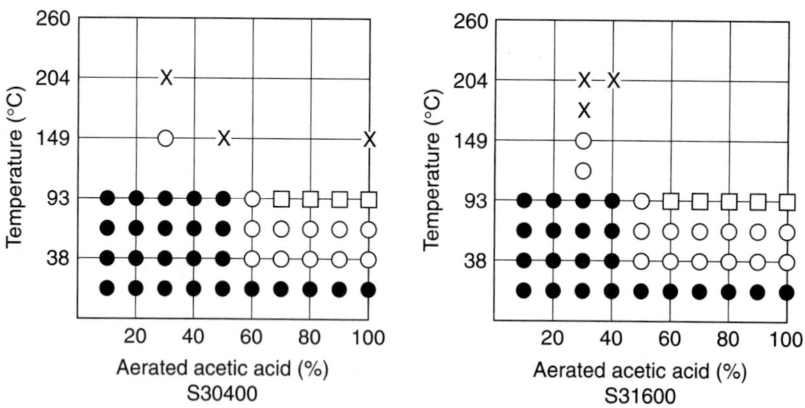

Figure 11.4 Corrosion behavior of S30400 and S31600 stainless steels in aerated pure acetic acid.

corrosion resistance of S30400 and S31600 stainless steels in aerated acetic acid service.

11.2.3 Precision of Corrosion Data

Corrosion data are overwhelmingly empirical, often widely scattered, and come in a variety of forms. Additionally, corrosion data from the literature can rarely be used to predict corrosion rates in field applications. There are many factors that explain why corrosion test results are typically more scattered than many other types of testing, an important one being the effect on corrosion rates due to minor impurities in the materials themselves or in the testing environments [8].

The accuracy of data against testing time and number of factors is illustrated in a 3-D plot (Fig. 11.5) showing the relative difficulties associated with reproducing industrially realistic corrosion problems. This intrinsic complexity has made the transformation of corrosion testing results into usable real-life functions for service applications a difficult task [9].

Corrosion behavior is often the result of complicated interactions between the conditions of a metallic surface and the adjacent environment to which it is exposed. Therefore, there is no universal

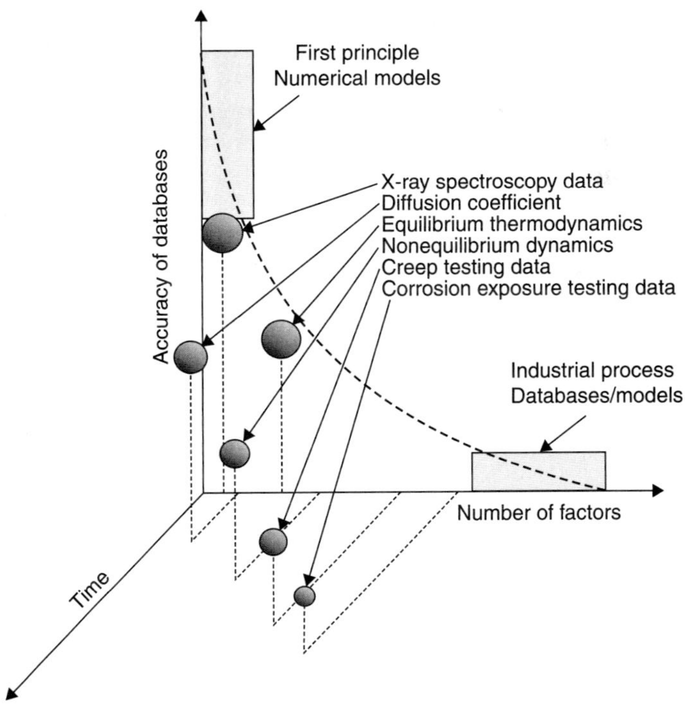

FIGURE 11.5 Time and environment dependency of databases and models [9].

corrosion test for all purposes. Corrosion data are also quite specific to materials and environments where they were taken. The type of corrosion and the rate at which it occurs may be highly influenced by even small changes in a material composition/processing history or by minor changes in the testing environment. For these reasons, the most reliable predictor of practical performance is service experience, followed closely by field testing, because both are based on the actual environment and its complex variations.

Many investigators have traditionally avoided the use of statistical techniques because the added reliability did not seem to offset the effort and time required to become familiar with the methodology or to perform the necessary calculations. Most often, the results must be interpreted in terms of relative rankings. Such interpretations can be subjective and depend much on the judgment of the investigator.

11.2.4 Complexity of Materials/Performance Interactions

Data may simply not be available for all potential forms of corrosion that might pertain to a specific application/environment combination. Additionally, the inherent corrosion resistance of a material is largely determined by its elemental composition, by its processing history, surface morphology, geometry, and in some cases its size. The following example illustrates how exfoliation corrosion cracking, a particular form of intergranular corrosion associated with high-strength aluminum alloys, may be minimized during the manufacturing process by accepting a deliberate materials loss.

Alloys that have been extruded or otherwise worked heavily, with a microstructure of elongated, flattened grains, are particularly prone to this damage. Figure 11.6 illustrates the anisotropic grain structure

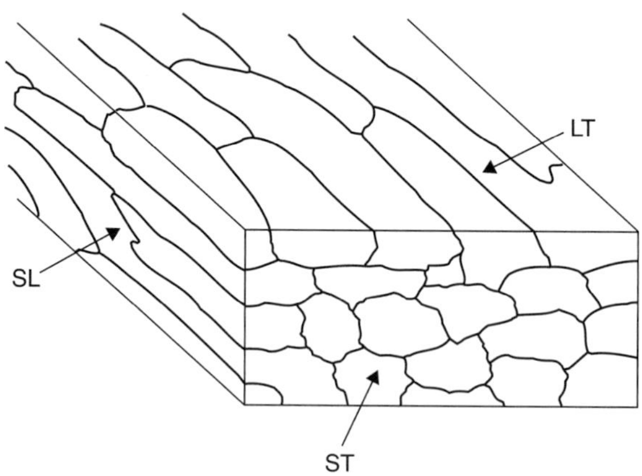

Figure 11.6 Schematic representation of the anisotropic grain structure of wrought aluminum alloys.

FIGURE 11.7 Lamellar corrosion exfoliation of an aluminum alloy (A97075). (Courtesy of Kingston Technical Software)

typical of wrought aluminum alloys where corrosion products building up along these grain boundaries exert pressure between the grains and the end result is a lifting or leafing effect (Fig. 11.7). The damage often initiates at end grains encountered in machined edges, holes, or grooves and can subsequently progress through an entire section. Figure 11.8 shows how a fraction of material is often sacrificed to alleviate the impact on the susceptibility to corrosion cracking of the short transverse sections of a component.

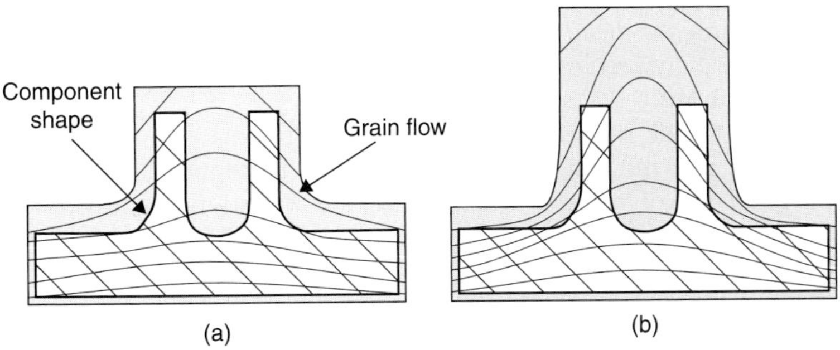

FIGURE 11.8 Machining for neutralizing the effects of grain flow on corrosion resistance: (*a*) saving on material and loosing on lifetime, and (*b*) loosing on material for increased lifetime.

11.3 Selection Compromises

One common reason designers pay little attention to corrosion prevention and control (CPAC) is that they have minimal experience in considering corrosion issues. Engineers must not only balance potential benefits against other performance parameters, but must also consider the longer-term attributes such as service life, anticipated maintenance and repair, and disposal.

As it is becoming obvious, selecting the *best* material, from a corrosion standpoint may not be a simple task. Numerous factors need to be considered and there is no single path to making an informed selection. Ultimately, designers must rely upon their best *engineering judgment* to select the optimum material, considering corrosion resistance in balance with other performance requirements, as well as other tangible factors, including cost, availability, and maintainability.

The choice of a material is the result of several compromises. For example, the technical appraisal of an alloy will generally be a compromise between corrosion resistance and some other properties such as strength and weldability. And the final selection will be a compromise between technical competence and economical considerations. In specifying a material, the task usually requires three stages:

1. Listing the requirements
2. Selecting and evaluating the candidate materials
3. Choosing the most economical material

Some particular requirements and typical selection considerations are presented in Table 11.2. The material selection process is also influenced by the fact that the materials are either considered for the construction of a new system, or for the modification or repairs in an existing facility. For the construction of new equipment, the selection procedure should begin as soon as possible and definitively much before the design is finalized since the optimum design for corrosion resistance may greatly affect how the system will be assembled or constructed.

In a repair application, there is usually less opportunity for redesign, and the main decision factors will be centered on delivery time and ease of fabrication in the field. It is also advisable to estimate the remaining life of the equipment so that the repair is not grossly overdesigned in terms of its corrosion allowance.

It is important to realize that one will seldom find the *perfect* choice, since most affordable commercial materials will corrode under some particular conditions. Instead, designers should identify the material and associated corrosion prevention and control practices that meet budgetary constraints specific to each program.

• **Requirements to be met**
Properties (corrosion, mechanical, physical, appearance)
Fabrication (ability to be formed, welded, machined, etc.)
Compatibility with existing equipment
Maintainability
Specification coverage
Availability of design data
• **Selection considerations**
Expected total life of plant or process
Estimated service life of material
Reliability (safety and economic consequences of failure)
Availability and delivery time
Need for further testing
Material costs
Fabrication costs
Maintenance and inspection costs
Return on investment analysis
Comparison with other corrosion-control methods

TABLE 11.2 Material Selection Checklist

11.3.1 Life-Cycle Costing

Life-cycle costing utilizes universally accepted accounting practices for determining the total cost of asset ownership or projects over the service life. The economic analysis is usually performed for comparing competing alternatives. Since the initial capital outlay, support, and maintenance over the service life and disposal costs are considered, the time value of money assumes major importance in life-cycle costing. Discounting future cash flows to present values essentially reduces all associated costs to a common point in time for objective comparison.

Economic calculations used to be laborious procedures that could only be properly performed by those who understood the interrelationship of all the factors involved. Relatively good estimates of initial costs, life, rate of depreciation, taxes, and value of money to the owner/employer have been combined into easily used equations in a NACE International standard that can be used without requiring a great knowledge of economics [10]. This report advocates the use of the discounted cash-flow method, which provides readily comparable calculations of net present worth. Factors that need to be considered in calculating net present worth include

1. Initial cost
2. Best estimate of expected life
3. Length of typical shutdown for emergency repair
4. Cost of planned maintenance during scheduled shutdowns
5. Effect of failure on total plant operation

When proposals are made with recommendations based on economic calculations, those controlling the purse strings are more inclined to listen. Calculated options provide them with an objective basis on which to make a decision regarding how and where to spend the money available, and thus add more credibility to a proposal.

The generalized economic analysis expression in Eq. (11.1) is particularly adapted to corrosion engineering problems. Given the numerous uncertainties associated with most corrosion problems this equation can provide fairly good estimates of prevention and control alternatives. This equation takes into account the influence of taxes, straight-line depreciation, operating expenses, and salvage value in the calculation of present worth and annual cost [11].

$$\mathrm{PW} = -P + \left(\frac{t(P-S)}{n}\right)\left(\frac{P}{A}, i\%, n\right) - (1-t)(X)\left(\frac{P}{A}, i\%, n\right) + S\left(\frac{P}{F}, i\%, n\right) \quad (11.1)$$

where A represents the annual end-of-period cash flow
F represents a future sum of money
$i\%$ represents the interest rate
n is the number of years
PW is the present worth referred also as Net Present Value (NPV)
P is the cost of the system at time 0
S represents salvage value
t is the tax rate expressed as a decimal
X represents the operating expenses

1. *First term* ($-P$): This term represents the initial project expense, at time zero. As an expense, it is assigned a negative value. There is no need to translate this value to a future value in time, as the PW approach discounts all money values to the present (time zero).
2. *Second term* $[(t(P-S)/n)(P/A, i\%, n)]$: The second term in this equation describes the depreciation of a system. The portion enclosed in braces expresses the annual amount of tax credit permitted. The portion in parentheses translates these as equal amounts back to time zero by converting them to PW.

3. *Third term* $[-(1-t)(X)(P/A, i\%, n)]$: The third term in the generalized equation consists of two parts. $(X)(P/A, i\%, n)$ represents the cost of items properly chargeable as expenses, such as the cost of maintenance, insurance, and the cost of inhibitors. Because this term involves expenditure of money, it also comes with a negative sign. The second part, $t(X)(P/A, i\%, n)$, accounts for the tax credit associated with this business expense and because it represents a saving it is associated with a positive sign.
4. *Fourth term* $[S(P/F, i\%, n)]$: The fourth term translates the future value of salvage to present value. This is a one time event rather than a uniform series and therefore it involves the single payment present worth factors. Many corrosion measures, such as coatings and other repetitive maintenance measures, have no salvage value, in which cases this term is zero.

PW can be converted to equivalent annual cost (A) by using the following formula:

$$A = (\text{PW})(A/P, i\%, n) \tag{11.2}$$

One can calculate different options by referring to interest tables or by simply using the formula describing the various functions. The capital recovery function (P/A) or how to find P once given A is provided by Eq. (11.3):

$$\left(\frac{P}{A}, i\%, n\right) \text{ where } P_n = A\frac{(1+i)^n - 1}{i(1+i)^n} \tag{11.3}$$

The compound amount factor (P/F) or how to find P once given F is provided by Eq. (11.4):

$$\left(\frac{P}{F}, i\%, n\right) \text{ where } F_n = P(1+i)^{-n} \tag{11.4}$$

The capital recovery factor (A/P) or how to find A once given P is provided by Eq. (11.5):

$$\left(\frac{A}{P}, i\%, n\right) \text{ where } A_n = P\frac{i(1+i)^n}{(1+i)^n - 1} \tag{11.5}$$

Several examples of the application of the generalized equation to corrosion engineering problems can be found in the literature [11;12].

11.3.2 Condition Assessment

A second major component of life-cycle asset management is systematic condition assessment surveys (CAS). The objective of CAS is to provide comprehensive information about the condition of an asset.

This information is imperative for predicting medium- and long-term maintenance requirements, projecting remaining service life, developing long-term maintenance and replacement strategies, planning future usage, and determining the available reaction time to damage. Therefore, CAS is in direct contrast to a short-term strategy of fixing serious defects as they are found. CAS includes three basic steps[12;13]:

1. A facility is divided into its systems and subsystems, forming a work breakdown structure (WBS).
2. Standards are developed to identify deficiencies that affect each component in the WBS and the extent of the deficiencies.
3. Each component in a WBS is evaluated against the standard or reference component.

CAS provides the maintenance managers the analytical information needed to optimize financial resource allocation for repair, maintenance, and replacement of assets. Through a well-executed CAS program, information will be available on the specific deficiencies of a system or component, the extent and coverage of that deficiency, and the urgency of repair. The following scenarios indicate a need for CAS in corrosion control strategies:

- Assets are aging, with increasing corrosion risks.
- Assets are complex engineering systems, although they may not always appear to be (for example, ordinary concrete is actually a highly complex material).
- Assets fulfilling a similar purpose have variations in design and operational histories.
- Existing asset information is incomplete and/or unreliable.
- Previous corrosion maintenance/repair work was performed but poorly documented.
- Information on the condition of assets is not transferred effectively from the field to management, leaving the decision makers ill-informed.
- Maintenance costs are increasing, yet asset utilization is decreasing.
- There is great variability in the condition of similar assets from poor to excellent. The condition appears to depend on local operating microenvironments but there is no knowledge on where the next major problem will appear.
- The information for long-term planning is very limited or nonexistent.
- Commitment within an organization to conduct long-term strategies and plans for corrosion control is limited or lacking.

11.3.3 Prioritization

Prioritizing maintenance activities is central to a methodical, structured maintenance approach, in contrast to merely addressing maintenance issues in a reactive, short-term mode. From the preceding sections, it should be apparent that life-cycle asset management can be used to develop a prioritization scheme that may then be employed in a wider set of funding decisions, not only maintenance "go/no-go" decisions. This entails the methodical evaluation of each maintenance action against preestablished values and attributes such as the key performance indicators described in Chap. 12.

Prioritization methodologies usually involve a numerical rating system to ensure that the most important work receives the most urgent attention. The criticality of equipment is an important element of some rating systems. Such an unbiased, "unemotional" rating will ensure that decisions are made for the best overall performance of an engineering system, rather than overemphasizing the performance of one of its parts.

Understanding the potential deterioration mechanisms that may lead to equipment failures, their likelihood of occurrence, and the potential consequences of the failures are key elements in a risk conscious environment. The challenge faced by corrosion engineers is to provide plant inspection personnel with a sound technical understanding of potential deterioration mechanisms for use in developing a practical and effective strategy to limit the risk of potential equipment failures. In this way a corrosion engineer can demonstrate the value of proactive corrosion control and use the results of a prioritization analysis to influence the decision makers and stakeholders in plant equipment integrity and reliability [14].

Using risk assessment techniques such as those described in Chap. 12 provides a sound basis for prioritizing and managing the inspection program for plant equipment. With this approach, the risk associated with the continued operation of each piece of plant equipment is ranked by assessing the likelihood of its failure versus the severity of failure consequences.

11.4 Materials Selection Road Map

A materials selection roadmap may be a good way to highlight the major steps needed to assess whether a new design may be susceptible to corrosion and how potential problems can be effectively mitigated or eliminated. In critical situations where corrosion can lead to high risk situations or serious safety concerns, it is usually preferable to involve a professional or a team of professionals that have the credentials to ensure that all aspects of the design are thoroughly investigated.

11.4.1 Identify Initial Slate of Candidate Materials

The flow chart in Fig. 11.9 illustrates a generic process for screening materials through a host of traditional means. Designers should initiate this process by first identifying a slate of candidate materials possessing the necessary physical, mechanical, thermal, and/or electrical properties. The number of materials chosen as initial candidates may vary depending with the criticality of the application and the design process employed.

The advantage of considering a larger number of materials at the first stage of a selection process is obviously to increase the likelihood

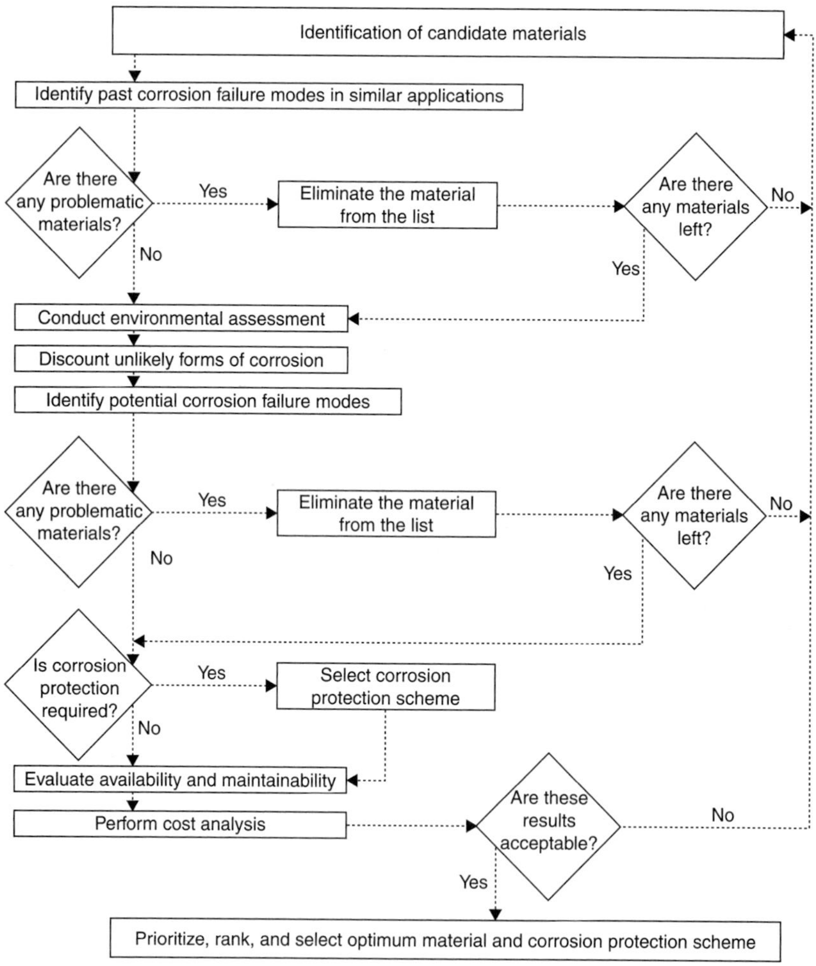

FIGURE 11.9 Generic process for selecting materials to improve corrosion resistance—a roadmap to life-cycle cost reduction.

of finding an appropriate candidate material that will meet traditional requirements while providing a required level of corrosion resistance. The disadvantage is that such an analysis is generally more costly. For highly important decisions with strict deadlines, analyzing multiple materials simultaneously would be the most efficient approach as it would minimize the impact on a project schedule.

11.4.2 Screen Materials Based on Past Experience

After choosing the material(s) to be initially considered, the next step is to look at whether there have been corrosion problems in applications similar to the system being designed, and to ascertain the nature of the causes of these problems. It is usually expedient to eliminate from consideration any material that has caused corrosion problems in a similar application in the past, especially those problems that could not be controlled using affordable corrosion prevention and control practices. The sources of information that should be consulted at this point will depend on many factors and vary greatly with organizations. Access to a good library with pertinent reference documents and online documentation is always a convenient starting point.

It is also important to consult any personnel that might have been involved with the materials considered for similar applications or consult maintenance records if these exist. Computer technologies are used extensively by large organizations to track and manage maintenance activities often supported by computerized technical drawings, parts lists, and detailed part descriptions. Figures 11.10 to 11.13 illustrate how such information technology can provide an easy graphical access to historical aircraft maintenance data [15].

11.4.3 Conduct Environmental Assessment

The next step in the materials selection process is to discount the forms of corrosion that are unlikely to occur. For example, if the application being designed will not be exposed to a flowing fluid, then erosion–corrosion would not be considered. To determine the more likely forms that will occur, the analyst should examine the factors responsible for initiating each corrosion mode and then determine which ones are present in the application. Chapters 6 and 7 provide descriptions that could be quite useful for this phase of the selection process.

After candidate materials have been selected, the next step is to analyze the environment in which a system will operate. When considering the operational environment many may automatically consider exposure to atmospheric, industrial, or marine conditions as the corrosion inducing factors. While important, such a global or "macro" view may be too limited since conditions imposed by the configuration and operation of a system may result in the formation of corrosion cells and of particularly corrosive microenvironments.

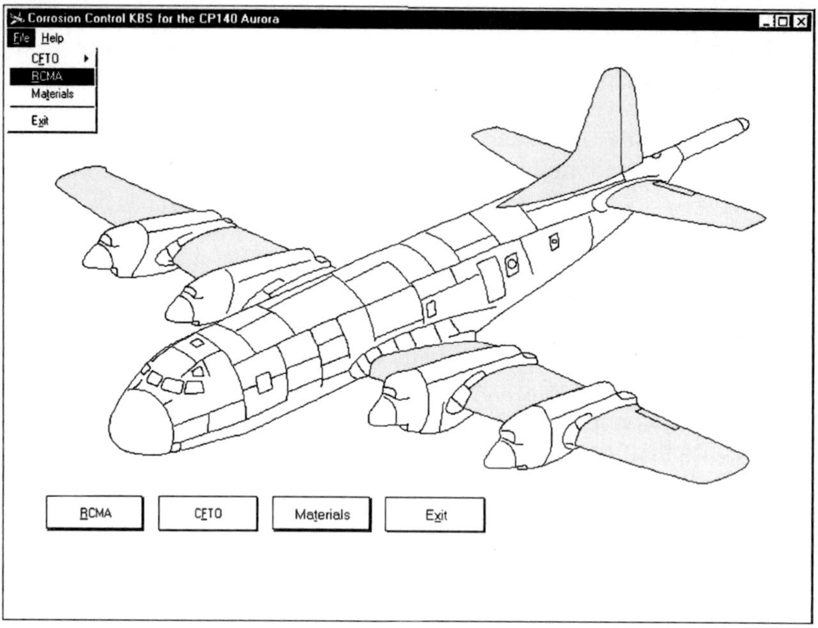

Figure 11.10 Main screen of a knowledge-based system (KBS) showing the areas of a patrol aircraft covered by an aircraft structural integrity program (ASIP).

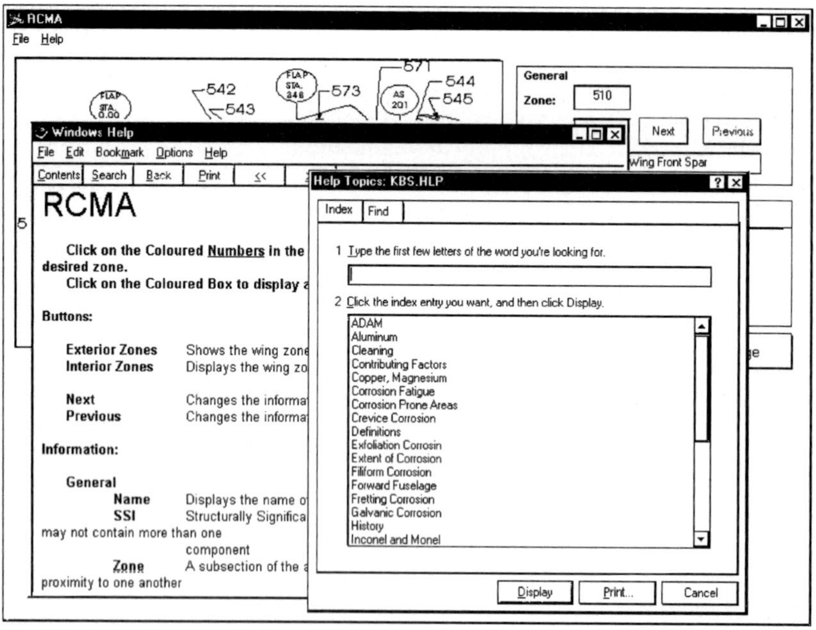

Figure 11.11 Example of integration of graphics and database information into a KBS for ASIP.

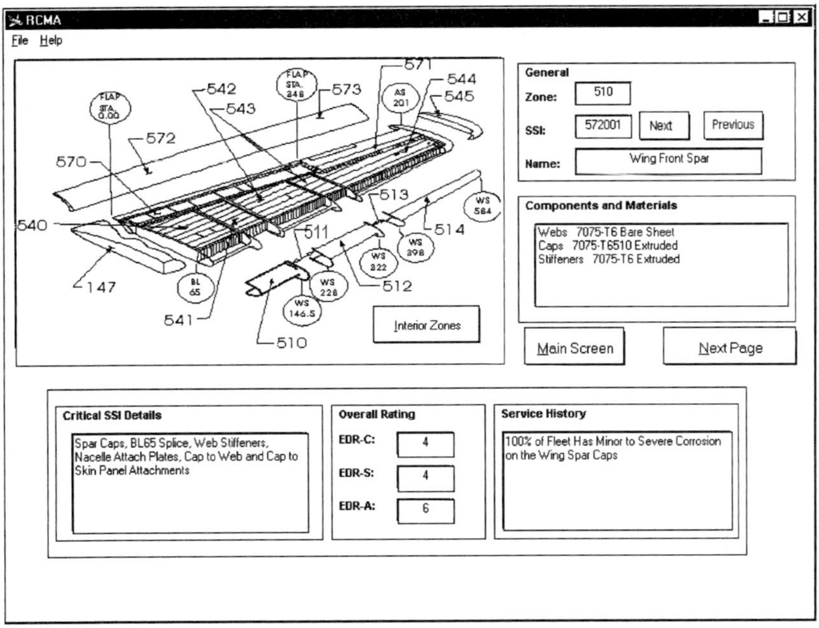

FIGURE 11.12 Example of context sensitive help in a KBS for ASIP.

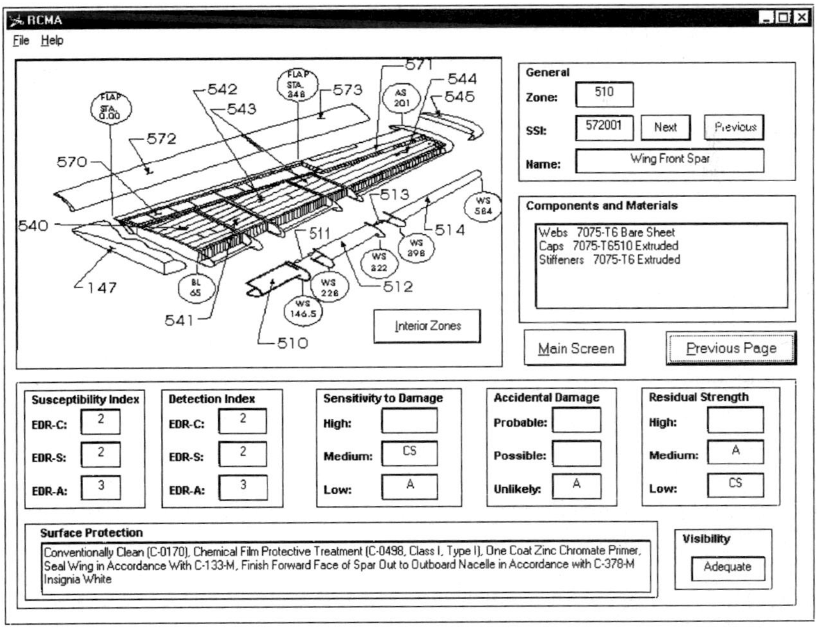

FIGURE 11.13 Display of some critical component information resident in a KBS for ASIP.

For instance, the buildup of scale within storage tanks and piping may result in situations where corrosive agents can accumulate and accelerate a very vicious hidden corrosion process. One particular example of such conditions that became extremely important after the energy crisis of the 1970s is what is called corrosion under insulation (CUI). However, the same problem exist in maybe a less severe fashion for structural details used in the construction of a system including lap joints, gaskets, and enclosed places that cannot be accessed.

Designers should thoroughly analyze the operating conditions the entire system will be subjected to and determine the potential combinations of design details, temperature, humidity, and chemical exposure that may be present at all points. The designer should also consider the conditions experienced by a system throughout its life-cycle including maintenance, storage, and transportation.

A comprehensive analysis of the operating environment can turn out to be an extensive undertaking. Therefore, depending upon the time and resources available, the extent of the analysis could range from general to detailed using a tiered approach. It is most important to first define the general operating conditions (e.g., atmospheric, immersed in liquid, buried, and so forth) which can be further defined with more precision as the information becomes available. A detailed environmental assessment could include specifics on pH, pollutants present, temperature, and other relevant factors. Design details should not be overlooked since they certainly can initiate the microenvironments responsible for triggering specific corrosion modes.

11.4.4 Evaluate Materials Based on Potential Corrosion Failure Modes

Once the candidate materials and the environmental conditions have been determined, the analyst must investigate the potential for one or more of the many forms of corrosion to become active. At this point in the process the designer should have at least one candidate material to consider and a listing of potential corrosion modes in the anticipated environmental conditions. Since the scope of the analysis has been defined, one can assess whether any of the potential forms of corrosion may become active when the candidate material is subjected to the operational environment.

The most cost-economical way to investigate both of these issues is to perform an analysis of the literature followed by carrying a test program if facilities are available and time permits. Following the completion of the literature analysis and/or testing program one should have adequate information to reject any problematic materials, thus yielding a listing of candidate materials that meet performance requirements while simultaneously possessing acceptable corrosion resistance.

11.4.5 Select Corrosion Prevention and Control Methods

Perhaps even more daunting than selecting the actual materials for a particular corrosion conscious design is choosing the corrosion prevention and control technologies that may be needed to protect the system or its components. Some materials require a minimum of corrosion control since they possess surface characteristics that make them resistant to corrosion. Stainless steels are a good example of such corrosion performance. However, most metals (and some other materials) need some form of additional protection from the corrosivity of their operating environment.

Several strategies can be employed to protect a metallic material from corrosion if and when necessary. One such strategy is to construct a barrier to prevent the corrosive environment from contacting the surface. Effective barriers (see Chap. 14) include coatings, surface treatments, and the use of corrosion preventive compounds (CPCs). Another strategy is to modify the environment itself. Now obviously this won't work when a structure is exposed to the elements themselves, but corrosion can also happen inside a structure or system, especially if it is totally enclosed. A good example is a heating plant where steam is circulated from a boiler to heat exchangers. The operational environment within this type of enclosed system contains water, which certainly will be laden with minerals and other contaminants that can initiate corrosion reactions. In such situations a chemical substance known as a corrosion inhibitor can be mixed with the fluid to reduce the corrosivity of the environment (Chap. 8).

Another approach for reducing corrosion is to employ mechanisms that can modify the electrochemical processes that consume materials. Cathodic protection, either through the use of sacrificial anodes or an impressed current system, can convert a material that normally will corrode quite readily into a material that resists corrosion. This approach, which is the topic of Chap. 13, works very well for protecting fixed assets in contact with potentially corrosive environments such as soils, seawater, or any other electrolytically conducting medium.

11.5 Design Considerations

Engineered structures should be designed to provide the desired functional qualities for the required period of service. This implies that a structure should neither be underdesigned (too risky) nor overdesigned (too costly). Thoughtful design requires more than the provision of adequate strength. The part must also last for a given period of time. However, no single text or academic course can give an engineer or technologist the answers to all corrosion problems. A checklist of some potential problems facing them when designing a new installation is provided in Table 11.3.

Plant Construction:
Prevailing wind direction OK?
Method of using cooling water defined (open recirculated, salt water, double exchange, ponds, air cooling)?
Cooling tower downwind?
Closed construction used (pipe or box stanchions, etc.)?
Exterior designs free of pockets?
Tank bottoms protected (cathodically or sealed, depending on type)?
Metal-to-metal contacts protected?
Equipment designed properly for coating?
Will hydrotesting corrode the steel or leave chlorides on stainless steels?
Is a system of material verification needed?
Will all steel be cleaned and primed before erection?
Are temporary preservatives needed for some items?
Should cathodic protection be installed now?
Is electrical equipment properly grounded to prevent stray currents?
Equipment Design:
No 'dead' spaces in lines, tanks, vessels?
All possible crevices sealed or flushed?
Inlets-outlets properly contoured? (If not, is protection provided?)
Offsets on mechanical equipment properly filleted?
Other potential fatigue sites?
Can equipment be cleaned adequately if needed?
Will end grain attack be a problem?
Homogenizing anneal needed?
Stress relief required to forestall SCC?
Are weld deposits compatible with parent metal in chemistry and hardness?
Corrosion allowances added to metal thickness?
Possibility of liquid metal embrittlement (e.g., zinc on stainless)?
Desiccating vents needed?
Temperature:
Evaporation area-vaporization or ebullition cause problem?
Hot spots-from stream injection point?
Hot spots-due to impinging flame?
Cold spots-due to attachment of external metal?
Cold spots-condensation zone a problem? (If so, keep + 10°C above dew point.)

Table 11.3 Some Design Considerations to Control Corrosion

Breakaway point exceeded (e.g., organic acids on stainless steels)?
Freeze-thaw zones properly protected?
Velocity:
Excessive linear flow?
Turbulent flow (from excess velocity; from design, such as raised tube ends)?
Insufficient flow? (Passive metals need oxygen (O_2). Vapor voids created?)
Adequate radii in bends?
Stream mixing tangential or in a tank?
Impingement and Cavitation:
Restrictions out of lines?
Adequate radii in bends?
Varying diameters in a line?
Pump problems analyzed?
Stream inlets baffled?
Bimetal Problems:
Galvanic couples in an aerated electrolyte?
Heavy metal (Fe, Cu) upstream of aluminum or steel?
Differential expansion problems?
Environmental Control:
Neutralization, dehydration, or deaeration of streams considered?
Parameters of operation to reduce corrosion registered with operators?
Two-phase liquid systems adequately evaluated?
Gas–liquid interfaces evaluated?
Solids in any gas or liquid stream?
Adequate alarms for breakthrough of corrosive agent (ppm's can be disastrous)?
Will atomic hydrogen be present? What effect?
Could bacteria create a corrosive situation?
Testing:
Initial testing defined required materials?
Extra nozzles available on equipment for later test rack insertion?
Resistance wire probes of help?
Linear polarization instrument of help?
Can side stream be built for critical test area?
Corrosion test program planned firmly?
Ultrasonic bench-mark readings made on critical equipment?

Table 11.3 *(continued)*

The best way to avoid corrosion is the imaginative use of knowledge. Proper analysis of failures, intelligent use of technical information, and creative use of engineering principles, combined with a deep appreciation of the economic aspects involved will result in improved techniques of materials selection and eliminate many needless and expensive corrosion failures. Much can be done in the detailed design to improve corrosion resistance and obtain better service from economical materials and affordable maintenance procedures. There are three important things to remember when designing a system for demanding service [16]:

1. Design for complete and free drainage
2. Eliminate or seal weld service
3. Make it easy to inspect

It cannot be overemphasized that corrosion control must start at the "drawing board" and that design details are critical for ensuring adequate long-term corrosion protection. The following sections provide some examples of drainage and joining to avoid corrosion problems.

11.5.1 Designing Adequate Drainage

Drains should always be located at the lowest point in a tank and the joint between the tank and the drain pipe should be designed to permit free drainage (Fig. 11.14). Such design avoids the presence of protruding weld beads that could trap residues or moisture inside the tank. The use of sumps in large tanks is a useful way to mitigate

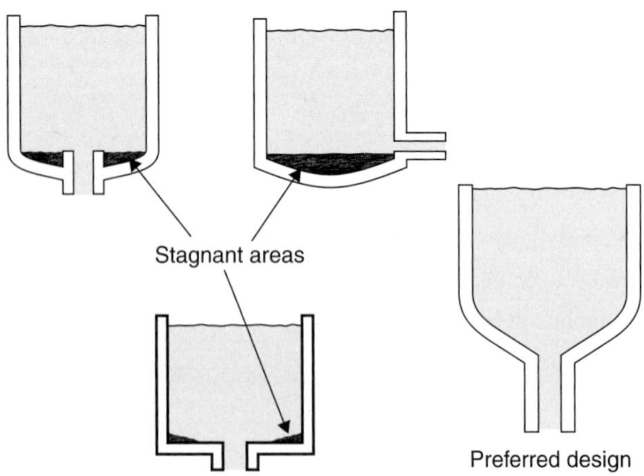

FIGURE 11.14 Design for drainage (adapted from [17]).

FIGURE 11.15 Outdoor oil tank with a poorly coated rain jacket and collecting sump.

drainage problems, however, special care must be made in the design of sumps to be sure that they are also free draining or can be inspected and cleaned easily. Inspection flanges are convenient for such purposes [17]. If rainwater is allowed to collect in a collecting sump as shown in Fig. 11.15, then the sump material itself should be protected accordingly, which is not obviously the case in Fig. 11.15.

Large flat-bottom tanks present a particular drainage problem because of the difficulty of avoiding areas that are difficult to drain. The design should incorporate some positive means of providing drainage, perhaps by sloping the bottom. Rain water or other moisture on the exterior of tanks can be drawn under flat-bottom tanks by capillary action. Use of drip skirts around the edges of such tanks will prevent water from clinging to the tanks and being drawn under the tanks (Fig. 11.16).

It is good practice to support small to moderate size tanks on stanchions rather than on pads. Metals sitting on concrete pose a special problem because the concrete is porous and holds moisture that eventually may corrode the base material in through crevice corrosion and pack rust causing loss of supporting strength as shown in Fig. 11.17 and Fig. 11.18.

The metal should sit off the floor so that spills and cleaning water contact the metal for only a brief time. Concrete should slope away from metal for rapid drainage [Fig. 11.19(*a*)]. Thin-walled tanks or any

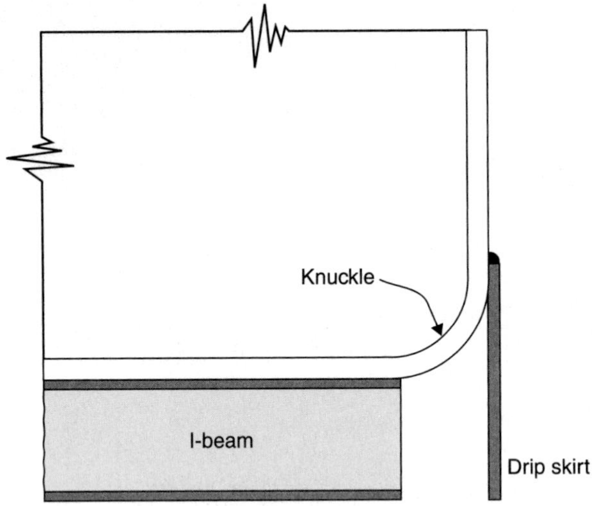

Figure 11.16 A drip skirt to reduce moisture collection under flat-bottom tanks(adapted from [17]).

Figure 11.17 City lamppost fallen due to corrosion of its base. (Courtesy of Bruce Hector, Rust Check Kingston)

Figure 11.18 Steel flag posts with steel base sitting on concrete with close-ups of pack rust area and missing post due to corrosion of the base.

other steel-based fixtures should be supported so that they do not contact concrete at all as in Fig. 11.19(*b*) [18] or in Fig. 11.20. If horizontal tanks must sit on a concrete saddle, as in Fig. 11.21, a support plate should be welded to the vessel to give an extra thickness allowance for the expected corrosion. A metal saddle would be even better.

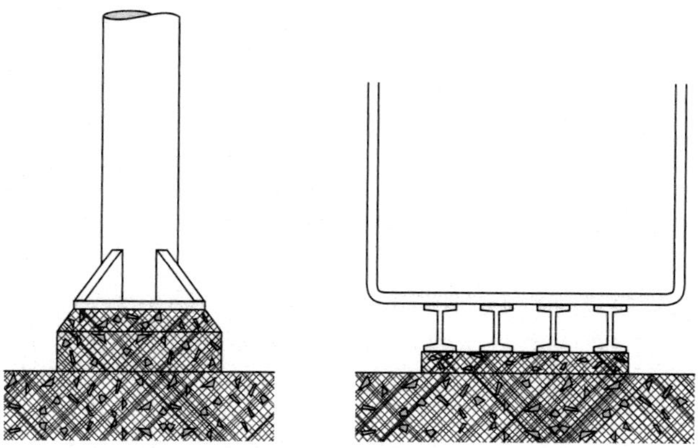

(a) (b)

Figure 11.19 (*a*) Support post with bottom plate fastened above floor level. (*b*) Support tanks above floor level (adapted from [18]).

Figure 11.20 Lamppost galvanized steel base with a gap over the concrete to prevent crevice corrosion and pack rust.

Crowned pads for flat bottomed tanks will encourage free drainage of contents, particularly where a sump is provided. Crowning also encourages free drainage of liquids from under the tank floor under variations of hydrostatic head, which can pull corrosive liquids under the tank. Use of pads slightly smaller in diameter than the tank base will reduce the danger of moisture and chemicals being

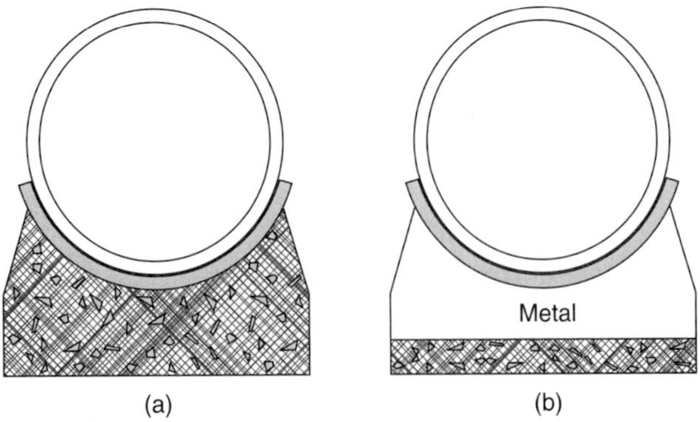

Figure 11.21 (*a*) Support pad protects horizontal tank from concrete saddle, (*b*) better design: metal saddle with support pad (adapted from [18]).

drawn under tanks by surface tension effects. Six common tank bottom arrangements are illustrated here with comments from a corrosion engineering point of view.

1. *Worst [Fig. 11.22(a)]:* The square corner flat bottom arrangement invites early failure from the inside at the corner weld where sediment will collect, increasing the probability of crevice attack. Moisture penetrating the flat bottom to pad support invites rapid crevice corrosion from the underside.
2. *Good corners, poor outside [Fig. 11.22(b)]:* The rounded bottom shown here is much more resistant from the inside, but is actually worse from the outside as condensation is funneled directly into the crevice between the tank bottom and pad support.
3. *Poor inside, poor outside [Fig. 11.22(c)]:* The grout used here to divert such condensation does help initially but soon shrinks back and becomes a maintenance demand itself.
4. *Good inside, good outside [Fig. 11.22(d)]:* The drip skirt shown here is the best arrangement for flat-bottom tanks.
5. *Good inside, good outside, fatigue resistant [Fig. 11.22(e)]:* The concave bottom shown here and the dished-head bottom on supports below are very good and superior to all flat-bottom tanks not only in corrosion resistance but also in fatigue. Fatigue stresses from filling and emptying are seldom considered in design, but can be significant and have led to failures in flat-bottom tanks. The concave- and dished-head arrangements can withstand much greater fatigue loadings than can flat bottoms.
6. *Best inside, best outside, fatigue resistant [Fig. 11.22(f)]:* Dished-head bottom.

11.5.2 Adequate Joining and Attachments

All attachments create potential crevice sites. Figure 11.23(*a*) shows a tray support angle with intermittent welds adequate for strength. There is, however, a severe crevice between the angle and the inside wall of the vessel which will become filled with debris and invite premature failure from crevice corrosion.

Figure 11.23(*b*) shows the same tray support with a continuous seal weld at the top preventing unwanted material from finding its way down the wall and into the crevice. However, the tray support is still open from the bottom but this is a much less severe crevice situation. Figure 11.23(*c*) shows a full seal weld at the top and bottom of the tray support angle. Here the potential crevice is fully stifled.

When the side wall of bottom-corner welds forms a right angle with the bottom, the fillet weld is seldom as smooth as shown in Fig. 11.24(*a*). It is usually rough and frequently varies in width

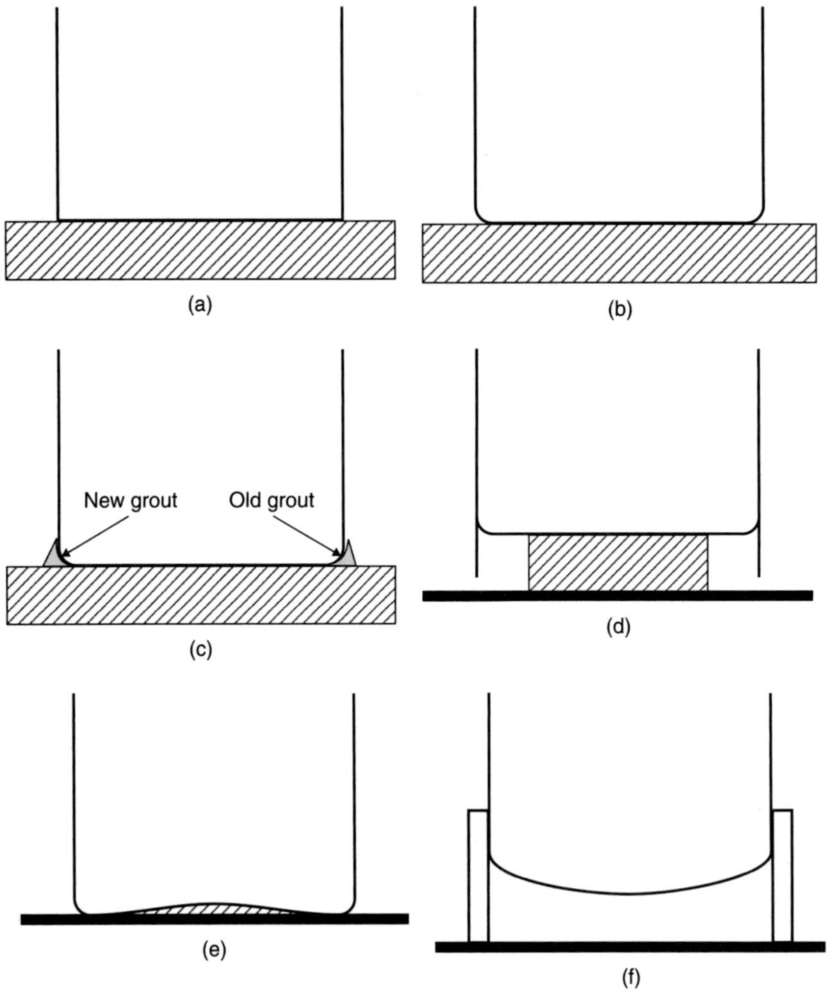

FIGURE 11.22 Tank bottoms: (*a*) worst; (*b*) good corners, poor outside; (*c*) poor inside, poor outside; (*d*) good inside, good outside; (*e*) good inside, good outside, fatigue resistant; (*f*) best inside, best outside, fatigue resistant.

compensating for variations in fit up. Because of the location, it is very difficult to grind and blend the weld into the adjacent sides. Debris tends to collect and is difficult to remove, leading to under sediment type crevice attack. Unless welded from the outside as shown in Fig. 11.24(*b*), the corner is vulnerable to crevice attack. However the inside weld area still is a cause for concern.

Rounding the corner and moving the weld to the side wall overcomes both short comings as shown in Fig. 11.24(*c*). This construction has much improved corrosion resistance and also has better fatigue resistance.

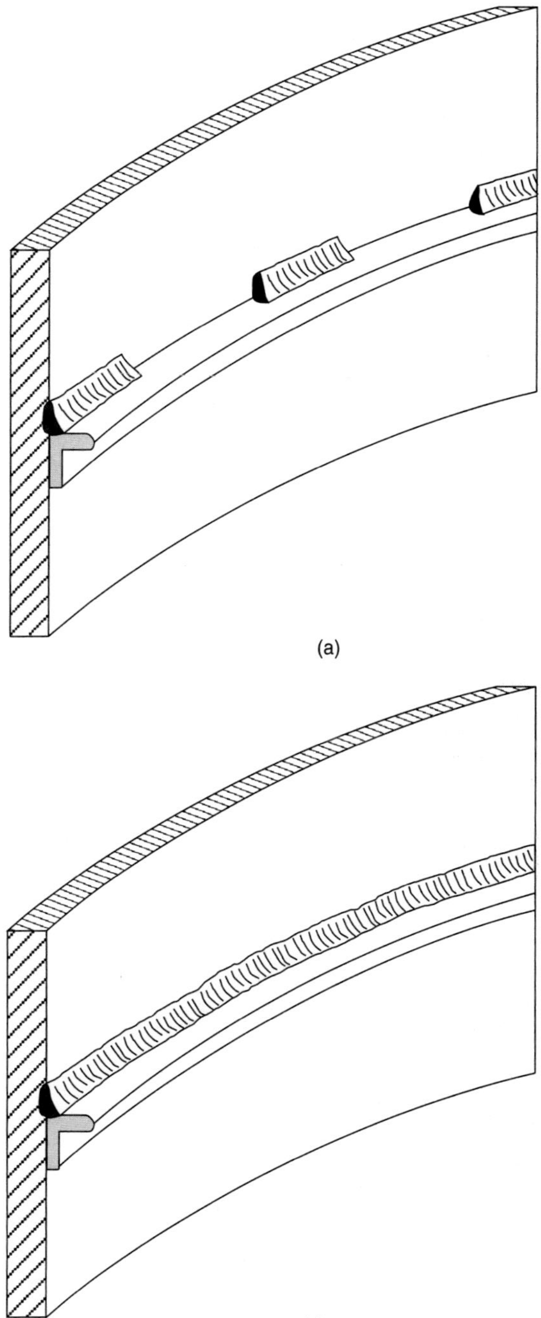

(a)

(b)

Figure 11.23 Structural elements: (*a*) adequate support, severe crevice; (*b*) good crevice resistance; (*c*) best crevice resistance.

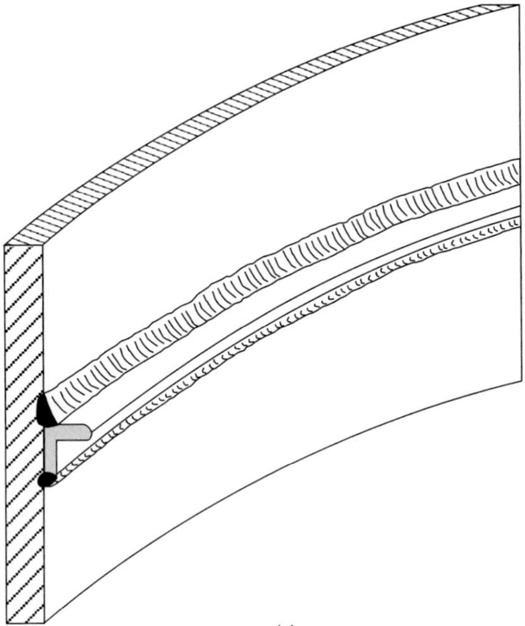

(c)

Figure 11.23 (*Countinued*)

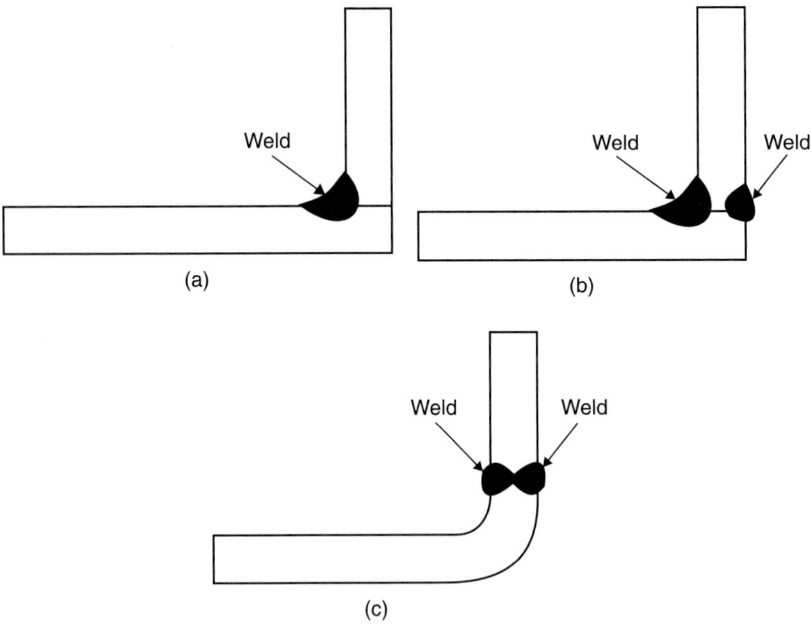

Figure 11.24 Bottom corner welds: (*a*) poor inside, worst outside; (*b*) poor inside, good outside; (*c*) best inside, good outside, fatigue resistant.

11.6 Testing Considerations

Part of the testing philosophy involves whether the test is intended to reproduce a certain environment accurately or whether it is more advisable to use a corrosive environment that represents a worst-case situation. In either case, the corrosion investigator must do everything possible to make the test reproducible by exercising explicit control over environmental factors such as concentration of reactants and contaminants, solution pH, temperature, aeration, velocity, impingement, and bacteriological effects [19].

11.6.1 Test Objectives

Properly conducted corrosion tests in a materials selection context scheme may provide important savings by preventing the use of a metallic material under unsuitable conditions or the use of a more expensive material than is required. Corrosion tests also help in the development of new alloys that perform more inexpensively, more efficiently, longer, or more safely than the alloys currently in use. Corrosion testing programs can be simple and completed in a few minutes or hours, or they can be complex, requiring the combined work of a number of investigators over a period of years. Preliminary time and effort spent considering metallurgical factors, environmental variations, statistical treatment, and the interpretation of accelerated test results is often the most useful effort of a testing program. Typical corrosion test objectives include

- Determining the best material to fill a need (material selection)
- Predicting the probable service life of a product or structure
- Evaluating new commercial alloys and processes
- Assisting with development of materials with improved resistance to corrosion
- Conducting lot-release and acceptance tests to determine whether material meets specifications (quality control)
- Evaluating environmental variations and controls such as corrosion inhibitors
- Determining the most economical means of reducing corrosion
- Studying corrosion mechanisms

Probably the most common corrosion test method is immersion in a liquid. Obvious differences in test procedures are the solutions used, agitation rates, and temperature. The environmental conditions that must be simulated and the degree of acceleration that is required often determine the choice of a laboratory test. In immersion testing, acceleration is achieved principally by

- Lengthening the exposure to the critical conditions that are suspected to cause corrosion damage. For example, if a vessel is to be batch-processed with a chemical for 24 hours, then laboratory corrosion exposure of 240 hours should be considered.
- By intensifying the conditions in order to increase corrosion rates, that is, increasing solution acidity, salt concentration, temperature or pressure, and so forth.
- By provoking localized environmental changes that can result in the formation of acidic/high chloride conditions furthering the initiation and propagation of crevice corrosion of susceptible alloys, for example, ASTM G78 [20] (Fig. 11.25).

Once the environmental conditions have been determined, and the test designed, then it should be repeated a sufficient number of times to determine whether it meets the desired standard for reproducibility.

With such a long list of considerations and options, it is important to simplify the design of test plans by adopting a testing strategy that relates requirements to the main test parameters. For example, the decision tree presented in Fig. 11.26 has been developed to facilitate the selection of tests designed to verify the susceptibility of steels to various forms of SCC. The strategy would be to start with the most severe and least expensive SCC test, that is, the slow strain-rate test, in which a bar made from the relevant material is exposed to the environment of interest and slowly monotonically strained to fracture [21]. When cracks are found, the susceptibility of the material should then be further evaluated by performing a battery of other tests designed to differentiate among the various mechanisms leading to SCC and hydrogen embrittlement.

Test methods for determining corrosion resistance are quite specific and must be based on the conditions prevailing and the materials to be used, including coatings and other protective measures planned for a given application. All these details, including the specification ranges for significant variables, must be determined from individually formulated tests based on the desired service life and other requirements of an application.

11.6.2 Test Standards

There are a multitude of organizations around the world dealing with the production of test methods and standards related to the overall behavior and performance of materials. Organized in 1898, the American Society for Testing and Materials (ASTM) has grown into one of the largest institutions of this kind. ASTM is a not-for-profit

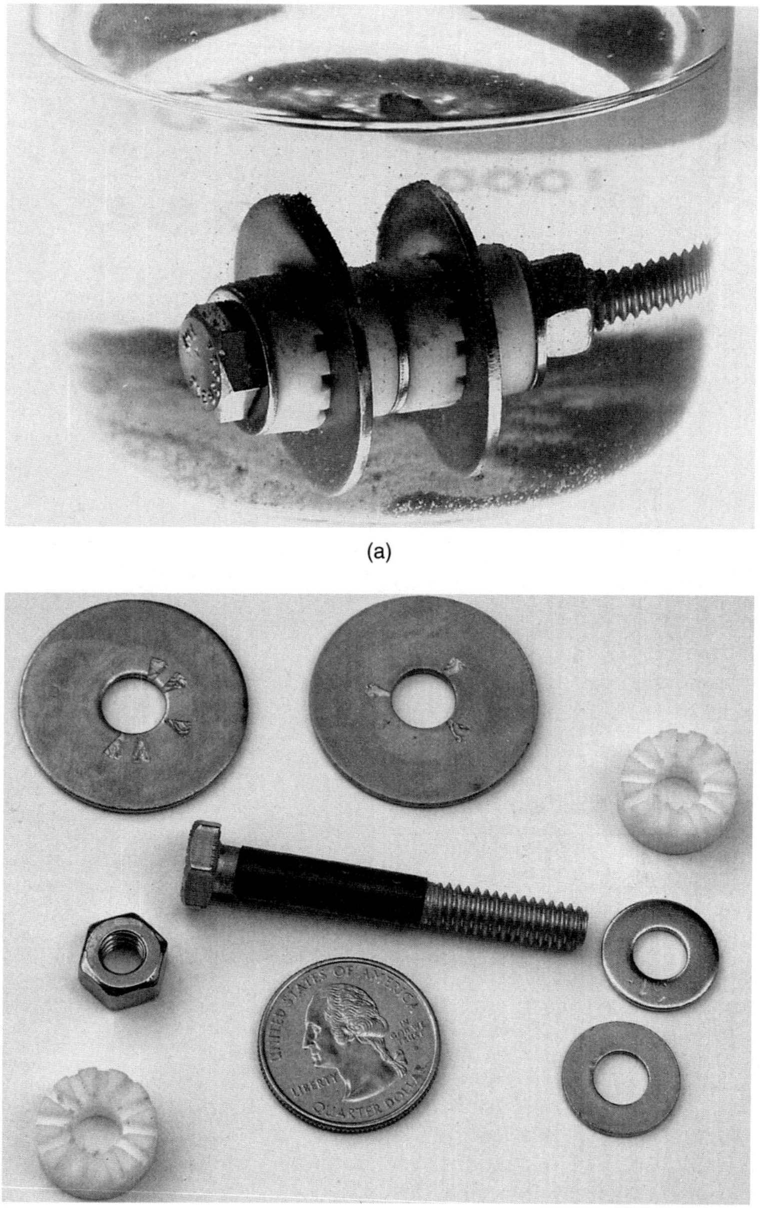

(a)

(b)

FIGURE 11.25 Crevice forming assembly in a beaker immersion test (*a*); components of the assembly and S30400 washer coupons after 30 days in a solution containing 4 percent NaCl and 8 percent $FeCl_3$ (*b*).

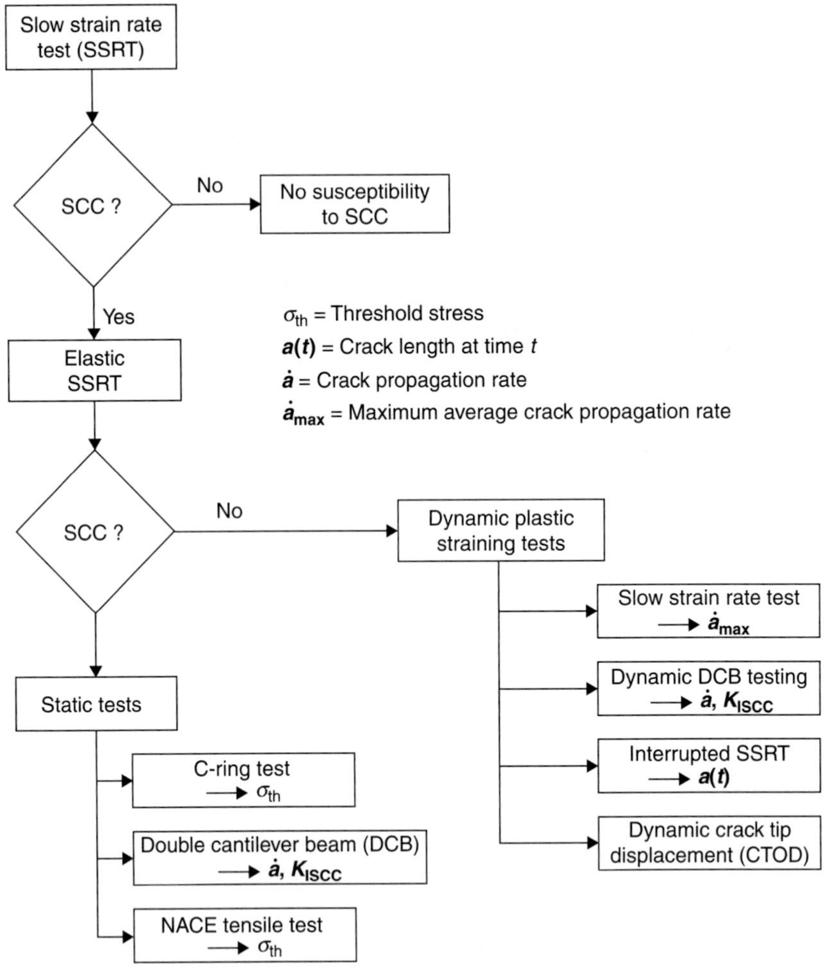

FIGURE 11.26 Decision tree for the selection of corrosion tests to verify the susceptibility of steels to SCC.

organization that provides a forum for producers, users, ultimate consumers, and those having a general interest to meet on common ground and write standards for materials, products, systems, and services. There are over 130 ASTM technical committees, and each has the same basic three-tiered structure (i.e., main committee, subcommittees, and task groups). Committee G01, Corrosion of Metals, is thus subdivided into the following subcommittees:

- G01.02 Terminology
- G01.04 Atmospheric Corrosion
- G01.05 Laboratory Corrosion Tests

- G01.06 Environmentally Assisted Cracking
- G01.09 Corrosion in Natural Waters
- G01.10 Corrosion in Soils
- G01.11 Electrochemical Measurements in Corrosion Testing
- G01.12 In-Plant Corrosion Tests
- G01.14 Corrosion of Metals in Construction Materials
- G01.14.01 Corrosion of Metals in Treated Lumber

Besides its regular standard-development meetings, the G01 committee has sponsored an impressive series of highly focused technical symposia that have led to the publication of many special technical publications (STP). Committee G-1 has also produced some generic reference documents summarizing state-of-the-art information related to corrosion testing. One such publication, *Corrosion Tests and Standards*, is a very valuable source of information for planning corrosion tests [22].

The information contained in that publication summarizes the efforts of over 400 experts in the field of corrosion testing and evaluation. The ASTM corrosion test handbook is highly redundant by design, and its users may find considerable overlap of subject matter (Fig. 11.27). For example, a specific type of corrosion can be thoroughly discussed in the section Testing for Corrosion Types and in the section Testing in Environments. If a specific metal or alloy is susceptible to that type of corrosion, the subject would also be discussed in the appropriate chapter in Materials Testing. And when a specific industry is involved, the appropriate chapter under Testing in Industries would include a discussion on testing for that type of corrosion in that industry. The test handbook is divided into the following five main sections:

1. *Types of Tests:* Each chapter includes basic principles, describes test techniques and important variables, discusses testing considerations such as specimen preparation and evaluation, and includes pertinent standards used.
2. *Testing for Corrosion Types:* Each chapter provides an overview and includes a description of the basic principles and factors controlling the type of corrosion.
3. *Testing in Environments:* The chapters in this section provide a description of each environment, including factors and variables affecting corrosion rates and mechanisms, and the unique characteristics of testing in the specific environment.
4. *By Materials:* This section includes a discussion of the nature of each material, such as the effects of composition, alloying, metallurgical treatments, microstructure, surface effects, and natural protective films on the corrosion behavior.

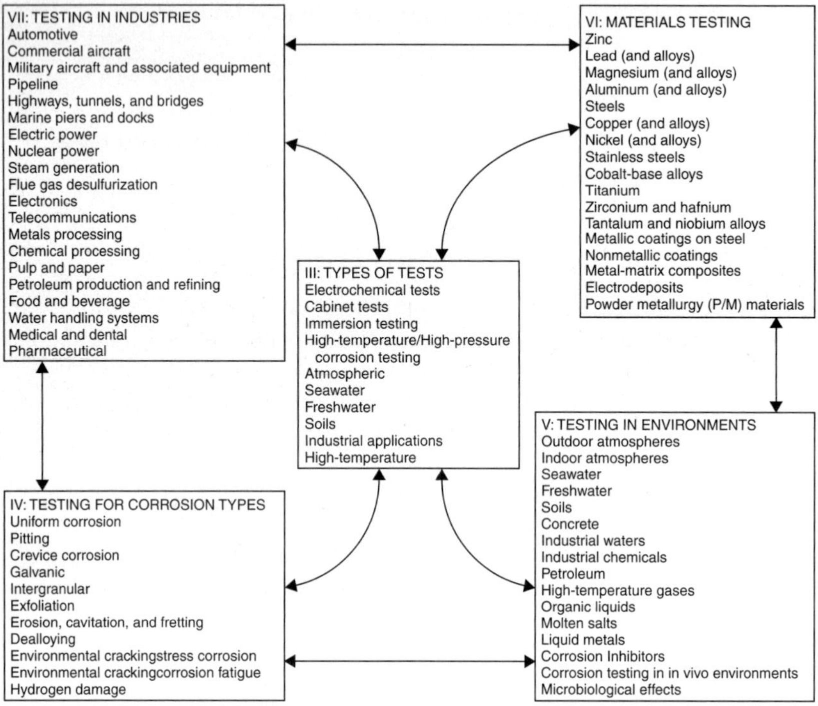

FIGURE 11.27 A graphical representation of the highly redundant index of the *Corrosion Tests and Standards Handbook*.

5. *Testing in Industries:* The chapters in this section provide an overview of the unique situations encountered by various industries, and how corrosion tests are used to combat the corrosion problems faced in these industries.

An industrial example for which standards have been feverishly developed in recent past is the oil and gas production handling sour (H_2S rich) fluids, an extremely corrosive environment for even the most noble alloys. Over the past 20 years a number of laboratory test methods to predict the behavior of steels in sour service have been developed, for example, in Fig. 11.28.

One of the best ways of assessing pipeline steels is to stress a full-ring specimen of the pipeline in a sour environment [23]. This can be achieved by welding end caps onto the sample and pressurizing it with a suitable gas or liquid medium (Figs. 11.29 and 11.30) or by applying external load as illustrated in Fig. 11.31. The advantage of the full-ring test is that it is not necessary to pressurize the pipeline's full-ring specimen in order to achieve the required stress loading. Equivalent stresses can be produced using mechanical means to deform the pipe by ovalization.

Figure 11.28 Bank of 50 NACE TM0177 method, a constant strain test facility with uniaxially loaded tensile specimens. (Courtesy of Bodycote Materials Testing)

Figure 11.29 Full-ring test in preparation for a 45-in pipe. (Courtesy of CAPCIS Ltd.)

Figure 11.30 24-in diameter full-ring test as per OTI 95-635 containing 300 L of NACE TM0177-96 solution saturated with H_2S. (Courtesy of Bodycote Materials Testing)

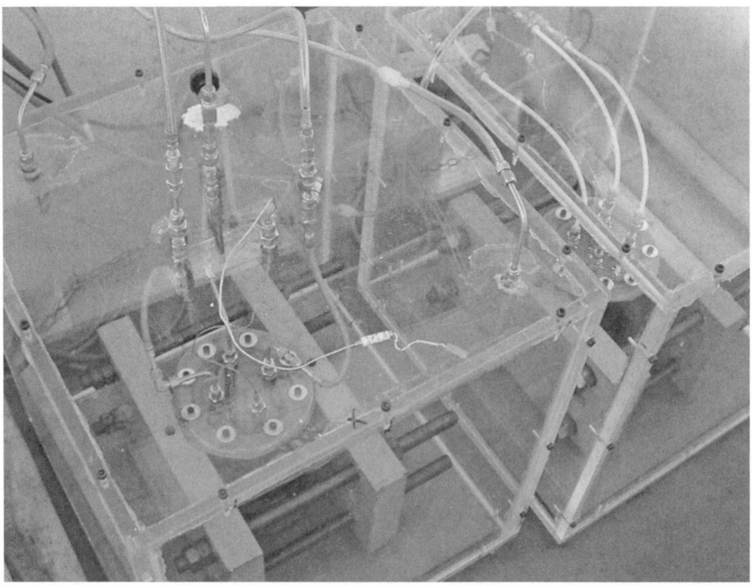

Figure 11.31 Super 13 percent Cr weldable pipe with external load in secondary containment tanks to totally exclude oxygen ingress. (Courtesy of Bodycote Materials Testing)

FIGURE 11.32 Controlled humidity test chamber.

11.6.3 Cabinet Testing

There are numerous options for the production of fog and humidity testing in cabinets in order to asses the corrosion resistance of a broad spectrum of products. The basic humidity test is most commonly used to evaluate the corrosion behavior of materials or the effects of residual contaminants. Cyclic humidity tests are conducted to simulate exposure to high humidity and heat typical of tropical environments.

A cabinet performing such tests should be equipped with a solid-state humidity sensor reading the current humidity condition and a feedback controller. The mechanism that controls the humidity typically moves chamber air via a blower motor and passes it with an atomizer nozzle over a heater coil at the bottom of the chamber (Fig. 11.32).

Hot, humid air is created by bubbling compressed air through a bubble (humidifying) tower containing hot, deionized water. Salt solution is typically moved from a reservoir through a filter to the nozzle by a gravity-feed system (Fig. 11.33). When the hot, humid air and the salt solution mix at the nozzle, it is atomized into a corrosive fog. This creates a 100 percent relative humidity condition in the exposure zone. For a low humidity state in the exposure zone of the chamber, air is forced into the exposure zone via a blower motor that directs air over the energized chamber heaters (Fig. 11.34).

The oldest and most widely used cabinet test is ASTM B 117 [*Test Method of Salt Spray (Fog) Testing*], a test that introduces a spray in a closed chamber where the test specimens are exposed at specific locations and angles [24]. The concentration of the sodium chloride (NaCl) solution has ranged from 3.5 to 20 percent. There is a wide

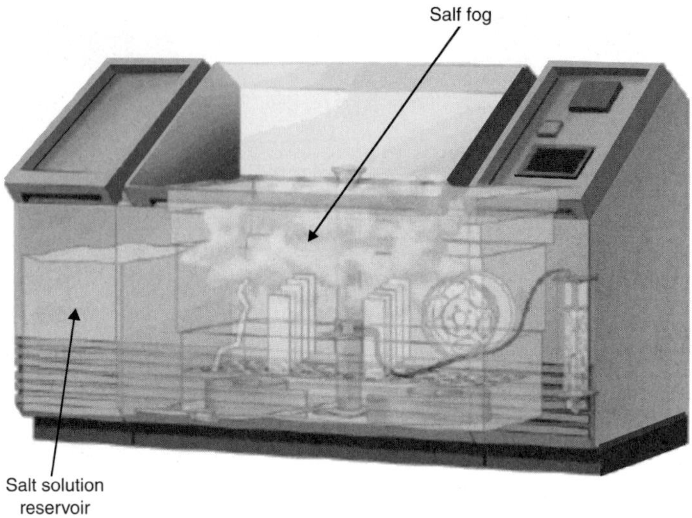

FIGURE 11.33 Controlled salt fog test chamber during a humid cycle.

range of chamber designs and sizes including walk-in or even garage-size rooms (Fig. 11.35) that are capable of performing this test.

In spite of serious shortcomings in providing guidance for service-life prediction, the continuous salt-spray test, which was originally designed to test coatings on metals, has been widely used to evaluate the resistance of metals to corrosion in marine service or on exposed shore locations [25;26]. Extensive experience has shown that, although salt-spray tests yield results somewhat similar to those in marine

FIGURE 11.34 Controlled salt fog test chamber during a dry cycle.

(a)

(b)

FIGURE 11.35 Salt corrosion/temperature/humidity chamber is designed to provide accelerated corrosion life testing of body, frame, and undercarriage components of passenger cars and trucks (*a*) as well as full vehicles (*b*). (Courtesy of Despatch Industries)

exposure, they do not reproduce all the factors causing corrosion in marine service. The reviews made by F.L. LaQue on this subject a few decades ago indicated that the salt-spray test cannot realistically be used, for example, for parts with complicated shapes because salt-spray particles fall in vertical patterns, creating a strong orientation dependency [27;28]. Another major inadequacy of the test is the variable sensitivity of different metallic materials to the ions present in various service environments.

Of considerable interest in recent years is the laboratory simulation and acceleration of atmospheric pollution effects to evaluate degradation of materials. These tests involve injection of a variety of gases, vapors, and particulate matter into a controlled pressure chamber with controlled condensation on metal specimens. Sometimes, ultraviolet light is used to produce photochemical changes in the gases or vapors to replicate smog conditions.

The ASTM B775 (*Test Method for Porosity in Gold Coatings on Metal Substrates by Nitric Acid Vapor*) and B 799 (*Test Method for Porosity in Gold or Palladium Coatings by Sulfurous Acid/Sulfur-Dioxide Vapor*), for example, employ very high concentrations of corrosive gases to amplify the presence of pores in gold or palladium coatings. A very sophisticated variation of these tests is the mixed-gas test (ASTM B 827), that consists in introducing parts per billion levels of pollutants such as chlorine, hydrogen sulfide, and nitrogen dioxide in a chamber at controlled temperature and humidity. This test is particularly adapted to the needs of the electronics industry.

References

1. Ashby MF. *Materials Selection in Mechanical Design*. 3rd edn. Oxford, U.K.: Elsevier, 2005.
2. Rose D. Corrosion conscious design. *AMPTIAC Quarterly* 9(3). 2005.
3. *External Corrosion—Introduction to Chemistry and Control*, 2nd Report M27. Denver, Co., American Water Works Association, 2004.
4. Komorowski JP, Krishnakumar S, Gould RW, Bellinger NC, Karpala F, Hageniers OL. Double pass retroreflection for corrosion detection in aircraft structures. *Materials Evaluation* 1996; 54: 80–6.
5. Henthorne M. Materials selection for corrosion control. *Chemical Engineering* 1971; 1139–146.
6. *Corrosion Data Survey Metals*. 6th edn. Houston, Tex.: National Association of Corrosion Engineers, 1985.
7. *Corrosion Data Survey Non-Metals*. 5th edn. Houston, Tex.: National Association of Corrosion Engineers, 1975.
8. Standard guide for applying statistics to analysis of corrosion data. In: *Annual Book of ASTM Standards*. Philadelphia, Pa.: American Society for Testing of Materials, 1999.
9. Tomiura A. Lessons for a case study of property databases in materials development. In: Nishijima S, Iwata S, Newton CH, eds. *Computerization/Networking of Materials Databases*, STP 1311. Philadelphia, Pa.: American Society for Testing and Materials, 1996; 3–20.
10. *Economics of Corrosion*. NACE 3C194. Houston, Tex., NACE International, 1994.

11. Verink ED. Corrosion economic calculations. In: *Metals Handbook: Corrosion. Metals Park*: ASM International, 1987; 369–74.
12. Roberge PR. *Handbook of Corrosion Engineering*. New York, N.Y.: McGraw-Hill, 2000.
13. Coullahan R, Siegfried C. Life cycle asset management. *Facilities Engineering Journal.*
14. Horvath RJ. The Role of the corrosion engineer in the development and application of risk-based inspection for plant equipment. *Materials Performance* 1998; 37: 70–5.
15. Townley NJ, Roberge PR, Little MA. Corrosion maintenance through knowledge re-use: the CP140 Aurora. *Canadian Aeronautics and Space Journal* 1997; 43: 189–94.
16. Avery RE, Tuthill AH. *Guidelines for the Welded Fabrication of Nickel-containing Stainless Steels for Corrosion Resistant Services. NiDI Report 11007*. Toronto, Canada: Nickel Development Institute (NiDI), 1992.
17. Verink ED. Designing to prevent corrosion. In: Revie RW, ed. *Uhlig's Corrosion Handbook.* New York, N.Y.: Wiley-Interscience, 2000; 97–109.
18. Bradford SA. *Corrosion Control*. Edmonton, Canada: CASTI Publishing, 2001.
19. Roberge PR. Designing, planning, and preparing corrosion tests. In: Cramer DS, Covino BS, eds. Vol. 13A: *Corrosion: Fundamentals, Testing, and Protection*. Metals Park, Ohio: ASM International, 2003; 420–4.
20. *Standard Guide for Crevice Corrosion Testing of Iron-Base and Nickel-Base Stainless Alloys in Seawater and Other Chloride-Containing Aqueous Environments*. West Conshohocken, Pa.: American Society for Testing of Materials, 2001; 03(02):G78–01.
21. Erlings JG, de Groot HW, Nauta J. The effect of slow plastic and elastic straining on sulphide stress cracking and hydrogen embrittlement of 3.5% Ni steel and API 5L X60 pipeline steel. *Corrosion Science* 1987; 27: 1153–67.
22. Baboian R. *Corrosion Tests and Standards: Application and Interpretation. 2nd edn*. Philadelphia, Pa.: American Society for Testing of Materials, 2005.
23. *A Test Method to Determine the Susceptibility of Cracking of Linepipe Steels in Sour Service*. OTI 95 635. Sudbury, U.K.: Health and Safety Executive (HSE), 1996.
24. *Standard Practice for Operating Salt Spray (Fog) Apparatus. Annual Book of ASTM Standards*, Philadelphia, Pa.: American Society for Testing of Materials, 1997; 03(02): B 117–97.
25. Capp JA. A rational test for metallic protective coatings. *Proceedings American Society for Testing of Materials* 1914; 14: 474–81.
26. Finn AN. Method of making the salt-spray corrosion test. *Proceedings American Society for Testing of Materials* 1918; 18: 237–8.
27. LaQue FL. *Marine Corrosion: Causes and Prevention*. New York, N.Y.: John Wiley & Sons, 1975.
28. LaQue FL. A critical look at salt spray tests. *Materials & Methods* 1952; 35: 77–81.

CHAPTER 12

Corrosion as a Risk

12.1 Risk Assessment

Industrial risk assessment is a careful examination of potential hazards that may affect the operation of a system, be it large and complex or relatively simple. These risks may be associated with the risks to the safety and integrity of physical assets, risks to the environment, financial risks from various decisions, and also risks from corrosion or poor corrosion mitigation procedures. Corrosion-related failures may be a major source of risk in many industry sectors. In the offshore oil and gas installations, for example, corrosion can be a life-limiting cause of deterioration by various corrosion damage attacks of plant items.

In the offshore processing equipment, vessels and pipework are fabricated from carbon-manganese steel. This is an economic choice, based on life-cycle costing at the design stage of a project. The use of carbon steels means potential hazards are present due to internal corrosion damage from aqueous produced fluids that contain acidic gases, carbon dioxide, and hydrogen sulfide [1]. In this industry, corrosion related failures may result in hydrocarbon releases and significant loss of production, as well as increased costs for maintenance, repair, or replacement.

A considerable catalyst to the advancement of corrosion inspection and monitoring technology has therefore been the exploitation of oil and gas resources in extreme environmental conditions. Many techniques that have been accepted in the oil and gas industries for years are only now beginning to be applied in other industries such as transportation, mining, and construction. Work in these conditions has necessitated enhanced instrument reliability and the automation of many tasks, including inspection.

In addition to the usual uncertainty of the onset or progression of corrosion of equipment, the oil industry has to face ever changing corrosivity of processing streams. The corrosivity at a well head, for example, may oscillate many times during the life of an exploitation field, between being benign to becoming extremely corrosive [2].

Such changes require more corrosion vigilance in terms of inspection and monitoring.

12.2 Risk Analysis

Risk analysis refers to techniques for identifying, characterizing, and evaluating hazards. The identification of risk, defined in Eq. (12.1), and risk analysis found their way into many applications where they can add value in prioritization and management processes. The application of general risk analysis principles to help prioritize and manage the inspection program for plant equipment, now commonly referred to as risk-based inspection (RBI), is one of the newest applications of risk principles [3]. Some examples of risk criteria and associated units are shown in Table 12.1.

$$\text{Risk} = \text{Probability of Failure (POF)} \times \text{Consequence of Failure (COF)} \quad (12.1)$$

In this equation the POF is either based on failure frequency or on remaining lifetime, while COF is usually related to safety, health, environment, and economics issues.

Data required for developing a risk assessment program are often acquired during the analysis of failed components and systems. However, conducting a failure analysis is not an easy or straightforward task. Early recognition of corrosion as a factor in a failure is critical, since much important corrosion information can be lost if a failure scene is altered or changed before appropriate observations and tests can be made.

To avoid these pitfalls, certain systematic procedures have been proposed to guide an investigator through the failure analysis process. Materials Technology Institute (MTI), for example, has produced an *Atlas of Corrosion and Related Failures* that maps out the process of a failure investigation from the request for the analysis to the submission of a report [4]. The section that relates the origin(s) of failure to plant or component geometry in this atlas is illustrated in Figs. 12.1 and 12.2.

Risk criteria	Expressed in
Financial risk (business impact)	Outage cost / day
Investment risk (asset damage)	Equipment cost / m^2
Safety	Injury cost / year
Environment	Cost / year
Potential of loss of life (PLL)	Events / year
Probability/likelihood of failure	Events / year

TABLE 12.1 Examples of Risk Criteria and Their Units

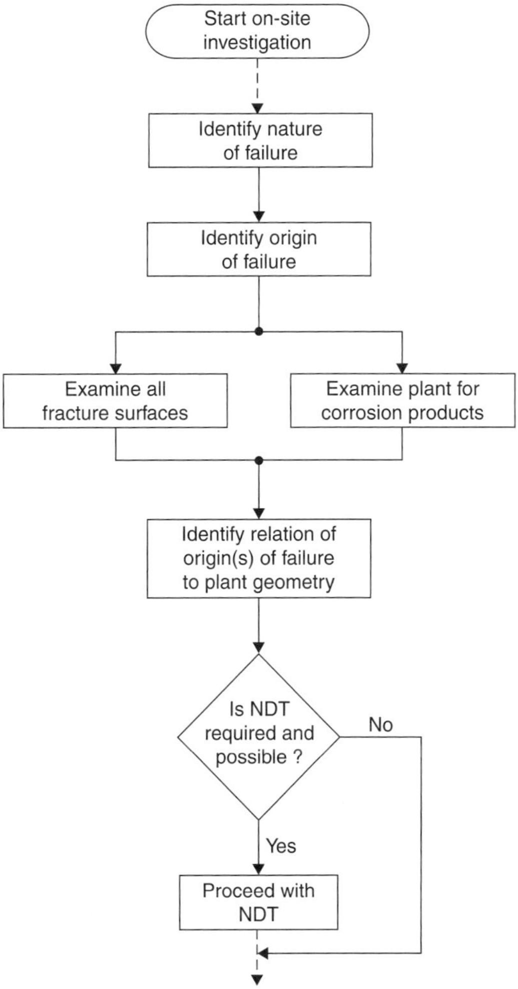

FIGURE 12.1 Decision tree to guide on-site investigations dealing with corrosion damage.

The depth of the failure analysis into the roots of a failure is the key to accurately unearthing all the failure sources. Looking at machinery failures one finds that there are [5]

Physical roots: The physical reasons why the parts failed

Human roots: The human errors of omission or commission that resulted in the physical roots

Latent roots: The deficiencies in the management systems or the management approaches that allow human errors to continue unchecked

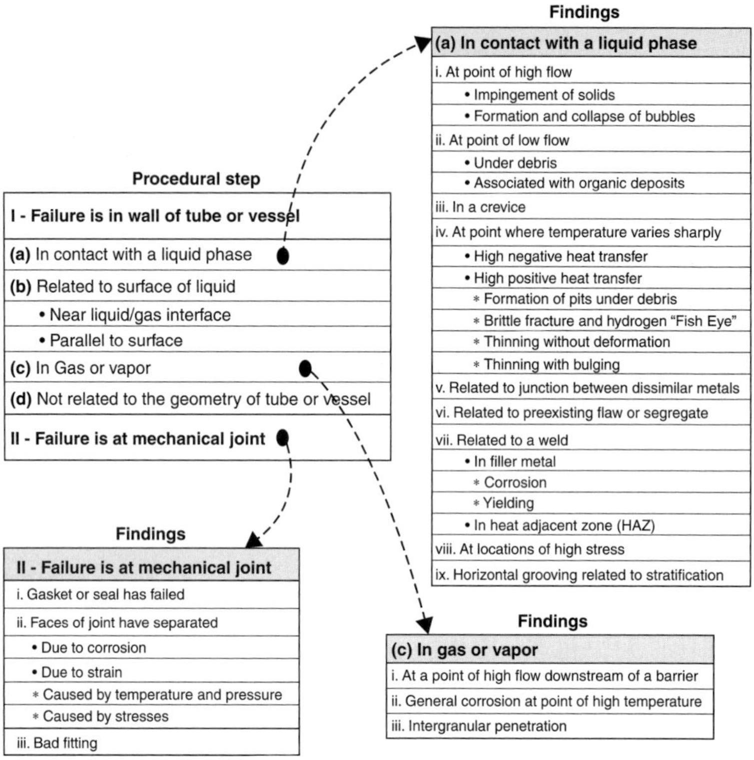

FIGURE 12.2 Recommendations for relating the origin(s) of failure to plant geometry.

The more detailed an analysis is, the better the events and mechanisms that contribute as the roots of the problem can be elucidated. The analyses can be further divided into three categories in order of complexity and depth of investigation:

Component failure analysis (CFA): It looks at the piece of the machine that failed, such as a bearing or a gear to determine the specific cause of the failure (fatigue, overload, or corrosion) and that there were these x, y, and z influences.

Root cause investigation (RCI): It is conducted in much greater depth than the CFA and goes substantially beyond the physical root of a problem to identify the human errors involved. It stops at the major human causes and does not involve management system deficiencies. RCIs are generally confined to a single operating unit.

Root cause analysis (RCA): It includes everything the RCI covers plus the minor human error causes and, more importantly, the management system problems that allow the human errors and other system weaknesses to exist. An RCA can sometimes extend to sites other than the one involved in the original problem.

Although the cost of an analysis increases as it becomes more complex, its benefits generally increase as well. Using a CFA to find the causes of a component failure answers why that specific part or machine failed and the results of the analysis can be used to prevent the occurrence of similar failures in the future. The cost of progressing to a RCI is 5 to 10 times that of a CFA, but the RCI adds a detailed understanding of the human errors contributing to the breakdown and can be used to eliminate groups of similar problems in the future. However, conducting a RCA and correcting the major roots has the potential to eliminate huge classes of problems.

12.3 Risk and Corrosion Control

In practice virtually all corrosion control is risk based. Typically in the first instance an attempt is made to identify the main mechanisms of corrosion attack, and thereafter a reasoned intervention program is considered. This may include a material change, the selection of a corrosion inhibitor, or a more drastic design change to the component or indeed the physical conditions expected. The critical step is invariably the corrosion analysis, which often progresses to the modeling approach. However, modeling unfortunately is often at odds with actual field experiences, mostly because corrosion analyses and solutions are highly judgmental and rely on imperfect sources of data [6].

In overall tight budgets, the selection and prioritization of the components, parts, or systems that should be inspected is of paramount importance. A good knowledge of intricate designs is therefore an important asset since corrosion factors are often related to the geometry of systems and components. This selection should also be based on a thorough knowledge of process conditions, materials of construction, geometry of the system, external factors and historical records. Historical data gathered during previous inspection and repairs cycles can be quite useful to determine the particular locations where future maintenance actions should be focused. Figure 12.3 illustrates some of the "hot spots" found for the P-3, a maritime patrol and antisubmarine warfare aircraft with 50 years of service.

Similarly there is a lot of knowledge in process operations. In the process industry, for example, units which start up and shut down frequently are more susceptible to stress corrosion cracking (SCC) than identical units in the same service which operate for long periods at a steady state. This is because start-up and shutdown produce thermal stresses and internal pressure surges which add a low-rate cyclic component to the static stress of normal operation [7].

In a single-train unit, obviously the greatest concern will be for capital intensive items such as major stills and heat exchangers and for items whose failure might affect production. Components within a given train which run hotter or that could experience more concentrated solutions than associated equipment, are worthy of extra attention. Reboilers, for example, frequently crack first when a

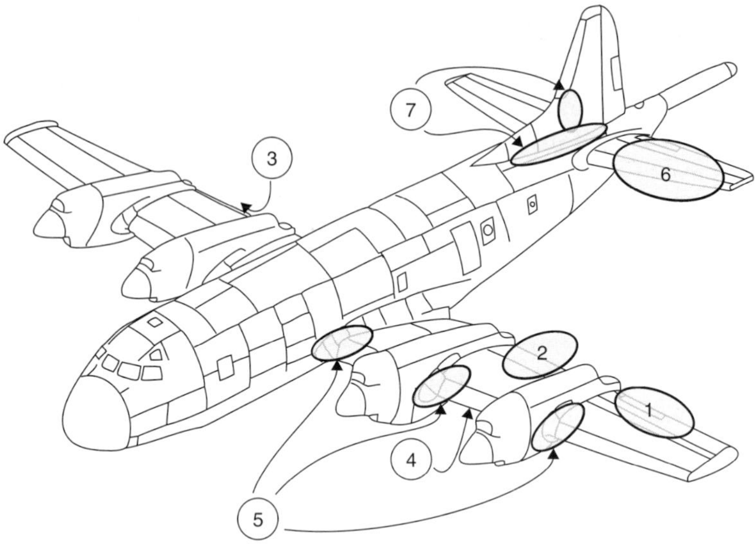

FIGURE 12.3 Corrosion areas found on P-3 maritime patrol aircraft over decades of service: 1) aileron bay bonded panels, 2) flap bay bonded panels, 3) main landing gear jack pad, 4) inboard nacelle structure, 5) fillet fairing, 6) horizontal stabilizers, and 7) vertical stabilizers.

corrosion problem surfaces in a unit, because reboilers normally run hotter than their associated stills, and their tubes are thinner and more highly stressed than the vessel walls.

There may be a multiplicity of welds in a piping system that may be susceptible to aggravated attack unless full penetration welds are achieved. In addition, cold-work due to bending operations, and especially that associated with flared pipe, provide areas highly susceptible to environmental cracking (Fig. 12.4).

All weldments in a reactor contain high residual stress and, unless they are suitably stress-relieved, they are focal points for environmental cracking. However, the circumferential welds by which nozzles are attached to the vessels are particularly vulnerable. The radius of a dished head also contains residual stress from the forming operation, and the top head is often the first point of failure when volatile species (e.g., traces of hydrogen or ferric chloride) are involved (Fig. 12.5) [7]. External jackets can cause problems, unrelated to the process, that arise from the use of cooling water or steam in the jacket which causes external SCC of the vessel proper.

In the particular cases of columns, weldments, especially circumferential nozzle welds, are primary areas for corrosion, as are the radii of dished heads. Nozzles themselves, if in the horizontal plane, may accumulate and hold chlorides or other corrosive agents, making them particularly susceptible to SCC (Fig. 12.6) [7].

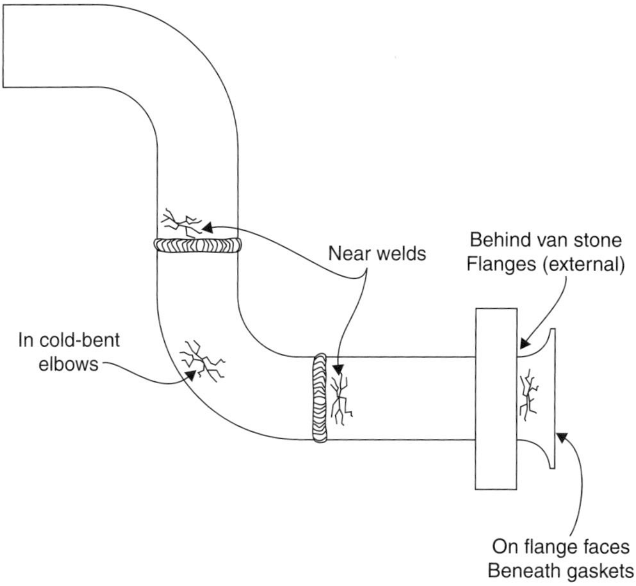

FIGURE 12.4 Typical cracking locations in piping [7].

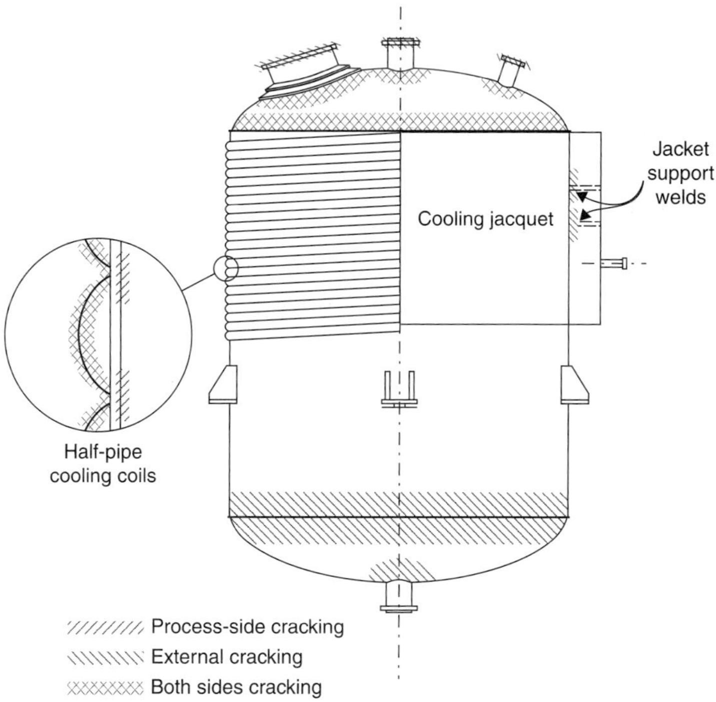

FIGURE 12.5 Typical cracking locations in chemical-process reactors [7].

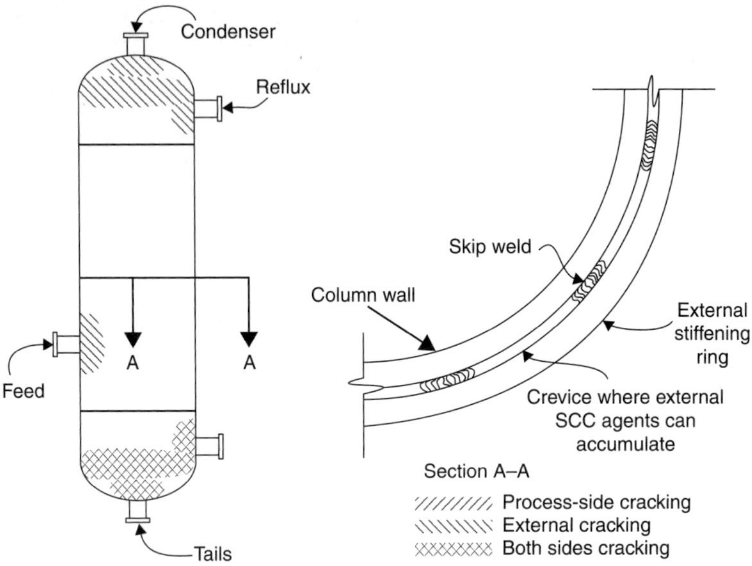

Figure 12.6 Typical cracking locations in distillation columns [7].

Column packing is also especially susceptible to SCC. Expanded metal packing or mesh contain a great deal of cold-work and very high stress levels, rendering it particularly susceptible to SCC. Although both types of packing can be stress-relieved, alone or with the column itself, a preferred method is to manufacture them from crack resistant alloys.

Other areas of corrosion in columns are trays, especially valve trays and bubble caps which may have high residual stress are also primary areas for environmental cracking. Tray posts, clips, fasteners, and other hardware may be highly stressed and susceptible to SCC. Sieve trays are usually manufactured by cold-punching, and bubble caps and valves are cold-formed. The latter particularly, unless stress-relieved or of resistant-alloy construction, may lose their restraining legs and fall off the trays, leaving only an extremely large "sieve" tray of inefficient characteristics.

12.4 Key Performance Indicators

The importance of meaningful management information is often highlighted when serious problems arise. However, it is operationally a much better practice to prevent these detrimental processes. Key performance indicators (KPIs) may be used in this context to optimize the need and sequence of inspection and other maintenance activities.

Management information is typically required on predicted costs of problems, the risks involved, the remaining life of the affected equipment and what can be done to improve or eradicate these problems.

The KPIs described in the following sections were developed specifically to estimate the effect of corrosion on the technical and financial performance of assets involved in oil and gas production facilities and to address the performance of particular critical corrosion-related systems [8]. The same approach can be generalized and adapted to many other high-risk systems.

12.4.1 Cost of Corrosion Key Performance Indicator

The cost of corrosion KPI allows converting the amount of corrosion damage sustained during a given period into a monetary figure in order to provide a clearer focus on corrosion management performance. Factors considered within this KPI are existing damage sustained prior to the period in question, the cost of repair or replacement, and the remaining service life of plant. Performance can then be computed in terms of the cost of damage in the last period examined, the annual damage cost, and/or life-cycle costs. The cost of corrosion damage (C_{corr}) sustained in a given period can be derived with the following Eq. (12.2):

$$C_{\text{corr}} = \left(\frac{N_C \cdot R_{\text{cost}}}{\text{FL}}\right)\left(\frac{D_p}{365}\right) \tag{12.2}$$

where C_{corr} is cost of corrosion damage in a specific time period
N_C is estimated number of replacement cycles to end of service life
R_{cost} is replacement cost (including lost product cost)
FL is required remaining field life (years)
D_p is days in monitoring period (days)

If, however, the calculated remaining life of a component (RL_C) as defined in Eq. (12.4) is greater than the field life (FL), C_{corr} can be assumed to be zero. This is based on the assumption that the KPI is a performance indicator reflecting the effect on operating costs (Opex) and does not consider depreciation against the initial capital cost (Capex). The number of replacement cycles (N_C) can be estimated with Eq. (12.3):

$$N_C = \left[1 + \left(\frac{\text{FL} - \text{RL}_C}{\text{RL}_R}\right)\right] \tag{12.3}$$

where RL_C is remaining life of current component (years)
RL_R is remaining life of replacement components (years)
RL_C and RL_R are respectively derived from Eqs. (12.4) and (12.5):

$$RL_C = \left(\frac{\text{CA} - \text{DT}}{\text{CR}}\right) \tag{12.4}$$

$$RL_R = \left(\frac{\text{CA}}{\text{CR}}\right) \tag{12.5}$$

where CA is corrosion allowance* (design or fitness for purpose) (mm)
DT is damage to date (mm)
CR is measured corrosion rate (mm/y)

The above formulae have been developed from an actual pipework monitoring and replacement program as a result of internal corrosion effects. Both replacement cost (R_{cost}), component remaining life (RL_C), and field life (FL) are key factors in this method and reflect the need to understand accurate costing of installation and replacement activities. Required field life is not necessarily the difference between installation and design lives but more appropriately the time remaining until end of field or production life.

12.4.2 Corrosion Inhibition Level Key Performance Indicator

This KPI is a measure of the availability of corrosion inhibitors to provide protection against corrosive processes. The inhibitor efficiency† itself should have been determined from a combination of previous laboratory and field testing to determine the optimum concentration

---

* Corrosion allowance depends on the type of defect anticipated, which needs to be identified by inspection. Once the defect geometry is known and the process parameters identified, the maximum allowable defect size may be calculated using fitness for purpose criteria to ensure that failure does not occur.

---

† The efficiency of an inhibitor is expressed as a measure of the improvement in lowering the corrosion rate of a system:

$$\text{Inhibitor efficiency (\%)} = \left(\frac{\text{CR}_{\text{uninhibited}} - \text{CR}_{\text{inhibited}}}{\text{CR}_{\text{uninhibited}}}\right) \times 100$$

where $\text{CR}_{\text{uninhibited}}$ = corrosion rate of the uninhibited system and
$\text{CR}_{\text{inhibited}}$ = corrosion rate of the inhibited system

of the chemical inhibitor in the produced fluids. There may be applications where it is justified to apply a degree of over injection to provide protection to downstream facilities where it is not practical to inject.

The KPI itself is derived from a measure of the produced fluids including water and hydrocarbon phases and the inhibitor injected in the produced fluid stream to provide a correlation between how much inhibitor should be in the produced fluid stream versus actual injected inhibitor concentrations. The KPI percent inhibitor availability (Inhibitor_{AV}) function is described by Eq. (12.6):

$$\text{Inhibitor}_{AV} = \left(\frac{C_{\text{actual}}}{C_{\text{required}}} \right) \times 100 \tag{12.6}$$

where C_{actual} is actual concentration of corrosion inhibitor (ppm) and
C_{required} is required concentration of corrosion inhibitor (ppm)

This KPI, when directly correlated with the corrosion cost KPI (C_{corr}), provides a clear indication of the corrosion performance of the asset and identifies clearly where effective action can be taken to improve performance if damage costs are seen to increase. A correlation between the cost of damage KPI and inhibition level KPI may indicate how the cost of damage and inhibitor dosage level trend in actual performance. The example shown in Fig. 12.7 illustrates how the availability of corrosion inhibitors relates to the cost of operation, that

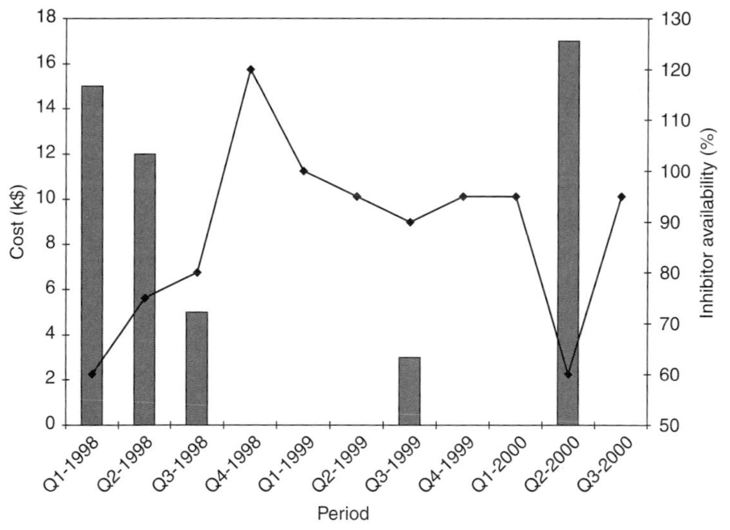

FIGURE 12.7 Example of corrosion cost and inhibitor availability KPIs for estimating corrosion management performance. [8].

is, the cost is down when the availability is high while it may become quite high when the availability decreases below 90 percent.

12.4.3 Completed Maintenance Key Performance Indicator

This KPI, which provides a measure of the reliability of the corrosion monitoring equipment, is determined from the asset's maintenance performance in repairing equipment faults reported during routine corrosion inspection visits. This measure reflects the recognized importance that equipment reliability is critical to the performance of corrosion inhibition systems and hence is also a key cost factor. The indicator is derived from a ratio of the number of maintenance actions raised versus the number of actions completed in a given monitoring period [Eq. (12.7)].

$$\% \text{ Completed maintenance} = \frac{\text{maintenance actions completed}}{\text{maintenance actions raised}} \times 100 \tag{12.7}$$

12.4.4 Selecting Key Performance Indicators

Measures of performance have been used by management for centuries to review current operational capabilities. Such measures have been used to assess both departmental and corporate performance, as well as trend performance achieved against plan. In many industrial facilities, these measurements are related to safety (e.g., number of incidents/accidents), environment (e.g., number of releases), costs, and productivity. These measures are needed in order to determine not only if resources and costs have been managed for the production achieved, but also whether the assets or plant remain in good health [9].

In order to define a complete set of performance measures, companies must ensure that simple, workable measures are in place. The real challenge is not only to select those indicators that satisfy budgetary goals, but also to build the activities needed to meet the levels of asset performance required. Selecting the right measures is crucial to ensure the effectiveness of an operation. Even more importantly, the metrics must be built into a performance measurement system that allows individuals and groups to understand how their behaviors and activities are fulfilling the overall corporate goals.

Asset Performance Metrics

An asset performance management initiative is comprised of business processes, workflows, and data capture that enable rigorous analysis to help define strategies based on best practices, plant history, and fact-based decision support. As with many other management issues the key to building a set of performance metrics is to do it in stages. Clear corporate goals are needed at this point, otherwise vague objectives will create impractical perspectives and metrics. By contrast, well-organized metrics and scorecards provide operational measures that have clear cause-and-effect relationships with the desired outcomes [9].

Each of these outcomes will build toward the goals of the perspective. And these metrics, if well chosen, will be the catalysts for change, providing warning signals to identify ineffective or failed asset performance strategies.

Tactical Perspectives

Table 12.2 illustrates a high-level map developed for a chemical company using operational excellence goals of managing risks and improving profitability. From this strategic goal, perspectives have been defined which are specific to four functions and their associated goals:

- *Operations:* It reduces operating costs/risks and maximizes output
- *Reliability*: It maximizes uptime and preserves plant and asset integrity
- *Work management:* It minimizes corrective work and restores asset condition
- *Safety and Environmental:* It provides controlled/audited environment and enhances safe/audited operational capabilities.

Within these perspectives, each discipline has been able to take charge of factors under its control by choosing the right metrics to measure its progress toward achieving the collective goals. Taking a methanol-producing chemical plant as example, the operations perspective would be to focus on delivering reduced operating costs and managing the risks inherent in the process and in operational activities, while maximizing methanol output.

In the safety and environmental perspective, the focus would be on providing the systems, procedures, and trainings which build operational awareness, skills, functional systems, and capabilities to prevent, manage, and eliminate safety and environmental incidents.

And in the work management perspective, the focus would be on efficiently completing maintenance work while minimizing the potential for future breakdowns and restoring assets to their operating condition. Finally, for the reliability perspective, the focus would be to build the analytics and skills required to increase and improve plant uptime while preserving the integrity and life of plant assets.

From each of these perspectives, tactical metrics can be set to stimulate new outcomes, build new processes, and build skill development and learning with clear links to the goals of each individual perspective.

Operations Perspective	Reliability Perspective	Work Management Perspective	Safety and Environmental Perspective
Strategic KPI			
• Plant availability • Number LPO events • Time operating outside deterioration limits (%) • Plant uptime (%) • Production target compliance	• Plant availability • Proactive work orders (%) • Emergency work orders on high critical systems (%) • Significant deterioration mechanisms improvements • Inspection compliance • Protective device schedule compliance • Quantified reliability target (total) • Predictive maintenance compliance	• Planning compliance • Work order complete (within 20 percent of planned costs) • Proactive work orders (%) • Scheduling compliance • Assessments of work order complete (total) • Quantified availability targets (total)	• Incident rate • Safety performance index • PHA/reviews completed (total) • PSM compliance audits (total) • Significant environmental aspects defined/quantified (total)
Operational KPI			
• Process availability variance • Utility variance • Product transfer indicator • Quality limit excursions (total) • Actual counter measures (total) • Startup indicators • Shutdown indicators • Offspec product • Scrap value • Inventory	• MTBF by equipment type • MTBR by equipment type • MTBM by equipment type • MTBF growth • Cumulative nonavailability of critical assets • Unscheduled maintenance events (total) • Completed work order records on significant failures (total) • Bad actor count • Current mechanical availability • Mechanical availability trend	• Emergency work orders (%) • Reactive work orders (%) • Backlog work orders (total) • Overtime hours (%) • Work orders planned (total) • Cumulative maintenance costs for standing order • Average direct cost per maintenance event • Work orders scheduled (total) • Rework (%) • Closed work orders within 2 days of schedule (%)	• Outstanding items from monthly safety inspection report (total) • A incidents (total) • B incidents (total) • C incidents (total) • Total days lost days due to injury • Process hazard analysis action items

Table 12.2 Performance Indicators for Managing Risk and Improving Profitability of a Chemical Plant

12.5 Risk Assessment Methods

Risk assessment methods can be either based on qualitative or on quantitative approaches. Qualitative methods provide a ranking of equipment and components based largely on experience and engineering judgment. Quantitative risk-based methods use several engineering disciplines to set priorities and develop maintenance programs. Some of the engineering disciplines include nondestructive examination, system and component design and analysis, fracture mechanics, probabilistic analysis, failure analysis, and operation of facilities.

The level of details required in risk assessment are usually directly related to the level of the intrinsic hazards. In general, the greater are the magnitude of the hazards and the complexity of the systems being considered, the greater should be the degree of rigor, robustness, and level of details to estimate that risks have been reduced *as low as is reasonably practicable* (ALARP). The level of risk arising from the undertaking should therefore determine the degree of sophistication needed in the risk assessment.

Qualitative approaches typically easier to apply but provide the least degree of insight. Conversely quantitative risk analysis (QRA) approaches are most demanding on resources and skill sets, but potentially deliver the most detailed understanding and provide the best basis if significant expenditure is involved. Semiquantitative approaches lie in between these extremes. The following sections will describe briefly some of the techniques that have been used in the context of corrosion risk assessment.

12.5.1 Hazard and Operability

The focus of a hazard and operability (HAZOP) study is to identify hazards that might affect safety and operability based on the use of guidewords. A team of experts in different aspects of the installation, under the guidance of an independent HAZOP leader, systematically considers each subsystem of the process typically referring to process and instrumentation diagrams. These experts use a standard list of guidewords to prompt them to identify deviations from design intent. Guidewords are simple words or phases used to qualify or quantify the intention and associated parameters in order to discover deviations (Table 12.3). For each credible deviation, they consider possible causes and consequences, and whether additional safeguards should be recommended. The conclusions are usually recorded in a standard format during the sessions.

Figure 12.8 illustrates how the HAZOP process is conducted on specific nodes of interest. In the HAZOP context, a node is a location on a process diagram (usually piping and instrumentation diagrams or P&IDs) at which process parameters are investigated for deviations. Nodes are also points where the process parameters have identified design intent. Nodes are usually pipe sections or vessels.

Guide Word	Meaning
Standard guidewords	
No	Negation of the design intent
More	Quantitative increase
Less	Quantitative decrease
As Well As	Qualitative increase
Part of	Qualitative decrease
Reverse	Logical opposite of the intent
Other Than	Complete Substitution
Auxiliary guidewords	
How	How is the step to be accomplished
Why	Is the step or operation really needed
When	Is timing of the step or operation important
Where	Is it important where the step is performed

TABLE 12.3 Standard and Auxiliary Guidewords Used to Conduct HAZOP Studies

Plant components (pumps, compressors, exchangers) are found within nodes. In the same fashion, a parameter is an aspect of the process that describes it physically, chemically, or in terms of what is happening (flow, level, temperature). Parameters are usually classified as specific or general to respectively describe aspects of the

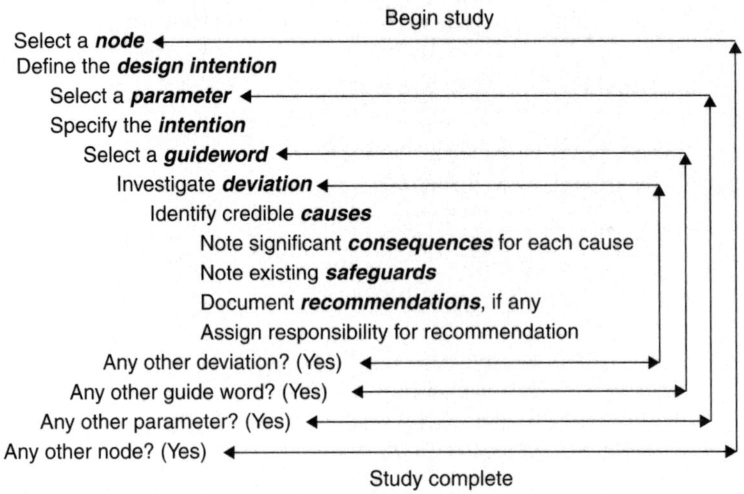

FIGURE 12.8 HAZOP process iterative loops.

process or aspects of design intent remaining after the specific parameters have been removed

The HAZOP procedure is a powerful tool for hazard analysis and its methodical approach ensures that weaknesses in the design intent are detected and acted upon. HAZOP is widely used in RBI processes. It is also applied at an early design stage through corrosion-risk assessments as a documented process for materials selection and corrosion prevention. The strengths of HAZOP are [10]

- It is widely used and its advantages and disadvantages are well understood.
- It uses the experience of operating personnel as part of the team.
- It is systematic and comprehensive and possibly identifies all hazardous process deviations.
- It is effective for both technical faults and human errors;
- It recognizes existing safeguards and develops recommendations for additional ones.
- The team approach is particularly appropriate to operations requiring the interaction of several disciplines or organizations.

The weaknesses of the HAZOP method are

- Its success depends on the facilitation of the leader and the knowledge of the team.
- It is optimized for process hazards, and needs modification to cover other types of hazards.
- It requires development of procedural descriptions, which are often not available in appropriate detail. However, the existence of these documents may benefit the operation.
- Documentation is lengthy (for complete recording).

12.5.2 Failure Modes, Effects, and Criticality Analysis

A failure modes, effects and criticality analysis (FMECA) (or its simpler form, FMEA) is a systematic method of identifying a system failure modes. FMEA is implemented by considering each equipment item and associated systems in the plant, detailing the possible failure modes (e.g. leak or break in the case of pressure equipment), and determining their resulting effect on the rest of the system. The analysis is more concerned with specifying the likely effects and criticality of different modes of failure rather than the mechanisms or events leading to a specific failure [11].

FMEA is relatively simple and easy to apply, yet it is a powerful tool that may be used to improve the quality of products and processes.

The method brings the focus of an analysis on consequences and additional safeguards to mitigate the effects of the failure. It is a common practice for individuals familiar with system functionality to perform FMEA, but teams of experts can produce greater insight into the mechanisms and wider range consequences. The analysis uses a form that begins with a systematic list of all components in the system:

- Component name
- Function of component
- Possible failure modes
- Causes of failure
- How failures are detected
- Effects of failure on primary system function
- Effects of failure on other components
- Necessary preventative/repair action
- Rating of failure frequency
- Rating of severity (i.e., consequence) of failure

Failures are rated as critical if they have high frequency or severity ratings. In these cases, special protection measures may be considered. The strengths of FMECA are [10]

- It is widely used and well understood.
- It can be performed by a single analyst.
- It is systematic and comprehensive, and identifies hazards.
- It identifies safety-critical equipment where a single failure would be critical for the system.

FMECA weaknesses are

- Its benefit depends on the experience of the analyst.
- It requires a hierarchical system drawing as the basis for the analysis, which the analyst usually has to develop before proceeding with the analysis.
- It is optimized for mechanical and electrical equipment, and does not apply to procedures or process equipment.
- It copes with multiple failures and human errors with difficulty.
- It does not produce a simple list of failure cases.

Most accidents have a significant human contribution, and FMECA is not well suited to identifying these. As FMECA can be conducted at various levels, it is important to decide before starting an analysis what level will be adopted as otherwise some areas may be examined in greater details than others. If conducted at too deep a level, FMECA can be time consuming and tedious, but it can also lead to great understanding of the system.

12.5.3 Risk Matrix Methods

Risk matrices provide a framework for an explicit examination of the frequency and consequences of hazards. The method may be used to rank the hazards in order of significance, screen out insignificant ones, or evaluate the need for risk reduction of each hazard. A risk matrix separates the dimensions of probability (POF) and consequence (COF) into typically three to six categories (A to E in Fig. 12.9). There is little standardization in matters such as the size of the matrix or the labeling of the axes.

Risk matrices may use quantitative definitions of the frequency and consequence categories or some numerical indices of frequency and consequence (e.g., one to five) before adding the frequency and consequence pairs to rank the risks of each hazard or each box on the risk matrix. The strengths of the risk matrix approach are [10]

- It is easy to apply and requires few specialist skills, and for this reason it is attractive to many project teams.
- It allows a consistent treatment of risks to people, property, environment, and business.
- It allows hazards to be ranked in priority order for risk reduction effort.

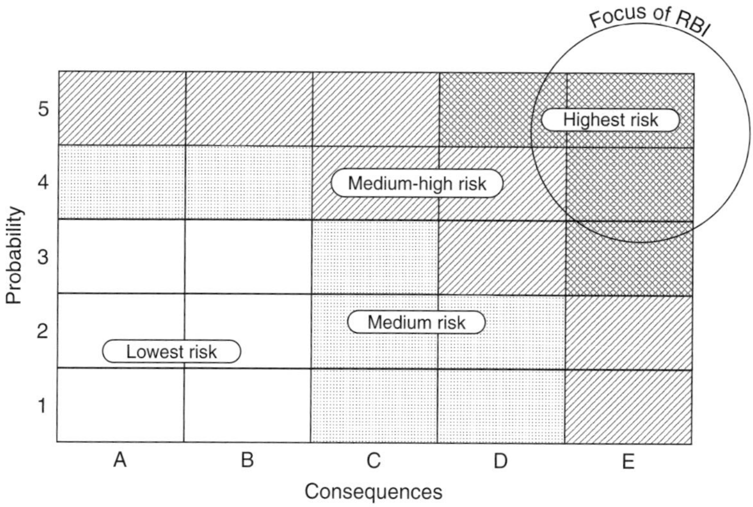

FIGURE 12.9 Five × five risk matrix.

However, there are several problems with this approach, which are less apparent:

- Many judgments are required on likelihood and consequence and unless properly recorded the basis for risk decisions will be lost.
- Judgments must be consistent among different team members, a condition difficult to achieve whether qualitative or quantitative definitions are used.
- Where multiple outcomes are possible (e.g., the consequence of a fall on a slippery deck can range from nothing to a broken neck), it can be difficult to select the *correct* consequence for the risk categorization.
- A risk matrix looks at hazards one at a time rather than in accumulation, whereas risk decisions should really be based on the total risk of an activity. Potentially many smaller risks can accumulate into an undesirably high total risk, but each smaller one on its own might not warrant risk reduction. As a consequence, risk matrix has the potential to underestimate total risk by ignoring accumulation.

12.5.4 Fault Tree Analysis

Fault tree analysis (FTA) provides a logical representation of many events and component failures that may combine to cause one critical event (e.g., pipeline explosion). It uses logic gates to show how *basic events* may combine to cause the critical *top event*. The top event would normally be a major hazard such as "pipeline SCC" as in the example shown in Fig. 12.10. The most commonly used tree symbols and gates used in the construction of fault trees are illustrated in Fig. 12.11 and briefly described here [12]:

- *Fault event (rectangle):* System-level fault or undesired event.
- *Conditional event (ellipse):* Specific condition or restriction applied to a logic gate (mostly used with inhibit gate).
- *Basic event (circle):* Lowest event of examination which has the capability of causing a fault to occur.
- *Undeveloped event (diamond):* Failure which is at the lowest event of examination by the fault tree, but can be further expanded.
- *Transfer (triangle):* The transfer function is used to signify a connection between two or more sections of the fault tree.
- *AND gate:* The output occurs only if all inputs exist (multiply probabilities on the input, therefore decreasing resulting probability).

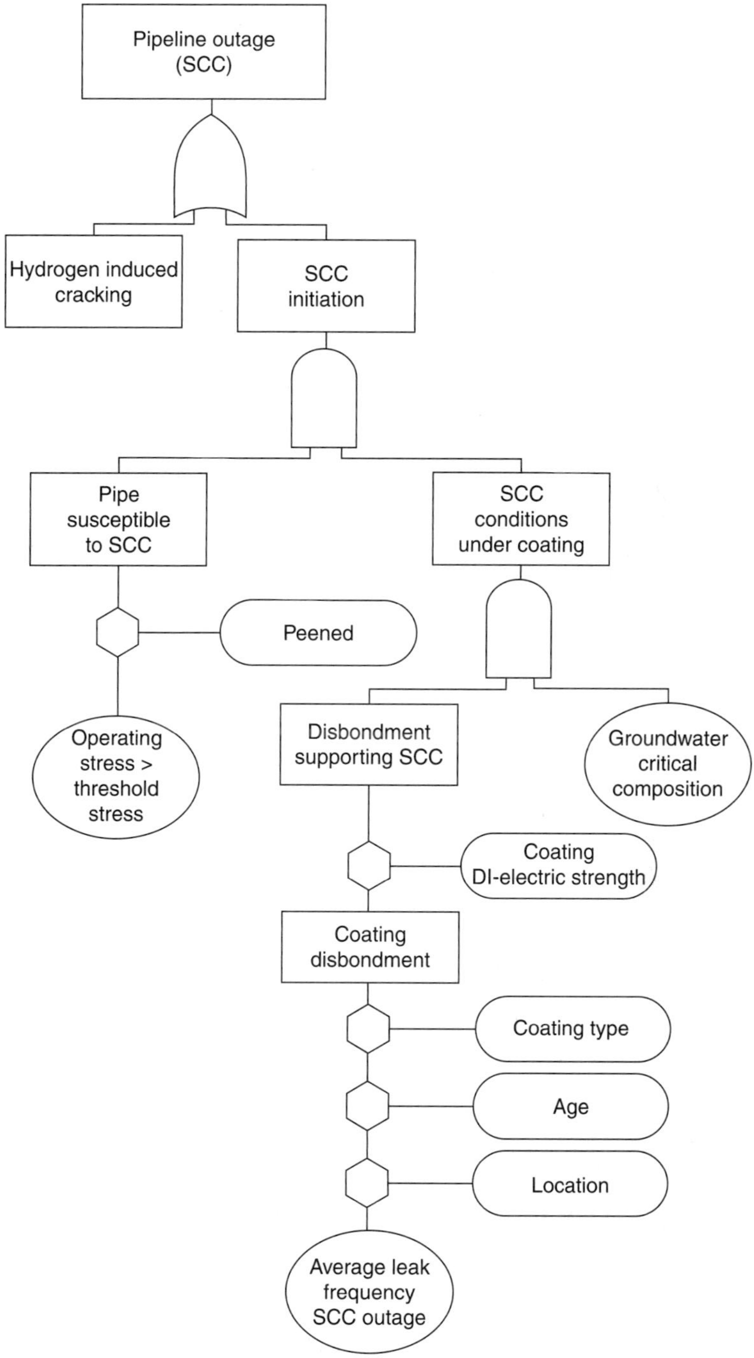

Figure 12.10 Fault tree for natural gas pipeline outage due to SCC.

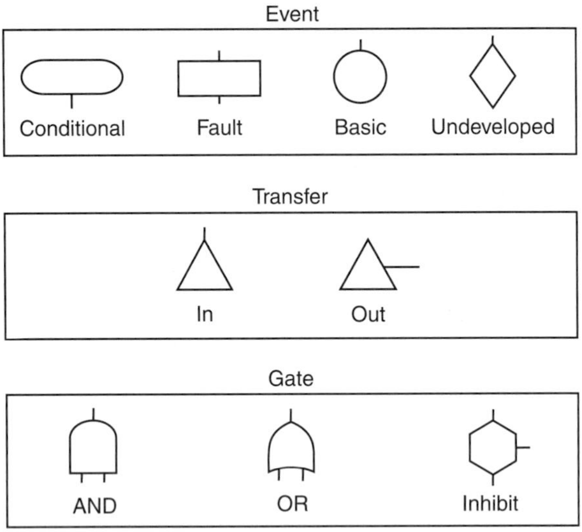

FIGURE 12.11 Fault tree symbols for gates, transfers, and events.

- *OR gate:* The output is true only if one or more of the input events occur (add probabilities on the input, therefore increasing resulting probability).
- *Inhibit gate (hexagon):* One input is a lower fault event and the other input is a conditional qualifier or accelerator (direct effect as a decreasing (< 1) or increasing factor (> 1)).

FTA emphasizes the lower-level fault occurrences that directly or indirectly contribute to a major fault or undesired event. The technique is one of "reverse thinking" where the analyst begins with the final undesirable event that is to be avoided and identifies the immediate causes of that event [11]. By developing the lower-level failure mechanisms necessary to produce higher level occurrences, a total overview of the system is achieved. Once completed, the fault tree allows an engineer to fully evaluate a system safety or reliability by altering the various lower-level attributes of the tree. Through this type of analysis, a number of variables may be visualized in a cost-effective manner.

Tracing the chain of events leading to the final outcome can indicate where extra monitoring, regular inspection, and protective schemes (e.g., temperature and pressure sensors, and alarms) would be most efficient at preventing an impending failure. Fault tree analysis is a very useful tool for studying the routes by which an accident can occur, and is particularly effective at identifying accident scenarios due to secondary and tertiary causes. However, it requires a great deal of skill and effort to implement. For this reason it is expected to be used only by industries where the consequences of failure might be very severe.

The strengths of fault tree analysis are [10]

- It is a widely used and well-accepted technique.
- It is suitable to quantify many of the hazards that arise from a combination of adverse circumstances.
- It is often the only technique that can generate credible likelihoods for novel, complex systems.
- It is suitable for technical faults and human errors.
- It provides a clear and logical form of presentation.

Its weaknesses are

- The diagrammatic format discourages analysts from stating explicitly the assumptions and conditional probabilities for each gate. This can be overcome by careful backup text documentation.
- FTA can become complicated very early in an analysis, therefore turning out to be time-consuming and difficult to follow for large systems.
- Analysts may overlook failure modes and fail to recognize common cause failures (i.e., a single fault affecting two or more safeguards) unless they have a high level of expertise and work jointly with the operator.
- All events are assumed to be independent.
- FTA may easily lose its clarity when applied to systems that do not fall into simple failed or working states (e.g., human error, adverse weather and so forth).

Figure 12.10 illustrates how a major gas transmission pipeline company adopted FTA for the risk assessment of SCC corrosion on its 18,000-km gas pipeline network [13;14]. The rupture risk FTA was normally performed for the review and analytical examination of systems or equipment to emphasize the lower-level fault occurrences. These results also served to schedule maintenance operations, conduct surveys, and plan research and development efforts.

Each element of the branch in Fig. 12.10 contains numerical probability information related to technical and historical data for each segment of the complete pipeline network. In some cases, it was simpler to assume some probability values for an entire system. The probabilities of operating at maximum permitted pressure and the presence of electrolyte were both set at value unity in Fig. 12.10, therefore forcing the focus on worst case scenarios. Other more verifiable variables can be fully developed as is shown in Fig. 12.12

```
Function PCPSH
Private PCPSH_Value

 PCPSH_Value = 0.05
 If COAT1 = "FBE"
   PCPSH_Value = if(DIST = 1, 0.05, PCPSH_Value)
   PCPSH_Value = if(DIST = 2, 0.05, PCPSH_Value)
   PCPSH_Value = if(DIST = 3, 0.05, PCPSH_Value)
   PCPSH_Value = if(DIST = 4, 0.05, PCPSH_Value)
 Endif

 If COAT1 = "EP"
   PCPSH_Value = if(DIST = 1, 0.05, PCPSH_Value)
   PCPSH_Value = if(DIST = 2, 0.05, PCPSH_Value)
   PCPSH_Value = if(DIST = 3, 0.05, PCPSH_Value)
   PCPSH_Value = if(DIST = 4, 0.05, PCPSH_Value)
 Endif

 If COAT1 = "AE". or. COAT1 = "CTE"
   PCPSH_Value = if(DIST = 1, 0.46, PCPSH_Value)
   PCPSH_Value = if(DIST = 2, 0.56, PCPSH_Value)
   PCPSH_Value = if(DIST = 3, 0.30, PCPSH_Value)
   PCPSH_Value = if(DIST = 4, 0.55, PCPSH_Value)
 Endif

 If COAT1 = "PT". or. COAT1 = "PT2". or.;
   COAT1 = "SRT". or. COAT1 = "HHT"
   PCPSH_Value = if(DIST = 1, 0.30, PCPSH_Value)
   PCPSH_Value = if(DIST = 2, 0.88, PCPSH_Value)
   PCPSH_Value = if(DIST = 3, 0.51, PCPSH_Value)
   PCPSH_Value = if(DIST = 4, 0.79, PCPSH_Value)
 Endif
```

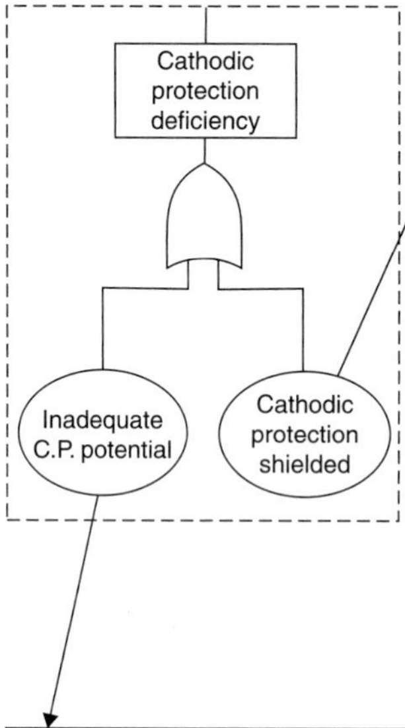


```
Function PCPINAD
Private PCPINAD_Value

 PCPINAD_Value = 0.0
 PCPINAD_Value = if(DIST = 1, 0.000152, PCPINAD_Value)
 PCPINAD_Value = if(DIST = 2, 0.007100, PCPINAD_Value)
 PCPINAD_Value = if(DIST = 3, 0.000134, PCPINAD_Value)
 PCPINAD_Value = if(DIST = 4, 0.000388, PCPINAD_Value)
```

FIGURE 12.12 Detailed code for the basic events leading to a natural gas pipeline cathodic protection deficiency.

for two basic events describing the probable impact of a cathodic protection deficiency on the pipeline network.

12.5.5 Event Tree Analysis

Event tree analysis (ETA) is a logical representation of the various events that may be triggered by an initiating event (e.g., a component failure). It uses branches to show the various possibilities that may arise at each step and it is often used to relate a failure event to various consequence models. It may also be used to quantify system-failure probabilities, where several contributory causes can only arise sequentially in time.

Like FTA, event tree analysis is also a logic-based methodology for identifying accident scenarios, but unlike FTA it is a "forward thinking" method. The analysis begins with a given initiating failure event and develops the resulting sequence of events, normally over a short time interval, making assumptions about the availability of safeguards and backup systems [11].

Event trees are valuable for examining the consequences of failure. However, they are less effective for the analysis of the causes of system failure and the short timescale over which events are considered may mask longer term consequences such as the gradual deterioration of equipment due to faults elsewhere.

Construction starts with the initiating event and works through each branch in turn. A branch is defined by a question (e.g., "Protective device fails?"). The answers are usually binary (e.g., "yes" or "no"), but there can also be multiple outcomes (e.g., 100, 20, or 0 percent in the operation of a control valve). Each branch is conditional on the appropriate answers to the previous ones in the tree.

Usually an event tree is presented with the initiating events on the left and the outcomes on the right. The questions defining the branches are placed across the top of the tree, with upward branches signifying "yes" and downward ones for "no." A probability is associated with each branch, being the conditional probability of the branch (i.e., the answer "yes" or "no" to the branch question) given the answers of all branches leading up to it. In each case, the sum of the probabilities of each branch must be unity. The probabilities of each outcome are the products of the probabilities at each branch leading to them. The sum of the probabilities for all outcomes must be unity as well. This provides a useful check on the analysis. The strengths of event tree analysis are [10]

- It is widely used and well accepted.
- It is suitable for many hazards in QRA that arise from sequences of successive failures.
- It a clear and logical form of presentation.
- It is simple and readily understood.

Its weaknesses are

- It is not efficient where many events must occur in combination, as it results in many redundant branches.
- All events are assumed to be independent.
- It loses its clarity when applied to systems that do not fall into simple failed or working states (e.g., human error, adverse weather, and so forth).

Figure 12.13 shows an event tree analysis that was performed on each process system of a fluid catalytic cracking unit (FCCU) gas plant using actual probabilities and consequences that are particular to that process system. In a span of seven months the refinery had experienced

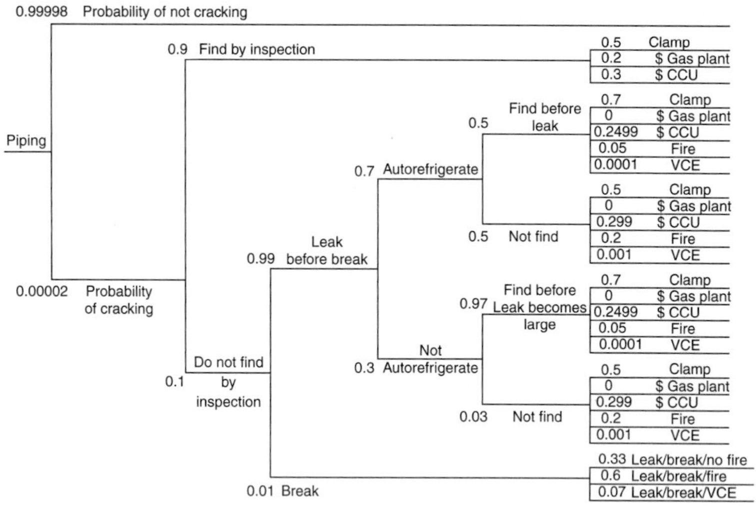

FIGURE 12.13 Event tree analysis that was performed on each process system of a FCCU gas plant ($ = cost of damage, CCU = catalytic cracking unit, VCE = vapor cloud explosion).

23 leaks in piping located in the fractionator's overhead and the wet gas compression sections of the FCCU. Follow-up with intensive ultrasonic (UT) shear wave inspection located an additional 73 carbonate cracking like indications in the gas recovery section of the FCCU. As with many forms of SCC, radiography was not considered a suitable inspection technique for carbonate cracking [15].

In petroleum refining, carbonate SCC can occur in FCCU gas plants where the process environment contains a significant amount of carbonate/bicarbonate ions, H_2S, free water, and ammonia. The equipment most likely to be affected in the FCCU gas plant are the main fractionator's overhead condensers and accumulators, wet-gas compressor's knockout drums and condenser, deethanizer (or other similar light-end fractionators), and all associated piping including the sour-water streams originating from these areas. In this example, probabilities were based on the occurrence of the event in any given year. Values assigned for the probabilities were determined using the number of welds that cracked divided by the total number of welds in the system. Consequences included

- Leaks/breaks
- Using clamps to contain the leak
- Shut down of the unit (SID)
- Fire
- Vapor cloud explosion (VCE)

Dollar values for each consequence were determined in agreement with operations, engineering, and inspection personnel based on costs of past failures and loss of production. The results of the risk assessment were then used to identify which equipment and piping were to be replaced during the next refinery FCCU turnaround.

12.6 Risk-Based Inspection

The application of RBI to refinery and petrochemical plant equipment inspection was started by several companies in the late 1980s and was first reported in the open literature in 1993. Since that time, RBI has become a widely accepted and mandated methodology. A 1995 survey of member companies from MTI found that most corrosion engineers were routinely involved with establishing the frequency of inspection, and about half established the inspection frequency based on the results of risk identification.

A joint industry project sponsored by 21 refining and petrochemical companies was set in 1993 working in conjunction with the American Petroleum Institute (API) to develop RBI methodologies for application within their industry. At the same time, the API was developing an industry consensus standard for RBI [16].

As mentioned earlier, there are several levels of risk-based assessments, usually described as qualitative, semiquantitative, and quantitative. These vary considerably in the amount of effort and input needed and the accuracy of the resultant assessment. The qualitative assessment is usually performed to determine risk associated with whole or large portions of process units. The semiquantitative and quantitative assessments are usually performed to determine risk associated with individual equipment items.

The end result of risk assessment is a risk rating for each equipment item, which may vary from low to high. More often than not, a large percentage of the risk (>80 percent) is found to be associated with a small percentage of the equipment items (<20 percent). Once identified, the higher-risk equipment becomes the focus of inspection and maintenance to reduce the risk, while opportunities may be found to reduce inspection and maintenance of the lower-risk equipment without significantly increasing risk. The main drivers for RBI in specific industries are [17]

- *Nuclear power industry:* Increased safety and availability
- *Fossil-fired power industry:* Life extension and cost reduction
- *Offshore petroleum industry:* Risk management, for example, safety, environmental, and economics, plus cost reduction of the inspection costs
- *Petrochemical industry:* Cost reduction, extension of inspection periods, and improvement of availability.

12.6.1 Probability of Failure Assessment

In the context of an RBI, two fundamental issues must be considered to determine the probability of a failure (POF): firstly the different forms and rates of corrosion and secondly the effectiveness of the inspection itself at detecting the corrosion. The input of corrosion experts is obviously required to identify the relevant forms of corrosion in a given situation and to determine the key variables affecting their propagation rate. It is also important to realize that full consensus and supporting data on the variables involved is highly unlikely in real-life complex systems and that simplification will invariably be necessary.

One semiquantitative approach for ranking process equipment is based on internal POF. The procedure is based on an analysis of equipment process and inspection parameters, following which the equipment is categorized on a scale of one to three, with one being the highest priority. The procedure requires a fair degree of engineering judgment and experience and, as such, is dependent on the background and expertise of the analyst. The procedure is designed to be both practical and efficient. The POF is intended to be a convenient and reproducible means for establishing equipment inspection priorities. As such, it facilitates the most efficient use of finite inspection resources when and where 100 percent inspection is not practical.

The POF approach is thus heavily based on a set of rules that depends on detailed inspection histories, knowledge of corrosion processes, and knowledge of normal and upset conditions. The equipment rankings may have to be modified and updated as additional knowledge is gained, process conditions change, and equipment ages. Maximum benefits of the procedure depend on fixed-equipment inspection programs that permit the capture, documentation, and retrieval of inspection, maintenance, and corrosion/failure mechanism information.

12.6.2 Consequence of Failure Assessment

It is important to obtain the input from experts in process engineering, safety, health, and environmental engineering to assess the consequences of a failure (COF). Considering the example of an uncontrollable fluid release, the following three factors would play a dominant role in the consequence assessment:

- The type of fluid that may be released and the hazards associated with a release.
- Its inventory available for release.
- The rate of release.

A determinant factor in the rate of release is the size of the breech of containment. Fortunately, there are many breeches of containment that are small leaks which, once detected, can be readily contained and mitigated without causing a significant incident. These are often the result of pinholes due to localized corrosion or small, tight cracks that only allow minor fluid seepage. On the other hand, some breeches of containment are the result of major ruptures that allow large quantities of hazardous fluid to escape in a short period of time. It is difficult and sometimes impossible to react quickly enough to contain and mitigate such releases without enduring significant incidents.

Corrosion and materials engineering expertise is required to estimate the size and nature of damage that could result in a plant item. As described in Chap. 6, different corrosion mechanisms can produce different damage morphologies. The difference in impact on the release rate created at a pinhole leak versus that of a large rupture is a good example of this aspect of consequence sensitivity.

Another important field covered by corrosion engineering is that of materials properties. For example, the risk of a catastrophic explosion from cracks in a brittle material associated with high release rates is obviously greater than in a material with higher fracture toughness. The toughness of a material is a key parameter in determining so-called "leak-before-break" safety criteria and the general tolerance toward defects. An understanding of how the toughness of a material can be reduced in service over time is thus obviously important [3].

12.6.3 Application of Risk-Based Inspection

The application of RBI becomes most effective when it involves the efforts of a multidiscipline team from operations, process engineering, equipment specialists, and maintenance in addition to inspection and corrosion engineers. As a team, these experts can exchange valuable information and use the team synergy to arrive at a mutually agreeable approach to risk reduction that may involve approaches other than increased inspection. The following sections describe three approaches to reduce the risk of operating plant equipment that are incorporated into the API RBI program [16].

Optimizing Inspection/Monitoring

Once the risk assessment has been completed, the results are used to evaluate the effectiveness of the present inspection/monitoring strategy while looking for ways to optimize the strategy and reduce the overall risk of continued operation. For equipment identified as high risk, changes in the detailed inspection plan that would reduce the risk should be considered. For equipment identified as low risk, changes in the detailed inspection plan that would reduce costs while not increasing risk appreciably should be considered [3].

The RBI assessment identifies the potential deterioration mechanisms that can lead to failure in each piece of equipment. It is very important to determine whether the existing inspection plan is addressing all of the potential deterioration mechanisms that were identified in that service. Are appropriate inspection methods being used? For example, it is unlikely that spot ultrasonic thickness measurements will find highly localized corrosion. Or perhaps there has been no inspection performed to detect a cracking mechanism that is highly likely to occur in that service. The fundamental understanding of a deterioration mechanism may be valuable input when evaluating the likelihood that an inspection method will actually detect the deterioration.

Materials of Construction Changes

The RBI approach can be used to evaluate the risk reduction associated with alternative alloys. Coupled with appropriate alloy cost information, the replacement alloy selection can be optimized on a risk reduction versus cost basis. Used in this manner, the RBI program becomes a valuable tool for the corrosion engineer. This type of output from an RBI program can be used when communicating with management about the justification of the added cost of a material upgrade [3].

Although it may not be readily evident, there is a great opportunity to use an RBI program when selecting materials for new construction projects. The corporate pressure for improving profitability has placed intense pressure to achieve the lowest initial cost for installation of new plants. Unfortunately, as a result, it is not uncommon to see low-cost materials with rather high expected corrosion rates chosen for original construction, therefore saving on initial plant cost while pushing the burden on future inspection and maintenance. The heavy reliance on inspection to maintain mechanical integrity more often than not comes with considerable risk that may not have been adequately quantified or addressed by the project leaders. When the new plant becomes operational, high risks require mitigation to ensure mechanical integrity, and this usually involves additional expenditures of inspection resources. As pointed out earlier, suitable risk reduction may not be achievable by increased inspection alone, and materials upgrades may be required sooner than expected. The risks associated with alternative materials selection for new projects should be quantified and considered, and an RBI program can provide an effective means of demonstrating the tradeoff between initial cost and risk.

Key Process Parameters

A real benefit derived from the development of an RBI program is the identification of key process parameters that most influence the equipment deterioration rate. These key process parameters often include fluid composition, temperature, pH, and fluid velocity. Once identified as such, a strong case can be made for routine monitoring of these highest-impact process parameters and maintaining them within prescribed limits. It is disturbing to find that process unit operating personnel sometimes do not fully understand the relevancy of certain process parameters to mechanical integrity, and therefore are not routinely monitoring them. This is often realized when there is a loss of containment incident and root cause investigation reveals that the failed equipment has been operating beyond one or more of the key process parameter limits [3].

An RBI program can also serve as a helpful tool for quantifying the risk associated with a change in the value of a key process parameter. This can be an invaluable resource to the corrosion engineer when addressing management of change issues in an operating process unit. It also can serve as an aid to reaching agreement on the desirable process monitoring and limits on operating conditions, and in communicating with process unit operators and management about the impact of changes on risk.

12.7 Industrial Example: Transmission Pipelines

The transmission pipeline industry is an irreplaceable component of modern infrastructures. Pipelines have historically been the safest means of transporting natural gas and hazardous liquids. However, recent pipeline failures have heightened the awareness of transmission pipeline systems. In particular, in many parts of the world cities have grown and are now located close to pipeline systems that have been operating for decades [18].

Many millions of kilometers of pipeline crisscross the globe, carrying oil and natural gas. In the United States alone there are approximately 3.5×10^6 km of transmission pipeline, 525,000 for the transmission and gathering of natural gas, 260,000 for the transmission and gathering of hazardous liquid and the balance for the distribution of natural gas. Each system is unique with respect to potential integrity threats and associated consequences in the unlikely event of a failure. The probability of a release can be reduced through an effective management program that addresses these integrity threats.

There are many causes and contributors to pipeline failures. The U.S. Department of Transportation (DOT) Research and Special Programs Administration, Office of Pipeline Safety (RSPA/OPS) compiles data on pipeline accidents and their causes. Tables 12.4 to 12.6 summarize the results collected during a two-year period (2002 and 2003) for respectively, natural gas transmission and gathering, hazardous liquid transmission, and natural gas distribution.

It is notable, in Tables 12.4 and 12.5, that corrosion was the most common cause (37 percent) of natural gas transmission and gathering pipeline incidents during that period, and the second most common cause (24 percent) of hazardous liquid pipeline incidents. During the same period over 60 percent of natural-gas distribution pipeline incidents were caused by outside forces (e.g., excavation by the operator or other parties, damage from natural forces, and the like) while only a small fraction (0.1 percent) of the property damage was attributed

Reported Cause	Number of Incidents	% of Total Incidents	Property Damages (k$)	% of Total Damages	Fatalities	Injuries
Excavation Damage	32	17.8	4,583	6.9	2	3
Natural Force Damage	12	6.7	8,278	13	0	0
Other Outside Force Damage	16	8.9	4,689	7.1	0	3
Corrosion	46	25.6	24,273	37	0	0
Equipment	12	6.7	5,337	8.0	0	5
Materials	36	20.0	12,131	18	0	0
Operation	6	3.3	2,286	3.4	0	2
Other	20	11.1	4,773	7.2	0	0
Total	180		66,351		2	13

Table 12.4 Natural Gas Transmission and Gathering Pipeline Incident Summary by Cause for 2002 and 2003

Reported Cause	Number of Accidents	% of Total Accidents	Barrels Lost	Property Damages (k$)	% of Total Damages	Fatalities	Injuries
Excavation	40	14.7	35,075	8,988	12	0	0
Natural Forces	13	4.8	5,045	2,646	3.5	0	0
Other Outside Force	12	4.4	3,068	2,063	2.8	0	0
Materials or Weld Failure	45	16.5	42,606	30,682	41	0	0
Equipment Failure	42	15.4	5,717	2,761	3.7	0	0
Corrosion	69	25.4	55,610	17,776	24	0	0
Operations	14	5.1	8,332	817	1.1	0	4
Other	37	13.6	20,022	9,060	12	1	1
Total	272		175,475	74,792		1	5

TABLE 12.5 Hazardous Liquid Pipeline Accident Summary by Cause for 2002 and 2003

Reported Cause	Number of Incidents	% of Total Incidents	Property Damages (k$)	% of Total Damages	Fatalities	Injuries
Construction/ Operation	20	8.1	3,086	6.7	0	16
Corrosion	3	1.2	60	0.1	2	9
Outside Force	153	62.2	32,334	70	6	48
Other	70	28.5	10,618	23	13	31
Total	246		46,098		21	104

TABLE 12.6 Natural-Gas Distribution Pipeline Incident Summary by Cause for 2002 and 2003

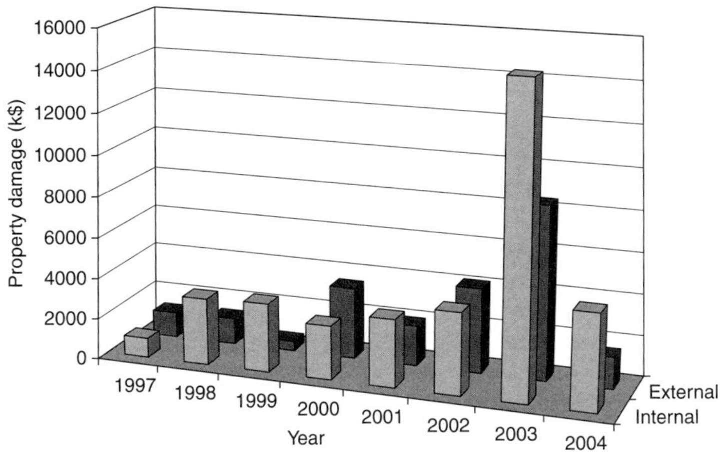

FIGURE 12.14 Property damage costs attributed to external and internal corrosion of transmission and gathering pipeline during 1997 to 2004.

to corrosion. Figures 12.14 and 12.15 illustrate the property damage costs associated with external and internal corrosion of natural gas (Fig. 12.14) and hazardous liquid (Fig. 12.15) transmission and gathering pipelines during a period of eight years (1997 to 2004). The data in these figures clearly indicate that both external and internal corrosion are serious contributors to pipelines property damage.

The pipeline industry uses considerable resources to minimize the likelihood of failures. A study recently completed by the DOT Federal Highway Administration (FHWA) estimates that the pipeline industry

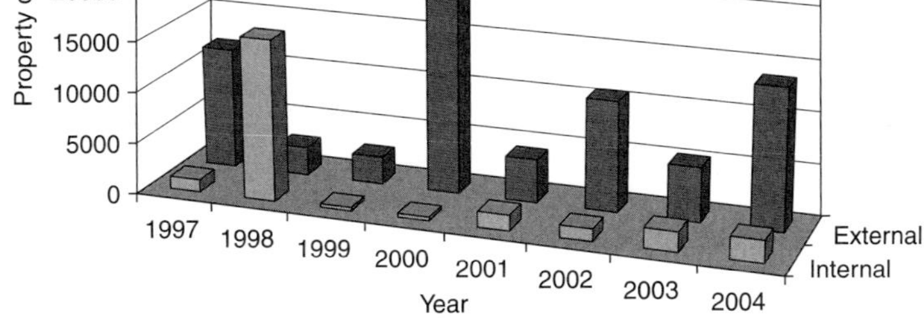

FIGURE 12.15 Property damage costs attributed to external and internal corrosion of hazardous liquid transmission pipeline during 1997 to 2004.

spends approximately seven billion dollars per year on corrosion control. This figure includes operations and maintenance activities, capital expenditures, and corrosion failure repairs [18]. The tools most commonly used by operators to verify pipeline integrity include both external and internal direct assessments, hydrostatic testing, and in-line inspection (ILI). Each of these tools can be used under various circumstances for baseline assessments and future reassessments.

12.7.1 External Corrosion Damage Assessment

External corrosion direct assessment (ECDA) is a structured process that consists of four key steps: preassessment, indirect examination, direct examination, and post assessment (Fig. 12.16). ECDA is intended to assist pipeline operators in establishing the integrity of pipelines. The process uses aboveground cathodic protection (CP) survey methods, many of which have been used in the pipeline industry for decades. ECDA further defines the process, validation, and data integration for these survey methods [18]. The ECDA process incorporates standard techniques for compiling historical information, pipeline and soil surveys, external pipeline inspections, and data analyses.

Preassessment

Preassessment is the first step in the ECDA process. All pertinent historical information is compiled, falling generally into five categories [19]:

- Pipe data
- Construction data

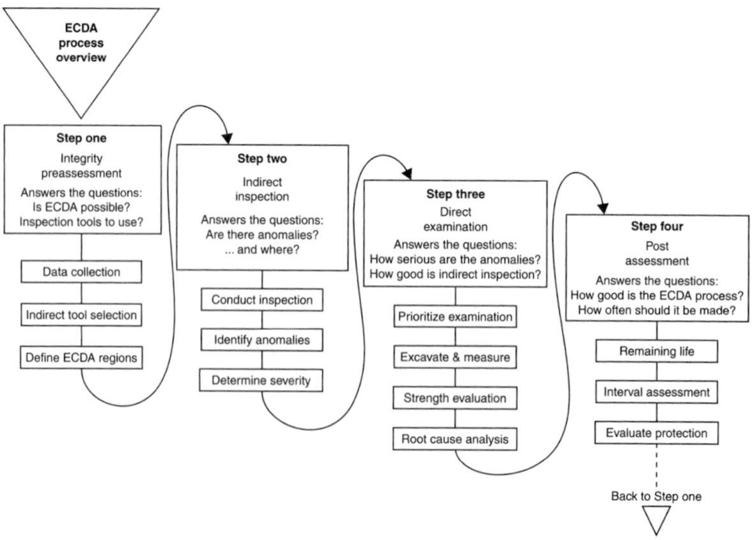

FIGURE 12.16 Overview of the ECDA process.

- Soil/environmental conditions
- Corrosion protection data
- Operating parameters/history

The data collected for preassessment are similar to the information required for risk assessment, which may be performed concurrently. The feasibility of using ECDA must be evaluated based on available data. Other factors that, besides the lack of information, could preclude ECDA from being the assessment method of choice include

- Electrical shielding from disbonded coating
- Rocky terrain and backfill
- Paved surfaces
- Inaccessible locations
- Paralleling buried metallic structures

If an operator determines that ECDA is an appropriate assessment methodology, the pipeline then must be divided into ECDA regions and two indirect inspection techniques need to be selected for each region. For example, on a cathodically protected pipeline with good coating, a close-interval potential survey (CIPS) will provide information concerning the level of protection, interference, and current shielding. A direct current voltage gradient (DCVG) survey performed in conjunction with the CIPS will detect specific locations of damaged coating (see Chap. 13 for additional details). However, it is crucial that, independent of the two techniques selected, the ECDA be performed along the entire length of the specified section of pipeline with both techniques.

Indirect Inspection

The purpose of indirect inspection is to identify the locations of coating faults, insufficient cathodic protection, electrical shorts, interference, geologic current shielding, and other anomalies along the pipeline. Indirect inspection also permits to identify the areas where corrosion may be occurring or has occurred. Typical pipeline survey techniques that may be used are [19]

- Close-interval on/off potential surveys (see Chap. 13)
- DCVG surveys
- Electromagnetic current attenuation surveys
- Alternating current voltage gradient surveys

All four survey techniques are well established. Each has its advantages and disadvantages, depending upon the type of anomaly that needs to be detected and the condition of the pipeline right-of-way. However, it is important that the inspections with the two chosen techniques be carried out at approximately the same time and that the data be tied to all permanent features along the pipeline route so that the two data sets can be aligned and compared. The data should then be analyzed and indications of corrosion activity identified. The data analysis should reveal all discrepancies in the two data sets, resolve these differences, and compare the results to those obtained in the preassessment.

Direct Examinations

Direct examination requires excavating the pipe and performing physical inspections and tests on the pipe surface and surrounding soil/water electrolytes. Direct examination includes

- Ranking and prioritizing indications identified during indirect inspections
- Excavating to expose the pipe and collecting data where corrosion activity is most likely
- Measuring coating damage and corrosion defects
- Performing a root-cause analysis (RCA)
- Evaluating the process

Each indication obtained with indirect inspections is then categorized for an action level as either

- Immediate action required
- Scheduled action required
- Suitable for monitoring

Subsequently, dig locations to expose the pipe are identified and given a priority. At least one excavation and direct examination is required for each ECDA region. When performed for the first time, a minimum of two direct examinations are required. Before exposing the pipe, the procedures for data collection and record keeping should be defined to ensure consistency. The procedures should include [19]

- Photographic documentation
- Pipe-to-soil (P/S) potential measurements
- Soil and groundwater tests
- Coating assessment
- Undercoating liquid pH
- Mapping and measurement of corrosion defects
- Data for other analyses such as microbiologically influenced corrosion and stress corrosion cracking

Once corrosion defects have been found, the remaining strength of the pipe should be estimated using a standard calculation routine (e.g., ASME B31G [20], RSTRENG*, or DNV RP-F101 [21]) following which a RCA should be carried out for all significant corrosion activities before implementing an appropriate corrosion mitigation program.

Postassessment

The final step in the ECDA process is postassessment, which provides an opportunity to assess the overall effectiveness of direct assessment and define reassessment intervals based on remaining-life calculations. Various methods can be used to calculate the remaining life from the corrosion-growth rate, wall thickness, calculated failure pressure, yield pressure, maximum allowable operating pressure, and appropriate safety margin.

Under some conditions, corrosion rates can be measured with a corrosion monitoring technique (e.g., linear polarization resistance (LPR), corrosion coupons) [22]. In the absence of other data, the NAGE ECDA document [23] recommends to use a pitting rate of 0.4 mm/y, which can be reduced by 24 percent in cases where the pipeline has consistently been cathodically protected by a minimum of 40 mV. The maximum reassessment interval is then established as one-half the calculated remaining life. For hazardous liquid pipelines, the maximum inspection interval is 10 years according to the high consequence area (HCA) regulations. The rule for gas pipelines is a maximum interval of seven years or five years when using direct assessment.

12.7.2 Internal Corrosion Damage Assessment

As previously discussed in reference to Fig. 12.14, internal corrosion is a serious problem for natural gas and transmission pipelines. Internal corrosion direct assessment (ICDA) is also a four-step method (Fig. 12.17) developed to determine the presence of internal corrosion in pipelines and ensure their integrity. ICDA is applied to natural gas systems that normally carry dry gas but experience intermittent dropout of liquid water. It is believed that internal corrosion occurs at regions where water accumulates along the pipeline. This observation forms the basis for the ICDA process to conduct a detailed examination at those regions [24].

Preassessment

Preassessment involves the collection of essential data required to conduct the assessment, feasibility determination of ICDA process, and classification of ICDA regions. Data to be collected include pipe operating history, pipe design, features with inclination and elevation data, maximum flow rates, temperature and pressures, corrosion

* RSTRENG uses the same basic equation for predicting failure stress as B31.G, but calls for different definitions of three key variables (i.e., flow stress, Folias factor, and area of missing metal). The last has a direct bearing on the way the results of a metal loss inspection are interpreted.

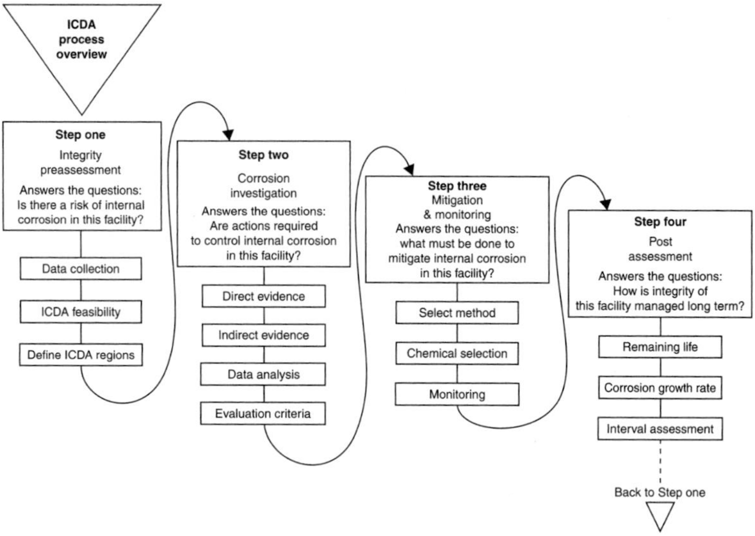

FIGURE 12.17 Overview of the ICDA process.

monitoring, and other internal corrosion-related data. Once the required data have been collected, the requirement to carry out the ICDA is assessed based on the following conditions:

- The gas in the pipeline should contain less than 110 g/m^3 of water as determined by the chemical analysis of collected fluids.
- The pipeline should not have a past history of transporting crude oil or products.
- The pipeline has a history of inefficient coating practices.
- The pipeline is not treated with corrosion inhibitors.
- Pipeline is not subjected to regular pigging.

The ICDA regions are then classified based on changes in temperature, pressure, and flow rate (in and out pressures).

Indirect Inspection

The main objective of the indirect inspection step is to predict regions highly susceptible to internal corrosion. These regions are further selected for detailed examination. For this reason, critical degree of inclination and pipeline inclinations are calculated and compared. This step is applicable to pipelines in which stratified flow* is the

* Stratified flow is defined as a flow pattern with liquid flowing at the bottom of the pipe and gas at the top.

dominant flow regime. A flow modeling calculation using Eq. (12.8) is applied to determine critical inclination angles (θ) [24]:

$$\theta = \sin^{-1}\left(\frac{\rho_g}{\rho_l - \rho_g}\right) \times \left(\frac{V_g^2}{g \times d_{\text{id}}}\right) \times F \tag{12.8}$$

where ρ_l is liquid density
ρ_g is gas density
g is acceleration due to gravity (9.8 m/s²)
d_{id} is pipe internal diameter
V_g is superficial gas velocity
F is modified Froude number

These calculations are bound by the following conditions:

- The maximum superficial gas velocity in the pipeline should be ≤ 8 m/s.
- The nominal pipe diameter should be between 0.1 m and 1.2 m.
- Pressures should be between 3.4 MPa and 7.6 MPa unless flow modeling has been performed outside this range.

The actual pipeline inclination angle is calculated based on Eq. (12.9):

$$\text{Inclination} = \tan^{-1}\left(\frac{\Delta\ \text{elevation}}{\Delta\ \text{length of pipe section}}\right) \tag{12.9}$$

Regions with water accumulation, which are most susceptible to internal corrosion, are identified based on the comparison between critical degree of inclination and pipeline inclination angles.

Detailed Examination

This step involves comparison of field-measured performance data with the pipeline historical and existing operating data to assess the effect of internal corrosion on the pipeline. Detailed examination is performed on the predicted high corrosion regions to confirm the presence of internal corrosion. If the pipeline inclination angle exceeds the critical degree of inclination, detailed examination is performed at pipe locations with inclination greater than the critical inclination within the ICDA region [24].

If there is no instance where the critical angle is exceeded, then detailed examination is performed at the angle of greatest inclination within the ICDA region. While performing a detailed examination, accurate measurements of wall thickness within wall loss areas need to be identified. The remaining strength of the corroded areas of the pipeline should also be determined.

Once the site has been exposed, corrosion monitoring devices (e.g., coupons, electrical resistance probes) can be installed to identify and monitor corrosivity in the pipeline. Inline inspection results may also provide information to assess the downstream conditions of the pipe. Once regions most susceptible to corrosion are free of damage, it is then considered that pipeline integrity to a large extent is assured.

Postassessment

Postassessment determines the effectiveness of ICDA for dry-gas transmission lines based on the correlation between detected corrosion obtained during detailed examination and ICDA predicted regions. Reassessment intervals are determined depending on the remaining life of the pipe and rate of corrosion growth in the pipe.

12.7.3 Hydrostatic Testing

Hydrostatic testing is an effective verification method for pipelines. This technique is used to conduct strength tests on new pipes while in the manufacturing process, as well as at the completion of pipeline installation in the field prior to being placed in service. Hydrostatic testing is also used, at times, for integrity assurance after a pipeline is in operation. Hydrostatic tests are generally the preferred integrity assessment method when the pipeline is not capable of being internally inspected or if defects are suspected that may not be detectable by ILI tools.

The hydrostatic test establishes the pressure carrying capacity of a pipeline and may identify defects that could affect integrity during operation. Testing is done to a pressure that is greater than the normal operating pressure of the pipeline. This provides a margin of safety. The test stresses the pipeline to a predetermined percentage (e.g., 93 to 100 percent) of its specified minimum yield strength (SMYS) generally for eight hours. If SCC is suspected, the test pressure may be increased to 100 percent of SMYS or higher for 30 minutes to an hour. Axial flaws such as SCC, longitudinal seam cracking, selective seam corrosion, long narrow axial (channel) corrosion and axial gouges, which are difficult to detect with magnetic flux leakage (MFL) pigs, are better detected with a hydrostatic test.

Although hydrostatic testing is an effective method for validating near-term integrity, it provides limited information about the status of integrity threats, creates environmental problems related to water treatment, and results in potentially significant downtime, affecting deliveries.

12.7.4 In-Line Inspection

In-line inspection (ILI) tools, also commonly called "smart" or "intelligent pigs," are cylinder-shaped electronic devices used by the pipeline industry to detect loss of metal and in some cases deformations in the pipeline. Inserted into the pipeline and propelled by the fluid or gas pressure, ILI tools record certain physical data about the pipeline integrity (e.g., location of reduced pipe wall thickness, dents, and so forth) as they move through the pipeline. Evaluation of the collected data allows the pipeline operator to make integrity decisions about the pipeline and mitigate potential problem areas before they become a problem.

Since their development in the 1960s, these tools have undergone several generations of technological advancements. ILI tools now include caliper/deformation tools to characterize dents and changes in ovality, and magnetic flux leakage (MFL) and ultrasonic testing (UT) tools to characterize corrosion-caused metal loss. Other technologies, such as ultrasonic crack-detection tools, are also used by pipeline operators to address specific integrity threats. Figure 12.18 presents a decision tree to help an operator choose which type of ILI tool would be the most adequate for a specific job [25].

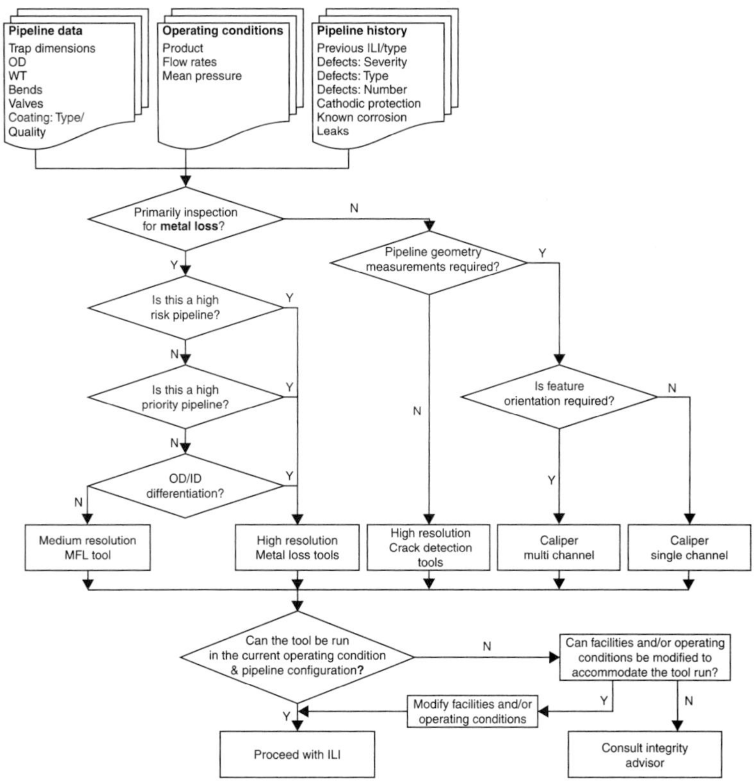

FIGURE 12.18 ILI tool selection decision tree [25].

Metal Loss (Corrosion) Tools

Metal loss tools are used to detect defects that have resulted in wall thinning in a pipeline. They can discriminate to some extent between manufacturing defects, corrosion defects and mechanical damage. There are two main types of tools:

1. The most commonly used and established ILI tool is without any doubt MFL (Fig. 12.19). Magnetic flux leakage tools use a circumferential array of MFL detectors embodying strong permanent magnets to magnetize the ppe wall to near saturation flux density. As the tool travels down the pipe it records magnetic flux leakages that are converted into information revealing anomalies i the pipe wall, such as corrosion pits (Fig. 12.20). The leakage fluxes are detected by Hall probes or induction coils that are part of the MFL ILI tool.

   Two types of MFL tools are available, i.e. high resolution MFL and standard resolution MFL. The main difference between the two is in the number of sensors and the effective resolution. Most MFL tools can determine the location and o'clock position of the metal loss anomaly and detect if a corrosion anomaly is nternal or external to the pipe wall. MFL tools also provide data of each corrosion anomaly including its length and maximum depth to determine the pipe remaining srength. MFL tools are typically capable of detecting corrosion depth greater than 20% of the pipe wall thickness. Axially-oriented flaws such as SCC, selective seam corrosion and axial gouges are difficult to detect with MFL tools.

2. Ultrasonic (UT) wall measurement tools use the ultrasound echo time technique for measuring the remaining wall thickness. The UT tool transmits an ultrasonic pulse into the pipe

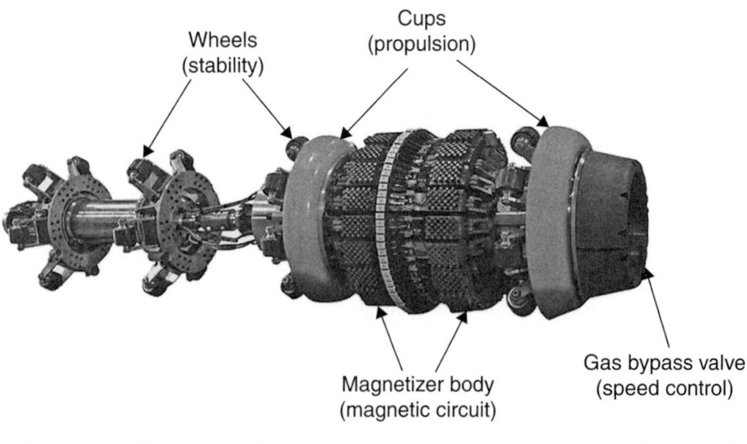

FIGURE 12.19 Magnetic flux leakage ILI tool designed for high-speed gas pipeline applications. (Courtesy of BJ Pipeline Inspection Services)

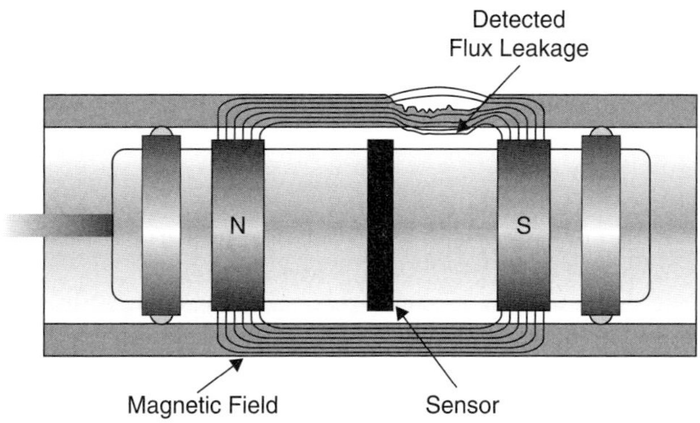

FIGURE 12.20 Pipe inspection using MFL.

wall and directly measures its thickness. Since this technology requires a clean pipe wall, it is generally not used for certain pipelines such as crude lines with a paraffin build-up. There are also wall-thickness limitations with the UT tool. It works well with heavy-wall pipe, but not as well with thin-wall pipe, and it is not as widely used as the MFL tool. This particular tool offers the following advantages in a pipeline environment:

- Direct measurement and high accuracy of wall thickness and defect depth;
- Precise distinction between internal and external defects with the exception of defects adjacent to welds;
- UT tools can also be used to approximate the remaining strength of affected pipe area;
- UT tools provide a more precise description of anomalies than MFL tools.

Crack Detection Tools

Crack detection tools are typically designed for longitudinal crack detection, but they can also be adapted to circumferential crack detection.

1. UT crack detection tool generates an ultrasonic signal into the pipe wall that is reflected off the internal and outer surfaces of the pipe. If a crack is detected, the signal reflects back along the same path of the tool. Since a liquid couplant is required between the sensors and the pipe wall, this tool works only with liquid pipelines.
2. Transverse MFL tools magnetize the pipe wall around its circumference to detect cracks, such as longitudinal seams cracks and longitudinal seam corrosion. This tool is similar to

the standard MFL tool but its induced magnetic field is in a transverse or perpendicular direction. This tool also has limitations, for example, cracks must have sufficient width, or gap, to be detected, and the severity of the crack is not determined.

3. Elastic wave tools operate by sending ultrasounds in two directions along the pipeline to locate and size longitudinally oriented crack and manufacturing defects.
4. Electromagnetic acoustic tools are particularly adapted to the detection of cracks in dry gas pipelines. This tool generates ultrasonic signals without requiring liquid couplant to transfer the ultrasound into the steel. These tools are new to the pipeline inspection industry, so their effectiveness has not yet been determined.

Geometry Tools

Geometry tools gather information about the physical shape, or geometry, of a pipeline. Geometry tools are primarily used to find "outside force damage," or dents, in the pipeline. However, they can also detect and locate mainline valves, or fittings. As with all ILI tools, geometry tools have limitations on their use and in the usefulness of the results obtained.

1. Caliper tools use a set of mechanical fingers or arms that ride against the internal surface of the pipe or use electromagnetic methods to detect dents or deformations.
2. Pipe deformation tools operate on the same principle of caliper tools with the addition of gyroscopes to provide the o'clock position of the dent or deformation in the pipe. This tool can also provide pipe bend information.
3. Mapping tools can be used in conjunction with other ILI tools to provide GPS correlated mapping of the pipeline and other physical details, for example, valves.

References

1. Dawson J, Bruce K, John DG. Corrosion risk assessment and safety management for offshore processing facilities. *Offshore Technology Report 1999/064*. Norwich, U.K.: Health and Safety Executive, 2001.
2. Moreland PJ, Hines JG. Corrosion Monitoring: Select the right system. *Hydrocarbon Processing*, 1978; 57: 251-255.
3. Horvath RJ. The role of the corrosion engineer in the development and application of risk-based inspection for plant equipment. *Materials Performance*, 1998; 37: 70–5.
4. Wyatt LM, Bagley DS, Moore MA, Baxter DC. *An Atlas of Corrosion and Related Failures.* St. Louis, Mo.: Materials Technology Institute, 1987.
5. Sachs NW. Understanding the multiple roots of machinery failures. *Reliability Magazine* 2002; 8: 18–21.

6. Singh B, Britton J, Poblete B, Smith G. The 3 R's—Risk rust and reliability. *CORROSION 2005*, Paper # 553. Houston, Tex.: NACE International, 2005.
7. McIntire DR, Dillon CP. *Guidelines for Preventing Stress Corrosion Cracking in the Process Industries*. Columbus, Ohio: Materials Technology Institute, 1985.
8. Queen DME, Ridd BR, Packman C. Key performance indicators for demonstrating effective corrosion management in the Oil and Gas Industry. *CORROSION 2001*, Paper # 056. Houston, Tex., NACE International, 2001.
9. McNeeney A. Selecting the right key performance indicators. *Maintenance Technology* 2005; 18: 27–34.
10. Marine Risk Assessment. Contract Research Report 2001/0631. Sudbury, U.K.: *Health and Safety Executive (HSE)*, 2002.
11. Wintle JB, Kenzie BW, Amphlett GJ, Smalley S. Best practice for risk based inspection as a part of plant integrity management. Contract Research Report 363/2001. Sudbury, U.K.: *Health and Safety Executive (HSE)*, 2001.
12. *Fault Tree Analysis Application Guide*. Rome, N.Y.: Reliability Analysis Center (RAC), 1990.
13. Roberge PR. *Handbook of Corrosion Engineering*. New York, N.Y.: McGraw-Hill, 2000.
14. Roberge PR. Modeling corrosion processes. In: Cramer DS, Covino BS, eds. *Vol. 13A: Corrosion: Fundamentals, Testing, and Protection*. Metals Park, Ohio: ASM International, 2003; 430–45.
15. Rivera M, Bolinger S, Wollenweber C. Carbonate cracking risk assessment for a FCCU gas plant. *CORROSION 2004*, Paper # 639. Houston, Tex.: NACE International, 2004.
16. *Risk-Based Inspection*. ANSI/API RP 580. 1st edn. Washington, D.C.: American Petroleum Institute, 2002.
17. *Report on Current Practices*. DNV Report No. 2004-0305. Hovik, Norway: DNV Library Services, 2005.
18. Vieth PH. Comprehensive, long-term integrity management programs are being developed and implemented to reduce the likelihood of pipeline failures. *Materials Performance* 2002; 41: 16–22.
19. Kroon DH. External corrosion direct assessment of buried pipelines: The Process. *Materials Performance* 2003; 42: 28–32.
20. ASME B31G—*Manual for Determining the Remaining Strength of Corroded Pipelines*. New York, N.Y.: American Society of Mechanical Engineers (ASME), 1992.
21. DNV-RP-F101 *Corroded Pipelines*. Høvik, Norway: Det Norske Veritas, 2004.
22. Roberge PR. *Corrosion Inspection and Monitoring*. New York, N.Y.: John Wiley & Sons, 2007.
23. NACE RP0502-2002—Standard Recommended Practice for Pipeline External Corrosion Direct Assessment Methodology. Houston, Tex.: NACE International, 2002.
24. Tadepally VP, Hendren ES. Internal corrosion direct assessment: Refining the method through new decision support models. *CORROSION 2005*, Paper # 178. Houston, Tex., NACE International, 2005.
25. Waker, S., Rosca, G, and Hylton, M. In-line inspection tool selection. *CORROSION 2004*, Paper # 168. Houston, Tex.: NACE International, 2004.

CHAPTER 13
Cathodic Protection

13.1 Cathodic Protection Historical Notes

The first application of cathodic protection (CP) can be traced back to 1824, when Sir Humphrey Davy, in a project financed by the British Navy, succeeded in protecting copper sheathing* against corrosion from seawater by the use of iron anodes. This limited use of CP on copper sheathing perdured and when wooden hulls were replaced by steel the fitting of zinc protector blocks on the sterns of naval vessels became traditional. These zinc slabs, although they offered some protection to steel hulls against local galvanic effects due to the presence of the bronze propellers, were generally not deemed to be effective.

This lack of efficiency was mainly due to the use of unsuitable zinc alloys and other factors such as insufficient appreciation of the technology of cathodic protection and the tendency to reduce the efficiency of the zinc material to zero by painting their surfaces [1]. From that early beginning, CP has grown to have many uses in marine and underground structures, water storage tanks, gas pipelines, oil platform supports, and many other facilities exposed to corrosive environments. More recently, CP has been proved to be an effective method for protecting and reinforcing steel from chloride-induced corrosion.

The CP effectiveness at protecting steel in soils has been demonstrated in the early 1940s when CP was applied to an old natural-gas piping network that had been developing leaks at a rapidly increasing rate, enough so that abandonment was seriously considered. The observed reduction in the number of leaks immediately after the CP installation was impressive [2]. A similarly impressive reduction in the frequency of leaks on a cast iron water main was achieved at about the same period as shown in Fig. 13.1 [3].

Modern specifications for the cathodic protection of active ocean-going ships were first described in 1950 [4]. Since that time progress has been rapid. Considerable advances in cathodic protection technology have been made, better sacrificial anode materials have

* Cladding with copper of the underwater hull of a ship was used on warships to prevent attack of the wooden hull by the Teredo worm in tropical waters.

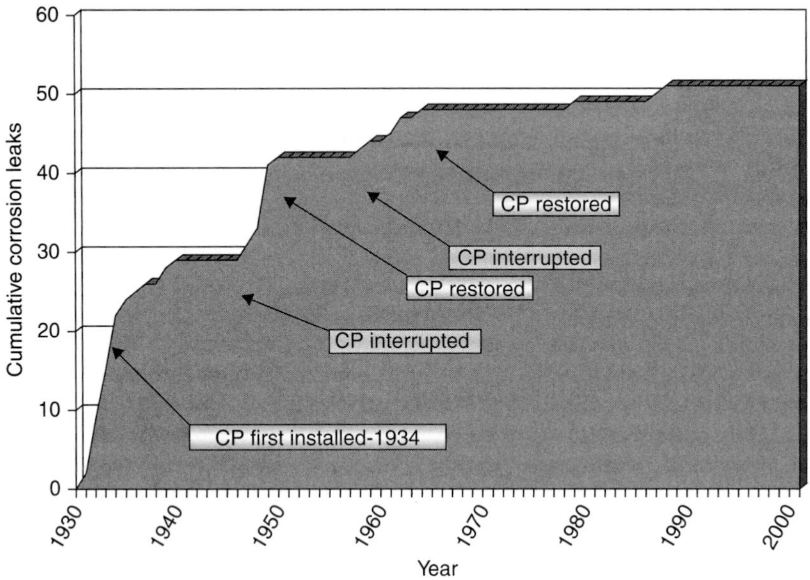

FIGURE 13.1 Leak history and cathodic protection on aqueduct no. 1. (Courtesy of East Bay Municipal Utility District, Oakland, California)

been developed, and circuits for the use of controlled applied current systems, using inert anodes, have been perfected.

The first reinforced concrete-impressed current CP system was an experimental system installed on a bridge support beam in 1959 [5]. A more advanced system was subsequently installed on a bridge deck in 1972 [6]. The anode system used in both applications was based on a conventional-impressed current CP system for pipelines, but "spread out" over a bridge deck. CP has since then become one of the few techniques that can be applied to control corrosion on existing structures.

13.2 How Cathodic Protection Works in Water

The basic principle of all CP techniques is that the unwanted anodic corrosion reactions are suppressed by the application of an opposing current forcing the local anodes to be polarized to the potential of the local cathodes therefore stifling corrosion cells. If less than this amount of cathodic current is supplied some corrosion would still occur, but the level of corrosion would be less without any CP. From a thermodynamics point of view, the application of a CP current basically reduces the corrosion rate of a metallic structure by reducing its corrosion potential toward its immune state (see Figs. 4.14 and 4.17). The two main methods of achieving this goal are by either:

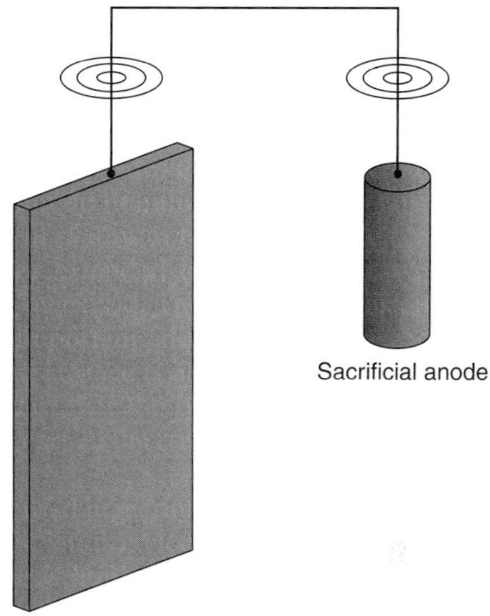

FIGURE 13.2 Schematic description of a sacrificial anode CP system.

- Using sacrificial anodes (Fig. 13.2) that have a corrosion potential lower than the metal to be protected. This ranking can be obtained by consulting the seawater galvanic series.
- Using an impressed current cathodic protection (ICCP) provided by an external current source (Fig. 13.3).

These simple schematics are neglecting one important component that may be a useful monitoring tool for sacrificial CP systems but is a must for ICCP for which controlling the potential applied to a protected structure is essential. This tool called a reference electrode is mentioned in the next sections, but details of its usage and functionality are described in Sec. 13.5.1.

13.2.1 Sacrificial Cathodic Protection

Sacrificial anodes are relatively inexpensive, easy to install, and in contrast to impressed current systems, can be used where there is no power supply. The method has the added advantage that there is no expensive electrical equipment to buy and current cannot be supplied in the wrong direction. Sacrificial anodes are very suitable in small-scale applications (Fig. 13.4), though they are also used extensively and with equal effect on large-scale structures (Fig. 13.5).

The use of sacrificial anodes to protect ship hulls has become less favored than impressed current techniques, but it is still found on smaller vessels, where the impressed current method is uneconomical. Zinc is the most common anode material for seawater applications while aluminum and magnesium provide a higher voltage for less

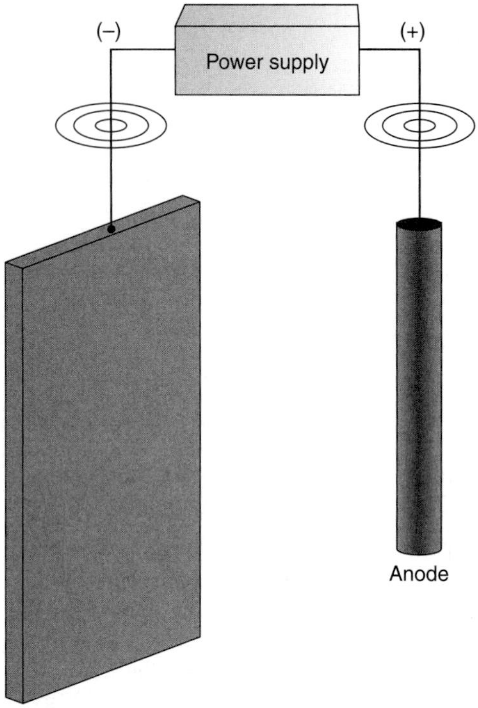

FIGURE 13.3 Schematic description of an impressed current CP system.

FIGURE 13.4 Sacrificial zinc anodes on a sailboat. (Courtesy of Kingston Technical Software)

FIGURE 13.5 Sacrificial zinc anodes on the interior wall of a drydock water handling pipe. (Courtesy of Defence R&D Canada-Atlantic)

conductive waters. Industry data have provided estimates for the 1998 sales of various hardware components totaling $146 million. The largest share of the cathodic protection market is taken up by sacrificial anodes at $60 million, of which magnesium has the greatest market share. Major markets for sacrificial anodes are the water heater and the underground storage tank markets [7].

The oil and gas industries have probably been responsible for the greatest seawater applications of sacrificial anodes in seawater applications. New technologies had to be developed to support the exploitation of deep sea resources which had expanded at a great pace since the mid 1970s. Corrosion protection of the expensive and intricate structures had to be based on CP systems, for which the available scientific data were sparse. Designs were often based more on inspired guesswork than on the application of science, particularly for impressed current systems. Designers preferred to use copious quantities of inexpensive zinc anodes in the belief that overprotection was safer than the risk of underprotection [8].

13.2.2 Impressed Current Cathodic Protection

ICCP of ships is always used in conjunction with protective coatings. The coatings are intended as the primary protection, and the ICCP is a backup in those areas where coating defects may be present. In the period immediately subsequent to the application of the coating, there is very little demand on the impressed current system. As the coating deteriorates during the operational life of the ship, the CP current demand increases. Eventually, the current demand placed may exceed the capabilities of the design, with high anode currents causing even more damage to the coating, particularly in areas adjacent to the anodes, where the concentration of hydroxyl ions may be very high. In addition, the high local currents around the anodes may reduce the protection supplied to the rest of the structure.

On a supertanker, the initial current of 10 A may rise to over 1000 A during the course of its operational life. Modern ICCP ship designs usually place anodes in symmetrical dispositions, but in bulk carriers, there is a need for internal access and for cable-runs to be away from anodes and reference electrodes. This usually precludes electrodes from being sited external to storage tanks. Instead electronics are placed either well forward or well aft, where the adjacent machinery spaces provide convenient access to the various pieces of equipment (Fig. 13.6). There are several reasons why potentiostatic systems frequently fail to provide adequate protection [8]:

- The galvanic effect of unpainted bronze propellers
- Varying seawater velocities
- Increasing exposure of bare steel as a result of paint damage
- Reference electrodes monitoring only in their own vicinity
- Poor positioning of zones, reference electrodes versus anodes

The underwater area of a ship is a large complex cathode with at least three components: painted steel, bare steel, and bronze (Fig. 13.7).

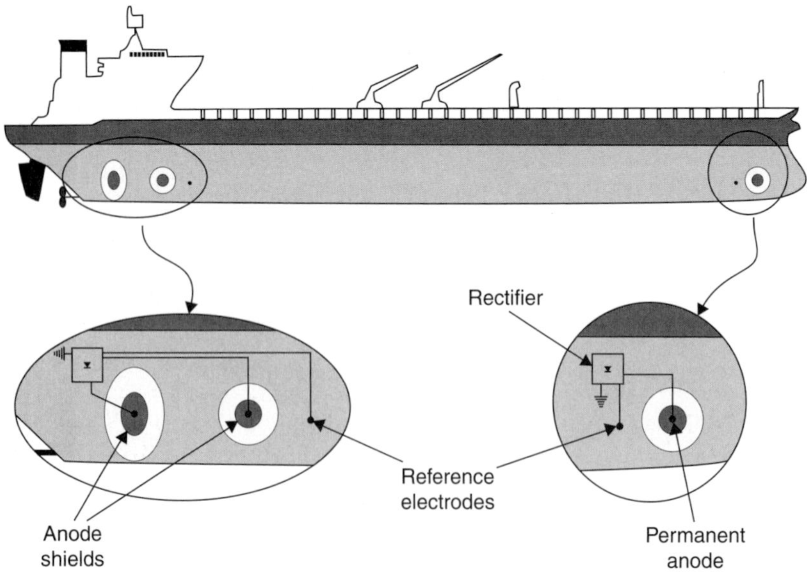

FIGURE 13.6 Typical ICCP system design for a commercial crude oil supertanker.

Figure 13.7 A naval ship during refit. (Courtesy of Kingston Technical Software)

These have different current density requirements for achieving the desired polarization and respond differently to changes in operational conditions, particularly seawater flow.

It was found, for example, that a protection current density of 30 mA/m^2 in quiet seawater rises to over 110 mA/m^2 at 15 knots and above. Ship ICCP anodes are few in number and very small in relation to the cathodic surface area, thus the distribution of potential from them is inevitably uneven. In the design process, it is the difficulty of knowing the expected potential distribution over the structure that leads to reliance on current density measurements as a means of assessment. Yet many ship ICCP designs utilize no more than two or three reference electrodes and these electrodes can only measure the potential in their adjacent vicinity.

The most significant part of a ship in determining the potential distribution over the hull is the stern area where propellers are located. Modeling techniques have produced accurate simulations of data measured on a real warship under static conditions. In conditions which simulated cruising speed, the protection system was found unable to provide the required potential at the stern [Figure 13.8(*a*)]. By adding an additional control electrode and repositioning the others, good protection levels could be obtained over the whole of the model, under both static and dynamic conditions. This demonstrates the critical nature of the reference position to the effective operation of the system [Figure 13.8(*b*)].

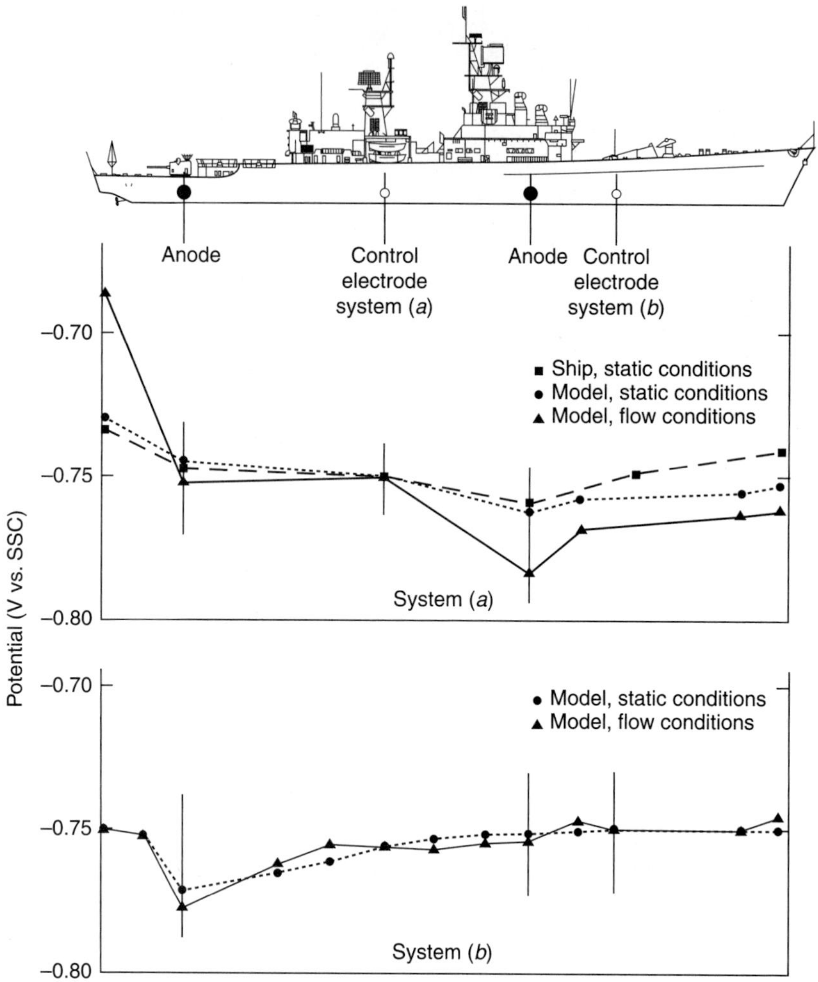

FIGURE 13.8 Effect of reference electrode location on potential distribution on a ship hull in both static and flow conditions: (*a*) with one reference electrode amidships; and (*b*) with an extra reference electrode aft and the original electrode repositioned ahead of the forward anode.

13.3 How Cathodic Protection Works in Soils

Concentration cells, such as shown in Fig. 13.9 are responsible for much of the corrosion which takes place in soils. Pipelines, for example, pass through different soils on their way across the land. Oil and gas well casings penetrate various strata of different compositions. In every case where two different soils are in contact with a single piece of metal, a concentration cell is possible.

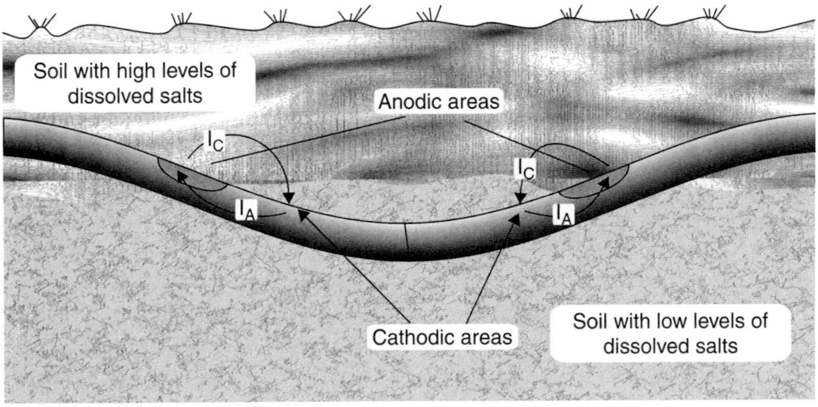

FIGURE 13.9 DC (I_C) and (I_A) flow, respectively, through earth from anodic area to cathodic area and through the pipe steel from cathodic area back to anodic area to complete the circuit. The anodic area, where current leaves the steel to enter the surrounding earth is where corrosion occurs.

In most cases occurring in naturally occurring soils, that portion of pipe lying in the more conductive soil is the anode and the portion in the less conductive soil is the cathode. Moist soils themselves act as the electrolyte and the pipe provides the connecting circuit. As shown in Fig. 13.9, current flows from the anodic area to the soil, then through the soil toward the cathodic area where it comes back through the pipe itself to complete the circuit. For a given situation, the amount of current flowing is limited by such factors as the resistivity of the environment and the degree of polarization at anodic and cathodic areas.

Corrosion occurs where the current discharges from the metallic structure into the soil at anodic areas. Where current flows from the environment onto the pipe (cathodic areas), there is no corrosion. In applying CP to a structure, then, the objective is to force the entire structure surface exposed to the environment to collect current from the environment. When this condition has been attained, the exposed surface becomes a cathode, and corrosion is then successfully mitigated. This is illustrated in Fig. 13.10, which shows how the originally corroding section of pipe used in Fig. 13.9 becomes a cathode with cancellation of all current-discharging areas on the pipe surface.

It can be seen in Fig. 13.10 that the CP current must flow into the environment from a special ground connection (usually called a groundbed) established for the purpose. By definition, the materials used in the groundbeds are anodes, and material consumption (corrosion) must occur. Corrosion has therefore not been really eliminated by the application of CP, but more or less transferred from the structure to be protected to known locations (the groundbeds) designed to discharge the CP current for a reasonably long time, and which, when consumed, may be replaced without interruption of the

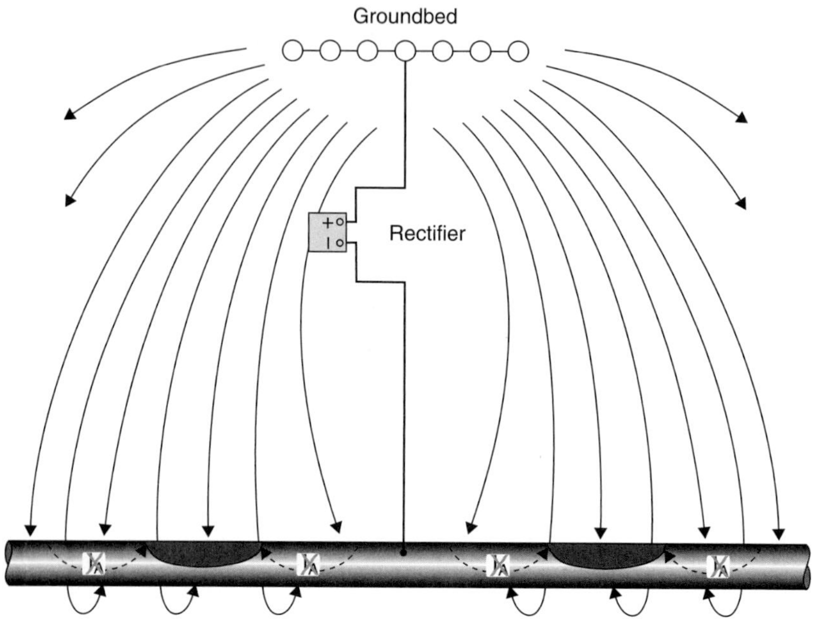

Figure 13.10 Basic impressed current cathodic protection installation. Shaded area on pipeline indicates anodic areas prior to application of CP and dotted lines represent current discharge from anodic areas which have been eliminated.

normal function of the structure being protected. While the basic principle of applying CP is quite simple, some practical questions have to be resolved, such as

- How much minimum current is needed to protect a given structure?
- What source of DC should be used?
- How should the installation actually be designed?
- How can it be ascertained, on a completely buried structure, whether or not the entire structure surface has, in fact, been made a cathode?

There are limitations to what CP can do. The CP current must be able to flow through a conducting environment onto the metallic surface being protected. Using the same pipeline example, a basic CP installation such as shown in Fig. 13.10 would protect only those external surfaces of the pipe that are in contact with the conducting environment. The internal surfaces of the pipe would receive no protection. This would remain true even if the pipe contained a conductive material since the metallic pipe would intercept the protective current and simply carry it back to its source. Similarly, any portions of the pipeline system in air (such as aboveground portions, valves, aboveground header manifolds) cannot receive any of the CP protection current since air will not carry current to the unburied surfaces.

Another example where CP has important limitations is for congested systems for which the current flow to the innermost structures is limited. The amount of protective current reaching hidden structures may not be sufficient to adequately protect them, even though the outermost surfaces would be fully protected. This is referred to as electrical shielding since the metallic structure surfaces on the outermost portions of the congested area intercept the flow of protective current and therefore shield the innermost structures. Piping in congested areas, such as pumping stations and tank farms, may encounter such shielding.

Figure 13.11 illustrates a condition in which a network of piping in a restricted area is protected by a remote-type groundbed. The remote groundbed and suitably sized rectifier may change the potential of the entire structure sufficiently to give an indication of full protection when measured with respect to a remote reference electrode, such as the −1.5 V indicated in Fig. 13.11. But, if a measurement is made between the remote earth and the earth in the midst of the congested area, the difference in potential between these two earths can be significant as indicated by the −0.8V reading in Fig. 13.11 [2].

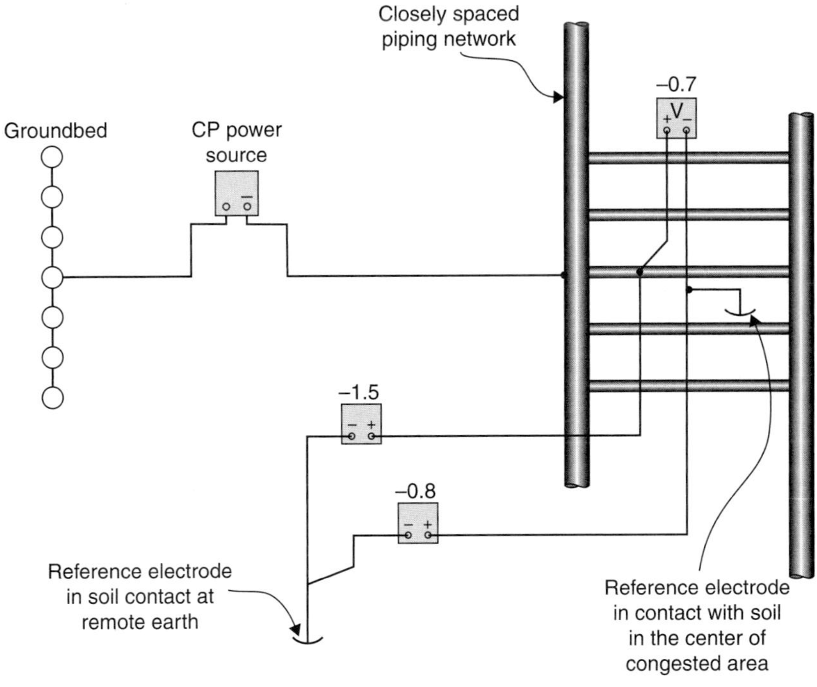

Figure 13.11 Electrical shielding in congested areas.

When this occurs, there may be relatively low potentials between the pipe and adjacent earth. This is indicated by the –0.7 V reading in Fig. 13.11, which is less than the full protection criterion of –0.85 V versus a copper sulfate reference electrode. The shielding effect will tend to be greatest near the center of the congested area.

Where congested area shielding is a problem, it may not be practical to rely solely on a remote CP system. However, CP can still be attained by the use of additional anodes (either galvanic or impressed current) distributed strategically throughout the congested area in such a way that the areas of influence surrounding the anodes overlap sufficiently to permit development of protective potentials on the innermost structures.

13.3.1 Sacrificial Cathodic Protection

Figure 13.12 illustrates the essential differences between sacrificial CP and ICCP systems for soil applications. As shown in Fig. 13.12, a voltage can be measured between the corroding structure and the material to be used as an anode. The structure must be positive (+) with respect to the anode before the anode can discharge current. There are certain requirements for a metal to be a practical galvanic anode material:

- The potential between the anode and corroding structure must be large enough to overcome the anode-cathode cells on the corroding structure.
- The anode material must have sufficient electrical energy current to permit reasonably long life (10 to 15 years plus) with a practical amount of anode material.
- Anodes must have good efficiency, meaning that a high percentage of the electrical energy content of the anode should be available for useful CP current output. The balance of the energy that is consumed in self-corrosion of the anode itself should be ideally small.

13.3.2 Impressed Current Cathodic Protection

With the ICCP system shown in Fig. 13.10, the groundbed anodes are not really providing the driving power. Instead, some external source of direct electric current is connected (or impressed) between the structure to be protected and the groundbed anodes. The positive terminal of the power source is connected to the groundbed, which is then forced to discharge as much CP current as is desirable. If a mistake is made and the positive terminal is connected erroneously to the structure to be protected, the structure will become an anode instead of a cathode and will corrode actively.

Groundbed anodes forced to discharge current will corrode. It is important to provide anode materials which are consumed at

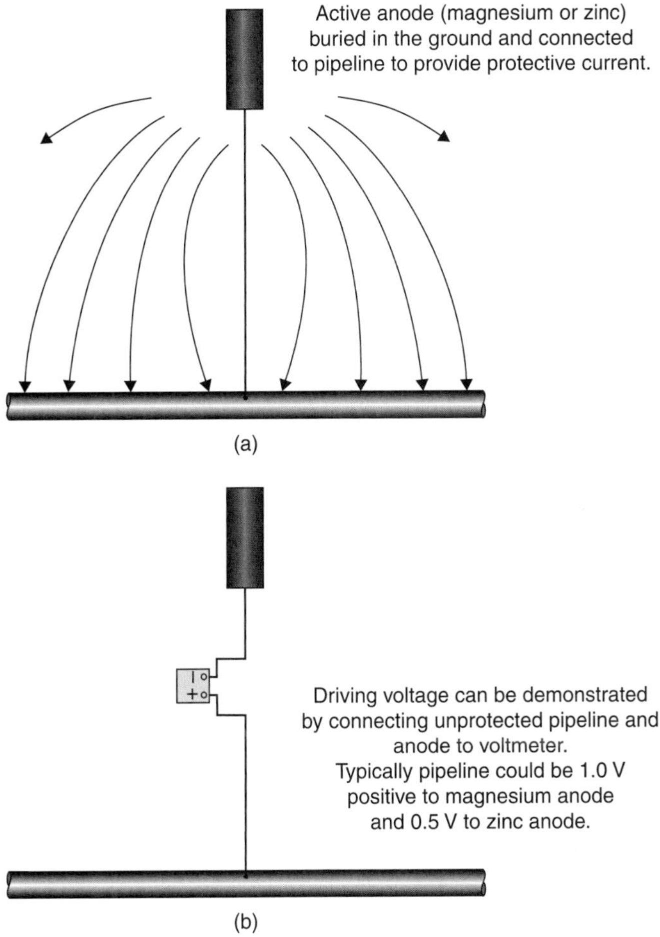

Figure 13.12 CP with galvanic anodes. (*a*) active galvanic anode buried in earth and connected to pipeline to provide protective current; (*b*) driving voltage measured between unprotected pipeline and anode with voltmeter.

relatively low rates and thus permit designing groundbeds that can discharge large amounts of current and still have a long life expectancy. While scrap steel pipe, rail, rod, or other similar iron or steel material may be used, such materials are consumed at the rate of about 44 kg/A-y (1 A flowing for 1 year). This means a relatively large amount of material is needed to achieve the desired operating life for an installation, something that was justified only when scrap steel was available at a low cost. Additional details of ICCP anode materials are provided later in the chapter.

13.3.3 Anode Beds

The type of anode bed used in an application is a key element of the ICCP design. Anode beds are located either close to the structure being protected or remote in order to distribute the current over the area containing the structures to be protected. Anode beds are classified as follows [9]:

- Local distributed anode-bed system
- Vertical distributed anode-bed system
- Horizontal distributed anode-bed system
- Deep anode-bed system
- Remote anode-bed system

If the anodes are installed vertically in 3 to 6 m deep holes about 5 to 6 m away from the structural elements to be protected, the system is called a vertical distributed anode-bed system. When the anodes are installed in a cluster, away from the structures that are being protected, the system is called a remote anode-bed system. A deep anode-bed system is a variation of a remote anode-bed system in which anodes are located remotely from the structures being protected in a vertical plane.

The location of the anode bed relative to the structure being protected and other structures in the area is the single most important factor in minimizing shielding issues. In power plants, for example, the current discharge from remote anode beds is shielded from spreading over a large area because of the shielding effects of reinforcements in the building foundations, duct runs, and other buried structures. When installing a new CP system, the influence on foreign structures should be a primary consideration in choosing the location of anode beds.

Local distributed anode-bed systems are the most common design. This type of bed can be installed either vertically or horizontally depending on their position in the soil. Horizontal installations may be necessary when obstructions or other soil conditions make digging holes difficult. Horizontal installations are also used where soil resistivities are very low and the increased resistance of the horizontal installation is not significant. A minimum of 0.6 m of burial for all cables and 0.9 m of burial for the anode is recommended. Figure 13.13 shows a typical local horizontal distributed anode bed.

Vertical installations are used when there is a need to obtain lower resistivities, when there are difficulties with trenching for a horizontal bed, or when the physical surface area is limited. Generally, due to their length, vertical bed anodes are brittle and must be carefully handled to prevent breakage. Figure 13.14 shows a typical local distributed vertical anode system. The anode cable is prone to failure if the insulation is damaged in any way. As a result, particular care should be exercised in handling the anode leads.

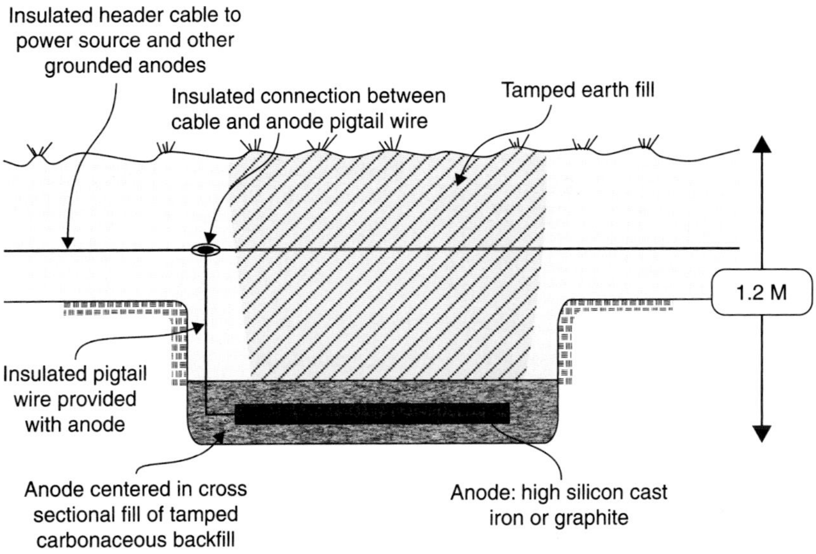

FIGURE 13.13 Typical horizontal impressed current anode installation.

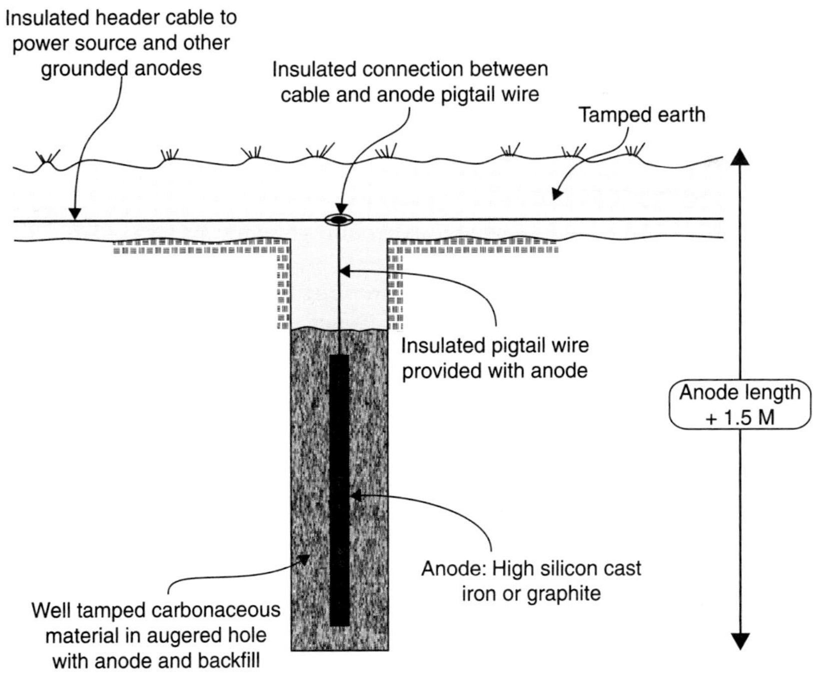

FIGURE 13.14 Typical vertical impressed current anode installation.

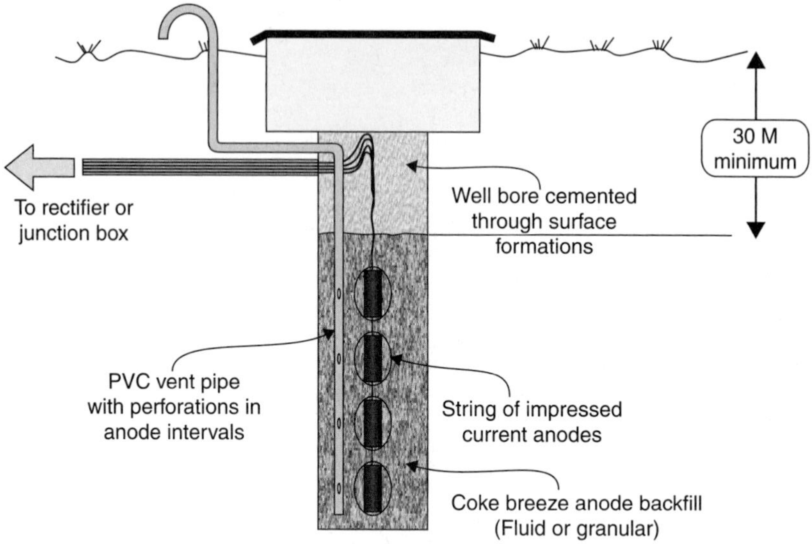

FIGURE 13.15 Typical deep bed anode in normal soil strata.

Anode beds are sometimes installed deep below the surface (deeper than 30 m) (Fig. 13.15), a design that is particularly useful for installations where electrical interference problems are severe or where the resistivity of the soil near the surface is high. The deep installation causes the current flowlines to become parallel and thus provide a more uniform distribution of the protective current.

13.3.4 Anode Backfill

Chemical backfill is widely used to surround galvanic anodes when they are buried in earth. There are several advantages to this. The special backfill provides a uniform environment to promote uniform anode consumption and maximum efficiency. It isolates the anode material from direct contact with the earth thus preventing negative effects by soil minerals that might otherwise build up high-resistance films on the anode surface. It has a low electrical resistivity that provides a lower anode-to-earth resistance and greater current output for cases where the surrounding soil is of higher resistivity.

A typical backfill material for magnesium anodes consists of 75 percent hydrated gypsum, 20 percent bentonite clay, and 5 percent sodium sulfate. A mixture of 50 percent molding plaster (plaster of Paris) and 50 percent bentonite clay works well with zinc anodes. These backfill mixtures are available from suppliers or are provided in prepackaged containers as shown in Fig. 13.16. Figure 13.17 illustrates how these anodes are installed in their prepackaged backfill containers.

Figure 13.16 Ductile iron municipal watermain protected with tape wrap and sacrificial magnesium anodes. (Courtesy of Dean Rookes, West Coast Corrosion Prevention Ltd.)

(a)

Figure 13.17 Holes being drilled for sacrificial anode installation (*a*); thermit welding of anode connection on watermain (*b*); anode cluster ready to be buried and connected to fire hydrant tap. (Courtesy of Drinking Water Services, City of Ottawa)

(b)

(c)

Figure 13.17 *(continued)*

For ICCP anode systems in soil, it is common practice to use a coke or graphite backfill to surround anodes (Fig. 13.18). There are two advantages to this. One is that graphite or suitable coke has a very low electrical resistivity (lower than most soils) and helps to produce low resistance between anode and earth. The second advantage is that if the coke or graphite is solidly packed around the impressed current anode, the greater part of the current discharged by the anode passes from the anode to the backfill through direct contact between the anode material and the backfill particles. This reduces the consumption rate of the anode. The current discharging from the outer surfaces of the backfill column will, however, consume the backfill material at a slow rate. The net result is a lower resistance, longer life anode installation than without the presence of backfill.

Suitable coke or graphite backfill materials are available from suppliers. Coke should be either coal coke or calcined (heat-treated) petroleum coke. Non-calcined petroleum coke can have high electrical resistivity and be unsuitable for groundbed backfill use.

FIGURE 13.18 Conventional groundbed using petroleum coke backfill and 5 cm cast iron anodes. (Courtesy of Dean Rookes, West Coast Corrosion Prevention Ltd.)

13.4 How Cathodic Protection Works in Concrete

The main causes of corrosion of steel in concrete are chloride attack and carbonation. These two mechanisms are unusual in that they do not attack the integrity of the concrete. Instead, aggressive chemical species pass through the pores in the concrete and attack the steel. This is unlike normal deterioration processes due to chemical attack on concrete.

Most problems associated with corrosion of steel in concrete are not due to the direct loss of steel but to the growth of the corrosion oxide products which eventually can lead to cracking and spalling of the concrete cover (Fig. 13.19). Structural collapses of reinforced concrete structures due to corrosion are rare. The most common problem caused by corrosion is spalling of concrete cover, which has caused a few deaths and accidents in recent history. It has been suggested that less than 100 μm of steel-section loss is needed to start cracking and spalling of the concrete. The actual amount needed will depend on the geometry in terms of cover, proximity to corners, rebar spacing, bar diameter, and the rate of corrosion [10].

A number of fundamental measures can be taken to address the problem of reinforcing steel corrosion. However, an important

FIGURE 13.19 Corrosion ravaged columns of a highway bridge in Illinois. (Courtesy of Vector Corrosion Technologies)

distinction has to be made in the applicability of remedial measures between new and existing structures. Unfortunately, the options in the following list of remedial measures are more limited for fixing existing aging structures than for designing new constructions [11]:

- Repairing the damaged concrete
- Modifying the external environment, for example, choosing a different deicing agent or changing deicing salt practices (however this will not remove salt already in the concrete).
- Modifying the internal concrete environment, for example, realkalization of concrete, chloride extraction, addition of corrosion inhibitors, and so on.
- Creating a barrier between the concrete and the external environment.
- Creating a barrier between the rebar steel and the internal concrete environment, for example, epoxy coating and galvanization.
- Applying CP to the rebar structure.
- Alternatively, more corrosion-resistant rebar materials, for example, stainless steel.
- Alternative methods of reinforcement, for example, fiber-reinforced polymer composites.

The following sections describe how CP can be used to protect reinforced concrete structures against corrosion by using either an ICCP system or sacrificial anodes.

13.4.1 Impressed Current Cathodic Protection

ICCP consists in polarizing the rebar with a DC current source (rectifier) to a cathodic potential where anodic dissolution of the rebar is minimized (Fig. 13.20). Three basic methods are available for controlling the output of a rectifier:

- *Constant current mode:* In this mode the rectifier maintains a constant current output. The output voltage then varies with changes in the circuit resistance. The potential of the reinforcing steel can be measured with a reference cell as a function of the applied current to ensure that certain protection criteria are met.
- *Constant voltage mode:* In this mode a constant output voltage is maintained by the rectifier. The applied current then changes with variations in circuit resistance. Low concrete resistance, often associated with increased risk for corrosion damage, can result in increased current output. It should be noted that in this mode, the rebar potential is not necessarily constant. It can again be monitored with a reference cell.

Figure 13.20 Impressed current cathodic protection being retrofitted on an old bridge structure: note vent tubes and drill holes anodes. (Courtesy of Vector Corrosion Technologies)

- *Constant rebar potential mode*: In this mode the current output is adjusted continuously to provide a constant (preselected) rebar potential. A method that may be used for immersed concrete where potentials are relatively uniform and stable, the rebar potential, measured continuously with reference electrodes, is fed back to the rectifier unit. Successful operation in this mode depends on minimizing the IR drop error in the rebar potential measurements and on the accuracy and stability of the reference electrodes over time.

An important issue in ICCP of reinforcing steel is how much current should be impressed between the reinforcing steel and the anode. Too little current will result in inadequate corrosion protection of the rebar, while excessive current can result in problems such as hydrogen embrittlement and concrete degradation. Hydrogen embrittlement of high-strength steel is a particularly severe problem for pre- and post-tensioned concrete systems. The current densities involved in meeting commonly used protection criteria are typically around 10 mA/m^2 of rebar surface. Uniformity in the current distribution is obviously a desirable feature.

Unfortunately, the current requirement cannot be measured directly and various indirect criteria have been proposed (Table 13.1). The ICCP current requirements are often expressed in terms of the potential of the reinforcing steel relative to a reference electrode, or alternatively as a shift in this potential when the ICCP system is activated or deactivated. Reference electrodes can be located externally by being placed in contact with the outside concrete surface, or they can be embedded directly in the concrete with the rebar.

Criterion	Details	Comments
Potential shift	100 mV shift of rebar potential in the positive direction when system is depolarized	Depolarization occurs when CP current is switched off. Time period required for rebar to depolarize is debatable. The potential reading before interrupting the CP current should be "IR" corrected.
Potential shift	300 mV shift of rebar potential in the negative direction due to application of CP current	The potential reading with the CP current on should be "IR" corrected. The method relies on a stable rebar potential before the application of CP current.
E-log(i) curve	The decrease in corrosion rate due to the application of CP current can be determined provided the relationship between rebar potential (E) and current (i) can be measured and modeled. A simple model is Tafel behavior with a linear relationship between E and log(i).	This methodology is structure specific and the measurements involved are relatively complex and require specialist interpretation. Ideal Tafel behavior is rarely observed for steel in concrete.
Current density	Application of 10 mA/m^2 of rebar surface area	Empirical approach based on limited experience. Does not consider individual characteristics of structures and environments.

TABLE 13.1 Cathodic Protection Criteria for Steel in Concrete.

13.4.2 Sacrificial Cathodic Protection

One alternative to impressed current CP is galvanic or sacrificial anode CP. For steel in concrete, sacrificial anodes are typically made of zinc or a zinc based alloy. While ICCP systems have been widely used since the 1980s, sacrificial CP systems have only become available around the end of the 1990s. Various configurations exist, including

- Thermal sprayed zinc, a few hundred microns in thickness applied directly to the concrete surface (Fig. 13.21). A humectant may be added for increasing moisture, reducing concrete resistance, and increasing the current level.
- Thin sheets of zinc with a conductive adhesive backing applied to the concrete surface.
- Expanded zinc mesh in a glass-reinforced plastic permanent form filled with a proprietary cementitious grout, usually applied to marine exposed piles in the splash and tidal zone.
- Zinc disks in a proprietary high alkaline mortar jacket embedded in patch repairs to prevent the ring anode or incipient anode effect discussed earlier in this section (Fig. 13.22).
- Zinc plugs in a cylinder of a proprietary high alkaline mortar for installation in cored holes in the concrete (Fig. 13.23).

FIGURE 13.21 Application of activated arc sprayed zinc on a reinforced concrete structure. (Courtesy of Vector Corrosion Technologies)

FIGURE 13.22 Zinc anode being connected to the rebar structure to control on-going corrosion and prevent the initiation of new corrosion activity in concrete structures. (Courtesy of Vector Corrosion Technologies)

FIGURE 13.23 Embedded zinc anodes used in concrete rehabilitation to prevent the formation of new corrosion sites adjacent to completed patch repairs. (Courtesy of Vector Corrosion Technologies)

The main advantage of sacrificial CP over ICCP is that it does not require a power supply or control system and therefore has low monitoring requirements. Its disadvantages are that it cannot be controlled to ensure complete corrosion control and that the anode types and anode lives are more limited.

13.5 Cathodic Protection Components

13.5.1 Reference Electrodes

The main criterion for monitoring and controlling the cathodic protection of most metals is based on a measure of the potential of the metal relative to a half cell or reference electrode. As described in Chap. 4, Corrosion Thermodynamics, reference electrodes consist of a metal electrode immersed in a specific solution of its own ions and this solution is connected to the corroding electrolyte through a salt bridge. This bridge may consist of some electrolyte-holding medium such as a gel, a sintered or porous ceramic, or a small wood plug saturated with the reference electrode solution, the corroding electrolyte, or a liquid compatible with both. An ideal reference electrode would possess the following properties [12]:

- Its potential should be constant irrespective of the electrolyte or environment in which it is used and it should not vary greatly with changes in electrolyte temperature or other parameters.
- Any changes should be predictable and should have no hysteresis effect.
- The device should not polarize at small currents either when these are cathodic or anodic.
- It should have a low internal resistance.
- These properties should be constant with time.
- A robust construction.
- It should be easy to replenish in the field if necessary.

The reference electrode generally used for monitoring cathodic protection in soils is the copper sulfate electrode. This consists of an electrode of electrolytic copper in a solution saturated with copper sulfate crystals. However, the most universal reference electrodes are of the silver chloride type. These consist of a silver wire, plate, or mesh coated with silver chloride so that both silver and silver chloride are in contact with a chloride ion rich electrolyte. The potential of the Ag/AgCl reference electrode will depend upon this chloride ion concentration.

A commonly used method for preparing the silver chloride electrode is by dipping a clean silver wire spiral or welded mesh into molten silver chloride, which had been heated some 50°C above its melting point in a pure silver or quartz crucible. The silver is removed from the melt and the majority of the molten salt shaken off leaving only a thin silver chloride film. This is then made the cathode in a solution of sodium chloride and some of its surface is reduced to silver.

The accuracy of the silver chloride reference electrode depends upon the amount of chloride ion that is present in the solution and the accuracy with which it is controlled. In practical applications two concentrations of chloride ion are popular. In the first the electrolyte surrounding the silver chloride element is seawater. The other type of reference electrode uses a saturated sodium or potassium chloride solution. Both silver chloride cells have large temperature coefficients but there is no hysteresis or other effects so that these variations can be calculated. These reference electrodes typically have a higher resistance than the copper sulfate cells and this resistance is a function of the thickness of the silver chloride layer.

A piece of zinc may also be used as a reference electrode by immersing it in an electrolyte that contains some quantity of chloride or sulfate ions or it can be used in the ground if surrounded by a suitable backfill. The anode backfill for zinc is usually bentonite, sodium sulfate, gypsum, and sodium chloride. Zinc electrodes made from high purity material usually provide a stable potential that is constant *t* within *a*5 mV bracket. The electrode may slowly polarize and may have marine growth on it in seawater but this is generally prevented if the zinc is allowed to act as an anode at a very low current density.

Often it is necessary to measure potentials where the electrolyte is not easily accessible as, for example, in the water box of a condenser or on the inside of a steel pipe carrying seawater. Permanent reference electrodes could be placed inside the box or pipe but this is expensive and no maintenance can be carried out on the electrode.

In such cases, the reference electrode can be extended by a long salt bridge which consists of a flexible plastic tube with a porous plug at its end and filled with a high conductivity solution compatible with the reference electrode. The size of the plug will be decided by two considerations, the resistance of the plug through the electrolyte and the uniqueness of the potential that the cell measures.

Most electrodes will need inspection and recharging of the electrolyte and some will need to have the metal cleaned or treated; preferably the reference electrode should be designed so that it can be taken apart for inspection and cleaning. For use on high resistivity ground, and particularly on pavements, or concrete, there is a need to enlarge the contact area. This can be done by the use of sponges following prewetting of the surface.

Measuring the potential of a buried structure is done by contacting the earth (or other conducting environment) with a reference electrode. Figure 13.24 shows how the measurement is usually taken. It is not necessary to separate the two half-cell potentials since only changes in the structure-to-earth measurements are of interest and these changes show up well enough in the total value measured in this fashion.

If we are talking about a steel structure, a value of –0.85 V or more on the voltmeter of Fig. 13.24 would indicate full cathodic protection. This is considered to mean that the structure potential is –0.85 V with respect to a copper sulfate electrode and is based on the value –0.80 V assigned to the most highly anodic steel found in practical situations.

When metals other than steel are to be cathodically protected in soils, different potentials to a copper sulfate reference may be used as a criterion of protection. For lead-sheathed cables, a customary value is –0.70 V. For aluminum, the potential should be held between the limits of –1.00 and –1.20 V.

Notice the maximum negative limit given for aluminum. This is because alkalinity, which is corrosive to aluminum, builds up at the aluminum surface due to the forced cathodic current. Aluminum and

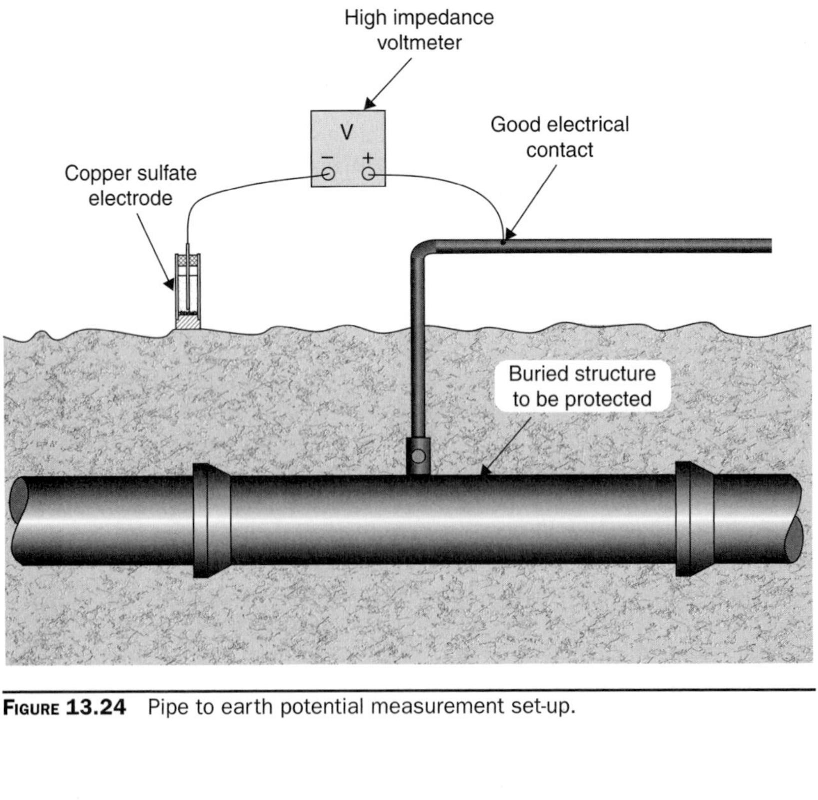

FIGURE 13.24 Pipe to earth potential measurement set-up.

other amphoteric metals such as zinc and magnesium may actually corrode more rapidly if subjected to excessive cathodic protection than if not protected at all. Steel is not subject to this effect. Lead may be to some extent, but usually is not a problem unless the CP current source is interrupted for an extended period. In this case, the alkalinity developed at the lead surface will have an opportunity to corrode the lead.

13.5.2 Anodes

Sacrificial Anodes

Materials suitable for use as galvanic anodes include aluminum, magnesium, and zinc. According to Faraday's law, the amount of electrical energy that can be obtained from a galvanic anode depends on the electrochemical equivalent of the metal used and the efficiency of the working anode. A useful unit to characterize the capacity of an anode is the ampere-hour (Ah) that corresponds to 1 A flowing for 1 hour, or any combination that will be equal to this, such as 0.5 A flowing for 2 hours, 2 A flowing for 0.5 hour, and so forth.

The efficiency of a galvanic anode is the ratio of an anode weight sacrificed for CP purposes divided by the total theoretical ampere-hours or capacity of the material actually consumed. Galvanic anodes are subject to self-corrosion which uses some of their capacity. This is why the efficiency is less than 100 percent.

For example, pure zinc has a theoretical maximum capacity of 820 Ah/Kg. This means that if a zinc anode were to discharge 1 A continuously, 1 kg would be consumed in 820 hours. If this kilogram was discharging 1/10 A, it would be totally consumed in 8200 hours or 48 weeks. Actually, zinc anodes operate, typically, at about 95 percent efficiency. This means that the energy content available for useful current output would be 820 × 0.95, or 779 Ah/kg.

Another useful way of expressing the anode consumption is in terms of kg per ampere-year. At 779 Ah/kg useful output for zinc, the conversion would be

$$\frac{8760\ (\text{h year}^{-1})}{779\ (\text{Ah kg}^{-1})} = 11.2\ (\text{kg A}^{-1}\text{year}^{-1}) \tag{13.1}$$

This equation indicates that 11.2 kg of zinc would be consumed from an anode discharging 1 A for 1 year.

Zinc anodes are made of high-purity zinc (99.99 percent purity or better) for soil use. For marine use, small amounts of aluminum and cadmium may be added to ensure maximum efficiency. If zinc of lesser purity is used, efficiency may suffer and the anodes may tend to become passive and cease to discharge useful amounts of protective current. Zinc anodes' working potential with respect to a copper sulfate electrode is in the order of −1.10 V.

Aluminum anodes have a theoretical capacity of 2960 Ah/kg. In recent years, aluminum anodes for seawater use have been developed that can operate at 95 percent efficiency with a useful output of approximately 2800 Ah/kg. Working potential is about the same as that of zinc, although there are variations depending on the alloy used. Aluminum anodes may have small amounts of alloying elements and are sometimes heat treated in order to obtain desired characteristics.

Magnesium anodes have a theoretical capacity of 2200 Ah/kg. Efficiency varies with current output density in terms of amperes per square meter of anode surface, but typically can be about 50 percent (useful output of 1100 Ah/kg) at a current output density of 0.3 A/m^2. Anodes' working potentials to copper sulfate electrodes can range from approximately −1.45 V for standard alloy magnesium to approximately −1.70 V with proprietary alloys also called *high-potential anodes*. Standard magnesium anodes have appreciable amounts of alloying materials. Typically, these may be 6 percent aluminum, 3 percent zinc, and 0.2 percent manganese. Anode materials are cast in numerous weights and shapes to meet CP design requirements (Fig. 13.25).

(a)

FIGURE 13.25 Magnesium anodes being prepared for use in soil: (*a*) electrical connection being soldered; (*b*) insulating tar added to protect the electrical connection; (*c*) sulfate rich powder added to the porous container to provide good anode current efficiency. (Courtesy of Drinking Water Services, City of Ottawa)

(b)

(c)

Figure 13.25 *(continued)*

Impressed Current Cathodic Protection Anodes

There are basically three broad categories of ICCP anodes: consumable, semiconsumable, and nonconsumable anodes. For consumable metals and alloys such as scrap steel or cast iron, the primary anodic reaction is the anodic metal dissolution reaction. Metal dissolution is negligible on inert or nonconsumable anodes and the main reactions are the evolution of gases. Oxygen can be evolved in the presence of water, whereas chlorine gas can be formed if chloride ions are present (seawater, deicing salts). On partially inert anodes, both the metal dissolution and gas evolution reactions are important. Graphite and other carbon-based anodes can additionally produce carbon dioxide gas as these anodes are partially consumed.

The anode consumption rates within each category of anodes depend on the level of the applied current and also on the operating environment (electrolyte). For example, the dissolution rate of platinized titanium anodes is significantly higher when buried in soil compared with their use in seawater. For actively corroding (consumable) materials approximate consumption rates are of the order of grams per ampere-hour (Ah), whereas for nonconsumable materials the corresponding consumption is on the scale of micrograms. The consumption rates for semiconsumable anode materials lie somewhere in between these extremes.

A wide range of materials can be considered for impressed current anodes, ranging from inexpensive scrap steel to high-cost platinized material. The following properties would all be ideally desirable in an impressed current anode material [13]:

- Low consumption rate, irrespective of environment and reaction products
- Low polarization levels, irrespective of the different anode reactions
- High electrical conductivity and low resistance at the anode-electrolyte interface
- High reliability
- High mechanical integrity to minimize mechanical damage during installation, maintenance, and service use
- High resistance to abrasion and erosion
- Ease of fabrication into different forms
- Low cost, relative to the overall corrosion protection scheme

Consumable Anodes. Scrap steel and iron have often been used in the form of abandoned pipes, railroad, or well casings, as well as any other scrap steel beams or tubes. These anodes found application particularly in the early years of impressed current CP installations. Because the dominant anode reaction is iron dissolution, gas production is restricted at the anode. The use of carbonaceous backfill assists in reducing the electrical resistance to ground associated with the buildup of corrosion products. Periodic flooding with water can also alleviate resistance problems in dry soils.

Theoretical anode consumption rates are at 9 kg A^{-1} y^{-1}. For grey cast iron, consumption rates may be lower than theoretical due to the formation of graphite-rich surface films. A full utilization of the anode is rarely achieved in practice due to preferential dissolution in certain areas. For long anode lengths, multiple current feed points are recommended to ensure a reasonably even current distribution over the surface and prevent premature failure near the electric connecting points.

Limitations include the buildup of corrosion products that will gradually lower the current output. Furthermore, in high-density urban areas, the use of abandoned structures as anodes can have serious consequences if these are shorted to foreign services. An abandoned gas main could, for example, appear to be a suitable anode for a new gas pipeline. However, if water mains are short circuited to the abandoned gas main in certain places, leaking water pipes will be encountered shortly afterward due to excessive anodic dissolution.

Semiconsumable Anodes. Semiconsumable anodes such as graphite and high-silicon iron (HSI) have been in service since the first industrial electrochemical systems were developed. Graphite is a porous material and for this reason graphite is often impregnated with resins to reduce solution ingress and improve mechanical strength. Graphite anodes are inert when chlorine evolution is occurring, chlorine being produced efficiently at low polarization. But if oxygen formation predominates, as in low chlorine media, graphite is oxidized to carbon dioxide. Graphite deterioration also increases with decreasing pH and increasing sulfate ions concentration [13].

Graphite is not recommended for use in closed systems because of the possibility of galvanic corrosion caused by detached pieces. In addition, graphite suffers high consumption rates in water at temperatures above 50°C. Consumption rates measured for graphite depend on the environment and thus range from 0.045 in seawater to 0.45 kg A^{-1} y^{-1} in freshwater. Similarly the corresponding operating current densities vary from 2.5 to 10 A m^{-2}. The main disadvantages of graphite compared to other impressed current anodes are low operating current densities and inferior mechanical strength. Graphite is generally used in conjunction with carbonaceous backfills in soil-based ICCP systems [11].

Polymeric anodes are a different type of carbon-based anodes consisting of flexible wire anodes with a copper core surrounded by a polymeric material that is impregnated with carbon. The impregnated carbon is gradually consumed in the conversion to carbon dioxide, with ultimate failure by perforation of the copper strand. The anodes are typically used in combination with carbonaceous backfill, which reportedly increases their lifetime substantially. Because these anodes are typically installed over long lengths, premature failures are common when soil resistivity varies widely [11].

HSI anodes are widely used for ICCP (Figs. 13.26 and 13.27). These anodes contain approximately 14.5 percent silicon and certain alloys contain additional 4.5 percent chromium replacing molybdenum as an alloying element in such anodes. The chromium alloying additions are particularly useful in chloride-containing environments to reduce the risk of pitting damage while the high silicon content ensures that the alloy forms a protective film containing silicon dioxide (SiO_2). A prerequisite for the formation of the silicon dioxide film is that the anode must initially corrode during the first few hours of operation. Silicon dioxide is highly resistant to acids but it is readily dissolved in alkaline conditions.

HSI anodes are extremely hard and cannot be machined easily. They are generally cast and then stress relieved by annealing. Although they are brittle these anodes have superior abrasion and erosion characteristics compared to graphite. HSI anodes are widely used usually in conjunction with carbonaceous backfills in soils. They have also found limited use in marine and freshwater environments. The maximum operating current density is determined by the type of alloy and the environment. For instance, in groundbeds with backfills the current density is limited to between 10 and 20 A m^{-2} because of problems caused by gas entrapment. In marine environments, a high iron chromium anode can be operated up to 50 A m^{-2}.

FIGURE 13.26 Sleds with six cast iron anodes, integrated cable, and protective hose for an operation capacity of 75 amperes for fifteen years. (Courtesy of Dean Rookes, West Coast Corrosion Prevention Ltd.)

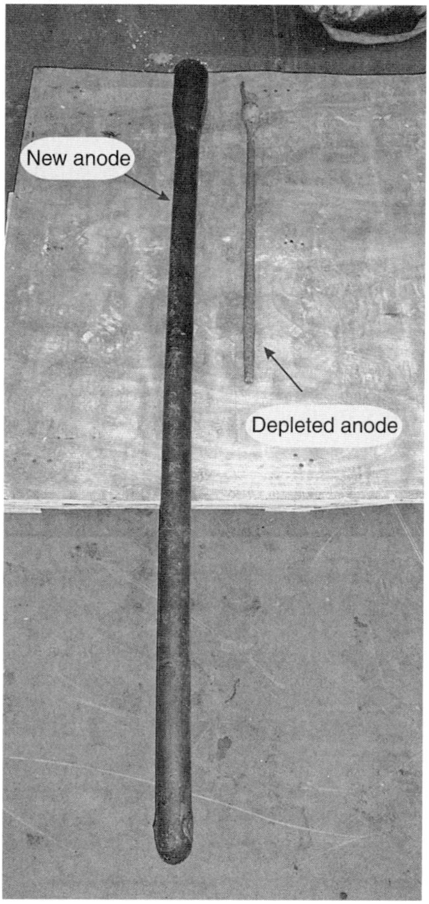

Figure 13.27 Impressed current cathodic protection anode made of high-silicon cast iron. A new anode is on the left and a depleted anode is on the right. (Courtesy of Dean Rookes, West Coast Corrosion Prevention Ltd.)

Nonconsumable Anodes. Platinized anodes utilize a few micrometers thick surface coating of platinum on titanium, niobium, or tantalum substrates. Restricting the use of platinum to a thin surface film has important cost advantages. For extended life, the thickness of the platinum surface layer can be increased. The corrosion resistance of the substrate materials is important due to the presence of discontinuities in the platinum surface coating that invariably will occur in practice. It is important that the potential of unplatinized areas on these anodes does not exceed the critical depassivation value for a given substrate material.

Platinized titanium is often used in marine environments. To avoid the dissolution of titanium at unplatinized locations on the surface, the operating voltage of the anode is limited by the anodic breakdown potential of titanium which is in the range of 9 to 9.5 V in the presence of chlorides. Hence the maximum recommended operating voltage of platinized titanium anodes is 8 V. The corresponding maximum current density output is approximately 1 kA m^{-2}. For ICCP systems where operating voltages are relatively high, niobium and tantalum based anodes with breakdown potentials greater than 100 V are generally selected in chloride containing electrolytes. The wastage rate of platinized anodes is approximately 8 mg A^{-1} y^{-1}.

Platinized anodes are fabricated in the form of wire, mesh, rods, tubes, and strips. For soil usage, they are usually embedded in a groundbed of carbonaceous material that provides a high surface area and lowers the anode/earth resistance [11]. Limitations include current attenuation in long sections of wire. Uneven current distribution results in premature localized anode degradation, especially near the connection to a single current feed point. Multiple feed points improve the current distribution and provide system redundancy in the event of excess local anode dissolution.

The rate of platinum consumption has been found to accelerate in the presence of AC current ripples. Most wastage has been observed to occur with AC frequencies of less than 50 Hz. The repeated oxidation/reduction processes result in the formation of a brownish layer of platinum oxide. To avoid the occurrence of this phenomenon, a single- or a three-phase full-wave rectification is recommended. The consumption rate of platinized anodes is also adversely affected by the presence of organic impurities such as sugar and diesel fuel.

Mixed metal oxide coated anodes, also called dimensionally stable anodes (DSA), are based on electrode technology developed in the early 1960s for the production of chlorine and caustic soda. The mixed metal oxide films are thermally applied to a noble metal such as titanium, niobium, and tantalum as substrate materials and are available in a variety of sizes and shapes. These oxide coatings have excellent conductivity, are resistant to acidic environments, are chemically stable, and have relatively low consumption rates. Groundbed installation in soils usually specifies that the anode be prepackaged in a canister with carbonaceous backfill material.

Electrode consumption is usually not the critical factor in determining anode life of mixed metal oxide anodes. It is rather the formation of nonconductive oxides between the substrate and the conductive surface film that limits the effective operation of these anodes. Excessive current densities accelerate the buildup of these insulating oxides to unacceptable levels. Mixed-metal oxide anodes have replaced most other types of impressed current anodes in seawater applications.

A variety of anode systems have evolved for ICCP of reinforcing steel, each with certain advantages and limitations. Continuous surface anodes have been based on conductive bituminous overlays and conductive surface coatings, the former being only suited to horizontal surfaces. In general, good current distribution is achievable with such systems. Discrete anodes have been used without overlays and with cementitious overlays.

For horizontal surfaces, anodes without overlay can be recessed in the concrete surface. Nonuniform current distribution is a fundamental concern in these systems. Anodes in the form of a titanium mesh, with proprietary surface coatings of precious metals are commonly used in concrete structures, in conjunction with cementitious overlays. These systems are applicable to both horizontal and vertical surfaces and generally provide uniform current distribution.

Magnetite is a cheap and naturally occurring material. It is a non-stoichiometric oxide with an electrical conductivity of 1.25 W^{-1} m^{-1}. Due to its brittleness, the anode is cast as a hollow cylinder and closed at one end. The inner surface is then copper plated and the cylinder is filled with polystyrene. Epoxy resin is used to fill any remaining space. The anode cable is soldered to the copper plate. Magnetite anodes have been successfully used in the cathodic protection of buried structures and those immersed in seawater. The maximum operating current density is 0.12 kA m^{-2} and the anode consumption rate is approximately from 1 to 4 g $A^{-1}y^{-1}$.

13.5.3 Rectified Current Sources

The most common ICCP power source is a rectifier that converts AC power from electric utility lines to lower voltage DC power by means of a step-down transformer and a rectifying circuit based on selenium or silicon elements (Fig. 13.28). These elements have a low resistance to current flow in one direction and a high resistance in the opposite direction. Figure 13.29 shows the simplified circuit of a rectifier utilizing a bridge connecting rectifying elements. The arrowhead symbols on the rectifying elements show the direction that current can flow readily. For the usual 60-cycle AC power source, the direction of current flow is reversed 120 times per second. Referring to Fig. 13.29, at some instant in time, current may be originating at connection (1) on the transformer secondary. The only path that this current can take will be through leg (C) of the bridge-connected rectifier, through the external circuit (groundbed to protected structure), and through leg (B) of the rectifier to return to the secondary winding at (2).

A 120th of a second later, the direction of the AC flow will have reversed and it will originate at connection (2). Under this condition, the only path that it can take to reach connection (1) will be through rectifier leg (D), the external circuit, and rectifier (A). For either direction of AC flow, the DC flow through the external circuit is in one direction only.

Rectifiers operate at less than 100 percent efficiency, meaning that the DC output power is less than the AC input power from the supply line. Rectifiers operate at maximum efficiency when operated at full-rated load. Where large rectifiers are used (as for large, bare structures), the rectifier rating should be selected so that it will operate at close to the full-load rating.

Figure 13.28 Five 300 ampere marine oil cooled rectifiers supported by an automatic dewatering oil spill containment pan. (Courtesy of Dean Rookes, West Coast Corrosion Prevention Ltd.)

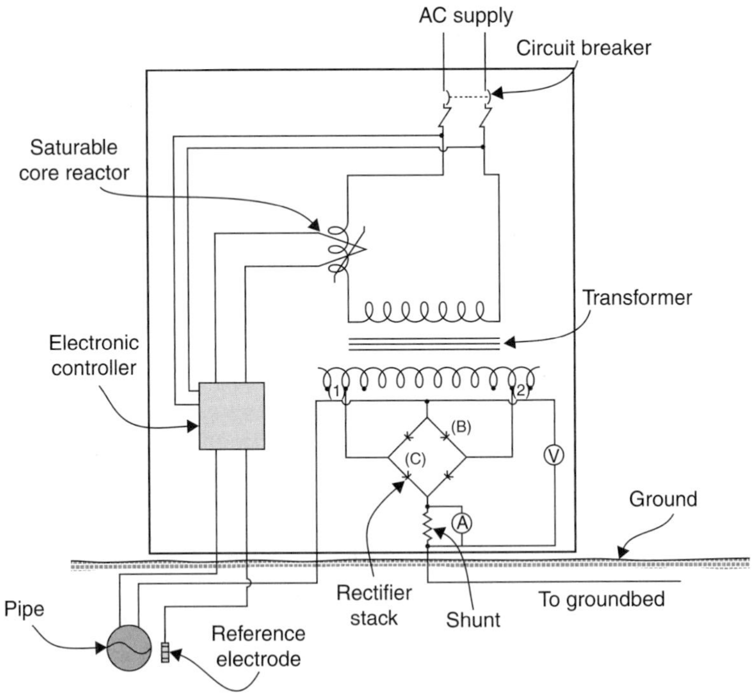

Figure 13.29 Typical constant potential rectifier diagram.

The efficiency of a CP rectifier may be determined by measuring the AC power input with a watt meter, measuring the DC volts and amperes output, and calculating the efficiency as follows:

$$\%\ \text{Efficiency} = \frac{\text{DC voltage (V)} \times \text{DC current (A)} \times 100}{\text{AC power (W)}} \tag{13.2}$$

For example, a rectifier operating at 20 A and 32 V DC output is found to have an AC input of 940 W. The efficiency will then be (20 × 32 × 100)/940 = 68 percent.

The power loss in the rectification process becomes heat and cooling is necessary to prevent the rectifying components from becoming too hot and eventually fail. One cooling method utilizes an air-cooled assembly with the rectifier cabinet arranged so that there will be a natural draft of air through the rectifier components. In another method, the rectifier components are immersed in electrical insulating oil in a large steel case. The rectifying elements give up their excess heat to the oil that is in turn cooled by radiation from fins on outer surfaces of the case.

13.5.4 Other Current Sources

Commercial AC power lines may not always be available. In such instances, alternate DC power sources may provide the necessary power [14]. Engine-generator sets, for example, may be used to provide the electrical energy for CP rectifiers if a large power source is needed. In such cases, gas or a petroleum product suitable for engine fuel taken directly from a pipeline can be used to power the engine of an ICCP system. Otherwise, fuel must be brought to the generator station periodically. Engine-generator installations must be designed with reliable equipment that will operate unattended for several weeks. The cost of operating such an installation will be relatively high, particularly if engine fuel must be brought in.

Closed cycle vapor turbogenerators (CCVTs) are commercially available as power sources for remote CP systems. Modern CCVT systems can supply up to 5 kW and 100 V. The system consists of a Rankine cycle turbine and an alternator. A burner heats an organic liquid that vaporizes and expands. The vapor is directed through the rotating turbine wheel, providing power to the alternator. Standard thermoelectric generator units are also available at power outputs up to 600 W and voltages up to 48 V (Fig. 13.30). Since thermoelectric generators contain no moving parts, maintenance is minimal. Annually, it is necessary to replace the fuel filter and clean the fuel orifice.

In areas where sunlight is abundant, a combination of solar cells and storage batteries may be a convenient power source to provide a continuous flow of current to a CP installation. Systems are now available that operate at power outputs of up to 1 kW, voltages up to 20 V, and currents up to 50 A. Battery storage capacities up to 3200 Ah (at 12 V) are available. Such a battery

Figure 13.30 Thermoelectric field generator. (Courtesy of Global Thermoelectric)

backup could supply a 10-A rectifier for almost two weeks with no recharging. Standalone batteries have also been used to power temporary CP installations to evaluate the current requirements at different locations of a system.

Wind-powered generators may be used as a source of power in areas where prevailing winds are of sufficient intensity and duration. Such units were used fairly extensively in the early days of pipeline CP. However, they are expensive and require extensive maintenance. Their use for this application has declined with the development of more cost-effective, reliable power sources, such as solar cells, CCVTs, and thermoelectric generators. Because the power output from a wind-powered generator will be neither steady nor continuous, some means must be used to assure a steady supply of current to the CP groundbed. This can be done by using storage batteries.

13.5.5 Wires and Cables

All ICCP underground or submerged cables from the positive DC terminal of the rectifier (or other power source) to the groundbed are at a positive voltage with respect to earth. For this reason, those cables must be very well insulated. It is important to remember that all buried parts of the groundbed assembly connected to the positive terminal of the power source may discharge current and corrode at any point where metal contacts the conducting environment in which the groundbed assembly is placed. This includes the cable from the rectifier to groundbed anodes and cable interconnecting anodes within the groundbed [11].

To prevent current discharge from the cables, wire having high-quality electrical insulation suitable for underground use must be provided, and all splices and connections must be insulated perfectly. Any defect in the cable insulation system may become a point of current discharge that will corrode the wire until it separates, thus breaking the electrical connection between the rectifier and all or part of the groundbed installation.

By contrast, all buried wire connecting self-powered galvanic anodes to a protected structure are subject to current collection from the environment and are free of corrosion. Insulation on such wires is used to prevent their picking up unnecessary current producing and associated loss of anode efficiency. This is also true of the insulated cable extending from the negative terminal of the impressed current power source to the protected structure.

13.6 Potential to Environment

The criteria used for the control of a CP system also provide a direct method to monitor the efficiency of that system and determine how well a structure supposedly under CP is actually protected against corrosion. Potential measurements are used most commonly as a criterion of protection. The basis for this is that if current is flowing onto a protected structure, there must be a change in the potential of the structure with respect to the environment. This is because the current flow causes a potential change, which is a combination of the voltage drop across the resistance between the protected structure and the environment, and the polarization potential developed at the structure surface.

The resistance between the protected structure and the environment includes the resistance of any electrically insulating paint or coating on the structure. This is illustrated in Fig. 13.31. detailing the various resistive components when a buried pipe is polarized by an ICCP system. In this example, the desirable place to measure potential would be across the interface between the pipe and the environment, as is represented by the terminals marked "polarization potential" on the equivalent circuit of Fig. 13.31.

Actually, in practice, this is rarely feasible on buried structures, and it becomes necessary to resort to measuring the potential between the structure and the environment surface directly above or at a point nearest to the structure. The measured potential now includes the polarization potential plus a portion of the voltage drop across the structure-to-remote-earth* resistance, as shown by the potential measured between the structure and point "A" on Fig. 13.31. The potential measured between the structure and remote earth, point "C," is not affected by the CP current lines and therefore provides a true estimate of the polarization shift produced on the structure by a CP system.

* The expression "remote-earth" used means the infinite conductor with practically no resistance which is the earth's mass.

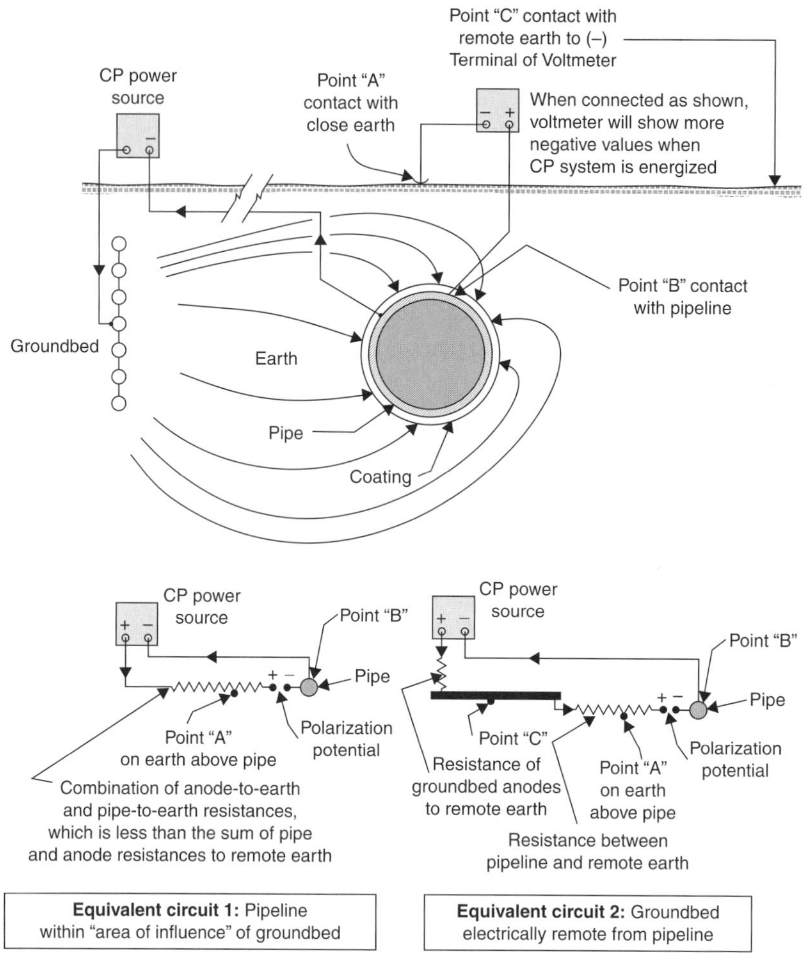

Figure 13.31 Pipe-to-environment potential change with flow of cathodic current.

13.7 Current Requirement Tests

A good practice prior to designing a CP installation is to conduct a field survey. Such a survey is made to assemble the data necessary to permit a workable design. During CP surveys, observations must be made regarding local conditions which may have a bearing on the CP design finally selected. These include the following:

- Availability of electric power for impressed current CP systems.
- Locations suitable for CP installations (subject to satisfactory soil resistivity) that will be accessible for routine inspection and maintenance. These should not be subject to construction

or other activities which may make the site untenable within the expected life of the installation.

- Presence of metallic structures, other than the one to be protected, which might be affected by the planned system.
- Presence of CP systems on other structures, which may have an effect on the structure to be protected.
- Presence of sources of stray DC such as DC railway systems.
- Presence of AC systems which can have an effect on the readings taken and on the criteria for full protection of the metal.
- Unusual environmental conditions such as acid-waste discharge from local manufacturing operations.

The current needed to protect a given buried metallic structure may vary over wide limits, depending on the nature of the environment and whether or not it has a protective coating, and if coated, on the quality and effectiveness of the coating as applied. For example, if it is assumed that a steel structure to be protected is buried in corrosive soil and has an exposed surface area of 90 m^2, the current required, assuming reasonably uniform current distribution, could range from about 3 A if the structure is bare, to as low as 30 μA or less if the structure has a superior coating. This means that a bare structure may require 100,000 times as much current as the same structure would if it were well coated.

It cannot be assumed, however, that just because a structure is coated, it will take a small amount of current to cathodically protect it. A poor coating material or an excellent material poorly applied can take much more current than the low value given above. For the same 90 m^2 structure, a relatively poor coating could result in a current requirement for CP in the order of 15 mA or more. A current of 15 mA may not sound like much, but it is 500 times as great as the 30 μA value given for a superior coating. The difference is considerable when working with large structures (such as large diameter cross-country pipelines).

13.7.1 Tests for a Coated System

When surveying a coated system, data on the electrical strength of the coating and on current requirements for CP can be taken concurrently. If the coating is in reasonably good condition, current requirements are obviously much smaller than on bare structures. This makes it possible to test large surfaces, distances, or structures with one test setup and a modest power supply. This may be done by actually applying current using a temporary test setup and adjusting the current from the power source until suitable protective potentials are obtained [15].

For a buried pipeline that covers large distances, while the same coating specification may have been used throughout its length, the effective electrical strength of that coating may vary considerably along the route due to the type of terrain, construction difficulties, changes in average soil resistivity, and different degrees of quality of pipeline construction work and inspection. When making the combined survey on such a system, the test arrangement shown in Fig. 13.32 may be used. At the test battery location, a current interrupter is used to automatically switch the current source on and off at a convenient time interval, for example, 10 seconds on and 5 seconds off, to gather the data needed for coating resistance calculations.

Testing section by section can be continued in each direction from the temporary CP location until the changes in the observed currents and potentials as the current interrupter switches on and off are no longer large enough to result in accurate data. The limits of the area that can be maintained above the protected criterion of –0.85 V or better can be established at this same time.

Coated systems polarize very rapidly and the better the coating, the faster the polarization. This means that conditions stabilize within the first few minutes and sometimes in a matter of seconds after the test current is applied.

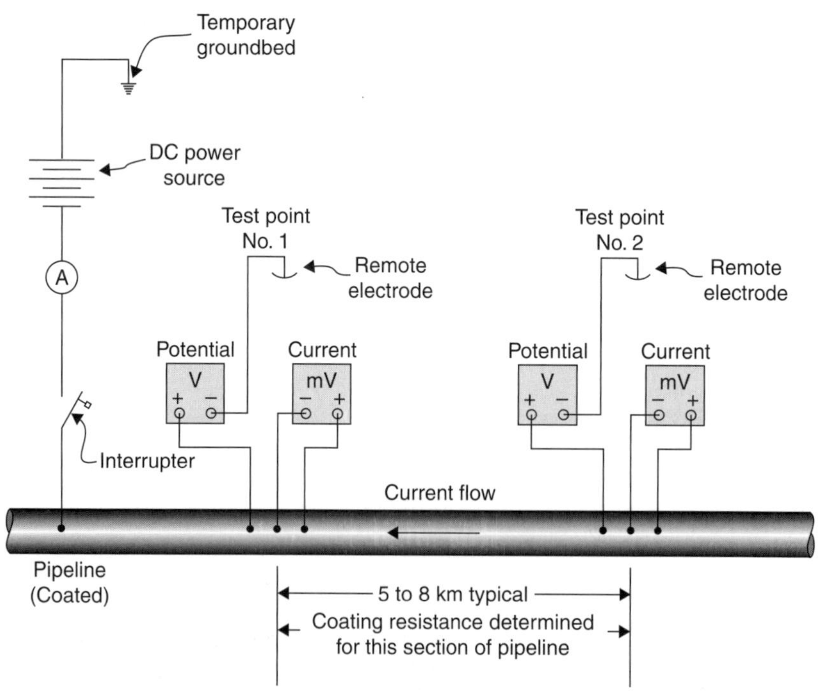

FIGURE 13.32 Coating resistance and CP current requirement tests.

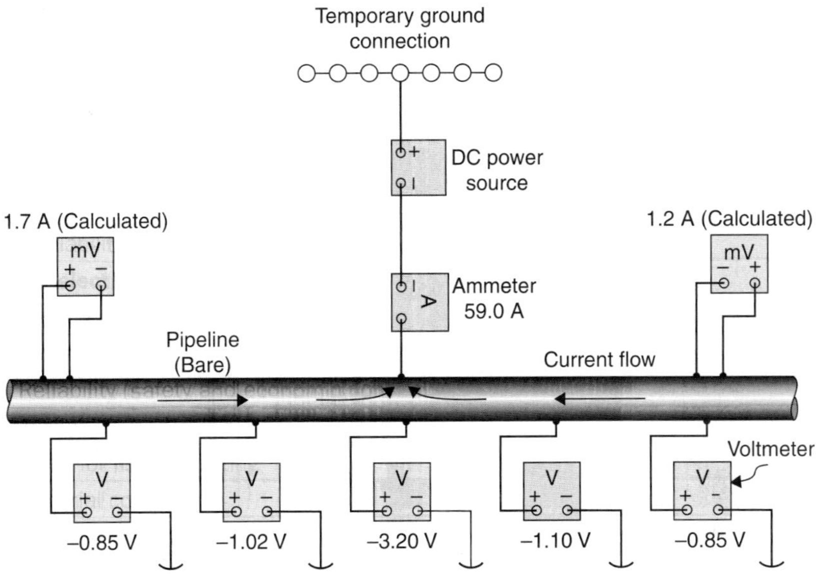

FIGURE 13.33 Evaluation of current requirement for CP.

13.7.2 Tests for a Bare Structure

When current requirement tests are made on an all-bare structure, results are typically quite different. Assume, as shown by the illustration in Fig. 13.33 that DC is forced to flow from a temporary ground connection to the system section being studied, and determining how much current will be needed to protect that section.

The output of the current source shown in Fig. 13.33 may be adjusted until protective potentials are attained at the ends of the section to be protected. When working with bare surfaces, the current should be allowed to flow steadily, which will permit the surface to polarize to some degree depending on the duration of the test. Full polarization on a bare surface may take weeks to achieve. However, if the increase in protective potential during the test is plotted versus time at a fixed current output, the curve can be extended to give a rough approximation of the potentials that would be obtained with complete polarization.

13.8 Stray Current Effects

Before preparing a CP design, it must be known whether or not stray currents are present. Currents which will most severely affect the structure to be protected are apt to be those of substantial magnitude that continually vary in quantity. These include stray currents from DC railway systems, mining operations using DC power, DC welding operations, and similar sources.

Depending on the severity of the stray current effect, the magnitude of the anodic potentials at areas where stray current is discharging from the affected structure and causing corrosion may be such that they cannot be counteracted readily with the usual CP system. This is particularly true where, at areas of current discharge, the structure may be a matter of several volts positive with respect to its environment. Such conditions necessitate special stray current control techniques involving metallic bonds from the affected structure to the source of the damaging current or other means of removing the stray current without harm to the affected structure.

The stray current sources described in Chap. 7 can result in very rapid corrosion which is usually much more severe than the corrosion caused by other environmental factors (Fig. 13.34). Another type of stray current which is variable in nature may be observed during periods of "magnetic storm" activity. Long structures such as pipelines or cables are most apt to be affected. During magnetic storms, the intensity of the earth's magnetic field can vary. When these variations occur, potentials are induced in the pipe or cable in much the same manner as potentials are induced in an electric generator.

Man-made variable stray currents usually demonstrate some patterns, as revealed by recording instrument records, showing rush hour peaks on transit systems, shift changes in mining operations, and so forth. However, variable stray currents resulting from magnetic disturbances rarely, if ever, demonstrate any pattern and may be active in a given area for a time without reappearing for a long time thereafter. Although sometimes intense for short periods, stray telluric currents resulting from magnetic induction will seldom cause as much corrosion as uncontrolled man-made stray current because the currents are of relatively short duration and usually are not concentrated in any given area for any appreciable lengths of time.

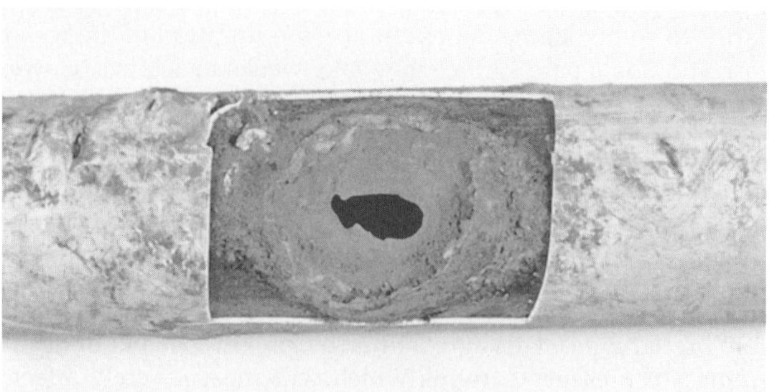

FIGURE 13.34 Pipeline perforated by a large fault current. (Courtesy of MACAW's Pipeline Defects, published by Yellow Pencil Marketing Co)

In addition to the variable-type stray currents described, steady-state stray currents may be encountered. These may be caused by an impressed current CP system on an adjacent, but electrically separate structure when the groundbed is too close to the unprotected structure. Under such a condition, the affected structure passes through the potential gradient field around the groundbed in which the soil is positive with respect to the system. This positive potential difference causes current to flow onto that part of the affected structure in the gradient field.

In severe cases, moving the groundbed that is causing the stray current condition may be necessary. Figure 7.29 illustrates the general features of typical stray current conditions that may be encountered in association with an electrified DC transit system. A pipeline is used, in this illustration, as a convenient means of illustration. Such situations may require the installation of metallic bonds between the possibly affected structure and the structure that is the source of the offending stray current. In the DC transit system, example, the metallic bound would be arranged to make a return path to the power house.

13.9 Monitoring Pipeline Cathodic Protection Systems

Monitoring pipeline CP systems can be quite complex and in many cases, condition-monitoring requirements are specified by regulatory authorities. Since CP systems are expected to operate in demanding environmental conditions over long time periods, the reliability requirements of the associated hardware are high. Regular monitoring of the equipment is therefore an important aspect of any CP program.

An increasing trend toward selective remote rectifier monitoring, using modern communication systems and computer networks is emerging to accomplish these tasks with reduced resources. Wireless cell phone and satellite communication systems are also available for interrogating rectifiers in remote locations. The other aspect of monitoring a pipeline CP system is monitoring the actual protection that is provided to the pipeline. The following sections briefly describe some of the techniques used to carry out this work.

13.9.1 Close Interval Potential Surveys

The principle of a close interval potentials surveys (CIPS) is to record the potential profile of a pipeline over its entire length by taking potential readings at approximately 1-m intervals. A reference electrode is connected to the pipeline at a test post and positioned in the ground over the pipeline at regular intervals for the measurement of the potential difference between the reference electrode and the pipeline (Fig. 13.35). The CIPS technique provides a complete pipe to soil potential profile and the interpretation of results, including the identification of defects, is relatively straightforward.

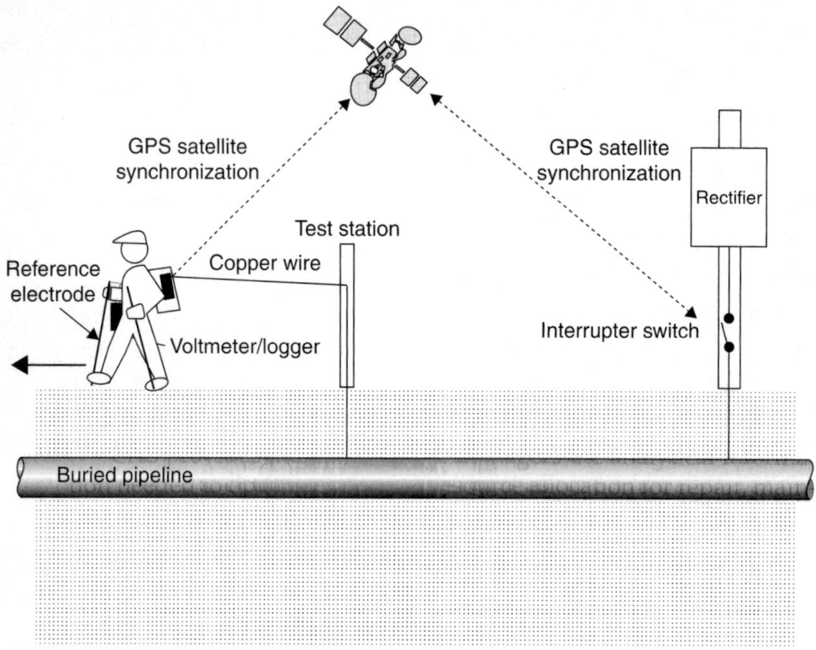

FIGURE 13.35 Schematic description of the CIPS methodology.

A three-person crew is typically required to perform these measurements. One person walking ahead locates the pipeline with a pipe locator to ensure that the potential measurements are performed directly above the pipeline. This person also carries a tape measure and inserts a distance marker (a small flag) at regular intervals over the pipeline. A second person carries a pair of electrodes connected to the test post by means of a trailing thin copper wire and the potential measuring instrumentation. The second person is also responsible for entering specific features as a function of the measuring distance. The third person collects the trailing wire, after individual survey sections have been completed.

CIPS measurements are quite demanding on field crews and require extensive logistical support from both the pipeline operator and the CIPS contractor. Field crews are typically required to move over/around fences, roads, highways, and other obstacles and difficult terrain. Breakage of the trailing copper wire is not uncommon and special strengthened wire has to be taped down onto road surfaces that are crossed.

An important consideration in the potential readings is the ohmic (IR) drop error that is included in the potential measurements when a CP system is operational. A commonly used method to correct for the IR drop, often called making instant-off measurements, can be a very accurate way to make potential measurements. In reality, however, making measurements completely free of voltage drop error is not always possible because some currents cannot be easily interrupted. Uninterruptible current sources may include: sacrificial anodes directly bonded to the structure, foreign rectifiers, stray currents, telluric currents, and long-line cells [16]. Modern interrupters are based on solid-state switches and are programmable to perform switching only when the survey is performed during the daytime. This feature minimizes the depolarization of the pipeline that may occur gradually due to the cumulative effects of the "off" periods.

When several rectifiers protect a structure, it is necessary that all rectifiers be interrupted at the exact same instant in order to obtain meaningful measurements. Pipeline operators usually specify that at least two rectifiers ahead of the survey team and two rectifiers behind the survey team have to be interrupted in a fully synchronized manner. The amount of time between current interruption and depolarization can vary from a fraction of a second to several seconds, depending upon details of the structure. In addition, capacitive spikes that occur shortly after current is interrupted may mask the instant-off potential. Measurements made with a recording voltmeter are preferred as they can be subsequently analyzed to determine the real instant-off potential [16].

An example of graphical CIPS data is presented in Fig. 13.36 [15]. In the simplest format, the "on" and "off" potentials are plotted as a function of distance. The usual sign convention is for potentials to be plotted as positive values. The difference between the "on" and "off" potential values should be noted. As is usually the case, the "off" potentials are less negative than the "on" values. When the relative position of these two lines is reversed, it indicates that some unusual conditions such as stray current interference may be at play.

13.9.2 Pearson Survey

Pearson surveys, named after its inventor, are used to locate coating defects in buried pipelines. Once these defects have been identified, the protection levels provided by the CP system can be investigated at these critical locations in more detail.

During a Pearson survey, an AC signal of approximately 1000 Hz is imposed on the pipeline by means of a transmitter (Fig. 13.37), which is connected to the pipeline and an earth spike as illustrated in Fig. 13.38. Two survey operators make earth contact either through aluminum poles or metal studded boots (Fig. 13.39). A distance of several meters typically separates the operators. Essentially, the signal measured by the receiver provides a measure of the potential gradient over the distance between the two operators. Defects are located by a change in the potential gradient, which translates into a change in signal intensity.

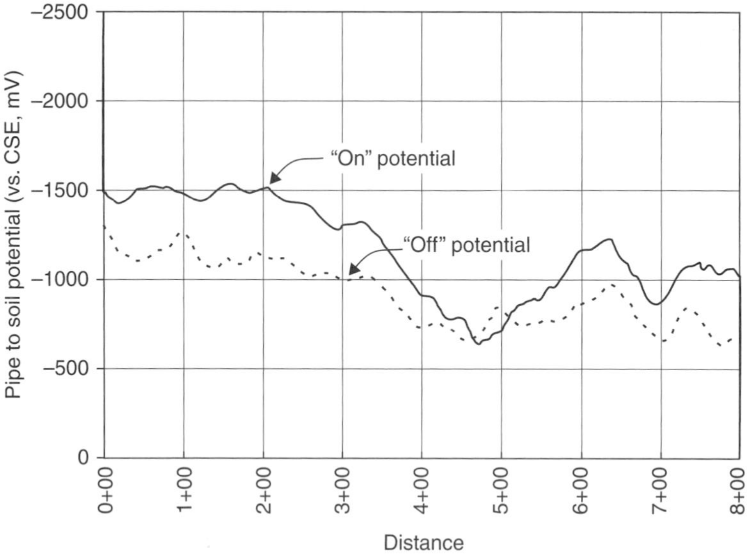

Figure 13.36 Over-the-line survey with cathodic protection.

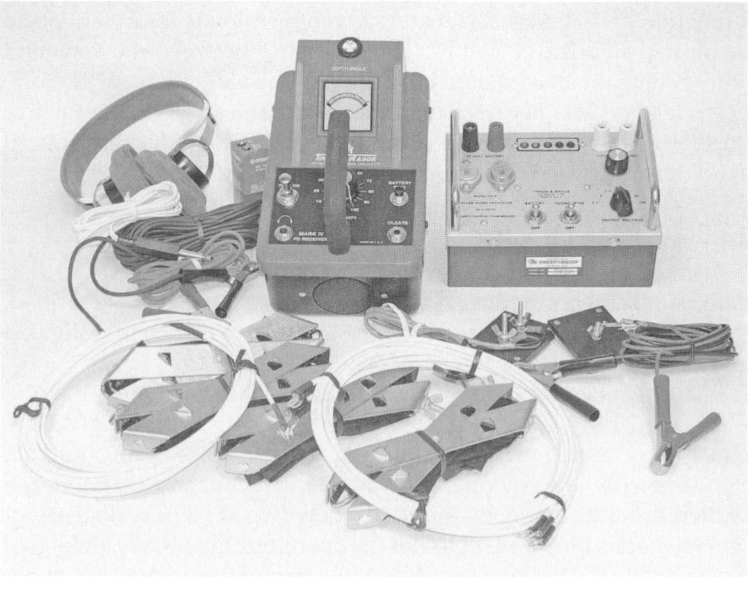

Figure 13.37 Instrumentation to carry out a Pearson survey. (Photo courtesy of Tinker & Rasor)

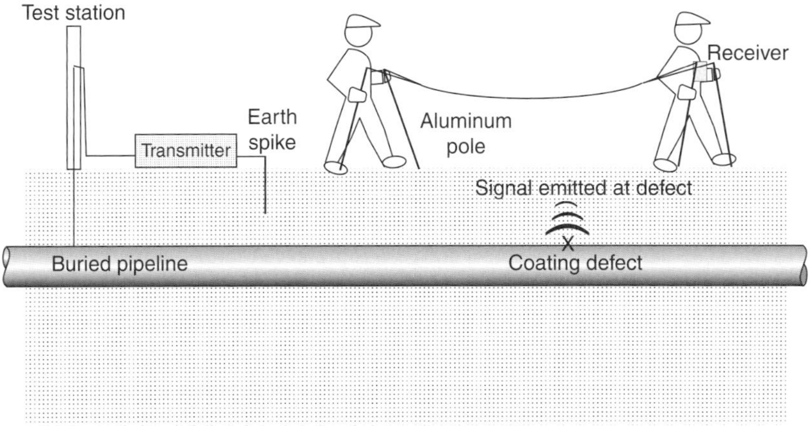

Figure 13.38 Schematic description of the Pearson survey technique.

As in the CIPS technique, the measurements are usually recorded by walking directly over the pipeline. As the front operator approaches a defect, increasing signal intensity is recorded. As the front person moves away from the defect, the signal intensity drops and later picks up again as the rear operator approaches the defect. The interpretation of signals may obviously be more difficult when several defects are located between the two operators.

Figure 13.39 Pipeline CP fault testing with Pearson-type detector. (Photo courtesy of Tinker & Rasor)

In principle, a Pearson survey can be performed with an impressed CP system still energized. However, sacrificial anodes should be disconnected since the signal from these may otherwise mask actual coating defects. A three-person team is usually required to locate the pipeline, perform the survey measurements, place defect markers into the ground and move the transmitters periodically.

By walking the entire length of a pipeline, an overall inspection of the right-of-way can be made together with the measurements. In principle, all significant defects and metallic conductors causing a potential gradient will be detected. There are no trailing wires and the impressed CP current does not have to be turned on and off.

The disadvantages associated with Pearson surveys are similar to those of CIPS, as the entire pipeline has to be walked and contact established with the ground. The technique is therefore unpractical in many areas such as roads, paved areas, or rivers.

13.9.3 Direct and Alternating Current Voltage Gradient Surveys

Direct current voltage gradient (DCVG) surveys are a more recent method to locate defects on coated buried pipelines and assess their severity. The technique again relies on the fundamental effect of a potential gradient being established in the soil at coating defects under the application of CP current. In general, the greater is the size of the defect, the greater is the potential gradient. The DCVG technique is particularly suited to complex CP systems, for example, areas with a relatively high density of buried structures. These are generally the most difficult survey conditions. The DCVG equipment is relatively simple and involves no trailing wires.

DCVG is measured by an operator between two reference electrodes, typically copper sulfate electrodes, separated by a distance of approximately 0.5 m. A pulsed DC signal is imposed on the pipeline for DCVG measurements. The pulsed input signal minimizes interference from other current sources, such as CP systems, electrified rail transit lines, telluric effects. This signal can be obtained with an interrupter on an existing rectifier or through a secondary current pulse superimposed on the existing steady CP current.

The operator walking the pipeline observes voltage deflections on a precision voltmeter to identify defect locations. The presence of a defect is indicated by an increased deflection as the defect is approached, no deflection when the operator is immediately above the defect and a decreasing deflection as the operator walks away from the defect (Fig. 13.40). The high precision in locating defects (~0.1 to 0.2 m) represents a major advantage in minimizing the work of subsequent digs if corrective action needs to be taken.

Alternating current voltage gradient surveys (ACVG) are similar to the DCVG technique, except that in an ACVG survey, a low-frequency transmitter is connected to the pipeline to create the AC signal.

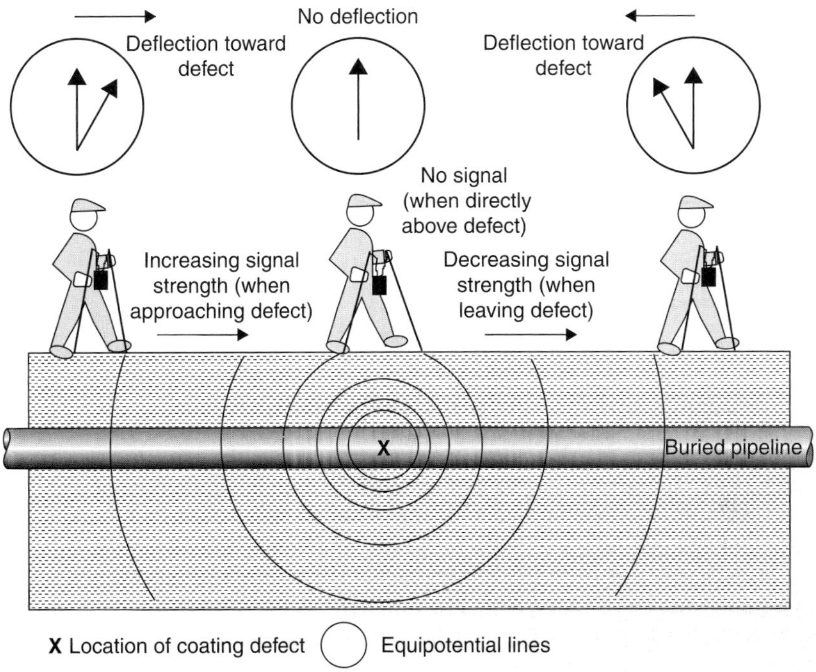

FIGURE 13.40 Schematic description of the DCVG methodology.

13.9.4 Corrosion Coupons

Cathodic protection coupons are now being used as an alternative method to make potential measurements that may be substantially free of voltage-drop error. A CP coupon is a small piece of metal that is electrically connected to the structure at a test station. The potential of a coupon will closely approximate the potential of any exposed portion of the structure (holiday) located in the vicinity of the coupon. By disconnecting the coupon from the structure at the test station, an "instant-off" potential measurement can be made on the coupon without having to interrupt any other current sources. However, these measurements are still not completely free of voltage-drop error. Any voltage drop occurring in the electrolyte in the distance between the reference electrode and the coupon surface will still be incorporated into measurements. Placing a reference electrode as close as possible to a coupon can minimize this error. However, the reference electrode must not be placed so close that it shields the coupon [16].

Perhaps the most important consideration in the installation of corrosion coupons is that a coupon must be representative of the actual pipeline surface/defect. The exact metallurgical detail and surface finish as found on the actual pipeline are therefore required on the coupon. The influence of corrosion product buildup may also be an important factor. The environmental conditions of the coupon have to be matched with those of the pipe being monitored, for example, temperature, soil conditions, soil compaction, and oxygen concentration.

Measurement of current flow to/from the coupon and its direction can also be determined, for example, by using a shunt resistor in the bond wire. Importantly, it is also possible to determine corrosion rates from the coupon. Electrical resistance sensors provide an option for in-situ corrosion-rate measurements, as an alternative to weight-loss coupons.

13.10 Simulation and Optimization of Cathodic Protection Designs

Early predictions of corrosion rates and estimates of adequate CP have traditionally been based on case studies and sample exposure tests. Applying these techniques to real structures usually involve extrapolations, use of large safety factors and ongoing corrections and maintenance of the system. In the late 1960s the finite element method was applied to the problem by discretization of the electrolytically conductive environment into a mesh and solving numerically with Laplace equations to define the intersection points, or nodes of this mesh [17].

However, creating a finite element mesh can be an extremely tedious and time-consuming process and, even when the mesh generation process is automated, it is difficult to perform a simulation when there are large geometry scale differences in the model, which is exactly the case found in most corrosion control problems where anodes are small compared to the size of the structure and the problem areas are most likely in corners and areas of complex geometry.

In the late 1970s, boundary element (BE) methods became available. As the name implies, the BE numerical method still requires mesh elements to be created, but now only on the boundary (or surfaces) of the problem geometry (Fig. 13.41). The main advantages of boundary elements for an ICCP analysis are [17]

- Mesh elements now describe only surfaces hence only two-dimensional elements are required. Mesh generators can be used with confidence, and models can be constructed extremely quickly and inexpensively once the geometry is defined.
- Models can be created with finer details for key or complex areas while modeling large and complex structures.

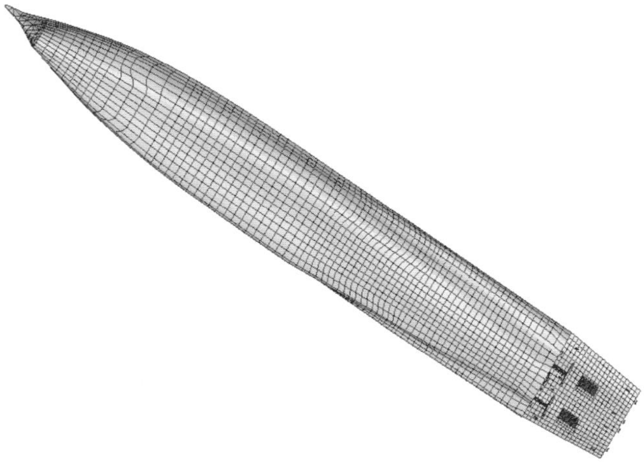

Figure 13.41 Ship geometry meshed with boundary elements. (Courtesy of Robert A. Adey, Computational Mechanics BEASY)

13.10.1 Modeling Ship Impressed Current Cathodic Protection

The design goal of an ICCP system is to produce an evenly distributed protection potential on a structure as well as to reduce the power consumption of the anodes to a minimum. The available design variables are the number of anodes, their location, and the location of the reference cells. The constraints on the design are the values of the potential on the structure. In order to provide adequate protection the potential must be less than a specified value, for example, −800 mV. In order to prevent overprotection the potential must be greater than a specified value, for example, −900 mV. An optimum solution can be obtained by combining an automatic optimization procedure with the BE model of the ICCP system. In this context Eq. (13.3) describes the electrochemical corrosion profile for the wetted surface of a ship hull is [18]

$$k\nabla^2\Phi = 0 \tag{13.3}$$

where $\mathbf{\Phi}$ is the potential and k is the conductivity of the electrolyte.

This equation is valid for homogeneous electrolytes and for situations where there are no stray currents, current sinks, or other interference. A shipboard ICCP system may be modeled in a way to meet these conditions. Current source points (Fig. 13.42) and exposed metal or damaged areas (Fig. 13.43) can be represented by boundary conditions eliminating the need to include sources and

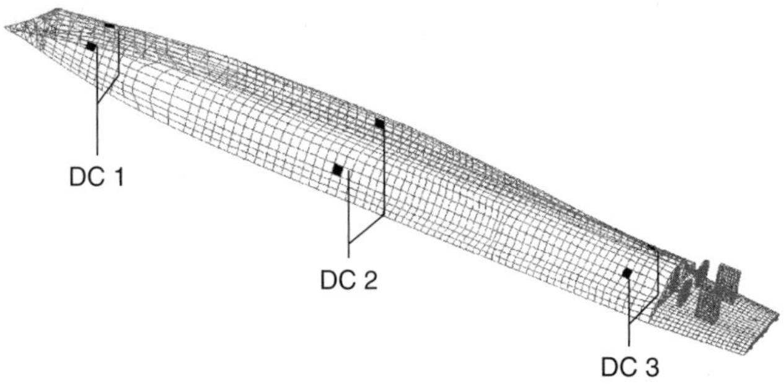

FIGURE 13.42 BE meshed ship hull with ICCP anode positions and power supplies. (Courtesy of Robert A. Adey, Computational Mechanics BEASY)

sinks in the model. Electroneutrality maintains charge equilibrium for the ship, surrounding water and ICCP system. The solution space for the problem defined by Eq. (13.3) used in the BE approach is the surface Γ which bounds the domain defined in Eq. (13.4):

$$\Gamma = \Gamma_A + \Gamma_C + \Gamma_I \tag{13.4}$$

where Γ_A is the anodic surface
Γ_C is the cathodic surface
Γ_I is the insulated surface

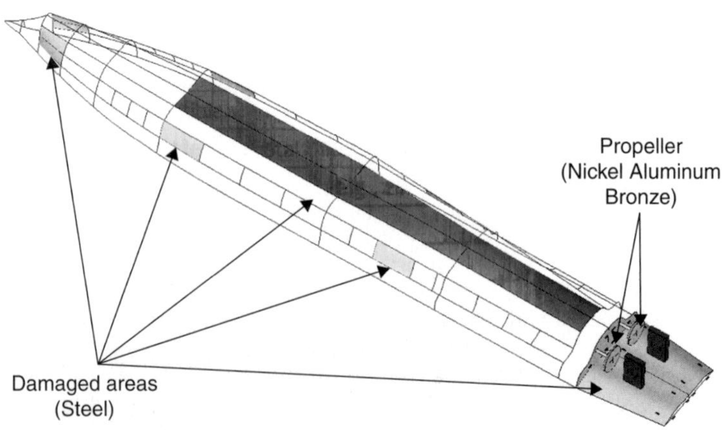

FIGURE 13.43 Schematic of ship hull showing damaged areas on steel and bare metal in propeller area. (Courtesy of Robert A. Adey, Computational Mechanics BEASY)

Γ must be continuous but all sections of one surface type do not have to be contiguous. An ICCP system consists of the surfaces to be protected, the anodes, the reference cells, and the power supply. Anodes are defined by maintaining the potential at a constant value, Φ_A: in Eq. (13.5):

$$\Phi_{(x,y)} = \Phi_A \tag{13.5}$$

defining the current density as a constant, q_A, on a surface in Eq. (13.6)

$$\frac{\partial \Phi(x,y)}{\partial n(x,y)} = q_A \tag{13.6}$$

where $\Phi(x, y)$ is the electrical potential at the point (x, y)
$n(x, y)$ is the normal to the surface at the point (x, y).

Reference cells are defined as specific points on the hull where the mathematical solution is obtained (Fig. 13.44) resulting in an optimal ICCP control with reference electrodes (Fig. 13.45).

It has been demonstrated that BE modeling can accurately predict experimental results. BE methods also can be used to evaluate the effect of a single parameter on system performance. In this way basic understanding of electrochemical corrosion and parameter interactions can be obtained. Several parametric studies have, for example, been published on damage levels in the propeller area, seawater conductivity, and paint resistance effects, as well as on the influence of stray current source on system performance [18].

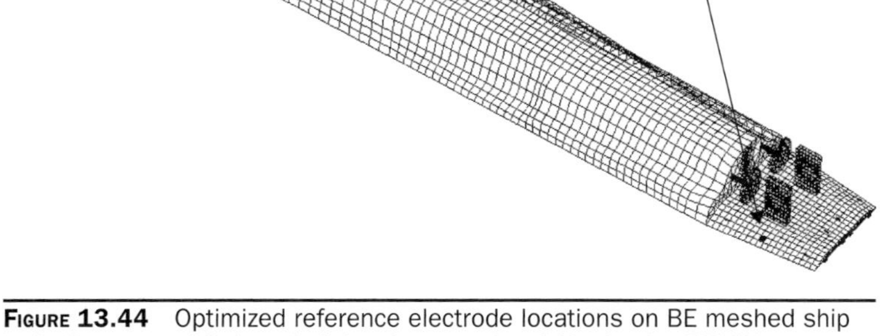

FIGURE 13.44 Optimized reference electrode locations on BE meshed ship hull. (Courtesy of Robert A. Adey, Computational Mechanics BEASY)

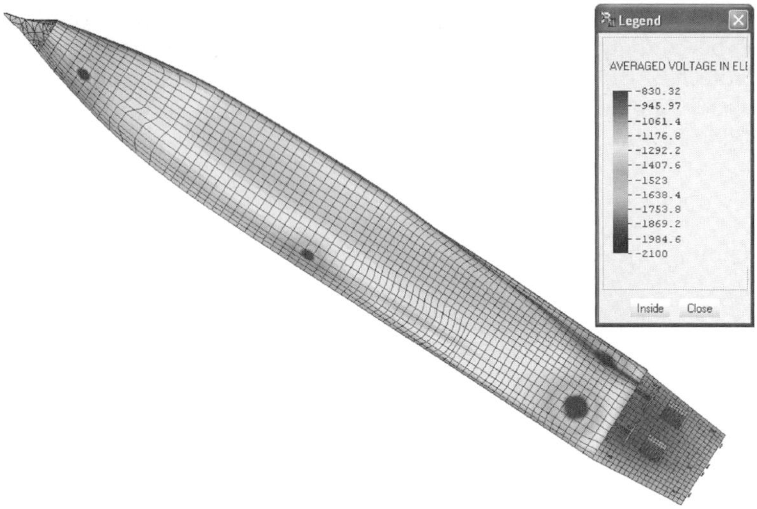

Figure 13.45 Automatic ICCP control with reference electrodes and power supplies set at 60 A (Stern, 27 A (middle), and 26 A (bow). (Courtesy of Robert A. Adey, Computational Mechanics BEASY)

13.10.2 Modeling Cathodic Protection in the Presence of Interference

Understanding the interactions between CP systems is becoming more complex particularly in the marine and offshore industry. The search for oil and gas in deeper waters and remote environments has resulted in engineering designs with many components on the seabed as well as at the sea surface. However, traditional design methods are inadequate in these situations due to the important interactions that may occur between CP systems [19].

It was common in the early CP models to make the assumption that the internal resistance of the structure through which the current returns was negligible compared with that of the electrolyte and the electrode kinetics. Therefore it was possible to ignore the metal resistance when formulating the modeling equations and hence assume that there was no IR drop in the return path. Pipelines present a typical situation where this assumption is not valid as over a long pipeline there is significant drop in the potential due to the internal resistance of the pipeline metal.

Another important situation for which the traditional assumptions are not valid is in new large-scale deep water oil and gas developments where not only there are long pipelines present but where hundreds of individual flowlines and connections introduce electrical connection paths that must be considered. Standard BE modeling tools are inadequate in these situations as interference will occur between the

CP systems, causing significant currents (and therefore voltage drops) to occur in the flowlines connecting the components.

The following example pertains to a floating production system (FPS) connected to a single point mooring (SPM) system by an offloading line (Fig. 13.46). The FPS may be protected by an ICCP system and the SPM by a sacrificial anode system. If these systems are not electrically connected then there is unlikely to be interference as the only electrical return path is through the seawater itself. However, if they are connected by the offloading lines an electrical circuit is created which allows current to flow from the ICCP anodes on the FPS to the SPM and current to return along the offloading lines. This can be expressed mathematically in Eq. (13.7) for the SPM and Eq. (13.8) for the FPS.

$$I_{\text{flowline connected to SPM}} = I_{\text{SPM anodes}} + I_{\text{SPM metal surfaces}} \tag{13.7}$$

$$I_{\text{flowline connected to FPS}} = I_{\text{FPS anodes}} + I_{\text{FPS metal surfaces}} \tag{13.8}$$

For equilibrium the current flowing along the flowline to either the SPM or the FPS must balance. The currents flowing into or out of the metal surfaces on the SPM and the FPS can be determined by integrating the normal current density calculated by the BE over the

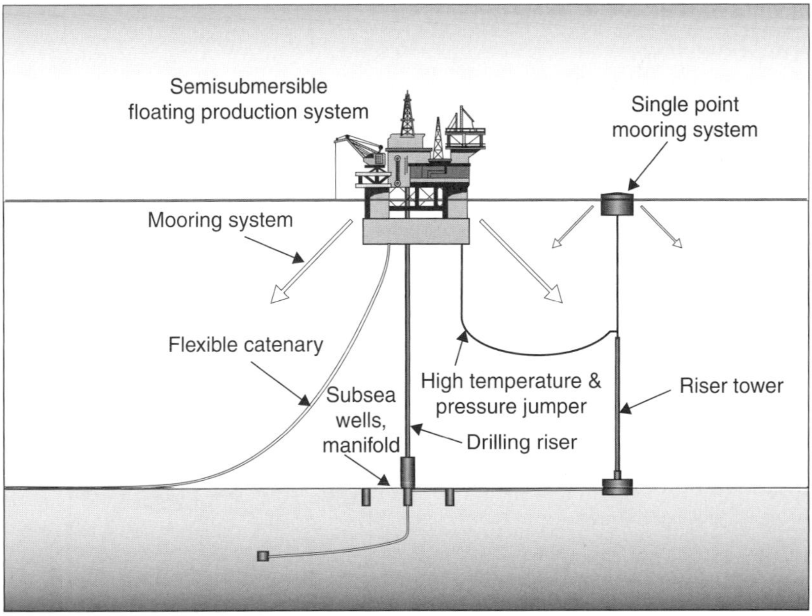

FIGURE 13.46 Example of a modern floating deepwater oil and gas production system.

metal surfaces of the structure. Therefore the internal circuit resistance equations and the BE equations representing the seawater can be solved together to predict the potentials and currents flowing in the overall system.

In the application example presented here the investigation focuses on an offloading SPM and its anchor chains. The anchor chains present an interesting CP problem because there is significant attenuation of the current flowing along the chains due to very high internal resistance caused by the chain link to chain link contact. Therefore near the SPM the chain will receive significant current from the anodes but this will drop off significantly within 25 to 30 m. The model of the SPM and its anchor chains is shown in Fig. 13.47.

The 3D model of the connectors, seen as cylinders in Fig. 13.47, needs to include the internal geometry because the anodes not only need to protect the outer metallic surfaces but also the internal structure. The connectors are hollow and the anchor chains enter into the connectors and are connected at the far end of the connector cylinder. In the model of the complete system the lines connecting to the FPS would also be included as well as the FPS and the subsea systems.

The predicted protection potentials provided by the CP system is shown in Fig. 13.48. One possible design change is to create holes in the connectors so the CP can provide better protection to the internal structure and chain. Computer modeling can help by predicting the effectiveness of different options.

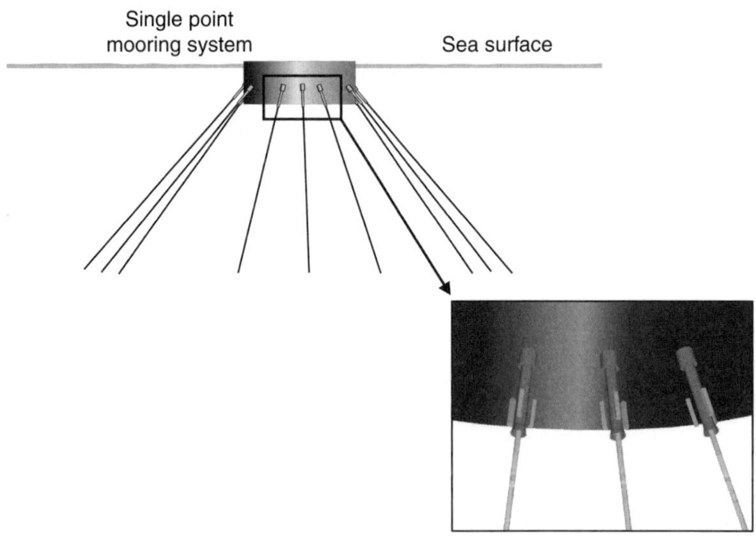

FIGURE 13.47 Model of the SPM and the anchor chains. (Courtesy of Robert A. Adey, Computational Mechanics BEASY)

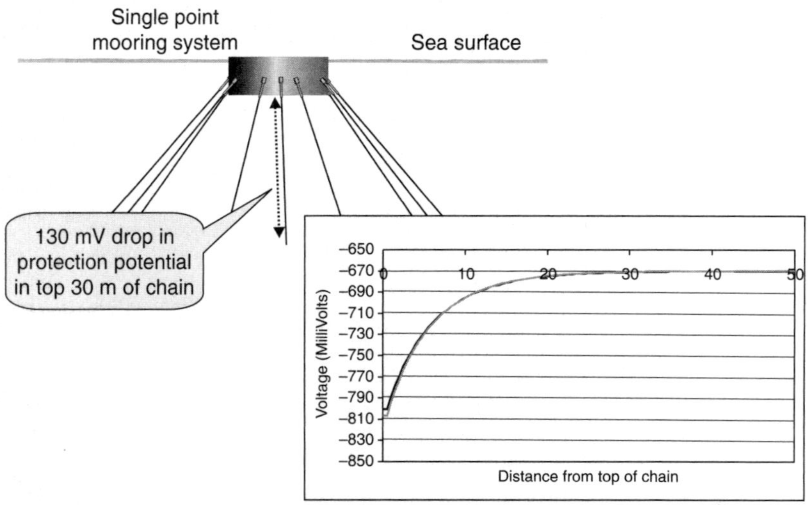

FIGURE 13.48 Estimated protection potentials on the SPM and its anchoring system. (Courtesy of Robert A. Adey, Computational Mechanics BEASY)

References

1. Rogers TH. *Marine Corrosion*. London, U.K.: Newnes, 1968.
2. Beavers JA. Cathodic protection—How it works. In: Peabody AW, Bianchetti RL, eds. *Peabody's Control of Pipeline Corrosion*. Houston, Tex.: NACE International, 2001; 21–47.
3. Stetler FE. Accelerating leak rate in cast iron water mains yields to CP. *Materials Performance* 1980; 19: 15–20.
4. Barnard KN, Christie GL. Cathodic protection of an active ship in sea water. *CORROSION 1950*; 6: 232–4.
5. Stratfull RF. Progress report on inhibiting the corrosion of steel in a reinforced concrete bridge. *CORROSION 1959*; 15: 65–9.
6. Stratfull RF. Experimental Cathodic Protection of a Bridge Deck. Transportation Research Record 500. Washington, D.C.: Transportation Research Board, 1974.
7. Koch GH, Brongers MPH, Thompson NG, Virmani YP, Payer JH. *Corrosion Costs and Preventive Strategies in the United States*. FHWA-RD-01-156. Springfield, Va.: National Technical Information Service, 2001.
8. Trethewey KR, Chamberlain J. *Corrosion for Science and Engineering. 2nd edn*. Burnt Mill, U.K.: Longman Scientific & Technical, 1995.
9. Johnson WE. *Cathodic Protection System Application and Maintenance Guide*. EPRI Report 1011905. Palo Alto, Calif.: Electric Power Research Institute, 2005.
10. Broomfield JP. *Corrosion of Steel in Concrete*. London, U.K.: E & FN Spon, 1997.
11. Roberge PR. *Handbook of Corrosion Engineering*. New York, N.Y.: McGraw-Hill, 2000.
12. Morgan JH. *Cathodic Protection. 2nd edn*. Houston, Tex.: NACE International, 1987.
13. Shreir LL, Hayfield PCS. Impressed Current Anodes. In: Ashworth V, Booker CJL, eds. *Cathodic Protection*. Chichester, U.K.: Ellis Horwood Limited, 1986.
14. Beavers JA. Cathodic Protection with other power sources. In: Peabody AW, Bianchetti RL. eds. *Peabody's Control of Pipeline Corrosion*. Houston, Tex.: NACE International, 2001; 201–10.

15. Bianchetti RL. Survey methods and evaluation techniques. In: Peabody AW, Bianchetti RL. eds. *Peabody's Control of Pipeline Corrosion*. Houston, Tex.: NACE International, 2001; 65–100.
16. Ansuini FJ, Dimond JR. Field tests on an advanced cathodic protection coupon. *CORROSION 2005*, Paper # 39. Houston, Tex.: NACE International, 2005.
17. Adey RA, Baynham J. Design and optimisation of cathodic protection systems using computer simulation. *CORROSION 2000*, Paper # 723. Houston, Tex.: NACE International, 2000.
18. DeGiorgi VG, Lucas KE. Computational design of ICCP systems: Lessons learned and future directions. In: Hack HP, eds. *Designing Cathodic Protection Systems for Marine Structures and Vehicles*. ASTM STP 1370, West Conshohocken, Pa., American Society for Testing and Materials, 1999; 87–100.
19. Adey R, Baynham J, Curtin T. Predicting the performance of cathodic protection systems with large scale interference. *Electrocor 2007*. Southampton, U.K.: Wessex Institute of Technology, 2007.

CHAPTER 14

Protective Coatings

14.1 Types of Coatings

The coating of surfaces is so ancient that its beginnings are lost in the mists of history. Paleolithic cave drawings exist which are reputed to be tens of thousands of years old. They were applied by fingers, a splayed twig, or, as some believe, were sprayed by blowing pigments through a hollow reed. The first evidence of coatings used for protection can be traced back to the Egyptians when they began using varnishes around 4000 BC. Polychrome Greek statues became common by 300 BC, and evidence exists that the Romans used coatings for both decorative and protective purposes. Coated Chinese artifacts are even more ancient than those in Western Europe.

The use of organic or metallic barrier coatings between a possibly aggressive environment and a material surely has become the most widely used method of protecting most engineered materials. There are three general classes of coatings: organic, inorganic, and metallic coatings. However, protective coatings often refer to integrated multifunctional systems that may combine more than one type of coatings.

Organic coatings protect more metal on a weight basis than any other means of corrosion protection. Besides providing protection by creating a physical barrier between the metal and the environment, these coatings may also contain corrosion inhibitors or other additives to stifle corrosion processes. Organic coatings include paints, resins, lacquers, and varnishes.

According to the U.S. Department of Commerce Census Bureau, the total amount of organic coating material sold in the United States in 1997 was 5.56×10^9 L, at a value of \$16.56 billion. The total sales can be broken down into architectural coatings, product original equipment manufacturers (OEM) coatings, special-purpose coatings, and miscellaneous paint products, with approximately one third of the total sales for the main purpose of corrosion protection [1].

Inorganic coatings include enamels, glass linings, and conversion coatings. Porcelain enamel coatings are inert in water and resistant to most weathers. They are common on appliances and plumbing fixtures. Glass-lined metals are used in process industries where there is concern over corrosion or contamination of the product. Conversion coatings are produced by intentionally corroding the metal surface in a controlled manner. This is done to produce an adherent corrosion product that protects the metal from further corrosion. Anodization of aluminum, one of the more commonly used conversion coating techniques, produces a protective aluminum oxide film on the aluminum metal [2].

Metallic coatings also create a barrier between the metal substrate and the environment. In addition, metallic coatings can sometimes provide cathodic protection when the coating is compromised. Metallic coatings and other inorganic coatings are produced using a variety of techniques, including hot dipping, electroplating, cladding, thermal spraying, chemical vapor deposition, or surface modification using directed energy (laser or ion) beams.

The most widely used metallic coating method for corrosion protection is galvanizing, which involves the application of metallic zinc to carbon steel for corrosion control purposes. Hot-dip galvanizing is the most common process, and as the name implies, it consists of dipping the steel member into a bath of molten zinc. According to the U.S. Department of Commerce approximately 8.6×10^6 t of hot-dip galvanized steel and 2.8×10^6 t of electrolytic galvanized steel were produced in 1997. The total market for metallizing and galvanizing in the United States is estimated at $1.4 billion.

14.2 Why Coatings Fail

Coating systems for the protection of materials need and probably demand more organized study than they have been given in many instances, and the benefits attainable from proper preliminary planning, job execution, on-site inspection, and postcompletion surveillance are well worth the effort and expense. Figure 14.1 describes how coating failure responsibilities were broken down in a throughout investigations made on 84 coating failures [3].

A coating may fail as a result of a large number of potentially adverse conditions. Some of these can be defined as mechanical, as when abrasion or impact removes the coating or more subtly and locally damage the coating and create a corrosion site that will show in a few months. A common reason protective coatings do not perform well is that they have not been considered as systems. Successful coating engineers approach a coatings project in much the same way that they approach any other engineering problem, beginning with the design of surfaces to be protected and ending with schedules for monitoring the completed work.

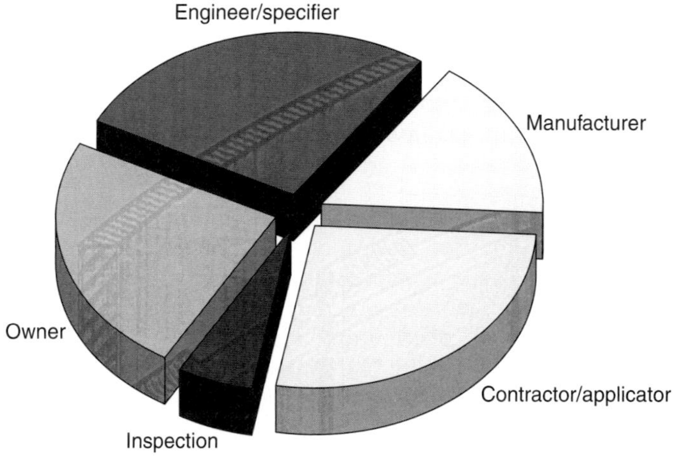

Figure 14.1 Breakdown of the responsibilities associated with 84 coating failures thoroughly investigated.

Prefabrication decisions and posterection corrections include elimination of sharp corners and hard to reach places, adoption of designs that will not collect or hold water or debris, provision for drainage of recessed zones, smoothing of rough surfaces and rounding of corners, removal of weld splatter, avoidance of skip welding, and caulking of crevices [4].

Tubular or box shapes should be used whenever possible to minimize edges and inside corners. If possible, the metallic materials selected should have the best resistance to corrosive attack or be alloyed to enhance coatings performance. A particular attention to design should be made to minimize the dwelling time of atmospheric condensation and provide special attention to surfaces shaded from the sun [5]. Large enclosed spaces that have not been totally sealed, such as the insides of box girders, should be coated, or ventilated, with fans if necessary. Decisions on coatings should not be left to the casual attention of inexperienced personnel. It is quite clear that

1. Decisions concerning effective protective coatings can be made properly only by experienced personnel.
2. The economic consequences of poor coating practices can be disastrous.
3. The number of coating materials available necessitates careful study if a proper material is to be selected.
4. Proper surface preparation is a prime economic factor for coatings in any environment.

5. A great variety of factors will determine the success or failure of an expensive coating job. These include such things as skill of the workmen, proper inspection of the job in progress and at its end, and provision of equipment to apply the coatings properly and promptly, among numerous others.

Continuous testing and evaluation of candidate systems and examination of systems actually in service more than pays its way in providing a basis for proper selection. The information thus obtained during the analysis of coating failures, either provoked during accelerated testing or more observed in service can be highly precious. Figure 14.2 provides a distribution by failure type of the 84 coating failures mentioned earlier [3]. The following list summarizes many of the factors that may have caused these failures and Table 14.1 briefly describes common coating failures with their causes and remedies:

1. *Water absorption:* All organic coatings absorb water and those used in corrosion protection are usually in the range of 0.1 to 3 percent by weight.
2. *Moisture vapor transfer:* This is the action of water vapor passing in molecular form through an organic substance. Generally, the lower the moisture vapor transfer rate, the better the protection that will be provided by that organic coating.
3. *Osmosis:* Osmosis refers to the passage of water through a semipermeable membrane, an organic coating in the present context, from a solution of lower concentration to one of greater concentration.

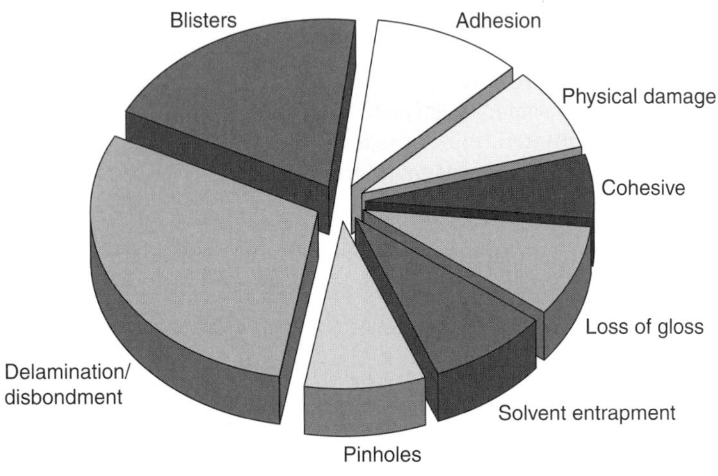

FIGURE 14.2 Breakdown of the root causes behind 84 coating failures thoroughly investigated.

Description	Causes	Remedies
Runs, sags, or curtains: excess flow of paint	Spray gun too close to work, too much thinner; surface too hard or glossy to hold paint	Before cure, brush out excess paint and modify spray conditions. After cure, sand and apply another coat.
Orange peel: hills and valleys in paint resembling skin of an orange	Paint too viscous, gun too close to surface, solvent evaporated too fast, air pressure too low for proper atomization	Before cure, brush out excess paint and modify spray conditions. After cure, sand and apply another coat.
Overspray	Paint applied to a surface beyond the intended surface; can be either wet or dry spray, but is usually dry	Change application method or equipment. Mask area behind surface intended to be painted.
Dry spray: dry, flat pebbly surface	Spray gun too far from surface, atomization too high, wrong solvent	Change application method; reduce atomization, change solvent.
Fish eyes: separation or pulling apart of wet film to expose underlying film or substrate	Application over oil, dirt, silicone, or incompatible coating	Remove by abrasive blasting or sanding. Apply a fresh coat by brush or spray.
Pinpoint rusting: rust in pinholes or holidays of film	Paint film too thin to cover the surface profile	Clean surface and apply additional coats as necessary.
Mud cracking: deep irregular cracks that resemble dried mud	A relatively inflexible coating applied too thickly (especially inorganic zincs)	Remove coating by abrasive blasting before applying at lesser thickness.
Lapping: each stroke of brush or spray gun visible after paint dries	Poor technique, fast-evaporating solvent, surface too hot	Change solvent; keep spray passes uniform, and paint before steel temperature peaks each day.

TABLE 14.1 Common Coating Failures, Their Causes, and Remedies [3]

4. *Electroendosmosis:* This phenomenon is caused by the presence of an electrical field that normally exists between a metal surface and its environment. Electroendosmosis can be defined as the forcing of water through a membrane by this electrical potential with the passage being governed by the polarity of the membrane.
5. *Blistering due to air or gas inclusion:* Air bubbles or volatile components of the coating may become incorporated in the film during the film formation and leave a void.
6. *Blistering due to phase separation during film formation:* A special type of osmotic blistering can occur when the formulation includes two solvents. The phase separation process occurs at a later stage in film formation and may occur at the coating/substrate interface.
7. *Surface contamination:* There are a number of surface contaminants that can be introduced before or during a coating operation. Water or water/oil emulsion can come from a faulty compressor. The blast abrasive can be either too acidic or too alkaline or may contain some small quantities of salts.
8. *Surface defects:* There are numerous substrate surface defects that occur during either the forming of the metal or the fabrication of the metal into a structure. These include hackles, laminations, needles, and weld splatter, among others.
9. *Soluble components in the film:* Some coating components, such as inhibitive pigments in primers, must be water soluble to some extent in order to passivate the substrate when water vapor permeates the coating film. However, if an inhibitive primer is used in immersion service, the water-soluble pigments may soon become blister sites because the continuous wetting action of the water dissolves the inhibitive pigment and blisters/delamination occur.
10. *Temperature differential or cold wall effect:* It is not uncommon to see blistering in a vessel at the uninsulated areas or areas that are heat sinks. The blistering caused by this temperature differential (ΔT) between the exterior and interior areas of the metal may be partly overcome by force curing to drive off the residual solvents in the film. However, the real solution may be to insulate the exterior surface in order to reduce the temperature differential ΔT.

14.3 Soluble Salts and Coating Failures

What happens when moisture and oxygen penetrate the coating? Perhaps nothing, if a truly good bond has been achieved, but a number of reactions may occur. Where the moisture contacts the steel, for example, the initiation or further oxidation of the iron can occur. A corrosion cell comparable to that in a pit on freely exposed iron may then develop. If ionic contaminants are present at the interface between the coating and the steel, the electrolyte would be more conductive and thus favor a more vigorous cell action.

Many contaminants, such as chlorides and other soluble salts, will actually enter into the chemistry of the cell to accentuate the attack. This is why many contractors are specifically requested to measure the level of these salts and continue washing a surface until the salt level is below an acceptable limit (Fig. 14.3).

The rate at which diffusion of water and contaminants is occurring through a coating is largely controlled by its thickness and formulation. Based on empirical data, economics, and some theoretical considerations, the thickness of a coating necessary to resist moisture permeation from the atmosphere and otherwise resist deterioration is approximately 125 μm.

The importance of adequate coating thickness cannot be overemphasized. If good flowout is not obtained during application, thin spots (particularly at the edges) or actual holidays (holes) may exist in the film. At times, the profile pattern (profile depth and uniform peak density) may not be matched to the wetability of the coating, resulting in thin areas over the peaks of metal and extended uneven distances between anchor points constituting the profile roughness.

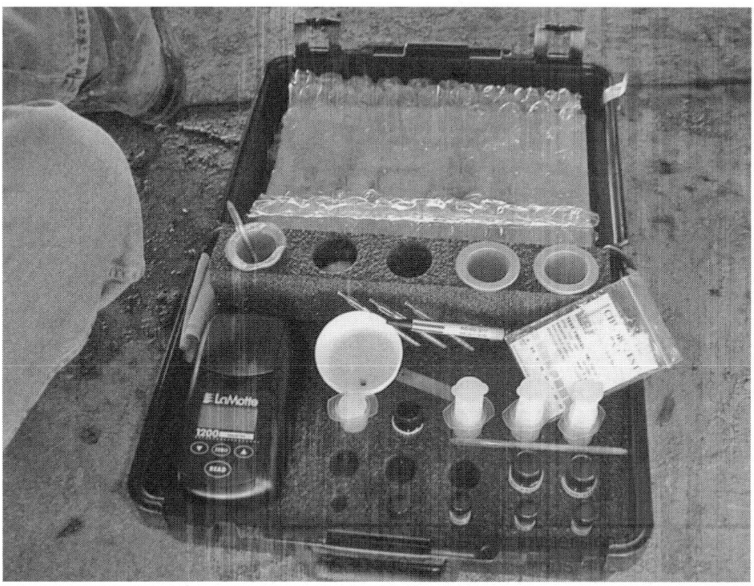

(a)

Figure 14.3 Soluble salts detector kit (*a*) and sample being taken on a surface in preparation (*b*). (Courtesy of Termarust Technologies)

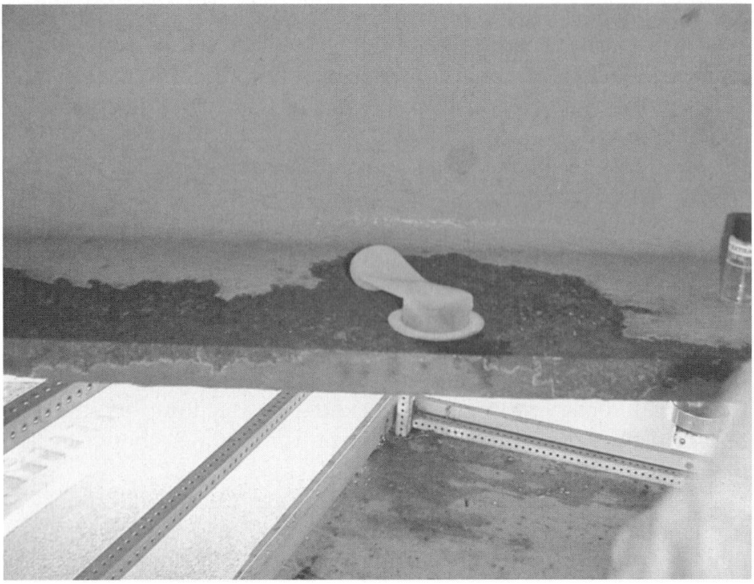

(b)

FIGURE 14.3 (*continued*)

All of these obviously produce a more rapid failure of the coating and decrease its service life.

In given conditions of humidity and oxygen permeation, the initial corrosion site may begin to move in a random manner as the corrosion product reduces the oxygen content at the surface and the area becomes highly anodic to the surrounding cathode area of oxygen saturation. The worm track of corrosion which then occurs is termed filiform corrosion (Fig. 14.4).

When a coating is exposed to an aggressive environment, such as that found in a marine application, the associated failure mechanisms of blistering and delamination are amongst the most important to consider. These two failure modes are often treated separately. However, from a mechanistic point of view, these two modes are quite similar and only differ in their degree of aggressiveness [6].

Due to the presence of imperfections in the coating, the steel substrate is directly exposed to its surroundings and may start corroding as described in Eq. (14.1). In order to maintain electroneutrality within the system, this reaction is balanced by at least one cathodic reaction. In most naturally occurring situations, this reaction will be the reduction of oxygen from ambient air, as illustrated in Eq. (14.2).

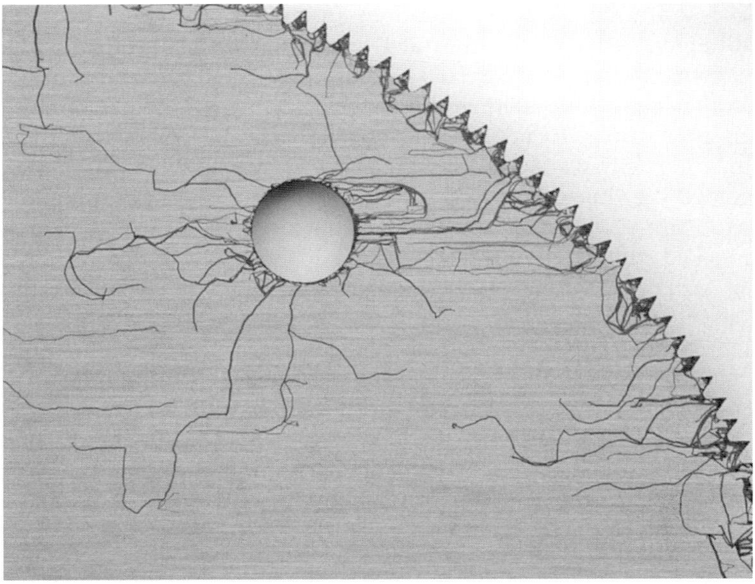

FIGURE 14.4 Filiform corrosion under the lacquer protecting the surface of a new hand saw (Courtesy of Kingston Technical Software).

These two reactions initially take place adjacent to each other but soon separate as the process continues with the cathode moving under the coating.

$$Fe(s) \rightarrow Fe^{2+} + 2e^- \quad (14.1)$$

$$O_2 + 2H_2O + 4e^- \rightarrow 4OH^- \quad (14.2)$$

The ferrous ions produced in Eq. (14.1) carry a positive charge while the cathode is producing hydroxyl ions (OH^-), resulting in an excess of negative charge. These local charge imbalances are unstable. Additionally, the ferrous ions produced by the corrosion reaction in Eq. (14.1) are not stable in the presence of oxygen as indicated in the E-pH diagram for iron shown in Fig. 4.15 where the pH-potential line for the reduction of oxygen is indicated as dotted line b. The ferrous ions thus soon react with water to form what is called rust or $Fe(OH)_3$ following the combination of oxidation and hydrolysis reaction described in Eq. (14.3).

$$Fe^{2+} + 3H_2O \rightarrow Fe(OH)_3 + 3H^+ + 1e^- \quad (14.3)$$

The electrons produced in Eq. (14.3) are also consumed by the reduction of oxygen. This scenario is not unique to coated steel and generally represents the corrosion of any iron or steel surface exposed to atmospheric humid environments with one important difference—that is, how the initial charge imbalance is subsequently satisfied by the movement of ions attracted to their opposite charges. In the case of uncoated steel, the route for the counter-cations and counter-anions to respectively the cathodic negative site or the anodic positive site is straightforward. However, when the steel is coated the situation becomes more complicated. The path from the exposure environment to these sites may be either restricted or blocked completely when the coating adheres well to the substrate. In such cases the corrosion process would be relatively stifled after the initial attack.

For less adherent coatings, available pathways (micronic dust, coating porosity, and holidays) for counter ions can be much less restrictive and the corrosion reaction would be allowed to proceed at a much faster pace. When this happens a second cathodic reaction can be triggered by the accumulation of protons at the anodic site to produce gaseous hydrogen [Eq. (14.4)]. Molecular hydrogen (H_2) is a highly active gas that can simply pry the coating loose.

$$2H^+ + 2e^- \rightarrow H_2(g) \tag{14.4}$$

Another aggravating factor is that organic coatings are generally poorly resistant to alkaline conditions and may be attacked by the hydroxyl ion causing a serious loss of surface adhesion. The reason this alkalinity causes such failure has been variously ascribed to saponification of the coating, dissolution of the oxide layer at the interface, and alteration of the ionic resistance of the film [6]. External cathodic currents provided by cathodic protection or internal currents produced by inorganic zinc additives during immersion service, for example, would increase the possibility of failure by the hydrogen and hydroxyl formation because much greater quantities of cathodic reaction products would be created.

In addition, external or internal cathodic currents on the steel can force more water through the coating than would be the case without such currents. The resulting forced diffusion of water through the coating is termed electro-osmosis (not to be confused with osmosis which occurs when water is drawn at a higher than normal rate through the film by a soluble salt lying beneath the coating). Thus, cathodic currents of sufficient magnitude can strip coatings from a steel surface. It has been demonstrated half a century ago that organic coatings were so permeable to water and oxygen that their rate of permeation to the cathodic region was greater than the amount required for corrosion to proceed [7].

One particular feature that has been identified by a number of workers is a delay time or initiation period between a coated substrate first being exposed to a corrosive environment and the start of the blistering or delamination process taking place. This delay can be partly explained by the time required to set up a steady-state diffusion of oxygen, water, and associated ions through the film or along the paint/metal interface [7].

The process starts with the initiation of blisters around the region of the point of damage. These blisters become larger as time progresses until they eventually coalesce. Once this has occurred, the delamination process takes over and corrosion progresses rapidly. As the coating delaminates from a fault, the pH at the interface behind it falls as the process continues.

These processes occurring under the film may be cyclic in nature, starting with an initiation time where there is no apparent activity. Figure 14.5 illustrate such a step process coating failure as it gradually occurred on a car door exposed to repeated winters of deicing salt exposure.

Blistering and delamination act in conjunction with each other in the failure of organic coatings. As blisters develop and grow, they coalesce to form a large disbonded region in the vicinity of a fault in the coating. This disbonded area provides the site for the delamination process to initiate and propagate.

FIGURE 14.5 Beach marks indicative of the cyclic progress of a car door coating failure.

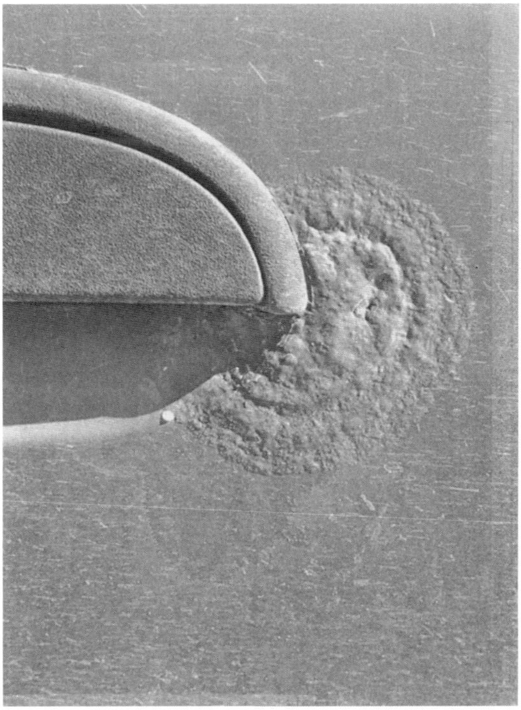

14.4 Economic Aspects of Coatings Selection and Maintenance

Expressions such as cost/m^2 or cost/m^2/y, which is better, percent of maintenance costs, percent of capital costs, and so forth, are often used to describe the cost of coating systems in economical terms. However, all these expressions provide an incomplete comparison on which to base a business decision. The cost of coatings materials for a maintenance overhaul only range between 5 to 21 percent of the total costs, while surface preparation is constant at about 45 percent of the total. Additionally, the performance of protective coatings has to be included in the overall decision matrix. High-performance epoxy polyamide, urethane, or zinc-rich systems, for example, may represent savings of nearly 40 percent over "less expensive" systems when a service life greater than 10 years is considered.

The number of coats to be applied, the compatibility of the various coats, and the requirements for their maintenance must all be considered to provide proper stability during the life of a project. Given the multitude of variables required for such analysis and the constant changing choices available, the identification of suitable protective coating systems for specific industrial environments can be a daunting task. Fortunately, general guidelines have been published for calculating approximate installed costs of coating systems with expected coating service lives for each system considered. The first guidelines have been updated every few years since they were first published in 1979 [8].

The latest version of the guide describes the basic elements of economic analysis and justification with guidance on the preparation of a *present value analysis* and on the effect of maintenance sequences on long-term costs and system performance [9]. Included in the most recent version of this guide are (1) most commonly used generic coating systems in typical service environments, (2) service life for each, (3) current material costs, and (4) current field and shop-painting costs. Guidelines for developing long-term life-cycle costs and number of paintings for the expected life of the structure are also included in the guide.

Tables 14.2 and 14.3 list the coating systems and their relative performance for respectively, an atmospheric exposure and an immersion service. In the new guide, immersion service has three categories: potable water, fresh water, and salt water, while the severity of atmospheric exposure uses a well-established scale based on the ISO 12944 standard [10]:

C_2: Low severity—atmospheres with low levels of pollution (rural areas)

C_3: Medium severity—urban and industrial atmospheres, moderate sulfur dioxide pollution, and coastal areas with low salinity

Type	Coating Systems for Atmospheric Exposure (primer/midcoat/to coat)	Surface Preparation	Number of Coats	DFT	C_2	C_3	$C_{5\text{-}I}$	$C_{5\text{-}M}$
Acrylic	Acrylic waterborne/acrylic WB/Acrylic WB	Hand/power	3	1.5	12	8	5	5
Acrylic	Acrylic waterborne/acrylic WB/Acrylic WB	Blast	3	1.5	17	12	9	9
Alkyd	Alkyd/alkyd	Hand/power	2	1.0	6	3	2	2
Alkyd	Alkyd/alkyd/alkyd	Blast	3	1.5	14	9	5	5
Alkyd	Alkyd/alkyd/silicone alkyd	Blast	3	1.0	11	6	3	3
Epoxy	Surface tolerant epoxy (STE)	Hand/power	1	1.3	12	8	5	5
Epoxy	Surface tolerant epoxy/STE	Hand/power	2	2.5	17	12	9	9
Epoxy	surface tolerant epoxy/STE	Blast	2	2.5	21	15	12	12
Epoxy	Surface tolerant epoxy/polyurethane	Hand/power	2	1.8	17	11	6	6
Epoxy	Surface tolerant epoxy/polyurethane	Blast	2	1.8	20	14	9	9
Epoxy	Surface tolerant epoxy/STE/polyurethane	Hand/power	3	3.0	23	17	12	12
Epoxy	Surface tolerant epoxy/STE/polyurethane	Blast	3	3.0	26	20	15	15
Epoxy	Epoxy 100% solids/penetrating sealer/epoxy	Hand/power	2	1.5	13	8	5	5
Epoxy	Epoxy 100% solids/penetrating sealer/polyurethane	Hand/power	2	1.5	13	8	5	5
Epoxy	Epoxy/epoxy	Blast	2	1.5	18	12	9	9
Epoxy	Epoxy/epoxy	Blast	2	2.0	20	14	11	11
Epoxy	Epoxy/epoxy/epoxy	Blast	3	2.5	23	17	14	14
Epoxy	Epoxy/polyurethane	Blast	2	1.5	17	11	8	8

TABLE 14.2 Estimated Service Life for Practical Maintenance Coating Systems for Atmospheric Exposure (in Years before First Maintenance Painting) [9]

Type	Coating Systems for Atmospheric Exposure (primer/midcoat/to coat)	Surface Preparation	Number of Coats	DFT	C_2	C_3	$C_{5\text{-}I}$	$C_{5\text{-}M}$
Epoxy	Epoxy/epoxy/polyurethane	Blast	3	2.0	20	14	11	11
Epoxy	Epoxy waterborne	Blast	3	2.3	18	12	9	9
Epoxy	Coal tar epoxy	Blast	2	4.1	21	17	14	14
Epoxy Zinc	Epoxy zinc/epoxy	Blast	2	1.8	26	18	12	12
Epoxy Zinc	Epoxy zinc/epoxy/epoxy	Blast	3	2.8	30	21	15	15
Epoxy Zinc	Epoxy zinc/polyurethane	Blast	2	1.8	24	17	12	12
Epoxy Zinc	Epoxy zinc/polyurethane/polyurethane	Blast	3	2.8	32	23	15	15
Epoxy Zinc	Epoxy zinc/epoxy/polyurethane	Blast	3	2.3	29	20	14	14
Inorganic Zinc	Inorganic zinc (IOZ)	Blast	1	0.8	27	17	12	12
Inorganic Zinc	IOZ/epoxy	Blast	2	1.8	26	18	14	14
Inorganic Zinc	IOZ/epoxy/epoxy	Blast	3	2.8	32	23	17	17
Inorganic Zinc	IOZ/polyurethane/polyurethane	Blast	3	2.8	32	23	17	17
Inorganic Zinc	IOZ/epoxy/polyurethane	Blast	3	2.3	30	21	15	15
Inorganic Zinc	IOZ/waterborne acrylic	Blast	3	1.5	24	17	12	12
Metalizing	Zinc metalizing (minimum 90% zinc)	Blast	1	1.3	33	22	16	16
Metalizing	Zinc metalizing/sealer	Blast	2	2.3	34	24	17	18
Metalizing	Zinc metalizing/sealer/polyurethane	Blast	3	3.3	39	27	22	22

TABLE 14.2 *(continued)*

Type	Coating Systems for Atmospheric Exposure (primer/midcoat/to coat)	Surface Preparation	Number of Coats	DFT	C_2	C_3	$C_{5\text{-}I}$	$C_{5\text{-}M}$
Moisture Curing Polyurethane (MCU)	MCU penetrating sealer/MCU/MCU	Hand/power	3	1.8	15	14	7	9
Moisture Curing Polyurethane	Zinc rich MCU/polyurethane/ polyurethane	Blast	3	2.3	30	21	15	15
Moisture Curing Polyurethane	Zinc rich MCU/MCU/MCU	Blast	3	2.3	29	21	14	15
Misc	Universal primer/epoxy	Hand/power	2	1.5	12	8	5	5
Misc	Universal primer/epoxy/polyurethane	Hand/power	3	2.0	14	9	6	6

Notes:

1. Service Life Estimates:
All estimates (in years) are for the "Practical" life of the system. Practical life is considered to be the time until 5 to 10% coating breakdown occurs (SSPC-Vis 2 Grade 4), and active rusting of the substrate is present.
2. Surface Preparation Definitions:
Hand/Power—Requires SSPC-SP 3 "Power Tool Cleaning" or SP 2 "Hand Tool Cleaning". Blast—Requires SSPC-SP 6 "Commercial Blast" or SP 10 "Near White Blast".
3. DFT is the dry film thickness in mm.
4. Service Life Environments per ISO 12944-2, "Classification of Environments":
C_2: Low—Atmospheres with low levels of pollution; mostly rural areas
C_3: Medium—Urban and industrial atmospheres, moderate sulfur dioxide pollution; coastal areas with low salinity
$C_{5\text{-}I}$: Very High, Industry—Industrial areas with high humidity and aggressive atmosphere
$C_{5\text{-}M}$: Very High, Marine—Coastal and offshore areas with high salinity
5. Typical Maintenance Painting Sequence:
Touch-Up painting occurs at Practical or "P" service life as listed in this Table
Maintenance Repaint occurs at "P" Life + 33% ("P" × 1.33)
Full Repaint occurs at Year of Maintenance Repaint + 50% of "P" Life (Maintenance Repaint Year + ["P" × 0.50].
6. Service Life for Hot Dip Galvanizing (0.1 mm minimum) as reported by American Galvanizers Association: Mild (rural) = 68 Years; Moderate (industrial) = 33 Years; Severe (heavy industrial) = 21 Years.

TABLE 14.2 *(continued)*

Type	Coating Systems for Immersion Service (primer/midcoat/topcoat)	Surface Preparation	Number of Coats	DFT	PW	FW	SW
Epoxy	Coal tar epoxy	Blast	2	4.1	—	17	14
Epoxy	Epoxy/epoxy	Blast	2	2.0	12	9	8
Epoxy	Epoxy/epoxy	Blast	2	1.5	10	8	6
Epoxy	Epoxy/epoxy/epoxy	Blast	3	2.5	15	12	11
Metallizing	Zinc metallizing/epoxy	Blast	2	2.3	20	17	15
Metallizing	Zinc metallizing/epoxy/epoxy	Blast	3	3.3	24	20	18

Notes:

1. Service Life Estimates:
   All estimates (in years) are for the "Practical" life of the system. Practical life is considered to be the time until 5 to 10% coating breakdown occurs (SSPC-Vis 2 Grade 4), and active rusting of the substrate is present.
2. Surface preparation definitions: Blast—Requires SP-10 "Near White Blast" for immersion service.
3. DFT is the dry film thickness in mm.
4. Service Life Environment Definitions:
   Potable Water (PW)—Immersion at an ambient temperature and pressure, coating requires NSF approval
   Fresh Water (FW)—Immersion at an ambient temperature and pressure
   Salt Water (SW)—Immersion at an ambient temperature and pressure
5. Typical Maintenance Painting Sequence:
   Touch-Up painting occurs at Practical or "P" service life as listed in this Table
   Maintenance Repaint occurs at "P" Life + 33% ("P" × 1.33)
   Full Repaint occurs at Year of Maintenance Repaint + 50% of "P" Life (Maintenance Repaint Year + ["P" × 0.50])

TABLE 14.3 Estimated Service Life for Practical Maintenance Coating Systems for Immersion Service (in Years before First Maintenance Painting) [9]

$C_{5\text{-}I}$: Very high industrial—industrial areas with high humidity and aggressive atmosphere

$C_{5\text{-}M}$: Very high marine—coastal and offshore areas with high salinity

Containment multiplier guidelines are included to aid in developing approximate costs for paint removal depending upon the containment type. These are to be used in conjunction with surface preparation costs.

Decisions involving whether or not a full repaint is required, as opposed to a maintenance repaint or touch-up, should be based on the results of an investigation. Generally, touch-up procedures are used when the amount of corrosion is limited or found in discrete areas. The success of maintenance repainting depends on the coating type, thickness, and adhesion of the existing coating system, as well as the substrate condition. The general procedures outlined in this guide may be used to determine installed costs and life-cycle costs for any combination of maintenance painting sequences.

Cost comparison examples are presented in the guide in the form of detailed worksheets following the basic economic principles of net present value and linear depreciation.

14.5 Organic Coatings

Organic coating materials have increased in complexity in parallel with developments in surface science and engineering, coating technologies, and environmental requirements. More can be achieved by the use of a proper coating material today than by simply covering a surface to reduce contact of an environment with the substrate. Of prime importance in the development of high-performance organic coatings was the petroleum industry, which produced most of the basic ingredients from which synthetic resins were developed.

The cracking of petroleum produced a multitude of unsaturated workable compounds that are important in the building of large resin polymers such as vinyls and acrylics. The solvents necessary for the solution of the resins were also derived from petroleum or natural gas [11]. The building blocks for epoxies and modern polyurethane coatings are other derivatives produced by refining petroleum products. Table 14.4 summarizes types of coatings, properties, and applications for a wide range of modern coating systems [12].

14.5.1 Coating Functionality

Some important considerations for designing corrosion-resistant coatings include coating protection, component design, component function, and coating formulation. Many coatings contain as many as 15 to 20 ingredients with their own range of functionality.

Coating type	% Solid	Thickness (μm)	Touch Dry	Time Before Overcoat	Uses and Properties
Alkyds					
Fast-dry enamel	32	38	1–2 hours(h)	6–8 h	Exterior exposure
Tank-finish aluminum	52	25	45 minutes(m)	6 h	Exteriors of steel storage tanks
Red oxide primer	42	38	2 h	24 h	Lead-free, general purpose, hard, water-resistant, anti-corrosive
Zinc phosphate primer	54	38	2 h	24 h	Lead- and chromate-free, general purpose for steel in atmospheric exposure can recoat conventional paint systems
General purpose gloss	55	38	1 h	16 h	Areas intermittently exposed to water immersion and atmospheric conditions
High gloss, high quality finish coat	54	38	1 h	16 h	High gloss enamel for interior and exterior surfaces
Aluminum finishing coat	48	25	45 m	6 h	For dry cargo holds, good impact resistance
All purpose primer	32	25	10 m	30 m	Fast-drying, anticorrosive
Coal Tar Epoxies					
Polyamide cured coal tar	80	400	7 h	24 h	Two components, one coat, high build, outstanding oil and water resistance, cures to 10°C and resistant to 93°C in dry exposure
Polyamine-adduct-cured coal tar	71	125–500	4 h	6 h	Two components, high build, cures down to −5°C when substrate ice-free

Table 14.4 Modern Organic Coating Materials, Applications, and Properties [12]

Coating type	% Solid	Thickness (μm)	Touch Dry	Time Before Overcoat	Uses and Properties
Epoxy Coatings					
Solvent-free epoxy (spray or trowel)	100	500–625	2 h	6 h	High resistance against tidal and impact action, good oil and water resistance, good adhesion
Weather-resistant epoxy acrylic	44	50	2 h	2–3 h	Two components, topcoat, excellent weathering resistance, color and gloss retention, isocyanate-free
Non-skid epoxy	72	500–1000	20 h	24 h	Helidecks, ramps, walkways and concrete floors, resistant to chemical spillage
High build, high solids epoxy	85	75–200	5 h	16 h	Potable water systems, resistant to cathodic disbondment
High solids epoxy primer	73	75–150	3 h	16 h	Tenacious adhesion to steel and concrete as primer for epoxies, urethanes and vinyls, good chemical resistance suitable for food contact
Armor epoxy	100	3000–5000	6 h	24 h	Solvent-free, quartz-reinforced
Zinc-rich epoxy primer	46	25–63	10 m	6 h	Two components, fast-drying, useful for a holding primer for rapid turnaround
High build, anticorrosive epoxy	61	125–150	2 h	3 h	Two components, for land and marine structures, unlimited overcoating time, good impact resistance and flexibility
Multipurpose epoxy primer	52	50–125	30 m	8 h	Two components, holding primer, can be overcoated with many other paints

Table 14.4 *(continued)*

Coating type	% Solid	Thickness (μm)	Touch Dry	Time Before Overcoat	Uses and Properties
Epoxy-based adhesion primer/ sealer	57	38	1 h	8 h	Two components for atmospheric and underwater systems, excellent adhesion to and sealing of aged zinc coatings, resists temperatures up to 175°C in dry exposure
Micaceous iron oxide epoxy primer	61	75–150	2 h	3 h	Two components, for land and marine structures
High build epoxy lining	78	250	3 h	8 h	Useful tank coating in two-coat system, resistant to wide range of chemicals
Epoxy high build/finish coat	55	75–150	3 h	10 h	Two components, topcoat on primed steel and concrete for atmospheric conditions, seals inorganic zinc coatings
Solvent-free epoxy	100	300	4 h	24 h	For crude oil/ballast tanks and aliphatic petroleum products
High build abrasion resistant epoxy	88	250–500	3 h	16 h	Amine-adduct-cured, glass-flake-reinforced
Acid-resistant epoxy	55	75–150	3 h	20 h	Two components, polyamine cured
High build, recoatable epoxy	62	150	2 h	12 h	Two components, general purpose
Solvent-free epoxy	100	> 300	5 h	40 h	One coat, long-life protection for steel structures and aliphatic petroleum products
High build, surface-tolerant epoxy	80	125–200	2.5 h	6 h	For maintenance where blast cleaning is not possible
Phenolic epoxy	60	50–100	45 m	18 h	Two components for low friction on inside of pipes, resistant to crude oil containing H_2S and CO_2

TABLE 14.4 *(continued)*

Coating type	% Solid	Thickness (μm)	Touch Dry	Time Before Overcoat	Uses and Properties
Phenolic epoxy lining	55	75–150		12 h	For sweet and sour crude, brine, processed petroleum products, CO_2- and H_2S-containing media
Phenolic epoxy tank lining primer/ intermediate and finish formulations	66	100	2 h	24 h	Two components, good resistance to chemicals and higher temperatures
Heat Resistant Coatings					
Heat-resisting aluminum	28	35	30 m	8 h	Two-coat system or overcoat for inorganic zinc primer, good up to 360°C
Heat-resisting aluminum finish	50	25	60 m	16 h	Good up to 175°C
Moisture-cured inorganic silicate	38	50	60 m	5 h	Finish coat over inorganic zinc primer, good from −90°C to 600°C
Inorganic Zincs					
Water-based inorganic zinc primer	65	100–125	15 m	8 h	Self-curing alkali silicate primer under most paint systems, good from −90°C to 400°C, withstands rain one hour after application, fair flexibility, good abrasion resistance when cured
Polyurethanes					
High solids aliphatic acrylic polyurethane	60	51–76	1h	12 h	Two components, unlimited recoatability, excellent resistance to chemicals such as water, oil and solvents, tough, flexible and abrasion resistant, excellent hiding power
Aliphatic acrylic polyurethane	41	50	30 m	4 h	Two components, excellent color finish, excellent resistance to atmospheric exposure, resistant to mineral oils, paraffin and gasoline

Table 14.4 *(continued)*

Coating type	% Solid	Thickness (μm)	Touch Dry	Time Before Overcoat	Uses and Properties
Waterborne Coatings					
Waterborne primer	43	50–75	10 m	2 h	Single-component primer as base for complete waterborne system on steel
Waterborne rust converting primer	43	50–75	30 m	6–8 h	Good for lightly rusted steel, eliminates the need for blasting, suitable for epoxy topcoats
Waterborne finish	42	50–75	30 m	1 h	For use over waterborne primer; for light to moderate chemical or high humidity exposure; good resistance to mechanical damage
Waterborne aluminum	45	50–75	15 m	2 h	For primer, intermediate or topcoats, excellent aluminum appearance and durability
Miscellaneous					
Polyvinyl butyral etch primer	10	13	5 m	1 h	Adhesion primer for ferrous and non-ferrous metals
Vinyl finish	35	50	60 m	16 h	Durable finish coat over high build chlorinated rubber and vinyl systems applied to interior and exterior surfaces in marine and industrial environments
Water-based decorative wall paint	42	38	15 m	2 h	Flat, decorative finish on walls and ceilings, for interior and exterior use
Low viscosity solvent-free epoxy sealer	100	—	—	16 h	Excellent penetrating sealer and adhesion promoter for concrete and prior to application of coatings, excellent chemical resistance

TABLE 14.4 *(continued)*

Some of the main variables used to design corrosion protective coatings are

- *Impermeability:* An ideal impermeable coating should be completely unaffected by the specific environment it is designed to block, be it most commonly humidity, water, or any other corrosive agent such as gases, ions, or electrons. Such coating should have a high dielectric constant and also have perfect adhesion to the underlying surface to avoid any entrapment of corrosive agents.
- *Inhibition:* In contrast with coatings developed on the basis of impermeability, inhibitive coatings function by reacting with a certain environment to provide a protective film or barrier on the metallic surface.
- *Cathodically protective pigments:* As with inhibition, cathodic protection in coatings is mostly provided by additives in the primer. The main function of these additives is to shift the potential of the environment to a less-corrosive cathodic potential. Inorganic zinc-based primers are good examples of this concept.

For serious corrosion situations, the coating system approach (primer, intermediate coat, and topcoat) provides all the ingredients for a long-lasting solution.

Primers

The original term "primer" implied a coating first applied to a substrate because it had a singular adhesive affinity for it and/or because it provided better adhesion for a subsequent coat than the subsequent coat could achieve if applied directly. A primer also functioned as a carrier for an inhibitor or for an anodic metal loading such as zinc, both of which will be discussed further in later sections of this chapter.

In modern usage and for many applications, primers usually are thin films (75 μm or less), and may be used on both metal and wood surfaces. On wood, their function is to seal the grain or to provide a smooth base for topcoats. On concrete, they should be compatible with the alkaline surface and help improve both adhesion and the life of subsequent coats.

Secondary or Intermediate Coatings

A second coat may be either the top or final coat, or an intermediate coat added when multiple thin films are required. The intermediate coat must serve as a tie coat between the primer and the subsequent coats, and may be of a different composition than either of them. When the three are of different formulations, it is the intermediate coat that is often used to provide the major film thickness for the coating.

Topcoats

A topcoat may be required to extend the life of the preceding coats. When designed as a topcoat, the film is normally more dense and hydrophobic than the remainder of the system to reduce the rate of moisture permeation to the underlying coats. A topcoat also may be selected to confer reflectivity, reduce photodegradation, or to affect color, among other reasons.

14.5.2 Basic Components

Coating materials consist, fundamentally of a binding resin, pigments, additives and a solvent or dispersant (except in the case of 100 percent solids coatings, where the latter two items are not used). The resinous material, plus the pigments, nonvolatile additives, and fillers, comprise the solids of the coating. These are carried in a vehicle (liquid solvent or dispersing medium) which may be varied in viscosity by the addition of a thinner. The vehicle also may ultimately be the binder of the coating (binding the pigment particles together and to the substrate).

Only the solids remain on a surface after the final coating has formed. However, a portion of, or the total vehicle may be reacted with the resin (100 percent solids coating) to produce the final film. Pigments are added to provide color or as inhibitors in a prime (first) coat. Fillers are selected to increase the bulk of the coating, improve the density, increase abrasion resistance, and increase the opacity of the film. Selection of the wrong filler or addition of excess filler can downgrade a coating by increasing the permeability or decreasing the cohesion in the film. Obviously, the solids content of a liquid coating determines the final film thickness of the coating, depending on the amount applied over a specific area:

$$\text{Film thickness } (\mu\text{m}) = \frac{\text{Vol liquid coating (mL)} \times \text{\% solids}}{\text{Surface area covered (m}^2\text{)}} \tag{14.5}$$

This theoretical coverage will not be precisely obtained because of the roughness of the surface and losses occurring during application of the wet film. Some 15 percent loss is to be anticipated for spray applications.

Binders

To perform in a practical environment, a coating must convert, after its application, into a dense, solid, and adherent membrane that has all the properties discussed previously. The binder is the material that makes this possible. It provides uniformity and coherence to the coating system. Not all binders are corrosion resistant, so only a few serve in the formulation of protective coatings. The binder's ability to form a dense, tight film is directly related to its molecular size and complexity. Binders that have the highest molecular weight will form films by the evaporation of the vehicle, whereas binders with smaller molecular weight will generally be reacted in situ. Binders can be classified according to their essential chemical reactions.

Oxygen-Reactive Binders. Oxygen-reactive binders are generally low molecular weight resins that are only capable of producing coatings through an intermolecular reaction with oxygen. This reaction is often catalyzed by metallic salts of cobalt or lead. Examples are

- *Alkyds:* Alkyds are produced by chemically reacting natural drying oils to form a synthetic resin with good film curability, chemical resistance, and weather resistance.
- *Epoxy esters:* Epoxy resins react chemically with drying oils to form epoxy esters. The drying oil part of the molecule determines the basic properties of the epoxy ester coatings. The coating dries by oxidation in the same manner as an alkyd.
- *Urethane alkyds:* Epoxy resins are also chemically combined with drying oils as part of the molecule that further reacts with isocyanates to produce urethane alkyds. Upon application as a liquid coating, the resin-oil combination converts by oxidation to a solid.
- *Silicone alkyds:* Alkyd resins are combined with silicone molecules to form an excellent weather-resistant combination known as silicone alkyds.

Lacquers. Lacquers are coatings that are converted from a liquid material to a solid film by the evaporation of solvents alone. Lacquers have generally a low volume of solids. Examples are

- *Polyvinyl chloride polymers:* This principal corrosion-resistant lacquer is made from polyvinyl chloride copolymers. The vinyl molecule is relatively large and will effectively dissolve in solvent in the 20 percent range.
- *Chlorinated rubbers:* To be effective, chlorinated rubbers have to be modified by other resistant resins to obtain higher solids, decreased brittleness, and increased adhesion.
- *Acrylics:* Acrylics are also of high molecular weight and may be combined with vinyls to improve exterior weatherability and color retention.
- *Bituminous materials:* Bituminous asphalts and coal tars are often combined with solvents to form lacquer-type films. They can provide good corrosion resistance but can only be applied where appearance is not a factor.

Heat-conversion binders. Examples of heat-conversion binders are

- *Hot melts:* Hot melts normally involve asphalt or coal tar and are melted and applied as 100 percent solids in the hot-liquid condition.

- *Organisols and plastisols:* These are high molecular resins (organisols) or vinyl materials (plastisols) that are dispersed in a solvent or plasticizer to solvate them into a filming material upon heating.
- *Powder coatings:* Powder coatings are high molecular weight thermoplastic resins or semithermoset resins applied to a substrate as a very fine powder that is melted to form a coating. Powder coatings can be applied by using an electrostatic field with the coating and substrate charged with opposing polarities. Such an application method is very efficient because the coated section of a substrate becomes insulating, therefore making the uncoated section the only one electrostatically attractive to the powder being applied.

Coreactive Binders. Coreactive binders are formed from two low molecular weight resins that are combined prior to the application to the substrate, where they react to form a very adherent and solid film. Examples are

- *Epoxies:* Epoxy binders are made of relatively low molecular weight resins in which the epoxy group is at the end of each molecule. The epoxy resins are then reacted with amines of various molecular weight and cured to form high molecular weight binders with good solvent and chemical resistance.
- *Polyurethanes:* Polyurethanes are coreactive binders in which low molecular weight resins containing alcohol or amine groups are reacted with di-isocynates into an intermediate resin prepolymer that is then capable of reacting with other groups containing amines, alcohol, or even water.

Condensation Binders. Condensation binders are based primarily on resins that interact to form cross-linked polymers when subject to sufficient thermal energy. These binders are also called high-baked materials and are commonly used as tank and pipe linings. Condensation is essentially the release of water during the polymerization process.

Coalescent Binders. Coalescent binders are coatings where binders of various resin types are emulsified to form a liquid binder. They are primarily emulsified with water or less commonly with some other solvent dispersions. When applied to the surface, the medium evaporates, leaving the coating in such a way that the binder resin gradually flows into itself, or coalesces, to form a continuous film.

Inorganic Binders. Inorganic binders are mostly inorganic silicates dissolved in water or solvent that react with moisture in the air after their application to a surface. The type of inorganic binder depends on the form of the silicate during the curing period. Examples are

- *Postcured silicates:* Soluble silicates are combined with zinc dust to form very hard rocklike films that are further stabilized by reacting them with an acidic curing agent.
- *Self-curing water silicates:* In this case, the soluble alkali silicates are combined with colloidal silica to improve the curing speed. Once applied to a surface they develop water insolubility by reacting with carbon dioxide and moisture from the air.
- *Self-curing solvent-based silicates:* These binders are organic esters of silica that are converted from a liquid form to a solid by reaction with moisture from the air, forming a very hard and corrosion-resistant binder. A major advantage of these materials is their conversion to rain- or moisture-resistant form shortly after their application.

Pigments

Pigments are essentially dry powders that are insoluble in the paint medium and that consequently need to be mixed in it by a dispersion technique. They range from naturally occurring minerals to artificial organic compounds. Pigments contribute several properties essential to the effective use of protective coatings. Several different pigments may be used within the same coating, all of them contributing to the coating's general characteristics to perform important functions such as providing

- Color
- Protection to resin binder
- Corrosion inhibition
- Corrosion resistance
- Film reinforcement
- Nonskid properties
- Sag control
- Increased coverage
- Hide and gloss control
- Adhesion

Zinc phosphates are now probably the most important pigments in anticorrosive paints. The selection of the correct binder for use with these pigments is very important and can dramatically affect their performance. Red lead is likely to accelerate the corrosion of nonferrous metals, but calcium plumbate is unique in providing adhesion to newly galvanized surfaces in the absence of any pretreatment and is claimed to behave similarly on other metals.

Solvents

Most coatings are made with multiple solvents and rarely with a single solvent. The choice of solvents influences the viscosity, flow properties, drying speed, spraying and brushing characteristics, and gloss. There is no universal solvent for protective coatings, the best solvent in one system being often impractical for another. Asphalts, for example, can be readily dissolved by hydrocarbons but are insoluble in alcohols. One of the most serious problems associated with coatings is the wrong choice of solvent because it can severely affect the curing and adhesion characteristics of the final coating. One convenient way to describe solvents is to regroup them into the following categories:

- *Aliphatic hydrocarbons:* Aliphatic hydrocarbons or paraffins such as naphtha or mineral spirits are typically used with asphalt-, oil-, and vinyl-based coatings.
- *Aromatic hydrocarbons:* Aromatic hydrocarbons, such as toluene, xylene, or some of the higher-boiling homologs, are typically used with chlorinated rubbers, coal tars, and certain alkyds.
- *Ketones:* Ketones such as acetone, methyl ethyl ketone, methyl isobutyl or amyl ketone, and many others, are very effectively used with vinyls, some epoxies, and other resin formulations.
- *Esters:* Esters such as ethyl, *n*-propyl, *n*-butyl, and amyl acetates are used commonly as latent solvents (a type of solvent that just swells the binder at room temperature) with epoxy and polyurethane formulations.
- *Alcohols:* Alcohols such as methyl, propyl, iso-propyl or butyl alcohols, and cyclo-hexanol are good solvents for highly polar binders such as phenolics. Some alcohols are used in connection with epoxies.
- *Ethers and alcohol ethers:* Ethers such as ethyl ether are excellent solvents for some of the natural resins, oils, and fats. The usual forms of ether used in protective coatings are alcohol ethers such as ethylene glycol mono methyl ether, known commonly as cellosolve. Cellosolve is a good solvent for many oils, gums, natural resins, and synthetic resins such as alkyds, ethyl-cellulose, nitro-cellulose, polyvinyl acetate, polyvinyl butyryl, and phenolics. Cellosolve is a slow solvent that is used in many lacquers to improve flow-out and gloss.

- *Water:* The recent regulations to reduce the emission of volatile organic compounds (VOCs) produced by organic solvents are forcing the coating industry to reconsider the applicability of water as a solvent. The most common water-borne coatings used for application to metals are air dried or force dried at temperatures below 90°C. A wide range of coating formulations falls into this category. The most commonly available technologies are water-reducible alkyds and modified alkyds, acrylic latexes, and acrylic epoxy hybrids.

14.6 Temporary Preservatives

A type of coating use which has grown steadily in use and diversity is temporary preservatives. The term temporary can mean intervals from hours to months or years during which metallic components are protected against specific indoor or outdoor environments. Such supplementary protection is often provided to surfaces that already have some form of permanent or semipermanent protection such as cladding or conversion coating. Some supplementary protection may be in the form of a fluid that can be easily applied or removed. Other temporary preservatives are low vapor pressure materials that will inhibit corrosion effects on systems stored in confined environments. All these products expected to be replaced or refreshed periodically during the life of a system. Jointing compounds, sealants, and corrosion prevention compounds are examples of this type.

14.6.1 Jointing Compounds and Sealants

Jointing compounds are used for protection at joints where they act by excluding dirt and moisture and by providing a reservoir of soluble corrosion inhibitors. Sealants are applied to joints to prevent the escape of fluids, such as fuel, but they also exclude moisture. Jointing compounds must remain flexible to allow easy disassembly of parts. Various synthetic resins are used for this purpose. The compounds harden sufficiently at edges to allow paint adhesion, but they remain tacky within the joint so that flexure does not cause cracking. The inhibitive sealants are very effective when used in faying surfaces and butt joints, for wet installation of fasteners, and over fastener patterns. They are also effective at insulating dissimilar metals.

14.6.2 Corrosion Prevention Compounds

Corrosion prevention compounds (CPCs) are fluids that can both prevent new corrosion sites from forming (Fig. 14.6) and, more importantly, suppress any corrosion that has initiated. CPCs are typically used as a postproduction treatment to provide cost-effective, temporary corrosion protection and to control existing corrosion on any metals. CPCs may be useful in providing supplementary protection for paint systems that have deteriorated or become damaged in service. They are applied as fluids by wiping, brushing, spraying, or dipping. These compounds are usually immiscible with water but may contain some water displacing components to remove water from surfaces and crevices.

A number of these fluids are based on lanolin and contain various solvents and inhibitors. The evaporation of the solvents leaves either thin soft films, semihard films, or hard resin films that provide varying degrees of short-term protection. There is a large range of CPCs that are commercially available ranging from water displacing to non-water displacing soft film and water displacing to non-water displacing hard film. The exact compositions of CPCs are not known due to their proprietary nature however information

(a)

FIGURE 14.6 Two steel washers exposed to a daily mist of water containing 3 percent sodium chloride (*a*) without CPC after two days and (*b*) with a commercial CPC after fifty days.

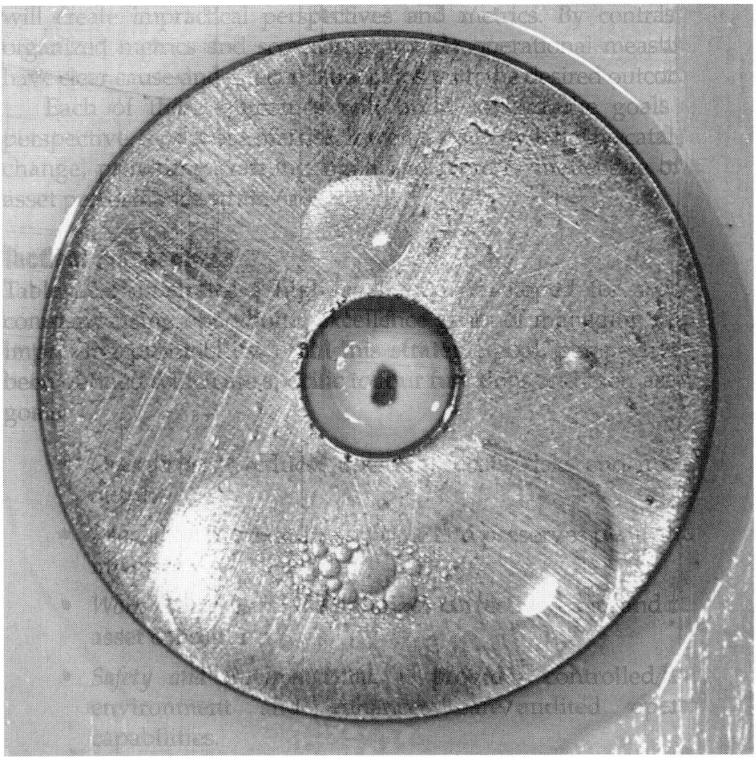

(b)

Figure 14.6 (*continued*)

contained in the Materials Safety Data Sheets (MSDS) reveal that they may include [13]

- An oil-, grease-, or resin-based film former
- A volatile, low surface tension carrier solvent
- A nonvolatile hydrophobic additive
- Various corrosion inhibitors, for example, sulphonates or surface active agents.

The water displacing CPCs act by spreading across surfaces, into cracks and crevices where they displace any moisture present, leaving behind a residue to act as a further barrier after the carrier solvent has evaporated. The hard film CPCs dry to a waxy or hard resin-like finish after application and provide a barrier film to corrosive environments.

CPCs have been used by the aircraft industry for many years as a means of protecting against corrosion. These treatments are often recommended by the manufacturer in the maintenance manuals as a way to help prevent the onset of corrosion in specific locations on the aircraft. The application of a CPC is relatively easy by using specially designed high-pressure pump system to convert it into a dense fog and literally fog the area or cavity that needs protection (Fig.14.7).

These products are also extensively used on road vehicles to reduce the severity of corrosion, particularly in cold-weather areas as roadways are coated with salt mixtures to break down snow and ice. There are plenty of locations in road vehicles where water can accumulate and rust form to cause premature failure of the equipment or perforation of the sheet metal. The underside of cars is especially vulnerable to rust and corrosion due to the exposure to high levels of moisture.

Since the 1970s, hot-melt wax thermoplastic CPCs have been used extensively during the construction of new vehicles to protect underbody structural components against corrosion and enhance vehicle durability. Hot melt waxes are usually applied through a dipping process. The wax is preheated to a temperature between 125 and 195°C. Following an alkali cleaning and water rinsing operation, parts are immersed in the molten wax. The thickness of the wax deposited on the parts is controlled through a preheat of the parts prior to dipping and the actual time of immersion in the hot melt wax. Following the immersion process, the coated parts are allowed to return to ambient temperature through a process that controls the uniformity and finish of the hot melt wax. Hot-melt wax thickness is commonly specified as 75 to 125 μm.

FIGURE 14.7 Fogging of aircraft with a commercial CPC.

Thin fluid film CPCs are typically used on road vehicles as a yearly maintenance application for added corrosion protection. The CPC treatment using air powered spray units is relatively inexpensive and can be provided by various rust-proofing centers (Fig. 14.8).

(a)

(b)

FIGURE 14.8 Fogging of car trunk (*a*) and underbody (*b*) with a commercial CPC.

No particular preparation is required. For outdoor spraying a NIOSH approved oil mist mask, as worn when spraying indoors, is recommended. Gloves are worn for comfort, as the aerosol cans or air powered spray guns quickly get cold. After treatment, any excess fluid may be wiped up with a cloth or paper towel.

A particular CPC package has been developed for very demanding applications where pack rust is threatening the integrity of structural components. The overbased crystalline calcium sulfonate (MOCCS) coating system made component consists of a penetrant/sealer designed to penetrate into pack rusted joints and connections displacing moisture, neutralizing acid, and sealing the joints. The MOCCS penetrant sealer stripe coat can be applied on large structure using an airless gun followed by brush touch-ups or simply by using a brush for smaller areas (Fig. 14.9).

14.6.3 Volatile Corrosion Inhibitors

Volatile corrosion inhibitors (VCIs), also called vapor phase inhibitors (VPIs), inhibit corrosion by being transported as a vapor from a permeable container to the site susceptible to corrosion. In boilers,

(a)

Figure 14.9 Brush application of a penetrant/sealer designed to stabilize pack rusted joints: (*a*) pack rust joint before and during application (*b*). (Courtesy of Termarust Technologies)

(b)

Figure 14.9 *(continued)*

volatile basic compounds, such as morpholine or hydrazine, are transported with steam to prevent corrosion in condenser tubes by neutralizing acidic carbon dioxide or by shifting surface pH toward less acidic and corrosive values. In closed vapor spaces, such as shipping containers, volatile solids such as salts of dicyclohexylamine, cyclohexylamine, and hexamethylene-amine are used.

On contact with the metal surface, the vapor of these salts condenses and reacts with moisture to liberate protective ions. It is desirable, for an efficient VCI, to provide inhibition rapidly and to last for long periods. Both qualities depend on the volatility of these compounds, fast action wanting high volatility, whereas enduring protection requires low volatility. A convenient partial vapor pressure for closed spaces VCIs will lie between 10^{-3} to 10 Pa (Table 14.5).

The first condition for good efficiency of a VCI is its capability to reach the metallic surface to be protected. The second is that the rate of transfer of the corrosion inhibiting molecules should not be too slow to prevent an initial attack of the metal surface by the aggressive environment before the inhibitor can act. These two conditions are related partly to the vapor pressure of the inhibitor, partly to the distance between the source(s) of the inhibitor and the metal surfaces, and partly to the accessibility of the surfaces [14].

It is significant that the most effective volatile corrosion inhibitors are the products of the reaction of a weak volatile base with a weak volatile acid. Such substances, although ionized in aqueous solutions, undergo substantial hydrolysis (reaction with water), the extent of

	Temperature	Vapor Pressure		Melting Point
Substance	(°C)	(mm Hg)	(Pa)	(°C)
Morpholine	20	8.0	1070	
Benzylamine	29	1.0	130	
Cyclohexylamine Carbonate	25.3	0.397	53	
Diisopropylamine Nitrite	21	4.84×10^{-3}	0.65	139
Morpholine Nitrite	21	3×10^{-3}	0.40	
Dicyclohexylamine Nitrite	21	1.3×10^{-4}	0.017	179
Cyclohexylamine Benzoate	21	8×10^{-5}	0.010	
Dicyclohexylamine Caprylate	21	5.5×10^{-4}	0.073	
Guanadine Chromate	21	1×10^{-5}	0.0013	
Hexamethyleneimine Benzoate	41	8×10^{-4}	0.110	64
Hexamethyleneamine Nitrobenzoate	41	1×10^{-6}	0.00013	136
Dicyclohexylamine Benzoate	41	1.2×10^{-6}	0.00016	210

TABLE 14.5 Saturated Vapor Pressures of Common VCIs

which is almost independent of concentration. In the case of the amine nitrites and amine carboxylates, the net result may be expressed by the following reaction:

$$H_2O + R_2NH_2NO_2 \rightarrow (R_2NH_2)^+{:}OH^- + H^+{:}(NO_2^-) \tag{14.6}$$

The nature of the adsorbed film formed at a metal-water interface is an important factor controlling the efficiency of VCIs. Metal surfaces exposed to vapors from VCIs in closed containers give evidence of having been covered by a hydrophobic-adsorbed layer. The contact angle of distilled water on such surfaces increases with time of exposure. Experimental studies on the adsorption of VCIs from the gas phase have confirmed that the VCIs indeed reacted with the metal surface to provide corrosion protection.

In these experiments, when a steel electrode was exposed to VCI vapors, the steady-state electrode potential shifted considerably into the region of positive values. The higher the vapor pressure was, the more important was the shift of the electrode potential in the positive direction [14]. This positive shift generally observed with most of the VCIs on ferrous metals is indicative of a preferentially anodic effect of the inhibitors. This anodic effect may be related either to a simple blocking effect of the anodic sites by the amine part of the inhibitors or to the contribution of the anionic component (i.e., the weak acid component).

In the case of nitrobenzoates, for example, it has been claimed that an acceleration of the cathodic partial process by reduction of the nitro group may lead, in addition to the effect of oxygen in the thin electrolyte layer, to a complete passivation of iron or ordinary steels. Contributions from the two parts of the dissociated molecule to the inhibitive effect are very likely.

14.7 Inorganic (Nonmetallic) Coatings

Inorganic coatings can be produced by chemical action, with or without electrical assistance, and include numerous classes of materials, among them the hydraulic cements that can set underwater, ceramics and clays, glass, carbon, silicates, and others. Some treatments to produce inorganic coatings can change the surface layer of a metal into a protective film of metallic oxide or compound that has better corrosion resistance than the natural oxide film and provides an effective base or key for supplementary protection such as paints. In some instances, these treatments can also be a preparatory step prior to painting.

14.7.1 Hydraulic Cement

Hydraulic cements are used to coat pipe inside and outside, especially pipes that are to be buried or submerged, for example, water or sewer lines. In the case of underwater piping to transport gas or liquid hydrocarbons, for example, it may be mixed with barites or other heavy materials to confer negative buoyancy. It also may be used in mixtures of organic materials where its ability to maintain a noncorrosive pH at the surface of steel is exploited.

Cast iron, ductile iron, or steel pipe may be coated at the mill by a process during which the pipe is spun on the center of its longitudinal axis while a mortar mixture is sprayed onto the inside surface in a uniform, dense layer. After proper curing, provided the pipe is handled carefully, this coating can protect the pipe interior against attack by water and many other liquid and gaseous corrosive environments.

A concrete coating on steel of any configuration can be protective to the steel as long as it does not crack or spall off, as in Fig. 14.10 because the alkaline reaction of hydraulic cement maintains a high pH at the steel surface, which effectively prevents corrosion. It does this when encasing reinforcing steel, for example, where it functions both as strength and protective component.

(a)

(b)

FIGURE 14.10 General view (*a*) and detail (*b*) of spalling concrete coating on water pipe. (Courtesy of Mark Lewis, East Bay Municipal Utility District)

14.7.2 Ceramics and Glass

Ceramics may also be used in much the same way as hydraulic cements, but they are usually applied to confer resistance to heat or attack by hot, high-velocity gases. Thus, they are used in exhaust passages emitting hot gases such as rocket nozzles, furnace linings, and other similar applications [5].

Also to be considered are the specialty mortars derived from the acid-resistant cements. These mortars can be applied to steel or concrete to provide remarkable acid, alkali, and heat resistance. They may be reinforced with wire mesh to sustain the heavy thicknesses normally applied. Coefficients of expansion close to that of steel have been attained to resolve one of the major problems when applying inorganic materials on steel. These can be used as floors, tank interiors, in scrubbers, and other process equipment of clean interior design.

Glass, in one formulation or another, is usually applied as a slurry and in a production line, but seldom for maintenance purposes. Once applied and as long as they are protected from mechanical abuse, glass surfaces are highly resistant to acidic and many mildly alkaline corrosives. They are also easily cleaned but difficult to repair. Glass-lined and -coated pipe, valves, pumps, and vessels of all kinds are widely used in the chemical, pharmaceutical, and food industries. As with the basic borosilicate glass from which the coatings are derived, the materials may be used up to about 175°C in most acid mixtures other than those containing fluorides.

14.7.3 Anodizing

Anodizing involves the electrolytic oxidation of typically magnesium or aluminum surfaces to produce an oxide scale that is thicker and more adherent than the naturally occurring film. An anodic electrical current passing through an electrolytic bath converts the metal surface to a durable aluminum oxide. The difference between plating and anodizing is that the oxide coating is integral with the metal substrate as opposed to being a metallic coating added by deposition. The oxidized surface is hard and abrasion resistant, and it provides some degree of corrosion resistance. Chapter 5 contains much more details on this topic.

17.7.4 Phosphatizing

One very successful treatment of steel has been by phosphatizing, also called phosphating or phosphate conversion coating. The coatings consist of a thick porous layer of fine phosphate crystals, tightly bonded to the steel. For many years, car bodies have been given this treatment prior to coating. The steel is pickled or otherwise well cleaned and immediately held in a bath of hot phosphoric acid containing zinc and perhaps manganese salts, plus other selected additives. A number of proprietary treatments such as parkerizing and bonderizing are available for use on steel.

The coatings do not provide significant corrosion resistance when used alone, but they provide an excellent base for oils, waxes, or paints, and they help to prevent the spreading of rust under layers of paint. Phosphating should not be applied to nitrided or finish-machined steel, and steel parts containing aluminum, magnesium, or zinc are subject to pitting in the bath. Some restrictions also apply to heat-treated stainless and high-strength steels. The requirements of a large vat to hold the hot acid salt solution also limits the process to production plants where large tonnages of steel are to be processed.

Cold phosphating of metal surfaces by spraying with a warm acid salt solution should not be compared with the phosphatizing treatment described above. Pickling of a metal surface can be obtained in this manner, and some residual phosphate formed, but not a full phosphate conversion coating.

14.7.5 Chromate Filming

The oxidation of a clean metal surface may greatly increase the corrosion resistance of the metal if a thin, continuous, dual metal oxide can be formed. Aluminum, zinc, and cadmium are often treated by a number of proprietary chromate filming treatments usually involving short-time immersion in strongly acid chromate solutions, but spraying or application by brushing or swabbing can also be used for touchup of parts. The resulting films are usually about 5 μm thick and are colored depending on the base alloy, being golden yellow on aluminum, dull gold on cadmium and zinc, and brown or black on magnesium.

The films contain soluble chromates that act as corrosion inhibitors that provide a modest improvement in corrosion resistance. As with phosphatizing, the film produced serves as an excellent coating base. Indeed, it can be difficult to make most polymeric coatings adhere to the nonferrous alloys without such a treatment. Epoxy primer, for example, which does not adhere well to bare aluminum, adheres very well to chemical conversion coatings. Recent concerns with the toxicity of the hexavalent chromium used in chromating have, however, limited its usage.

14.7.6 Nitriding

Steels containing nitride-forming elements such as chromium, molybdenum, aluminum, and vanadium can be treated to produce hard surface coatings with excellent, providing improved wear resistance. Many of the processes employed are proprietary, but typically they involve exposure of cleaned surfaces to anhydrous ammonia at elevated temperatures. The nitrides formed are not only harder but they are also more voluminous than the original steel which creates compressive residual stresses on the treated surface. Therefore, nitrided steels usually exhibit improved fatigue and corrosion fatigue resistance. Similar beneficial effects can be achieved by shot peening.

14.7.7 Passive Films

The bluing of steel has long been used as a process for producing a protective oxide on steel. When oiled and maintained, the thin oxide film prevents the rusting of steel in normal atmospheres. Such oxide coatings can be produced by exposing a clean steel surface to dry steam, strong alkalies, cyanides, or a great number of other chemicals. The films must be very thin and continuous to have the mechanical and physical properties desired.

Austenitic stainless steels and hardenable stainless steels such as martensitic, precipitation hardening, and maraging stainless steels are seldom coated, but their corrosion resistance depends on the formation of naturally occurring transparent oxide films. These films may be impaired by surface contaminants such as organic compounds or metallic or inorganic materials. Treatments are available to clean and degrease the surface of these materials to produce uniform protective oxide films. These usually involve immersion in an aqueous solution of nitric acid and a dichromate solution.

14.7.8 Pack Cementation

Certain materials can be diffused into the surface of a metal to produce a surface product entirely different from the metal. Powders of carbonaceous materials, aluminides, chromium salts, borides, or silicides are packed around the object in a container. The mass is taken to high temperature for sufficient time for the carbon, aluminum, chromium, or boron to diffuse into the metal surface. A hard, abrasion-resistant surface of some 20 to 50 μm in thickness can be thus produced. These coatings are used to impart abrasion and/or high-temperature resistance to gas turbine blades and vanes, among other similar uses. Aluminum coatings are frequently used on heat exchanger surfaces subject to attack by sulfur gases (H_2S, SO_2, and SO_3).

The sherardizing process involves diffusion of a zinc coating into steel by tumbling steel parts with zinc dust at high temperatures.

14.8 Metallic Coatings

For many rugged services, the choice of a metal coating is preferred. Where severe impact, abrasion, or high temperatures are a part of the materials selection consideration, the use of metallic coatings should be considered. However, for service in presence of an electrolyte it is imperative to examine if there are any galvanic incompatibility between the substrate and the metallic coating being applied. If there is a breach in the protective coating and a contact established between the two different metals by the presence of some aqueous electrolyte, corrosion of either may be accelerated or accentuated, depending on the existing conditions (Fig. 14.11).

14.8.1 Electroplating

Electroplating is achieved by passing an electrical current through a solution containing dissolved metal ions and the metal object to be plated. The metal object serves as the cathode in an electrochemical cell, attracting metal ions from the solution. Ferrous and nonferrous

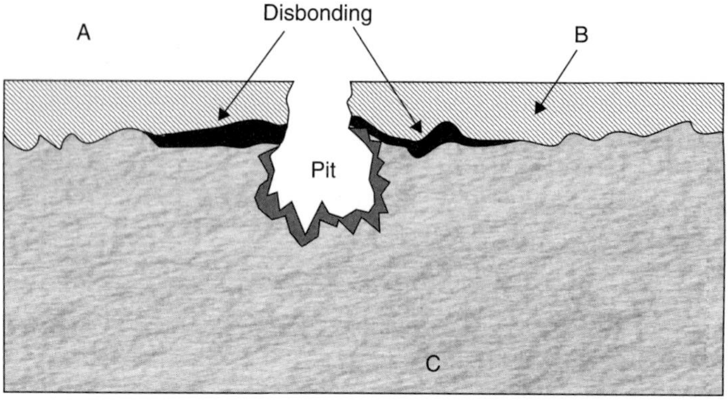

Figure 14.11 Shows how a holiday (hole) in a coating cathodic to a base may cause a pit in the base metal. (A) electrolyte; (B) cathodic layer; and (C) base metal. The indicated disbonding may or may not take place.

metal objects are plated with a variety of metals, including aluminum, brass, bronze, cadmium, copper, chromium, iron, lead, nickel, tin, and zinc, as well as precious metals, such as gold, platinum, and silver.

The process is regulated by controlling a variety of parameters, including the voltage and amperage, temperature, residence times, and the purity of bath solutions. Plating baths are almost always aqueous solutions; therefore, only those metals that can be reduced from aqueous solutions of their salts can be electrodeposited. The only major exception is aluminum that may only be plated from nonaqueous organic electrolytes. The sequence of unit operations in an electroplating operation typically involves various cleaning steps, stripping of old plating or paint, electroplating steps, and rinsing between and after each of these operations. Electroless plating uses similar steps but involves the deposition of metal on a substrate without the use of external electrical energy.

For small, intricate shapes such as bolts and screws, plating metal may be applied by barrel plating. A rotating barrel containing the parts to be plated is turned slowly in the electrolyte, while current is discharged to the tumbling parts inside.

Brush plating is also possible, particularly for touch-up of small surfaces. In this case, an absorbent pad holding the electrolyte is wrapped around a metal anode and the brush is moved over the cathodic surface.

Chrome plate, which was extensively used on automotive parts until the introduction of plastics, is actually a three-ply coating. A flash of copper is first laid down, followed by a nickel coating which comprises most of the finished thickness. A thin, hard coating of bright chromium is then applied as the topcoat. The total thickness is normally 25 to 50 μm.

Electroplated coatings normally fail by cracking from repeated expansion and contraction or impact. Another problem associated with plating is the presence of large quantities of hydrogen on the cathodic surface. Some of this hydrogen can enter the substrate metal (normally steel). For low-strength steels (hardness < R_C22), hydrogen permeation seldom poses a problem. However, the higher-strength steels can be badly embrittled and fail prematurely. When poisons for the recombination into molecular hydrogen, for example, S, CN, As, Cd, and so on, are present in the bath, the amount of atomic hydrogen absorbed can be quite high. If immediate damage is not done, the parts may be baked at temperatures around 200°C to drive most of the hydrogen out of the metal.

Thin electroplated films of gold, platinum, and silver are used on electrical contacts and in electronic equipment. Because gold and platinum resist oxidation, they are important in contacts where small currents are being switched or passed through connections, because any oxides on the contact surfaces impose unwanted resistances in the circuits.

14.8.2 Electroless Plating

Electroless nickel (EN) plating is a chemical reduction process that depends upon the catalytic reduction process of nickel ions in an aqueous solution (containing a chemical reducing agent) and the subsequent deposition of nickel metal without the use of electrical energy. Electroless plating processes are widely used in industry to meet the end-use functional requirements and are only rarely used for decorative purposes.

In the EN plating process, the driving force for the reduction of nickel metal ions and their deposition is supplied by a reducing agent in solution. This plating power is thus relatively constant at all points of the surface of the component, provided the agitation is sufficient to ensure a uniform concentration of metal ions and reducing agents. The electroless deposits are therefore very uniform in thickness all over the part's shape and size. This offers distinct advantages when plating irregularly shaped objects, holes, recesses, internal surfaces, valves, threaded parts, and so forth.

In a true electroless plating process, reduction of metal ions occurs only on the surface of a catalytic substrate in contact with the plating solution. Once the catalytic substrate is covered by the deposited metal, the plating continues because the deposited metal is also catalytic.

Nickel deposits have unique magnetic properties, except deposits containing more than 8 percent phosphorus are essentially nonmagnetic. In Ni-P coatings, phosphorus is present as supersaturated solution in fine microcrystalline solid solution, bordering on amorphous or liquid-like (glass-like) metastable structure, and this phosphorus is responsible for the nonferromagnetic behavior of as-plated Ni-P deposits with more than 8 percent phosphorus.

A second generation of EN plating has been developed by codepositing micrometer-sized particles of silicon carbide with the nickel, thereby creating an extremely wear- and corrosion-resistant coating. In this composite coating, the nickel alloy matrix provides corrosion resistance while the silicon carbide particles add wear resistance.

14.8.3 Hot-Dip Galvanizing

Hot-dip galvanizing is the process of applying a zinc coating to fabricated iron or steel material by immersing the material in a bath consisting primarily of molten zinc. Galvanizing can be found in almost every major application and industry where iron or mild steel is used. The simplicity of the galvanizing process is a distinct advantage over other methods of providing corrosion protection. The automotive industry depends heavily on this process for the production of many components used in car manufacturing, including fully assembled car bodies.

The electrochemical protection provided by zinc coatings is a vital element in the effectiveness of galvanized coatings in protecting steel from corrosion. All pregalvanized products rely on the cathodic protection provided by zinc to prevent corrosion of exposed steel at cut edges. In any situation where zinc is corroded sacrificially to protect exposed steel, the mass of available zinc will determine the corrosion protection performance. Corrosion rates of zinc coatings required to cathodically protect uncoated steel in aggressive environments (saltwater/marine) may be 25 times higher than the normal zinc corrosion rate.

14.8.4 Cladding

Metal layers of varying thicknesses can be applied to other metals by various methods designed to obtain a more corrosion-resistant surface while using a less expensive or a stronger structural material to fabricate the component or system of interest. Cladding is most commonly applied at the mill stage by the manufacturers of sheet, plate, tubing, or coins. Cladding by pressing, rolling, or extrusion can produce a coating in which the thickness and distribution can be controlled over wide ranges, and the coatings produced are free of porosity.

Although there is almost no practical limit to the thickness of coatings that can be produced by cladding, the application of the process is limited to simple-shaped articles that do not require much subsequent mechanical deformation. Among the principal uses are lead and cadmium sheathing for cables, lead-sheathed sheets for architectural applications, and composite extruded tubes for heat exchangers.

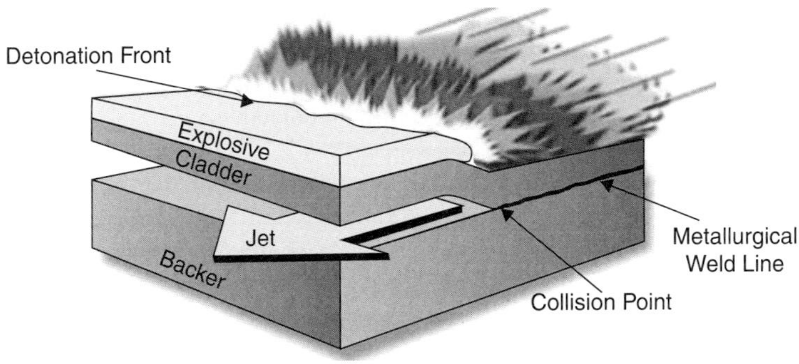

FIGURE 14.12 Manner in which explosives bond a layer of cladding metal to a base: explosive; cladding metal; base metal; and "jet" securing cladding layer to base.

In roll bonding, layers of two metals are mated by heavy rolling in a mill after the surfaces have been thoroughly cleaned and treated. Clad thicknesses of 5 to 10 percent of the base steel thickness are common. Some small areas of unbonded metal will be present. Two metals may also be coextruded through a die.

Cladding can also be accomplished by other methods such as arc or gas welding in which relatively thick layers of weld metal are deposited either by manual or machine methods on surfaces. The interior of pulp digesters or other pressure vessels requiring an alloy composition to resist the chemical conditions can be constructed in this manner.

Laser Welding Propeller

In explosion bonding, a base metal and a covering metal coating are contacted, placed in an appropriate enclosure with the coating material on top. A layer of explosive is placed over the coating metal. When the explosion is activated, the resulting shock wave merges the two materials, as shown in Fig. 14.12.

14.8.5 Metallizing (Thermal Spray)

The metallizing technique for applying metal coatings consists of thermally spraying one type of metal onto a metallic substrate, most often steel, or reinforced concrete to provide cathodic protection to the rebar material as discussed in Chap. 13. The technique is accomplished in several ways. Table 14.6 shows the main metallic materials that have been used for the production of spray coatings and Table 14.7 summarizes the limits and applicability of each technique.

Coatings can be sprayed from rod or wire stock or from powdered materials. The material (e.g., wire) is fed into a flame, where it is melted. The molten stock is then stripped from the end of the wire and atomized by a high-velocity stream of compressed air or other

Type Coating	General Qualities
Aluminum	Highly resistant to heat, hot water, and corrosive gases. Excellent heat distribution and reflection.
Babbitt	Excellent bearing wearability.
Brass	Machines well, takes a good finish.
Bronze	Excellent wear resistance; exceptional machineability; dense coatings (especially Al, Bronze).
Copper	High heat and electrical conductivity.
Iron	Excellent machining qualities.
Lead	Good corrosion protection fast deposits and dense coatings.
Molybdenum (Molybond)	Self-bonding for steel surface preparation.
Monel	Excellent machining qualities; highly resistant to corrosion.
Nickel	Good machine finishing; excellent corrosion protection.
Nickel-Chrome	High-temperature applications.
Steel	Hard finishes, good machineability.
Chrome Steel (Tufton)	Bright, hard finish, highly resistant to wear.
Stainless	Excellent corrosion protection and superior wearability.
Tin	High purity for food applications.
Zinc	Superior corrosion resistance and bonding qualities.

Table 14.6 Spray-Coating Materials

gas, which propels the material onto a prepared substrate or workpiece. Depending on the substrate, bonding occurs either due to mechanical interlock with a roughened surface, due to localized diffusion and alloying, and/or by means of van der Waals forces (i.e., mutual attraction and cohesion between two surfaces). There are three basic categories of thermal spray technologies:

- Feeding a wire or metal powder into oxyacetylene or oxyhydrogen flames (Fig. 14.13).
- Feeding a wire into the heated zone created by electrodes arcing alternating current from which molten metal is impelled by a stream of compressed air toward the surface to be coated (Fig. 14.14).
- Blowing metallic particles suspended in a gas through a blowpipe flame, twin-arc (Fig. 14.15) or plasma (Fig. 14.16), a process that can be used to coat inside pipes and tubes.

Combustion Torch/Flame Spraying
Flame spraying is noted for its relatively high as-deposited porosity, significant oxidation of the metallic components, low resistance to impact or point loading, and limited thickness (typically 0.5 to 3.5 mm). Advantages include the low capital cost of the equipment, its simplicity, and the relative ease of training the operators. In addition, the technique uses materials efficiently and has low associated maintenance costs.
Combustion Torch/High-Velocity Oxy-Fuel (HVOF)
This technique has very high-velocity impact, and coatings exhibit little or no porosity. Deposition rates are relatively high, and the coatings have acceptable bond strength. Coating thickness range from 0.000013 to 3 mm. Some oxidation of metallics or reduction of some oxides may occur, altering the coating's properties.
Combustion Torch/Detonation Gun
This technique produces the densest of the thermal coatings. Almost any metallic, ceramic, or cement materials that melt without decomposing can be used to produce a coating. Typical coating thickness range from 0.05 to 0.5 mm, but both thinner and thicker coatings are used. Because of the high velocities, the properties of the coatings are much less sensitive to the angle of deposition than most other thermal spray coatings.
Electric Arc Spraying
Coating thickness can range from a few hundredths of a mm to almost unlimited thickness, depending on the end use. Electric arc spraying can be used for simple metallic coatings, such as copper and zinc, and for some ferrous alloys. The coatings have high porosity and low bond strength.
Plasma Spraying
Plasma spraying can be used to achieve thickness from 0.3 to 6 mm, depending on the coating and the substrate materials. Sprayed materials include aluminum, zinc, copper alloys, tin, molybdenum, some steels, and numerous ceramic materials. With proper process controls, this technique can produce coatings with a wide range of selected physical properties, such as coatings with porosity ranging from essentially zero to high porosity.
Ion Plating/Plasma Based
This technique produces coatings that typically range from 0.008 to 0.025 mm. Advantages include a wide variety of processes as sources of the depositing material; in-situ cleaning of the substrate prior to film deposition; excellent surface covering ability; good adhesion; flexibility in tailoring film properties such as morphology, density, and residual film stress; and equipment requirements and costs equivalent to sputter deposition. Disadvantages include many processing parameters must be controlled; contamination may be released and activated in the plasma; and bombarding gas species may be incorporated in the substrate and coating.

Table 14.7 Limits and Applicability of the Main Techniques for Producing Thermal Spray Coatings

Ion Plating/Ion Beam Enhanced Deposition (IBED)
Advantages include increased adhesion; increased coating density; decreased coating porosity and prevalence of pinholes; and increased control of internal stress, morphology, density, and composition. Disadvantages include high equipment and processing costs; limited coating thickness; part geometry and size are limited; and gas precursors used for some implantation species are toxic. This technique can produce a chromium deposit 10 μm thick with greater thickness attained by layering. Such thickness is too thin for most hard chrome requirements (25 to 75 μm with some dimensional restoration work requiring 750 μm) and layering would significantly add to the cost of the process. IBED provides some surface cleaning when the surface is initially illuminated with a flux of high energy inert gas ions; however, the process will still require precleaning (e.g., degreasing).
Ion Implantation
Ion implantation can be used for any element that can be vaporized and ionized in a vacuum chamber. Since material is added to the surface, rather than onto the surface, there is no significant dimensional change or problems with adhesion. The process is easily controlled, offers high reliability and reproducibility, requires no posttreatment, and generates minimal waste. If exposed to high temperatures, however, implanted ions may diffuse away from the surface due to limited depth of penetration and penetration does not always withstand severe abrasive wear. Implantation is used to alter surface properties, such as hardness, friction, wear resistance, conductance, optical properties, corrosion resistance, and catalysis. Commercial availability is limited by general unfamiliarity with the technology, scarcity of equipment, lack of quality control and assurance, and competition with other surface modification techniques. Areas of research includes ion implantation of ceramic materials for high temperature internal combustion engines, glass to reduce infrared radiation transmission and reduce corrosion, as well as automotive parts (piston rings, cylinder liners) to reduce wear.
Sputtering and Sputter Deposition
This technique is a versatile process for depositing coatings of metals, alloys, compounds, and dielectrics on surfaces. The process has been applied in industrial hard and protective coatings. Primarily TiN, as well as other nitrides and carbides, has demonstrated high hardness, low porosity, good chemical inertness, good conductivity, and attractive appearance. Sputtering is capable of producing dense films, often with near bulk quantities. Areas requiring future research and development include better methods for in-situ process control; methods for removing deposited TiN and other hard, ceramic-like coatings from poorly coated or worn components without damage to the product; and improved understanding of the factors the affect film properties.

Table 14.7 *(continued)*

Laser Surface Alloying
This technique can be used to apply most of the same materials that can be applied via thermal spray techniques; the powders used for both methods are generally the same. Materials that are easily oxidized, however, will prove difficult to deposit without recourse to inert gas streams and envelopes. Deposition rates depend on laser power, powder feed rates, and traverse speed. The rates are typically in the region of 2×10^{-4} cm^3 for a 500-W beam. Thickness of several hundred microns can be laid down on each pass of the laser beam allowing thickness of several millimeters to accumulate. If the powder density is too high, this thermal cycling causes cracking and delamination of earlier layers, severely limiting the attainable buildup. Research has found that easily oxidized materials, such as aluminum, cannot be laser clad because the brittle oxide causes cracking and delamination. Some steels may be difficult to coat effectively. The small size of the laser's beam limits the size of the workpieces that can be treated cost effectively. Shapes are restricted to those that prevent line-of-sight access to the region to be coated.
Chemical Vapor Deposition
CVD is used mainly for purposes of corrosion resistance and wear resistance. CVD processes are also usually applied in cases where specific properties of materials of interest are difficult to obtain by other means. CVD is unique because it controls the microstructure and/or chemistry of the deposited material. The microstructure of CVD deposits depends on chemical makeup and energy of atoms, ions, or molecular fragments impinging on the substrate; chemical composition and surface properties of the substrate; substrate temperature; and presence or absence of a substrate bias voltage. The most useful CVD coatings are nickel, tungsten, chromium, and titanium carbide. Titanium carbide is used for coating punching and embossing tools to impart wear resistance.

Table 14.7 *(continued)*

Thorough surface preparation usually is required because the molten metal droplets must achieve adhesion before solidification. The coating achieved is somewhat oxidized and porous. Common applications are on steel surfaces exposed on offshore petroleum production platforms, chemical tanker compartments, stack breechings, and other structural steel locations. Metallized surfaces frequently are top-coated (Fig. 14.17).

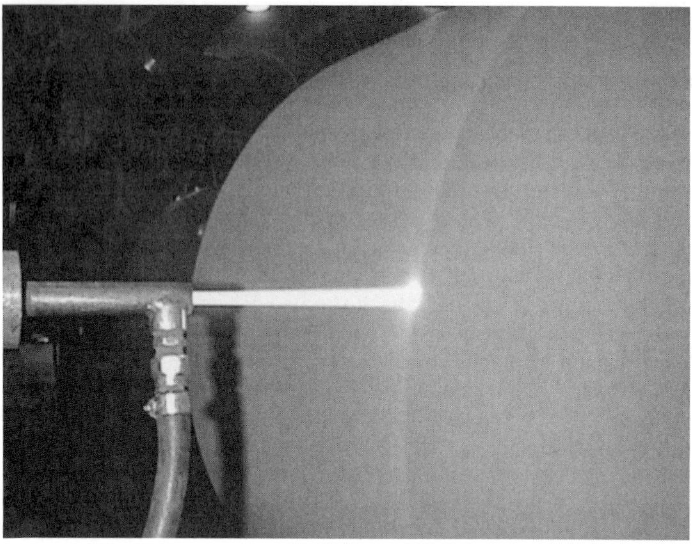

Figure 14.13 High-velocity oxygen fuel thermal spray process. (Courtesy of M. Gariepy, WEIR Engineering Services, Montreal, Canada)

Figure 14.14 Arc-spraying process. (Courtesy of M. Gariepy, WEIR Engineering Services, Montreal, Canada)

Figure 14.15 Twin arc blowing torch for metallizing inside pipes and tubes. (Courtesy of M. Gariepy, WEIR Engineering Services, Montreal, Canada)

Figure 14.16 Plasma-blowing torch for metallizing inside pipes and tubes. (Courtesy of M. Gariepy, WEIR Engineering Services, Montreal, Canada)

(a)

(b)

Figure 14.17 Pump casing before (*a*) being arc sprayed with stainless steel S43000 and top-coated with epoxy (*b*). (Courtesy of M. Gariepy, WEIR Engineering Services, Montreal, Canada)

14.9 Coating Inspection and Testing

Tremendous financial losses may be incurred as a result of premature failure of paints and coatings. Costs of such failures often far outweigh the initial costs of painting because of the complexity of repairs and the liabilities associated with downtime to correct the problems. It is therefore important to choose the right coating system for an application and in order to do so coating operations of almost any size should have some type of meaningful test program. Such programs are the only objective basis on which to make decisions relating to future work. These programs should evaluate materials, surface preparation, application, and inspection procedures.

It is also important to investigate coating failures when they do occur since this may provide critical clues on what should be corrected to avoid recurring problems. These investigations may be done in-house or contracted to the many consulting firms available to help those without the necessary personnel. A similar team of experts should also carry out condition surveys of a structure or a facility prior to beginning a protective coatings project by considering the following factors [3]:

- Determine the condition of the substrate.
- Determine the condition of the existing protective coating system.
- Identify the environment in which the protective coating systems will be required to protect the substrate.
- Estimate the surface area to be protected with coating systems.

A proper survey requires dividing the structure or facility into easily defined parts, zones, or sections before proceeding in a methodical fashion to establish exactly what has happened to the structures and the coating systems since its last painting program.

14.9.1 Condition of the Substrate

Determining the condition of the substrate is a critical part of any condition survey. Coatings can mask some rather severe corrosion for a period of time. Corrosion under insulation (CUI) is a particular example of the development of a very severe environment that may go unnoticed until a catastrophic leak occurs. Contaminants from process or environmental upsets can be difficult to find from a purely visual standpoint. The underlying substrate must be accessed in some way. If it is under insulation, strategically placed inspection windows may need to be cut into the insulation or its cladding so proper inspection or monitoring can be performed on the substrate to determine its true state.

If it is under an old protective coating system, strategically located sections of the coating system may need to be removed to provide access to the substrate for the same types of tests. Adhesion may need to be verified by following either ASTM D3359 [15] or ASTM D4541 [16]. If it is visually rusty, the extent of the rust and the depth of the pitting must be established. This can be accomplished using the SSPC-VIS 2 Guide [17] that provides comparative photographs in addition to the rust grades shown in Table 14.8.

Blisters should be identified using the ASTM D714 [18] in which blisters are categorized by size, shape, location, and density. The standard has a set of photographs depicting blister sizes from No. 2, which is the largest size to No. 8, which is the smallest. The photographs also depict the density of the blisters as few, medium, medium dense, and dense.

It is often critical to determine how much of the original wall thickness of the metal has been lost on visually rusty surfaces. This can be done with a number of instruments. A portable A-scan ultrasonic thickness gauge may be quite useful to determine and store in memory the remaining wall thickness at any given point of a structure. B-scan-type ultrasonic thickness gauges can measure and plot wall thickness on complete structures or sections of structures from which isolated corrosion pits, vessel corrosion rate, general wall thickness, and visual presentation of the section being measured can be produced.

If it is a tank or other vessel that has contained liquids, the type and temperature of the liquids must be identified. The substrate must

Rust Grades	Description	Standard
10	No rusting or less than 0.01% of surface rusted	Unnecessary
9	Minute rusting, less than 0.03% of surface rusted	No. 9
8	Few isolated rust spots, less than 0.1% of surface rusted	No. 8
7	Less than 0.3% of surface rusted	No. 7
6	Extensive spots, but less than 1% of surface rusted	No. 6
5	Rusting to the extent of 3% of surface rusted	No. 5
4	Rusting to the extent of 10% of surface rusted	No. 4
3	Approximately 1/6 of surface rusted	No. 3
2	Approximately 1/3 of surface rusted	No. 2
1	Approximately 1/2 of surface rusted	No. 1
0	Approximately 100% of surface rusted	Unnecessary

TABLE 14.8 Guide to SSPC-VIS 2: Standard Method of Evaluating Degree of Rusting on Painted Steel Surfaces.

then be analyzed for contaminants normally associated with those liquids. Coatings that have been otherwise properly applied over steel that appeared to be perfectly clean have failed due to the lack of identification and removal of such contaminants.

14.9.2 Condition of the Existing Coating System

Historical records of a coating system when it was first applied, as well as during periodic annual surveys, can provide a good indication of the relative life left in the coating system, particularly when compared with the life expectancy charts published by the Society for Protective Coatings (SSPC). These life expectancy charts are categorized both by the generic type of coating and the environment to which it is exposed. For example, epoxy resins are known to be fragile to sunlight UV rays that can degrade them to a powdery film easily removed by wind and rain, eventually exposing underlying coats or the substrate. By checking the historical records, the remaining life of a coating system can be estimated fairly well.

Simple visual scans do provide a broad indication of the areas of a structure that are deteriorating at a faster rate. The appearance of red iron oxide pigmented primers through the remaining colored topcoats is a very convincing indication of the need for immediate corrective action at the lowest maintenance cost of any maintenance practice. Unless corrosion is visually seen, a simple waterjet washing with fresh water, and a new topcoat of the same generic coating previously used, or of one that is compatible with the previous coating system, can extend the life of the current system by a few years.

If only small isolated areas of the structure show corrosion while the majority of the structure appears sound, it can prompt the maintenance engineer to schedule a spot blast or power tool white metal cleaning of the rusty spots, a spot prime with an inhibitive primer, and a new refresher coat of the topcoat of the system. Again, the economy comes in avoiding a total breakdown of the protective coating system, which would require total removal and replacement of the entire system. That is the most costly of all maintenance programs.

14.9.3 Coating Inspection

Characteristics of a properly prepared surface may include specification of a surface profile in terms of mils depth, as well as other criteria. This is important in some instances, especially when a thin film coating is to be applied. If the surface profile is such that an inadequate thickness of coating is deposited on the peaks, or the coating bridges the valley (Fig. 14.18), a coating failure may follow, or, at best, a greater volume of coating will be required for adequate

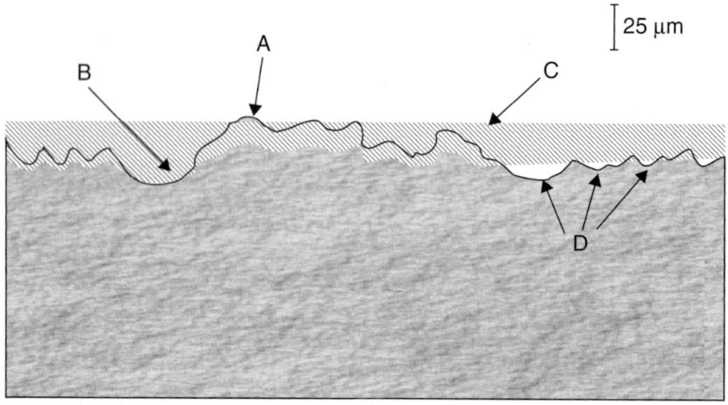

Figure 14.18 Diagram showing cross-section through a base which has been centrifugally blasted with G-50 grit (fine) and then coated. Note that peak A extends above coating surface, B indicates a valley, and C indicates coating surface. Coating fails first at peak A, while D indicates "bridging" of valleys where coating failure may start.

coverage of the peaks than would be the case if the profile were of proper depth and spacing.

Gauges and equipment are available for checking surface profiles (Figs. 14.19 to 14.24). In the hands of an experienced person, a good statistical estimate can be made in a short time of a properly prepared surface. Profilometers (useful laboratory instruments) give exact

Figure 14.19 Keane-Tator surface profile comparator disc. (Courtesy of KTA-Tator, Inc.)

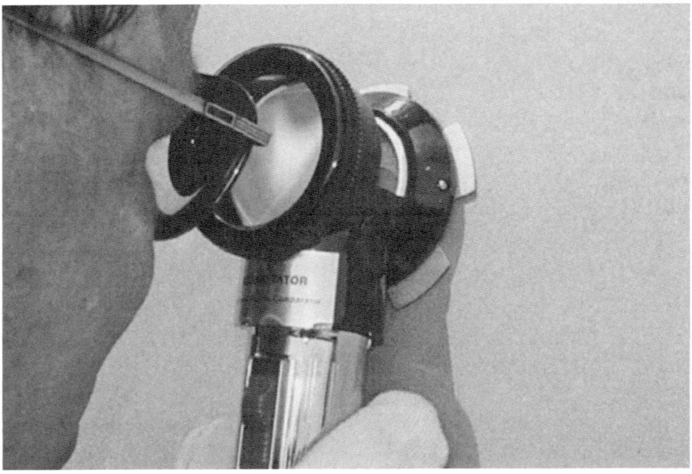

FIGURE 14.20 Use of Keane-Tator surface profile comparator method for measurement of surface profile depth. (Courtesy of KTA-Tator, Inc.)

FIGURE 14.21 Measuring surface profile depth using depth micrometer. (Courtesy of KTA-Tator, Inc.)

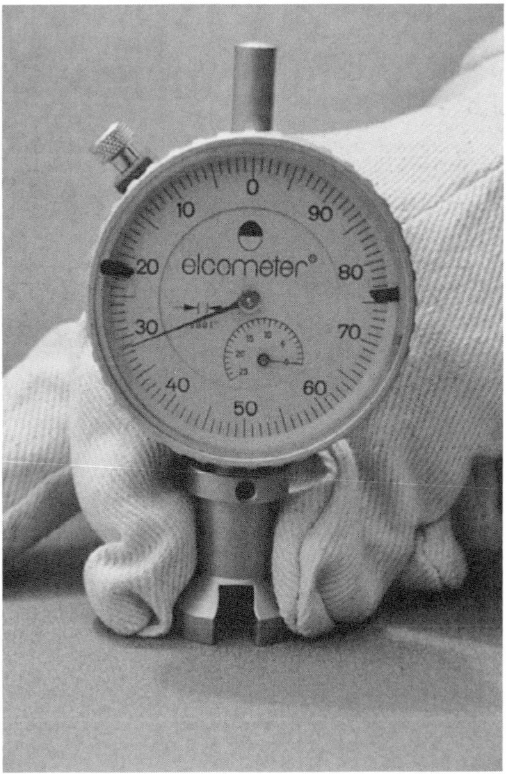

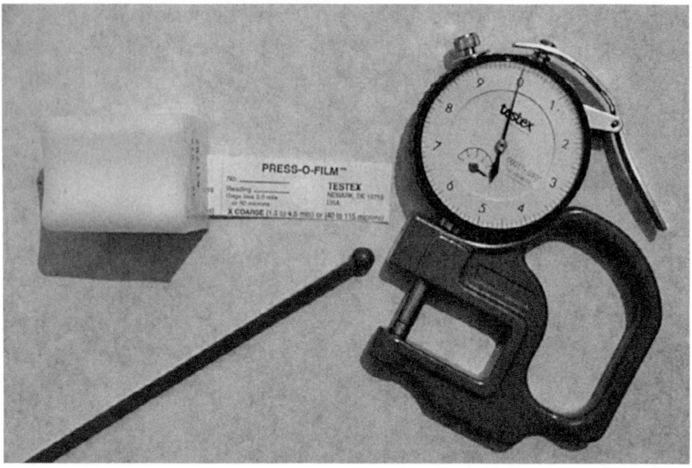

Figure 14.22 Testex Press-O-Film replica tape surface profile measurement equipment. (Courtesy of KTA-Tator, Inc.)

information on the profile, but it is difficult and usually impossible to use them in the field.

> *Wet film gauging:* Thickness of a wet film can be determined by any of several gauges which consist of a series of calibrated indentations along the edge of a handheld ruler (Fig. 14.25). Probable thickness of the dried film can be calculated if solids content by volume is known.

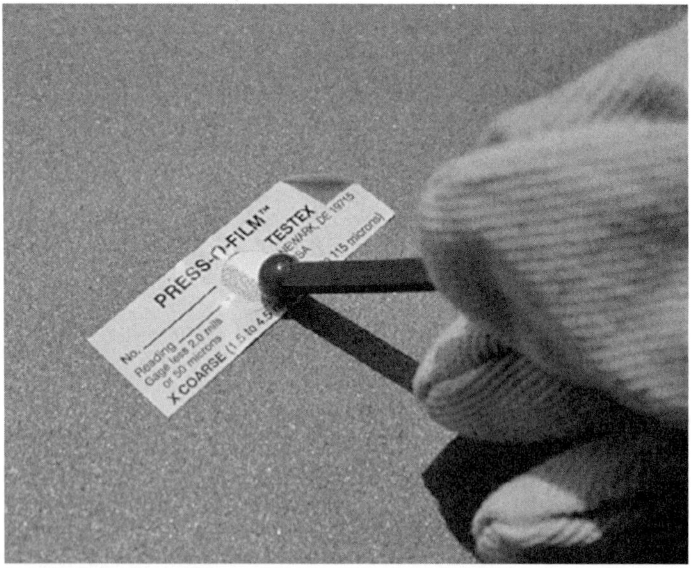

Figure 14.23 Burnishing Testex Press-O-Film replica tape. (Courtesy of KTA-Tator, Inc.)

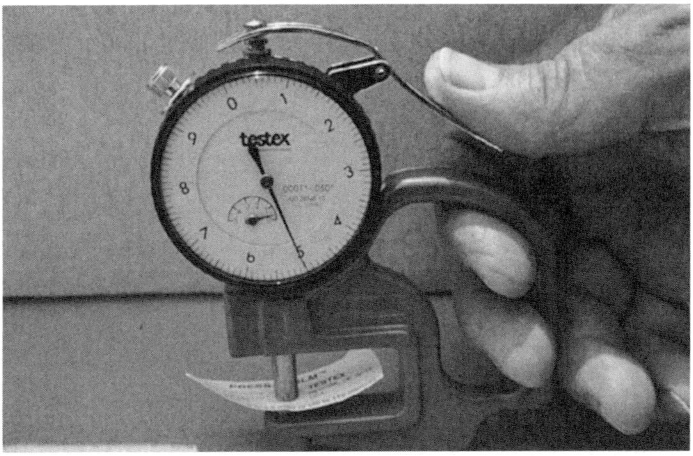

Figure 14.24 Measuring surface profile depth using Testex micrometer. (Courtesy of KTA-Tator, Inc.)

Dry film gauging: Electrical and magnetic nondestructive and visual destructive equipment for determining thickness are described in several places (Figs. 14.26 and 14.27).

Coverage completeness: Coverage completeness can be checked by using different colors of multiple coats or conductive and nonconductive coats.

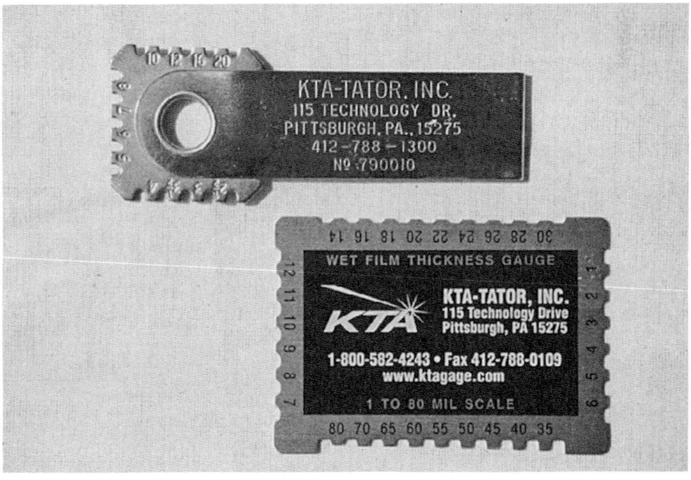

Figure 14.25 Notch-type wet film thickness gauges (top: stainless steel; bottom: aluminum). (Courtesy of KTA-Tator, Inc.)

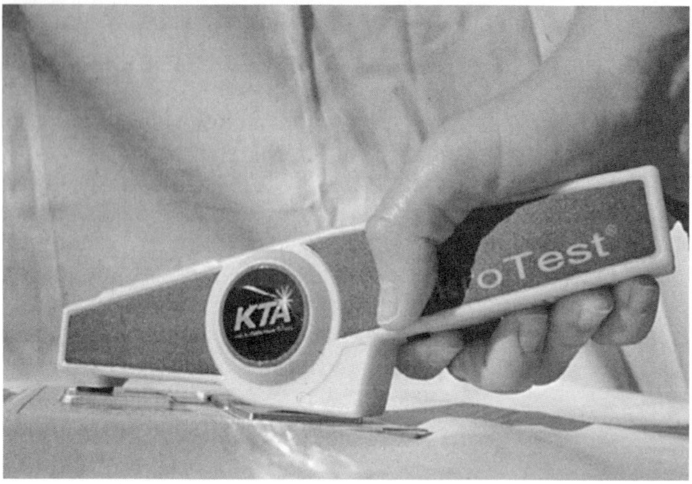

FIGURE 14.26 Magnetic pull-off dry film thickness measurement gauge. (Courtesy of KTA-Tator, Inc.)

FIGURE 14.27 Electronic dry film thickness measurement gauge. (Courtesy of KTA-Tator, Inc.)

FIGURE 14.28 Tooke gauge. (Courtesy of KTA-Tator, Inc.)

Visual inspection: Visual inspection can be conducted with or without magnification. The Tooke gauge, which makes a diagonal cut through a coatings system to the substrate, permits measurement and identification of the coating layers under magnification (Figs. 14.28 and 14.29). These gauges are less commonly used because they are slower to operate, and create a cut through the film.

14.9.4 Laboratory Testing

Whenever possible, standard test procedures and means of evaluation provided by organizations such as ASTM, NACE, or SSPC should be consulted and followed. Only in this manner can correlation with the results of other workers be established. SSPC and NACE certified coating/corrosion professionals are trained to investigate paint failures and know which analytical technique or investigation method may provide the most useful information.

For example, gas chromatography is excellent at detecting solvents, but would be of little use in determining why an alkyd resin failed. Another example is that only electrochemical impedance spectroscopy (EIS) and actual exposure tests can provide a distinction between storage-related problems and paint application/formulation problems when white rust is formed on painted galvanized steels. Traditional approaches cannot make such distinction. These insights could save unnecessary effort and expense, simply by selecting the right test.

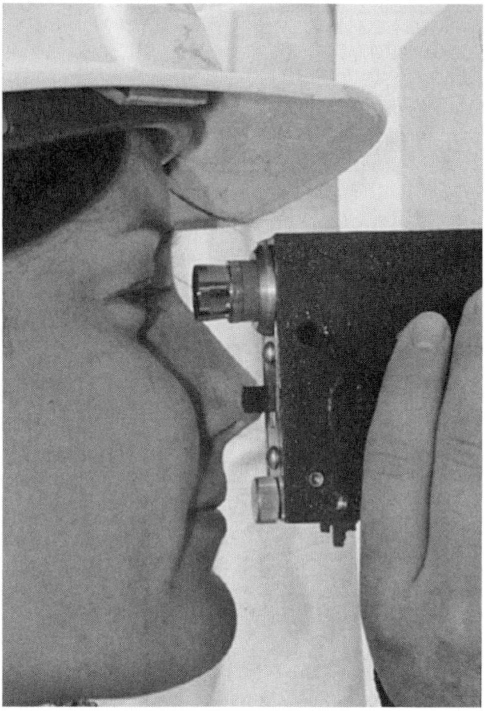

FIGURE 14.29 Use of Tooke gage for measurement of paint layers. (Courtesy of KTA-Tator, Inc.)

The number and types of tests that can be conducted in the laboratory are almost limitless. Table 14.9 lists some of the ASTM tests relevant to protective coatings. Because field testing of coatings is so costly in time, labor, and facilities, it has been the persistent aim of formulators, raw materials producers, and users of coatings to establish meaningful accelerated tests. These may involve a wide variety of comparatively simple, as well as highly sophisticated procedures.

These tests can define certain characteristics of a coating system. This discrimination usually is specific to the type of accelerated test, and frequently does not indicate in advance how a given coating will perform in service. However, the tests are invaluable when developing coatings or appraising new concepts in coatings application or use.

General immersion tests: General immersion tests can be as simple as the standard saltwater immersion test for coatings to be used in the atmosphere, or as complicated as cold wall effect immersion tests for heavy coatings to be used constantly in immersion service. It is essential to reproduce the anticipated conditions of the field exposure as closely as possible when evaluating coatings for immersion service.

ASTM Standard	Title
B 117	Test Method of Salt Spray (Fog) Testing
B 368	Method for Copper Accelerated Acetic Acid-Salt Spray (Fog) Testing (CASS Test)
B 457	Test Method for Measurement of Impedance of Anodic Coatings on Aluminum
C 536	Test Method for Continuity of Coatings in Glassed Steel Equipment by Electrical Testing
C 743	Test Method for Continuity of Porcelain Enamel Coatings
D 522	Test Methods for Mandrel Bend Test of Attached Organic Coatings
D 523	Test Method for Specular Gloss
D 610	Test Method for Evaluating Degree of Rusting on Painted Steel Surfaces
D 662	Test Method for Evaluating Degree of Erosion of Exterior Paints
D 714	Test Method for Evaluating Degree of Blistering of Paints
D 822	Practice for Conducting Tests on Paint and Related Coatings and Materials Using Filtered Open-Flame, Carbon-Arc Light, and Water Exposure Apparatus
D 823	Practices for Producing Films of Uniform Thickness of Paint, Varnish, Lacquer, and Related Products on Test Panels
D 870	Practice for Testing Water Resistance of Coatings Using Water Immersion
D 968	Test Methods for Abrasion Resistance of Organic Coatings by Falling Abrasive
D 1014	Test Method for Conducting Exterior Exposure Tests of Paints on Steel
D 1186	Test Method for Nondestructive Measurement of Dry Film Thickness of Nonmagnetic Coatings Applied to a Ferrous Base
D 1400	Test Method for Nondestructive Measurement of Dry Film Thickness of Nonconductive Coatings Applied to a Nonferrous Metal Base
D 1567	Test Method for Testing Detergent Cleaners for Evaluation of Corrosive Effects on Certain Porcelain Enamels
D 1653	Test Method for Water Vapor Transmission of Organic Coating Films
D 1654	Test Method for Evaluation pf Painted or Coated Specimens Subject to Corrosive Environments
D 1735	Practice for Testing Water Resistance of Organic Coatings Using Water Fog Apparatus
D 2197	Test Methods for Adhesion of Coatings by Scrape Adhesion

Table 14.9 Coatings Tests and Methods Pertinent to Corrosion Protection.

ASTM Standard	Title
D 2247	Practice for Testing Water Resistance of Coatings in 100% Relative Humidity
D 2248	Practice for Detergent Resistance of Organic Finishes
D 3258	Test Method for Porosity of Paint Films
D 3273	Test Method for Resistance to Growth of Mold on the Surface of Interior Coatings in an Environmental Chamber
D 3276	Guide for Paint Inspectors (Metal Substrates)
D 3359	Test Methods for Measuring Adhesion by Tape Test
D 3361	Practice for Operating Light- and Water-Exposure Apparatus (Unfiltered Carbon-Arc Type) for Testing Paint, Varnish, Lacquer, and Related Products Using the Dew Cycle
D 3363	Test Method for Film Hardness by Pencil Test
D 4145	Test Method for Coating Flexibility of Prepainted Sheet
D 4585	Practice for Testing Water Resistance of Coatings Using Controlled Condensation
D 4587	Practice for Conducting Tests on Paint and Related Coatings and Materials Using a Fluorescent UV-Condensation Light- and Water-Exposure Apparatus
E 376	Practice for Measuring Coating Thickness by Magnetic-field or Eddy-Current (Electromagnetic) Test Methods
G 8	Test Method for Cathodic Disbonding of Pipeline Coatings
G 12	Test Method for Nondestructive Measurement of Film Thickness of Pipeline Coatings on Steel
G 42	Test Methods for Cathodic Disbonding of Pipeline Coatings Subjected to Elevated Temperatures
G 53	Practice for Operating Light- and Water-Exposure Apparatus (Fluorescent UV-Condensation Type) for Exposure of Nonmetallic Materials
G 60	Method for Conducting Cyclic Humidity Tests
G 80	Test Method for Specific Cathodic Disbonding of Pipeline Coatings
G 84	Practice for Measurement of Time-of-Wetness on Surfaces Exposed to Wetting Conditions as in Atmospheric Testing
G 85	Practice for Modified Salt Spray (Fog) Testing
G 87	Practice for Conducting Moist SO_2 Tests
G 90	Practice for Performing Accelerated Outdoor Weathering of Nonmetallic Materials Using Concentrated Natural Sunlight
G 92	Practice for Characterization of Atmospheric Test Sites

Table 14.9 *(continued)*

Salt-spray testing: The salt-spray test, which was originally designed to test coatings on metals, has been widely used to evaluate the resistance of metals to corrosion in marine service or on exposed shore locations [19;20]. However, extensive experience has shown that, although salt-spray tests yield results somewhat similar to those exposed in marine environments, they do not reproduce all the factors causing corrosion in marine service. Salt-spray tests should thus be considered to be arbitrary performance tests and their validity dependent on the extent to which a correlation has been established between the results of the test and the behavior under expected conditions or service. Despite the current widespread use of continuous salt-spray methods, their unrealistic simulation of outdoor environments is a serious shortcoming.

Cathodic protection exposures: One accelerated test involves subjecting the material to which the coating has been applied to various voltages of impressed current, often in excess of that normally required to achieve cathodic protection, with the aim of demonstrating the ability of the coatings to resist the disbonding effect of the cathodic current or to resist electroendosmosis (Fig. 14.30).

Condensation apparatus: Relying on the known susceptibility of coatings to permeation by condensed water, a screening test is available in a laboratory-type cabinet which results in condensation of water on the painted surface. Disbonding of poorer coatings occurs.

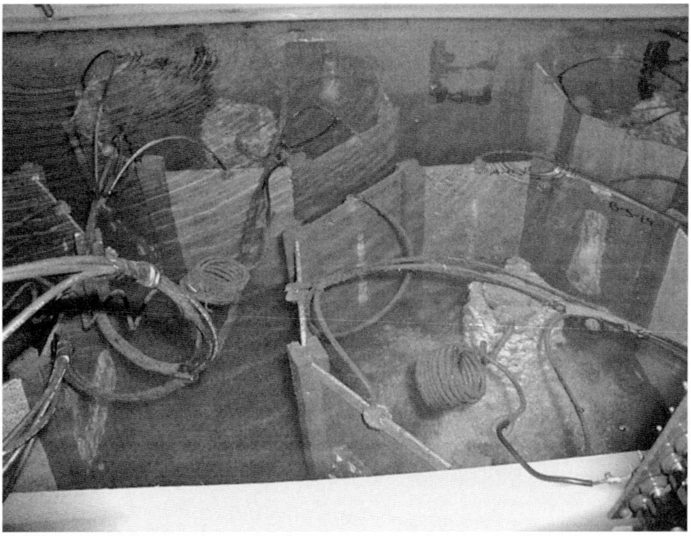

FIGURE 14.30 Laboratory setup to test for disbonding during the application of cathodic protection. (Courtesy of Corrpro)

Environmental test rooms: Environmental test rooms which permit various combinations of temperature, fogging, humidity levels, and shower effects to simulate rain can reproduce in an accelerated mode many of the factors present in an actual exposure. For the most part, these tests use small cut or machined specimens, but assemblies of simulated components and complete systems can also be tested as shown in Fig. 11.35.

14.9.5 Holiday Detection

Holidays are pinholes and voids in normally non-conductive organic coatings that allow current to pass through the protective coating to the metal base material. These discontinuities are usually not detectable with normal or even enhanced visual inspection techniques. A holiday detector will help to find coating defects by impressing an electrical voltage across the coating (Fig. 14.31). An electrode passes over the entire coating surface and as it crosses a defect, it gives off an electrical discharge, or spark, which signals the operator that a holiday has been detected. The operator can then mark the holiday for subsequent repair prior to acceptance.

Holiday detectors must be selected for the thickness and type of coating being tested. Excessive voltage can stress or damage thin coatings if the detector has been set too high. Adherence to the manufacturer's instructions is recommended, including both the manufacturer

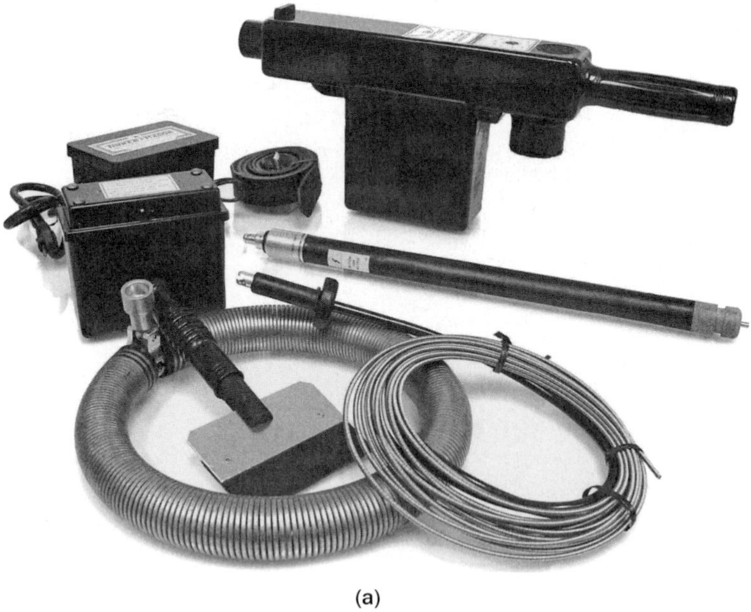

(a)

Figure 14.31 High voltage holiday detector kit (*a*); low-voltage holiday detector (*b*); holiday testing with high voltage detector (*c*). (Courtesy of Tinker & Rasor)

(b)

(c)

Figure 14.31 *(continued)*

of the coating and the manufacturer of the holiday detector. ASTM G62-07 Standard describes two test methods for holiday detection in pipeline coatings.

ASTM G62-0 Method A is designed to detect holidays such as pinholes and voids in thin-film coatings from 0.0254 to 0.254 mm in thickness using ordinary tap water and an applied voltage of less than 100 V DC. This method is effective on films up to 0.508 mm thickness if a wetting agent is used with the water and may be considered to be a non-destructive test because of the relatively low voltage. But Method A will not detect thin spots in the coating, even those as thin as 0.635 mm. However, this can be achieved with ASTM G62-07 Method B because of the higher applied voltages, between 900 and 20000 V DC. Method B is considered destructive because the high voltages involved generally destroy the coating at thin spots.

14.10 Surface Preparation

Surface preparation is a critical part of coating projects and must provide a surface that is compatible with the coating material to be applied. The main concerns in this aspect of protective coatings are the cleanliness of the surface required and the surface roughness or profile that essentially provide anchorage to the coating. It is believed that, of the cost of a coating job, as much as one-half to two thirds goes for surface preparation and labor.

In some cases it is impossible to provide the best prepared surface because there is insufficient money or time to do the job and/or it is not permitted because of possible product contamination, fire hazard, or some other reason. It must however be recognized that when there is a lesser surface preparation, the longevity of a coating system will likely be compromised.

The surface properties of the substrate to which a coating is applied will influence its performance. Surfaces in compression often increase coating performance because they tend to be attacked less than surfaces that are less stressed; on the other hand, surfaces in tension may cause coatings to fail earlier than they would otherwise. Differences in expansion rates between a coating and a substrate impose stresses on the coating which may lead to cracking and failure. Additionally, both inadequate and excessive coating thickness may lead to failure.

14.10.1 Principles of Coating Adhesion

Of the cleaning methods available to provide the cleanest and largest effective surface per unit area, the use of abrasive blasting produces a surface closest to the ideal surface profile. However, the best of blasted surfaces can still be badly contaminated with residual metal oxides, extraneous dirt, particles of the abrasive materials, and adsorbed gases on the metal. As a consequence, it is necessary to construct coating formulations with the maximum capacity to compete for and lock into the available bonding sites. For this reason, coating materials containing polar functionalities such as OH, and others, particularly C=O functional groups provide the best adhesion to a metal substrate.

Good wetting of the surface by liquid coatings is obviously a requisite for good adhesion. The coating material must remain stable to maintain adequate adhesion. If embrittlement occurs through oxidation, cross-linking, or volatilization of a portion of the coating, the resultant shrinkage stresses in the coating can pull it from the surface. Any loss of adhesion from such sources is obviously undesirable.

14.10.2 Abrasive Cleaning

The two main methods of abrasive cleaning before application of coatings are centrifugal blasting and air-pressure blasting. Centrifugal blasting is accomplished by machines that propel abrasives against the surface to be cleaned by imparting velocity to the particles by means of a rapidly rotating wheel. The abrasive material (commonly grit, and/or shot, but also specialized abrasives such as cut wire, various hard oxides, or carbides) impinges against the surface, removing both the surface contamination and making a pattern of indentations, such as shown in Fig. 14.18. This pattern, consisting of peaks and valleys, serves as anchors to the chemical bonding forces of materials applied to the surface. The total surface area exposed in a profile of this type is obviously much greater than that exposed by a smooth surface.

Centrifugal blasting machines are most economical when continuous or intermittent shipments of steel are cleaned by the wheel machine, such as in a steel fabrication shop. All steel, as it is received, can be run through the blasting machine and immediately primed with a preconstruction primer to provide an advantageous start on the coating system. Without this initial treatment of the steel, a more tedious, expensive, and difficult blast-cleaning operation may be required after erection. Even then, overlapping surfaces and many recesses in the erected structure cannot be properly cleaned and primed, and must be touched-up by air pressure blast cleaning.

Field blasting involves injecting a supply of abrasive into a rapidly moving air stream and expelling it through a nozzle of the proper configuration so that the solid particles impinge against the surface to be cleaned (Fig. 14.32). A standard commercial setup for pressure blasting is illustrated in Fig. 14.33 [5].

For smaller items, or for touch-up, venturi-abrasive blast cleaning cup guns may be used. The abrasive in this equipment is stored in a small container fastened beneath the blasting gun and ejected into the air stream. This equipment is used by hobbyists, by automotive repainting shops, and for small areas or inside shops for work on small pieces.

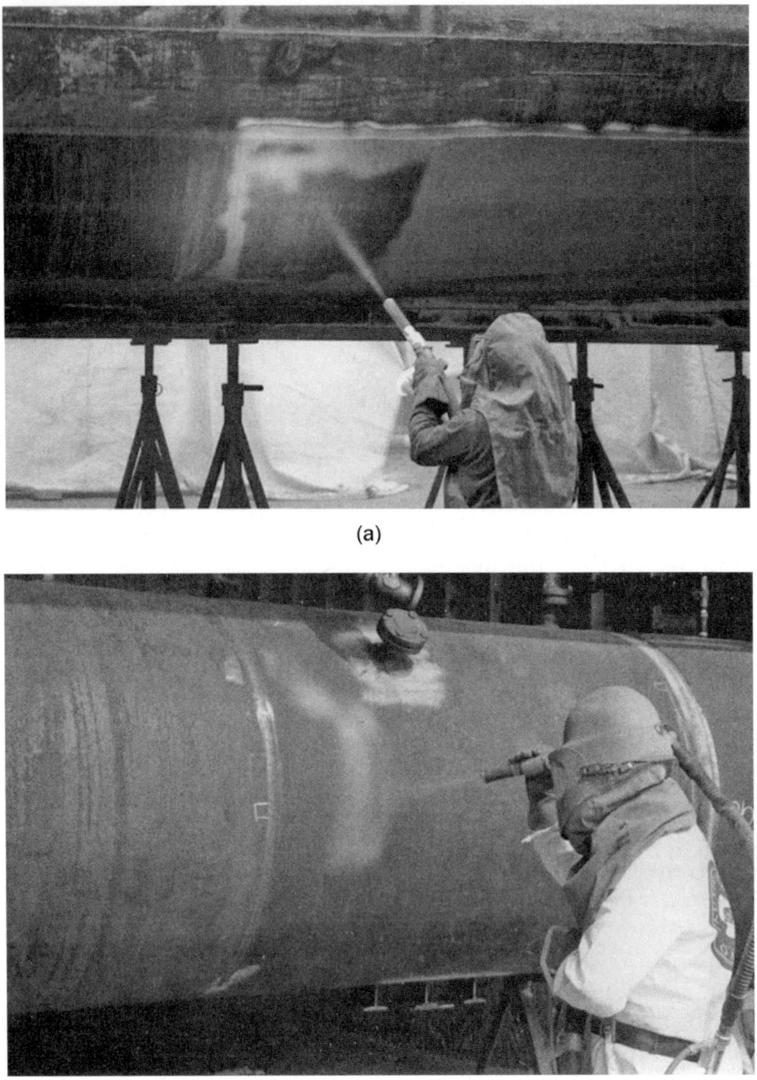

(a)

(b)

FIGURE 14.32 Dry abrasive blasting of (*a*) a ship hull; (*b*) a pipeline. (Courtesy of Barton Mines Co.)

The skill of the workman in keeping the blast stream at the right angle and at proper distance to the work while making sure that all areas are properly cleaned are a prerequisite to a good job. NACE International and the SSPC have jointly adopted a schedule of five

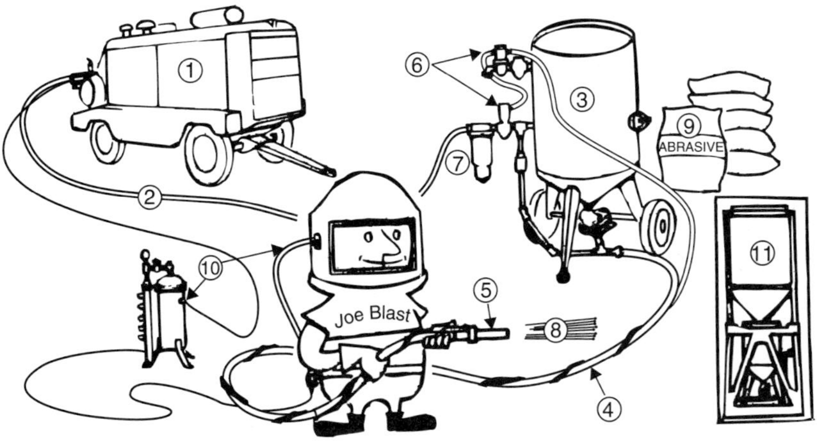

FIGURE 14.33 Typical rig for compressed air abrasive blasting: (1) an adequate and efficient air supply (compressor); (2) air hose, couplings, and valves of ample size; (3) a portable, high-production blast machine; (4) the correct size of antistatic pressure hose with externally fitted quick couplings; (5) high-production venturi nozzle; (6) pneumatic remote control valves for safety and cost savings; (7) an effective moisture separator; (8) high nozzle air pressure; (9) the proper blasting abrasive; (10) a safe air-fed helmet and air purifier; (11) inset showing bulk abrasive storage hopper over pressure blast vessel equipped with remote control valves-when pop-up filling valve opens, abrasive from the hopper fills the machine.

degrees of cleaning steel by blasting operations. These are described briefly as follows:

1. *NACE 1/SSPC 5:* White metal blast cleaned—produces a gray-white, uniform, metallic color with no obvious foreign material remaining
2. *NACE 2/SSPC 10:* Near-white blast cleaned—foreign matter is removed, but with differing shades of metallic gray surface allowable
3. *NACE 3/SSPC 6:* Commercial blast cleaned—rust and foreign matter is removed, except for tight specks of oxide or paint uniformly distributed over a minor portion of the surface; some residues in pits
4. *NACE 8/SSPC 14:* Industrial blast cleaned—similar to NACE 3/SSPC 6, except that evenly disbursed small islands of mill scale are also allowed to remain on the cleaned surface.
5. *NACE 4/SSPC 7:* Brush-off blast cleaned—light mill scale and tightly adhered rust is allowed if distributed over the surface.

14.10.3 Water Jetting

Water jetting involves playing a stream of water moving at a high velocity (~ 35 to 275 MPa) against a surface to be cleaned (Fig. 14.34).

(a)

(b)

FIGURE 14.34 Water-jetting operation (*a*) and pressurizing equipment (*b*). (Courtesy of Termarust Technologies)

The energy imparted to the water dislodges the scale or contamination from the surface to reveal the substrate profile underneath existing coatings. This method is especially effective in removing resilient materials which may not be removed efficiently by abrasives, or in cleaning surfaces of complex cross section such as gratings or sieves. Water also can be used when grit blasting is a fire or environmental hazard. Water jetting excels at removing water soluble salts from a surface. The coatings industry uses waterjetting primarily for recoating or relining projects where there is an adequate preexisting profile.

NACE International and the SSPC have jointly adopted a schedule of four degrees of cleaning steel with waterjetting to parallel the degrees of cleaning with abrasives. NACE No. 5 and SSPC SP-12 describes the degrees briefly as follows:

WJ-1: Clean to bare substrate—surface is free of all visible rust, dirt, previous coatings, mill scale, and foreign matter.

WJ-2: Very thorough or substantial cleaning—allows for randomly dispersed stains of rust, tightly adherent thin coatings, and other tightly adherent foreign matter. The staining or tightly adherent matter is limited to a maximum of five percent of each unit area of the surface.

WJ-3: Thorough cleaning—allows staining or tightly adherent matter to a maximum of 33 percent of each unit area of the surface.

WJ-4: Light cleaning—provides a lesser degree of cleaning than thorough cleaning (WJ-3). The objective of light cleaning is to allow as much of an existing coating or foreign matter to remain as possible and to roughen the surface prior to coating application.

14.10.4 Wet Abrasive Blasting

Abrasive wet blasting is variation of abrasive particle-air blasting that involves introducing abrasives into a rapidly moving stream of water and air. The combination obviously cuts down on the atmospheric pollution experienced when using the dry abrasive alone. Inhibitors are often added to the water to reduce rusting of the "clean" steel before the prime coat is applied.

Wet abrasive blast cleaning systems commonly encountered include air, water, or abrasive blast cleaning, which uses compressed air to propel the abrasive; pressurized water or abrasive blast cleaning, which uses water to propel the abrasive; and a combination of pressurized water or pressurized air with abrasives. The abrasive defines the profile.

14.10.5 Other Surface Preparation Methods

Although abrasive blasting and water jetting are widely used for plants, ships, bridges and other large systems there is a great variety of other methods that can be used to prepare a surface for the final coating application.

For certain surfaces, it is only necessary to wipe the surface with a selected solvent to remove oil, grease, and loose dirt. The use of detergent solutions, isopropanol, ketones, and aliphatic and aromatic hydrocarbons for this purpose can be effective. The major requirement for the process is that the wipe shall not leave a residual contaminant, but rather a clean, dry metal surface. Such a cleaning procedure is used on previously cleaned steel, stainless steel surfaces, and old paint films before overcoating, or on dirty steel prior to abrasive blasting.

A surface that is dry and free from oily residues can be obtained by suspending the item in a closed booth in which the vapors of a solvent are condensing. The vapors condense on the part, coalesce, dissolve the contaminant, and drip from the surface. 1,1,1 Trichloroethane (Trichlor) and perchloroethylene have been widely used solvents for this purpose. However, these otherwise excellent solvents are being phased out due to their high health and environmental risk. Many production line items are cleaned in this manner before coating. Baths of cleaning solvents may also be prepared to remove oily or loose contaminants. Aqueous detergent systems, emulsion systems, and alkali baths are also used with good mechanical or ultrasonic agitation to ensure scouring of the surface by the liquid.

Hot water heaters or steam generators (gennys or gensets) are available for use in the field to clean metals, coatings, or concrete surfaces. Strong detergents or alkalies are added to the water to emulsify oils and similar organic contaminants on contact. Impingement of the hot, strong emulsifying agent on the surface removes even heavy soil at a rapid rate. This method can often be used with good results prior to abrasive blasting.

The process of flame cleaning has found greater favor in Europe than in America. A wide oxy-acetylene flame is played on the surface of steel parts to produce two effects. First, the heavier portions of rust scale or mill scale will pop from the surface because of the thermal expansion difference between the scale and the substrate. Secondly, the surface is freed of moisture if sufficient heat is applied to the substrate. Removal of loose rust particles after the treatment, followed by prompt coating can produce an adequate coating procedure for certain applications. When using an open flame for cleaning, extreme caution must be used to prevent fires and explosions.

Manual or power tool cleaning of an oxidized surface is sometimes the least thorough and slowest means of surface preparation. However, cost, location of the part, or the availability of tools often dictates the use of such a technique. Normally, only the top loose layers of rust on a piece of steel are removed through this method. Moisture and other contaminants remain in the residual scale. It is important to consider the compatibility of the tool with the substrate when using brushes or grinding wheels. For instance, the use of steel or bronze brushes to clean aluminum can leave sufficient heavy metal (Cu or Fe) contamination on the surface to initiate severe pitting of the metal. An air-actuated needle gun is also available to clean small areas by mechanical action. Barrel cleaning, where parts are placed in a rotating drum containing an abrasive, is also used for small parts.

Great quantities of sheet, plate, coil stock, and other forms of metal are pickled in production mills or by metal recyclers. On steel products, the goal is the removal of the mill scale formed on the surface during hot processing (> 575°C). The bath is an aqueous hot hydrochloric, sulfuric, or less frequently phosphoric acid solution containing an inhibitor. The inhibitor must allow uniform attack on the metal, as little metal loss as possible, and when pickling steel, must leave a clean surface free of carbon smut.

The pickling process is conducted in large vats and does not lend itself readily to use in the field. Acid gels or washes based on phosphoric acid can be prepared which will lightly etch a steel surface when held on the surface once the heavy scale has been removed. This type of "pickling" is restricted to smaller articles where the acid can be managed properly.

References

1. Koch GH, Brongers MPH, Thompson NG, Virmani YP, Payer JH. *Corrosion Costs and Preventive Strategies in the United States*. FHWA-RD-01-156. Springfield, Va.: National Technical Information Service, 2001.
2. Natishan P. Introduction to methods of corrosion protection. In: Cramer DS, Covino BS, eds. Vol. 13A: *Corrosion: Fundamentals, Testing, and Protection*. Metals Park, Ohio: ASM International, 2003; 685–6.
3. Vincent LD. *The Protective Coating User's Handbook*. Houston, Tex.: NACE International, 2004.
4. *Coatings and Linings for Immersion Service*. Revised edn. Houston, Tex.: NACE International, 1998.
5. Roberge PR. *Corrosion Basics—An Introduction. 2nd edn.* Houston, Tex.: NACE International, 2006.
6. Greenfield D, Scantlebury JD. Blistering and delamination processes on coated steel. *The Journal of Corrosion Science and Engineering* (electronic) 2000; 2(26).
7. Mayne JEO. The mechanism of the protective action of an unpigmented film of polystyrene. *Journal of the Oil and Colour Chemists' Association*. 1949; 32: 481–7.
8. Brevoort GH, Roebuck AH. Simplified cost calculations and comparisons of paint and protective coating systems, Expected Life and Economic Justification. *CORROSION 1979*, Paper # 037. Houston, Tex: NACE International, 1979.
9. Helsel JL, Melampy MF, Wissmar K. Expected service life and cost considerations for maintenance and new construction protective coating work. *CORROSION 2006*, Paper # 318. Houston, Tex.: NACE International, 2006.
10. *Corrosion of metals and alloys—Corrosivity of atmospheres—Guiding values for the corrosivity categories*. ISO 9224: Switzerland, International Organization for Standardization (ISO), 1992.
11. Munger CG, Vincent LD. *Corrosion Prevention by Protective Coatings. 2nd edn.* Houston, Tex: NACE International, 1999.
12. Trethewey KR, Chamberlain J. *Corrosion for Science and Engineering. 2nd edn.* Burnt Mill, U.K.: Longman Scientific & Technical, 1995.

13. Salagaras M, Bushell PG, Trathen PN, Hinton BRW. *The Use of Corrosion Prevention Compounds for Arresting the Growth of Corrosion in Aluminium Alloys.* Fifth Joint NASA/FAA/DoD Conference on Aging Aircraft, Orlando, Fla.: September 10–13, 2001.
14. Fiaud C. Theory and practice of vapour phase inhibitors. In: Trabanelli, G. *Corrosion Inhibitors.* The Institute of Materials, 1994; 1–11.
15. ASTM D3359: *Standard Test Methods for Measuring Adhesion by Tape Test. Annual Book of ASTM Standards.* West Conshohocken, Pa.: American Society for Testing of Materials, 2002.
16. ASTM D4541: *Standard Test Method for Pull-Off Strength of Coatings Using Portable Adhesion Testers. Annual Book of ASTM Standards.* West Conshohocken, Pa.: American Society for Testing of Materials, 2002.
17. SSPC-VIS 2: *Standard Method of Evaluating Degree of Rusting on Painted Steel Surfaces.* Pittsburgh, Pa.: The Society for Protective Coatings, 2002.
18. ASTM D714: *Standard Test Method for Evaluating Degree of Blistering of Paints. Annual Book of ASTM Standards.* West Conshohocken, Pa.: American Society for Testing of Materials, 2002.
19. Capp JA. A rational test for metallic protective coatings. *Proceedings American Society for Testing of Materials.* 1914; 14: 474–81.
20. Finn AN. Method of making the salt-spray corrosion test. *Proceedings American Society for Testing of Materials.* 1918; 18: 237–8.

CHAPTER 15

High-Temperature Corrosion

15.1 Introduction

High-temperature corrosion is a form of corrosion that does not require the presence of a liquid electrolyte. Sometimes, this type of damage is called *dry corrosion* or *scaling*. The first quantitative approach to oxidation behavior was made in the early 1920s with the postulation of the parabolic rate theory of oxidation by Tammann and, independently, by Pilling and Bedworth.

All materials have their limitations and the solution to high-temperature problems is often a compromise between careful materials selection when the cause of a problem is known, process control in order to impose a safe limit for temperature or gas composition, for example, and better design specifications to recognize mechanical constraints at elevated temperature or resulting from thermal cycling. The ultimate choice will be a compromise based on what is available and how much it costs. In some cases it is rational to accept a short life expectancy with a high reliability factor where the component is replaced on a planned time schedule [1].

Alloys generally rely upon an oxidation reaction for the formation of a protective scale that will improve the corrosion resistance to sulfidation, carburization, and the other forms of high-temperature attack. The properties of high-temperature oxide films, such as their thermodynamic stability, ionic defect structure, and detailed morphology, therefore play a crucial role in determining the oxidation resistance of a metal/alloy in a specific environment.

In general, the names of the corrosion mechanisms are determined by the most abundant dominant corrosion products. For example: oxidation implies oxides, sulfidation implies sulfides, sulfidation/oxidation implies sulfides plus oxides, and carburization implies carbides [2]. Oxidizing environments refer to high oxygen activities, with excess oxygen. Reducing environments are characterized by low oxygen activities, with no excess oxygen available, a situation that may be detrimental to the stability of an oxide scale. It is for this reason that reducing industrial environments are generally considered to be more corrosive than the oxidizing variety.

High-temperature materials are used for many critical components in a wide number of industries, including power generation, chemical processing, and gas turbine. With ever-continuing demands for increased throughput and efficiency, there has been a trend toward higher service temperatures and pressures. This has resulted in continued corrosion problems, countered by continued improvements in material compositions such as minimizing detrimental trace elements, development of coating procedures, and improved fabrication, notably casting, forging, and welding. For example, in the gas turbine industry, alloys designed to cope with high stress-bearing/elevated temperature scaling are presently used for service temperatures in excess of 1100°C, compared with about 800°C 50 years ago [1].

Most common process temperatures are in the range 450 to 850°C or higher (Fig. 15.1). Materials of construction must withstand excessive metal loss by scale formation from oxidation and from penetration by internal oxidation products that could reduce the remaining cross-sectional area to a level that cannot sustain the load-bearing requirements. The component will then yield and may swell or distort. In some cases the internal fluid pressures can be sufficient to burst the component releasing hot, possibly toxic or flammable fluids.

There are several ways of measuring the extent of high temperature corrosion attack. Measurement of weight change per unit area in a given time has been a popular procedure. However the weight change/area information is not directly related to the thickness (penetration) of corroded metal, which is often needed in assessing the strength of equipment components. Corrosion is best reported in penetration units that reveal the actual loss of sound metal. A metallographic technique which is used to determine the extent of damage is illustrated in Fig. 15.2 [3]. The parameters shown in Fig. 15.2 relate to cylindrical specimens and provide information about the load-bearing section (metal loss) and on the extent of grain boundary attack that can also affect structural integrity.

Heating and cooling rates can also cause the buildup of invisible damage due to thermal stress and other fatigue effects. The need for a careful study of the properties of a heat-resistant alloy and its behavior in the anticipated environment is therefore of considerable importance in the selection of a suitable alloy for a particular service application. New alloys and nonmetallic materials, which are continually being made available to industry, are making it possible to make better selections and to establish safe working limits within which the material can be expected to give satisfactory performance over a reasonable length of time.

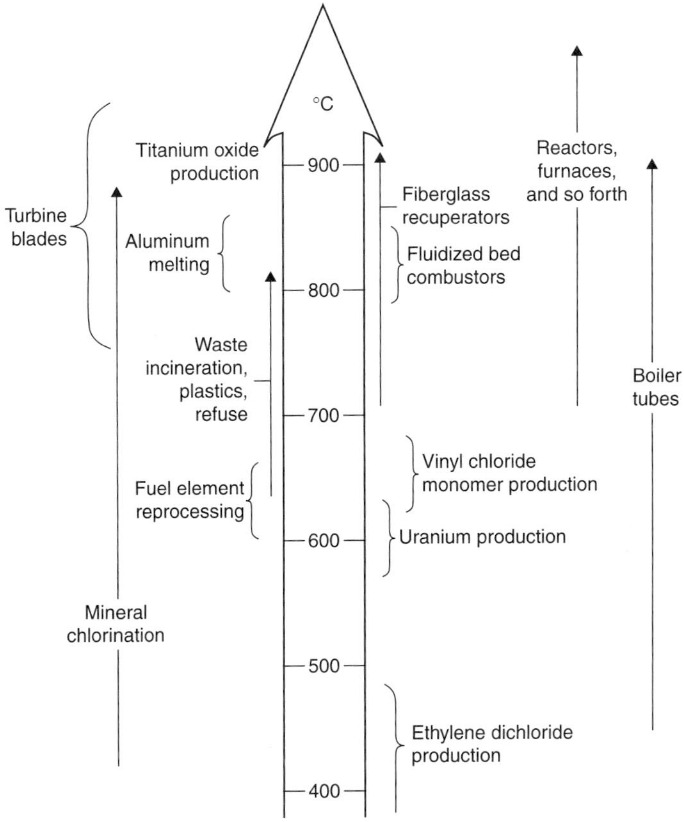

Figure 15.1 Representative process temperatures (adapted from [1]).

Figure 15.2 Metallographic method of measuring hot corrosion attack [3].

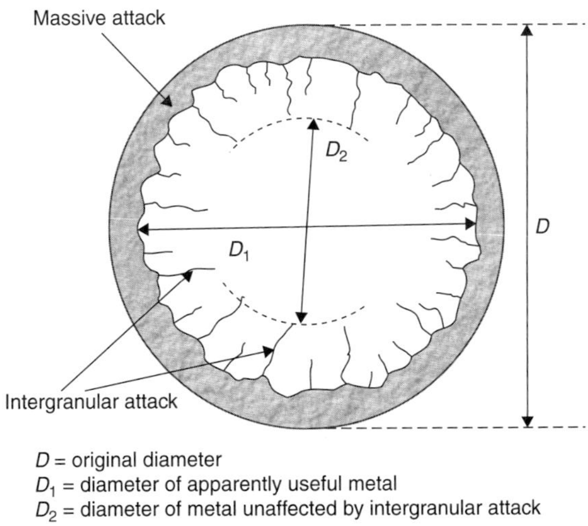

15.2 Thermodynamic Principles

15.2.1 Standard Free Energy of Formation

In a corrosion context, the thermodynamic properties of a metallic material describe the energy that is lost or gained in the form of heat as the metal reacts to form oxides or other compounds. The stability of the various oxides can also differ.

Two important points are involved. First, the more energy released when forming the compound, such as an oxide, the more stable the compound. Secondly, as the temperature is raised, more of the oxidizing species is required to maintain the oxide or other film.

As an illustration, let us consider a bar of copper and a bar of iron at 980°C in an atmosphere made up of 20 percent H_2 and 80 percent H_2O. This atmosphere is partly oxidizing and partly reducing. Under these conditions, copper will not corrode and will remain bright. Iron, on the other hand, will become coated with a layer of iron oxide.

In the case of copper, the reaction with water that produces copper oxide and hydrogen is thermodynamically not favored.

$$2\mathrm{Cu} + \mathrm{H_2O}\,(80\%) \leftarrow \mathrm{Cu_2O} + \mathrm{H_2}(20\%) \tag{15.1}$$

In fact, the reverse reaction would occur if some copper oxide were on the copper; that is, the copper oxide would be reduced by the hydrogen to form H_2O and metallic copper.

In the case of iron, however, the reaction with water to form iron oxide and hydrogen in the same conditions is thermodynamically favored and would therefore proceed.

$$\mathrm{Fe} + \mathrm{H_2O}\,(80\%) \rightarrow \mathrm{FeO} + \mathrm{H_2}(20\%) \tag{15.2}$$

The 80 percent H_2O-20 percent H_2 mixture would not permit H_2 to reduce FeO to form Fe and H_2O, as in the case of copper. Thus, we may say that copper is corrosion resistant under these circumstances, or is more noble than iron. Iron, however, is thermodynamically subject to corrosive attack.

It is possible to use plots of the free energy of formation of metal oxides versus temperature to predict the temperatures at which a metal is stable and the temperatures at which it will spontaneously oxidize. For temperatures at which the free energy of formation of the oxide is positive, the reverse reaction is favored and the oxide will spontaneously decompose to the metal.

Plots of the standard free energy of reaction (ΔG^0) as a function of temperature, commonly called Ellingham diagrams, can help to visualize the relative stability of metals and their oxidized products. Figure 15.3 shows an Ellingham diagram for many simple oxides. The ΔG^0 values on an Ellingham diagram are expressed as kilojoule per mole of O_2 to normalize the energy scale and compare the stability of these oxides directly, that is, the lower the position of the line on

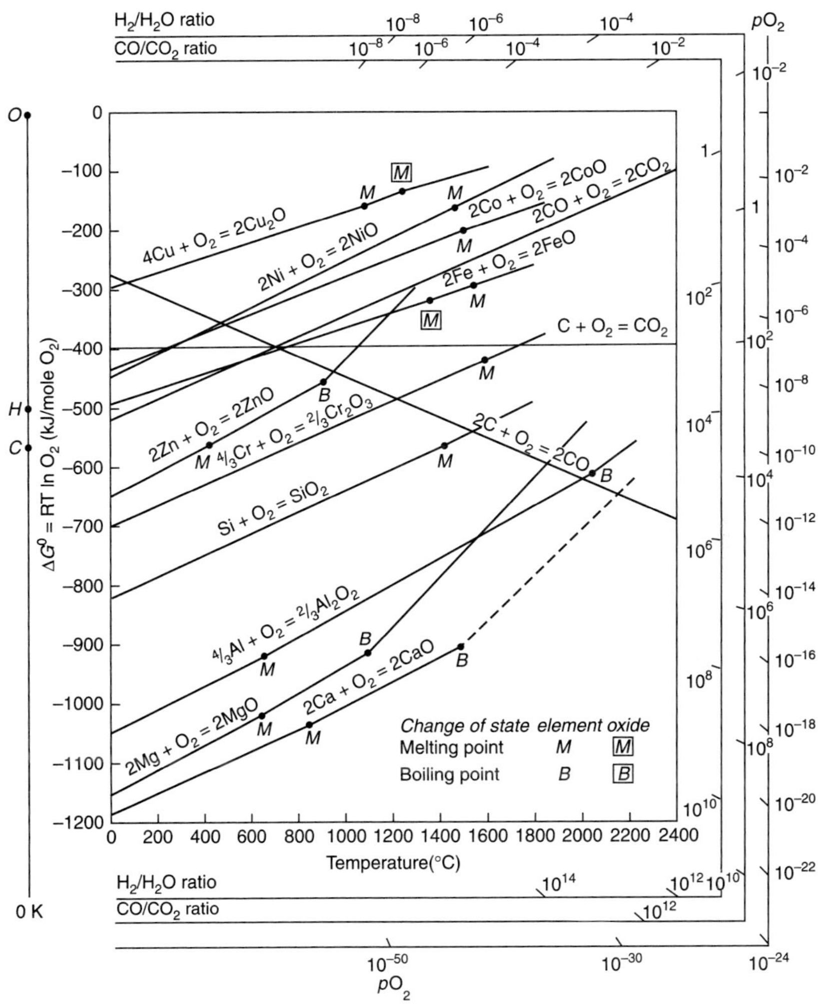

Figure 15.3 Ellingham diagram showing the standard free energy of formation of selected oxides as a function of temperature.

the diagram the more stable is the oxide [4]. For a given reaction, expressed in generic terms in Eq. (15.3), Eq. (15.4), or its logarithmic form Eq. (15.5) may be used to express the oxygen partial pressure (p_{O_2}) at which the metal and oxide are at equilibrium:

$$M + O_2 \rightleftarrows MO_2 \tag{15.3}$$

$$p_{O_2} = e^{\frac{\Delta G^0}{RT}} \tag{15.4}$$

$$\log(p_{O_2}) = \frac{\Delta G^0}{2.303 \cdot RT} \tag{15.5}$$

The oxygen partial pressures may be obtained directly from the Ellingham diagram by drawing a straight line from the origin marked *O* through the free energy line at the temperature of interest and reading the oxygen pressure from its intersection with the scale at the right side labeled p_{O_2}. Values for the pressure ratio H_2/H_2O for equilibrium between a given metal and oxide may be obtained by drawing a similar line from the point marked *H* to the scale labeled H_2/H_2O ratio and values for the equilibrium CO/CO_2 ratio may be obtained by drawing a line from point *C* to the scale CO/CO_2 ratio.

Note that environments other than oxygen, water, or CO_2 can produce metallic oxides because of their oxidizing capacity. Table 15.1 lists a number of common gases and environments encountered in high-temperature works, and the type of reaction to be expected when they predominate.

Referring back to the statements regarding copper and iron at 980°C in a 20 percent H_2-80 percent H_2O gas atmosphere, Fig. 15.3 indicates that although pure H_2O would oxidize copper (use point *H* on left vertical axis as a fulcrum), as little as about 0.1 percent H_2 (H_2/H_2O ratio of 10^{-3}) would be adequate to suppress copper corrosion.

This is important to remember and is in accord with the relatively high nobility of copper. In the case of iron, however, about 70 percent H_2 would be necessary to suppress iron corrosion at 980°C. It can be determined from Fig. 15.3 that the $2Fe + O_2 = 2FeO$ line at 980°C indicates an equilibrium gas mixture of about 2:1 for H_2/H_2O ratio. Since a gas made up of 2 parts H_2 and 1 part H_2O is a 67 percent H_2 gas, 70 percent H_2 or more would favor the reduction of oxide scale at 980°C, and iron would remain bright in this atmosphere. If the iron were to be cooled in this atmosphere, it would oxidize because drier hydrogen is necessary at lower temperatures.

By using the heavy line shown from fulcrum *O* we can determine that 10^{-7} or more volume percent of oxygen also will sustain the oxide on iron at 980°C.

Environment	
Oxidizing	**Reducing**
Oxygen	Hydrogen
Sulfur dioxide	Ammonia
Sulfur trioxide	Hydrogen sulfide
Water (steam)	Sulfur
Chlorine	Carbon
Carbon dioxide	Carbon monoxide
Molten salts	

Table 15.1 Oxidation-Reduction Conditions Existing at High Temperature

By using Fig. 15.3 in an appropriate fashion, one could determine the partial pressure of oxygen that would cause any of the listed metals to corrode, as well as the CO/CO_2 ratios or the H_2/H_2O ratios to produce a similar oxidation. If all these gases were actually mixed together, the situation would of course become more complicated and beyond the scope of the present introduction to the subject. One could further add to such complexity by introducing other gases such as sulfur dioxide, hydrogen sulfide, chlorine gas, or other contaminants which, in addition to oxides, tend to form sulfides, chlorides, and so forth.

15.2.2 Vapor Species Diagrams

Vapor species that form in any given high-temperature corrosion situation often have a strong influence on the rate of attack of the metallic substrate, the rate generally being accelerated when volatile corrosion products are formed. Gulbransen and Jansson have shown that metal and volatile oxide species are important in the kinetics of high-temperature oxidation of carbon, silicon, molybdenum, and chromium [5]. Six types of oxidation/corrosion rate-controlling mechanisms have been identified:

1. At low temperature, diffusion of oxygen and metal species through a compact oxide film
2. At moderate and high temperatures, a combination of oxide film formation and oxide volatility
3. At moderate and high temperatures, the formation of volatile metal and oxide species at the metal-oxide interface and transport through the oxide lattice and mechanically formed cracks in the oxide layer
4. At moderate and high temperatures, the direct formation of volatile oxide gases
5. At high temperature, the gaseous diffusion of oxygen through a barrier layer of volatilized oxides
6. At high temperature, spalling of metal and oxide particles

The diagrams most suited for presentation of vapor pressure data in oxide systems are $\log(p_{M_xO_y})$ versus $\log(p_{O_2})$ at constant temperature and Arrhenius diagrams of $\log(p_{M_xO_y})$ versus $1/T$ at constant oxygen pressure. The following example will illustrate the construction of the first type of these diagrams by considering the Cr-O system at 1200°C or 1473 K. Only one condensed oxide, Cr_2O_3, is formed under these conditions of high-temperature oxidation [4]. The thermochemical data for producing the vapor equilibria diagram shown in Fig. 15.4 are given in Table 15.2. The thermochemical data for Cr_2O_3(s), Cr(s), and the four possible gaseous species Cr(g), CrO(g), CrO_2(g), and CrO_3(g) were obtained from a commercial database [6].

		Enthalpy	Entropy	Heat Capacity (Cp)				Temperature Range	
Species	**State**	H	S	A	B	C	D	T_1	T_2
		(kJ mol^{-1})	(J mol^{-1} K^{-1})	(J mol^{-1} K^{-1})				(K)	(K)
Cr	gas	397.48	174.305	20.786	0	0	0	298.15	700
	gas	–	–	15.456	2.556	16.828	0.874	700	3000
Cr	solid	0	23.64	17.715	22.966	−0.377	−9.033	298.15	1000
	solid	–	–	18.067	15.531	−16.698	0	1000	2130
	liquid	16.933	7.95	39.33	0	0	0	2130	2945
Cr_2O_3	solid	−1139.7	81.199	104.307	24.451	7.59	−3.807	298.15	2603
CrO	gas	188.28	239.266	24.518	25.33	0.222	−11.201	298.15	400
	gas	–	–	34.292	3.979	−4.351	-1.138	400	1600
CrO_2	gas	−75.312	269.24	30.878	46.689	0	−15.782	298.15	400
	gas	–	–	45.309	18.552	−7.222	−7.632	400	1100
CrO_3	gas	−292.88	266.169	74.002	7.686	−18.393	−1.644	298.15	3000
O_2	gas	0	205.146	31.321	3.895	−3.105	−0.335	298.15	5000

Table 15.2 Thermochemical Data for the Cr-O System

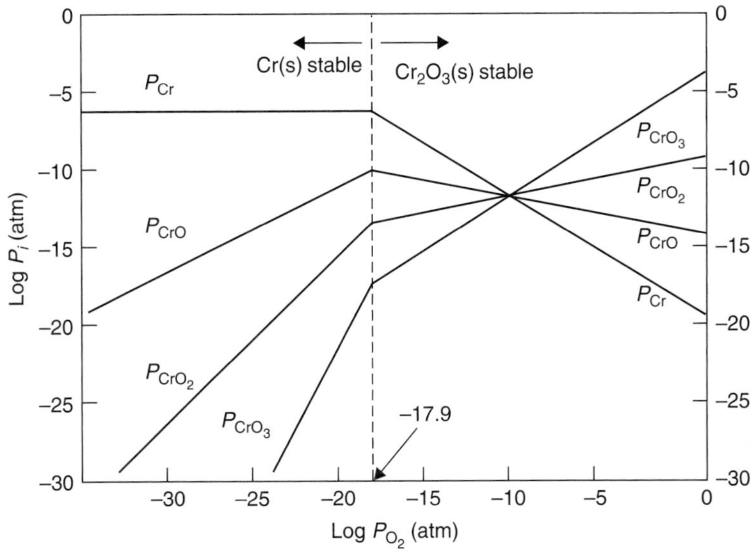

FIGURE 15.4 Vapor equilibria in the Cr-O system at 1250°C.

The enthalpy (H_T), entropy (S), and heat capacity (C_p) of each species considered were calculated using Eqs. (15.6) to (15.8) in which T was set at 1473 K (Table 15.3). In these equations H_{tr} and T_{tr} represent, respectively, the enthalpy and temperature associated with any phase transition encountered between the reference temperature (298.15 K) and the temperature of interest.

$$H_{(T)} = H_{f(298.15)} + \int_{298.15}^{T} C_p \cdot dT + \sum H_{tr} \tag{15.6}$$

$$S = S^0_{(298.15)} + \int_{298.15}^{T} (C_p / T) \cdot dT + \sum H_{tr} / T_{tr} \tag{15.7}$$

$$C_p = A + B \cdot 10^{-3} \cdot T + C \cdot 10^5 \cdot T^{-2} + D \cdot 10^{-6} \cdot T^2 \tag{15.8}$$

The free energy (G) for each species considered was then calculated with Eq. (15.9) and used to evaluate the stability of these species and the predicted energy of reaction for each equilibrium (Table 15.4).

$$G = H - T \cdot S \tag{15.9}$$

Vapor pressures of species at equilibrium with either the metal or its most stable oxide (i.e., Cr_2O_3) must then be determined. The boundary between these regions is the oxygen pressure for the Cr/Cr_2O_3 equilibrium expressed in Eq. (15.10).

$$2\,Cr(s) + 1.5\,O_2(g) = Cr_2O_3(s) \tag{15.10}$$

Species	State	H	S	G
		(kJ mol^{-1})	(J mol^{-1} K^{-1})	(kJ mol^{-1})
Cr	gas	422.02	207.58	116.25
Cr	solid	36.97	70.78	−67.29
Cr_2O_3	solid	−993.71	276.68	−1401.27
CrO	gas	230.37	295.28	-204.57
CrO_2	gas	−12.73	351.72	−530.81
CrO_3	gas	−204.60	381.78	−766.96
O_2	gas	39.67	257.73	−339.97

TABLE 15.3 Thermochemical Data for the Cr-O System at 1473 K

The equilibrium constant (K_p) is evaluated with Eq. (15.12), giving an equilibrium pressure of oxygen calculated with Eq. (15.12).

$$\log K_p = \frac{-\Delta G^0}{2.303 \cdot RT} \tag{15.11}$$

$$\log(p_{O_2}) = -\frac{2}{3}\log K_p^{Cr_2O_3} = -17.90 \tag{15.12}$$

The dotted vertical line in Fig. 15.4 represents this boundary. At low oxygen pressure it can be seen that the presence of Cr(g) is independent of oxygen pressure. For oxygen pressures greater than

Reaction	ΔG^0
	(kJ mol^{-1})
2 Cr(s) + 1.5 O_2 = Cr_2O_3	−756.72
Over Cr(s)	
Cr(s) = Cr(g)	183.54
Cr(s) + 0.5 O_2 = CrO(g)	32.71
Cr(s) + O_2 = CrO_2(g)	−123.54
Cr(s) + 1.5 O_2 = CrO_3(g)	−189.71
Over Cr_2O_3	
0.5 Cr_2O_3(s) = Cr(g) + 0.75 O_2	561.90
0.5 Cr_2O_3(s) = CrO(g) + 0.25 O_2	411.07
0.5 Cr_2O_3(s) + 0.25 O_2= CrO_2(g)	254.81
0.5 Cr_2O_3(s) + 0.75 O_2 = CrO_3(g)	188.65

TABLE 15.4 Standard Energy of Reactions for the Cr-O System at 1473 K

the Cr/Cr_2O_3 equilibrium, the Cr(g) vapor pressure may be obtained from the equilibrium expressed in Eq. (15.13).

$$0.5\ Cr_2O_3(s) = Cr(g) + 0.75\ O_2(g) \tag{15.13}$$

The other lines in Fig. 15.4 are obtained by using similar equilibrium equations (Table 15.4). The vapor equilibria presented in Fig. 15.4 show that significant Cr(g) vapor pressures are developed at low-oxygen partial pressure (e.g., at the alloy-scale interface of a Cr_2O_3-forming alloy) but that a much larger pressure of CrO_3(g) develops at high-oxygen partial pressure. This high CrO_3(g) pressure is responsible for the thinning of Cr_2O_3 scales by vapor losses during exposure to oxygen-rich environments.

15.2.3 2D Isothermal Stability Diagrams

When a metal reacts with a gas containing more than one oxidant, a number of different phases may form depending on both thermodynamic and kinetic considerations. Isothermal stability diagrams, usually constructed with the logarithmic values of the activities or partial pressures of the two nonmetallic components as the coordinate axes, are useful in interpreting the condensed phases that may form. The metal-sulfur-oxygen stability diagrams for iron, nickel, cobalt, and chromium are shown in Figs. 15.5 to 15.8. One important assumption in these diagrams is that all condensed species are at unit activity. This assumption places important limitations on the use of such diagrams for alloys.

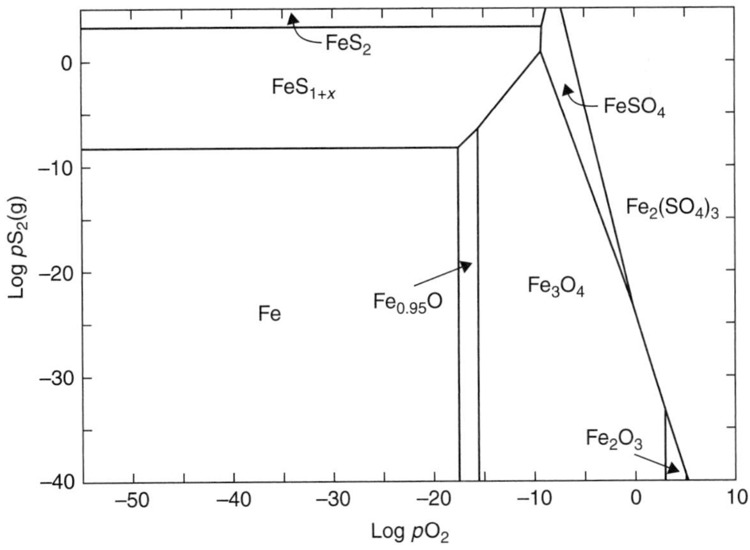

FIGURE 15.5 Stability diagram of the Fe-S-O system at 870°C.

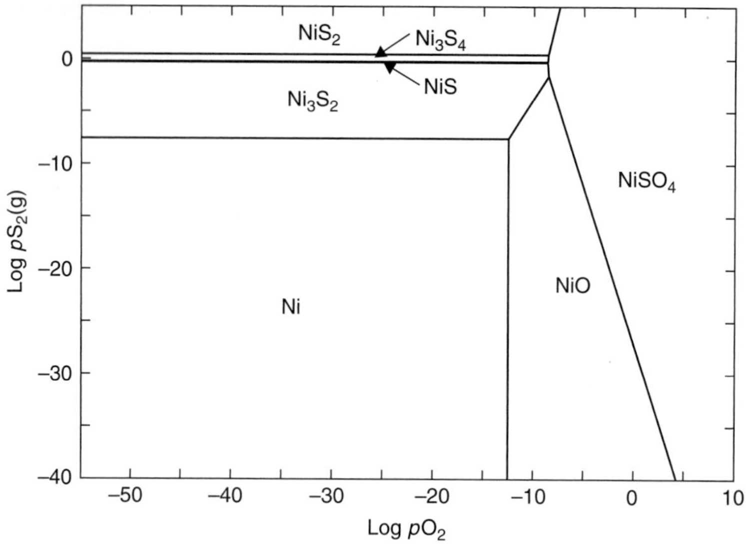

FIGURE 15.6 Stability diagram of the Ni-S-O system at 870°C.

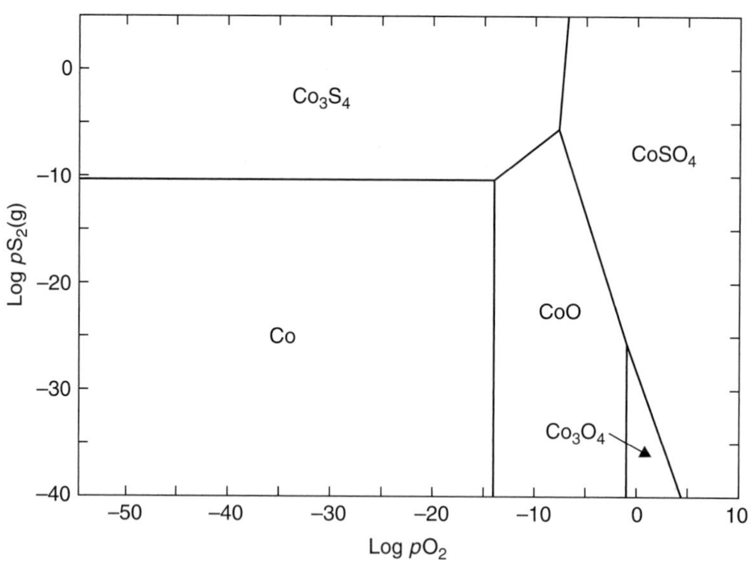

FIGURE 15.7 Stability diagram of the Co-S-O system at 870°C.

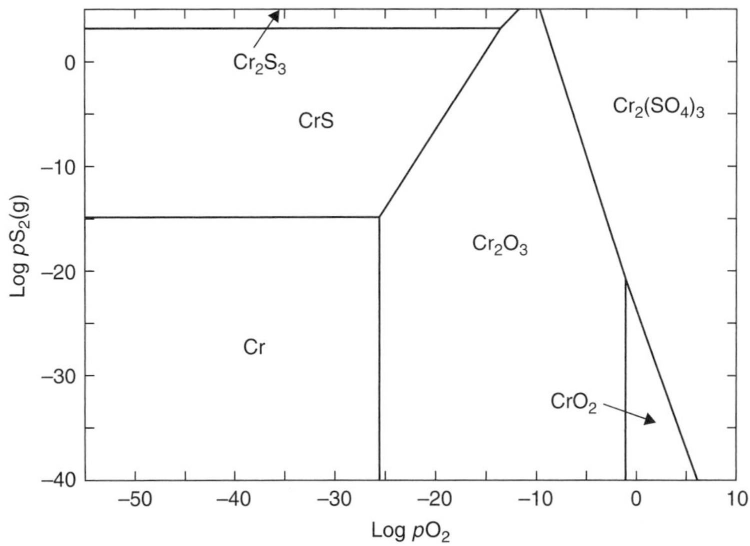

FIGURE 15.8 Stability diagram of the Cr-S-O system at 870°C.

15.3 Kinetic Principles

When a metal is oxidized at elevated temperatures, a stable scale of oxides or other compounds may buildup to cover and protect the exposed metal surface from further corrosion. The corrosion product layer may therefore act as a barrier between the underlying metal and the corrosive environment (air, flue gas, molten salt, or any other corrosive agent). The compound may be a solid, a liquid, or a gas. For instance, if chromium, vanadium, and molybdenum were exposed to air at 1100°C:

1. Chromium would form solid Cr_2O_3.
2. Vanadium would form liquid V_2O_5.
3. Molybdenum would form gaseous MoO_3.

In such case, only chromium would not be severely corroded. If the oxide formed serves as an effective barrier, corrosion can be strongly retarded and only a thin oxide layer will form. On the other hand, if the barrier is not very effective, corrosion nay continue at a moderate rate. It could then be concluded that the oxide scale was only slightly protective and that the metal was not corrosion-resistant under these conditions of exposure.

15.3.1 Scale as a Diffusion Barrier

Once a corrosion product layer is formed, the corrosion process may continue through the diffusion of at least one of the reactants through the corrosion product layer. Let us consider, for example, the case of nickel exposed to air at high temperature. Corrosion can theoretically continue through the nickel oxide layer by means of diffusion in either direction, alone or by counter-current diffusion, as illustrated in Fig. 15.9.

From the atomic viewpoint, diffusion through the oxide occurs by atomic movement which is made easier by existing atomic vacancies, as illustrated in Fig. 15.10 and 15.11. The schematic diagram in Fig. 15.10 represents the type of oxide that forms on nickel and on a number of other metals, including copper, cobalt, manganese, and chromium. In the case of nickel, only the NiO oxide will form in the presence of oxygen, but, as the figure shows, it is a metal-deficit or *p*-type* oxide; that is, a slight excess of oxygen atoms over nickel atoms will be found under normal conditions, and the charge balance in the scale that must remain electrically neutral will be maintained by a few of the nickel atoms assuming a higher valence state.

For this type of oxide, the controlling step in increasing scale thickness is the ability of the nickel ions to migrate outward toward the surface. If a small quantity of a higher valence ion (Cr^{3+}, for example) is added to the nickel, as in Fig. 15.11, charge balance will require a greater number of nickel vacancies and hence an easier,

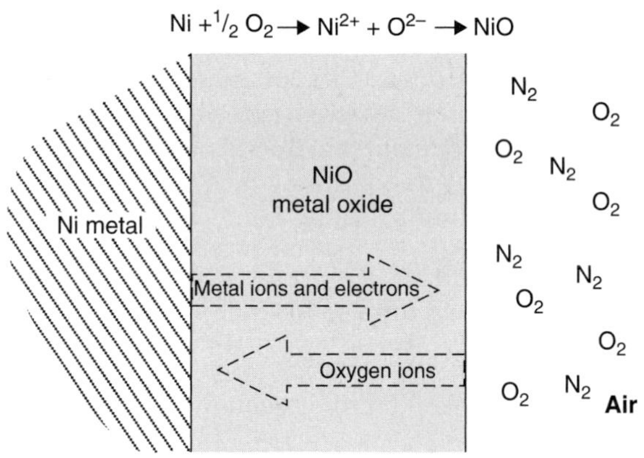

FIGURE 15.9 Schematic diagram showing countercurrent diffusion, that is, diffusion of oxygen inward and metal ions plus electrons outward. Thus, the anode is at the metal/ metal oxide interface where oxidation ($M \rightarrow M^{2+} + 2e^-$) occurs. The cathode is at the oxide/air interface where reduction ($\frac{1}{2}O_2 + 2e^- \rightarrow O^{2-}$) occurs.

* A *p*-type material is a semiconductor material in which the conduction is mostly assumed by holes having the equivalent of positive charges.

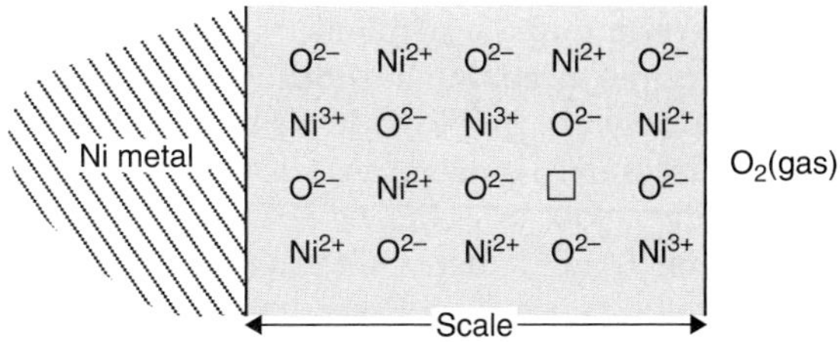

FIGURE 15.10 Nickel oxide as an example of metal-deficit scale. Schematic of NiO lattice shows a few Ni^{3+} sites among the Ni^{2+}. For every two Ni^{3+} sites, there is a vacant □ Ni^{2+} site.

quicker path for nickel ions to migrate to the surface. Conversely, if a lower valence metal ion (Li^+ is the best known example) is added, charge balance requirements will reduce the number of defects or vacancies, and the oxidation rate, conforming to the nickel ion diffusion rate, will decrease.

The opposite situation, a metal-excess or *n*-type* scale, is also possible (zinc is an example). In this case, extra metal ions are believed to occupy intermediate positions in the scale. The introduction of a higher valence metal ion will reduce the number of zinc ions in the scale and thereby lower the oxidation rate. A lower valence metal ion will, of course, have the opposite effect, causing the metal-excess scale to respond to the addition of other metal ions in a manner just the reverse of that in metal-deficit scales.

Some metals form more than one oxide. In the case of iron, for example, Fe_2O_3, Fe_3O_4, and FeO can form, depending upon the availability of oxygen, usually expressed as a partial pressure. Generally, all three types will form concurrently on the metal surface. Fe_2O_3, which requires the most oxygen, will be found on the outside, and FeO, which requires the least, on the inside, with Fe_3O_4 between them. The boundaries between the various oxides usually are quite well-defined and, provided the several layers remain intact, each will

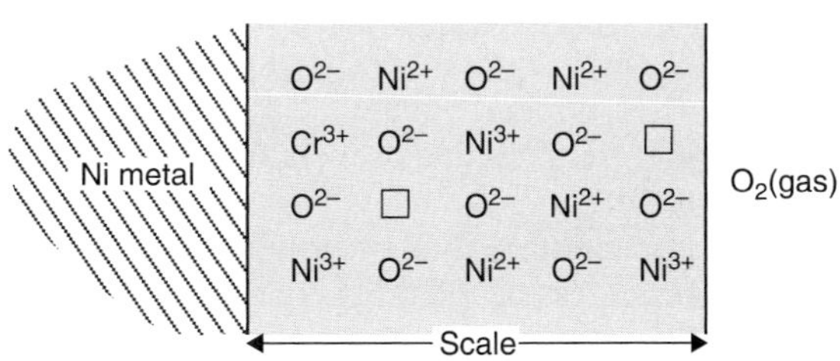

FIGURE 15.11 Nickel oxide containing small amounts of chromic oxide. Note that this increases the density of vacant □ Ni^{2+} sites.

* A *n*-type material is a semiconductor material in which the conduction is mostly assumed by electrons having negative charges.

grow according to the parabolic rate law as discussed later. In most instances, the predominant diffusion is that of metal ions plus electrons outward to the oxide-air interface shown in Fig. 15.9.

The migration of electrons should not be overlooked. If electron mobility was low, as in the case of oxides which behave as electrical insulators (Al_2O_3, BeO, Cr_2O_3), the metals developing these compounds as surface films could be expected to have good corrosion resistance.

Because nickel at high temperatures does not form an oxide layer with a good diffusion-resistant coating, alloying elements are generally added to nickel-based alloys, in order to provide added imperviousness to the oxide layer. Chromium is an excellent alloying element, and the 80 Ni-20 Cr alloy is one of several compositions commercially used for heating elements in high-quality domestic electric toasters or electric irons. This alloy generally develops a suitable NiO-Cr_2O_3 spinel (oxide) on the surface, but under suitable conditions, Cr_2O_3 forms an even more diffusion-resistant barrier, the *spinel*.

15.3.2 Basic Kinetic Models

Three basic kinetic laws have been used to describe the oxidation rates of pure metals. It is important to bear in mind that these laws are based on relatively simple oxidation models. Practical oxidation problems usually involve alloys and considerably more complicated oxidation mechanisms and scale properties than considered in these simple models.

Linear Behavior

If the oxide film or scale cracks or is porous, that is, if the corrosive gas can continue to penetrate readily and react with the base metal in a catastrophic manner, no protection will be afforded and attack will proceed at a rate determined essentially by the availability of the corrosive gas. In this case, the rate will not sensibly change with time, and, as is apparent from Fig. 15.12, the weight change or depth of penetration from oxidation is a straight line or linear function of time and may be expressed as

$$y = k_L t \tag{15.14}$$

where y is scale thickness or weight change

t is time

k_L is a constant that is dependent on temperature, the material being tested, and other conditions of the high temperature test

Logarithmic Behavior

The logarithmic rate behavior follows an empirical relationship, which has no fundamental underlying mechanism. This behavior, also shown in Fig. 15.12, is mainly applicable to thin oxide films

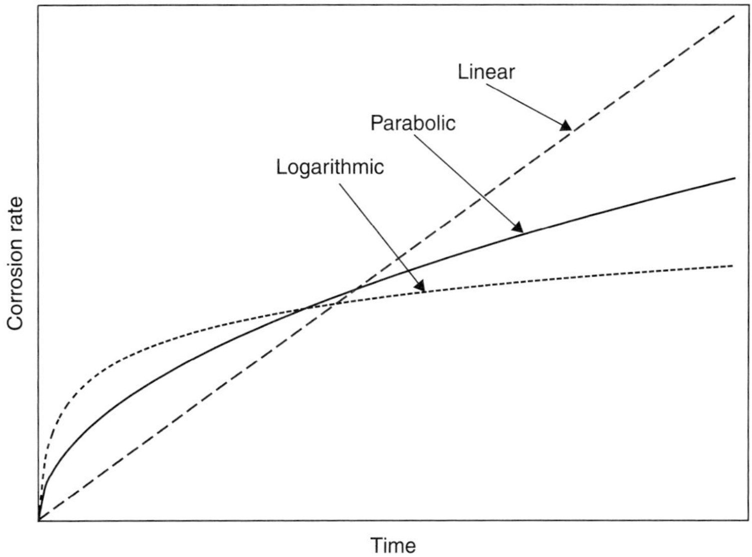

FIGURE 15.12 Linear, parabolic, and logarithmic relationships between time and extent of corrosion which may be measured as depth of penetration or weight change (gain or loss).

formed at relatively low temperatures, and therefore rarely applicable to high temperature engineering problems.

$$y = k_e \log(ct + b) \tag{15.15}$$

where k_e is the rate constant for the corrosive attack
c and b are empirical constants for the particular conditions of the test

Parabolic Behavior

If, on the other hand, the scale formed is continuous, adherent, and prevents easy access of the corrosive gas to the underlying base metal, a considerable measure of protection may occur, and the extent of protection will increase as the scale thickens. In this case, the availability of the corrosive gas will not determine the reaction rate.

Diffusion through the scale will be the slowest and hence, the rate-controlling step. The diffusing element that controls the reaction rate may be either the oxygen moving inward or the metal moving outward. The latter is more likely in alloys possessing good corrosion resistance. As the scale thickens, everything else remaining the same, the diffusion rate will decrease, slowing proportionally the rate of scale growth. A mathematical expression of the process is a parabola (Fig. 15.12) having the equation:

$$y = y_0 + k_p\sqrt{t} \tag{15.16}$$

where k_p is the rate constant

y_0 an empirical constant of the system studied

After an initial period of time, the rate of scale growth will decrease to a low level, that is, a considerable measure of oxidation resistance will be obtained and, provided the scale remains intact, will be maintained for very long times.

Under special conditions, particularly at intermediate temperatures, the oxidation process may essentially stop after the initial scale formation. In such cases, no change in scale thickness or weight would be observed after very long intervals. Presumably this occurs because, when the scale reaches some minimum thickness the diffusion rate is so low as to be essentially zero, thus creating a condition under which the metal should last almost indefinitely without further deterioration. Some investigators have chosen to call this type of behavior tarnishing and thereby to distinguish it from the higher temperature phenomenon of scaling. The latter will be of greater concern in the remainder of this discussion.

The parabolic rate law, although obviously an oversimplification of the actual state of affairs during oxidation, nevertheless provides a relatively accurate approximation of the behavior of the majority of metals and alloys practically used at high temperatures.

15.3.3 Pilling-Bedworth Ratio

Thus far, principally oxide scale formation was considered, and it was assumed that once formed, the scale would remain intact and protective. Scale thickness may be an important factor. In early stages, the scale is thin and reasonably elastic. It also will have little intrinsic strength and usually will remain tightly adherent to the base metal. As it thickens, however, less desirable properties may become manifest. Heavy scales tend to be brittle and will therefore crack and spall from the surface.

Figure 15.13 presents data on a stainless steel that reflects this type of behavior. At the lower temperatures, the scale thickens slowly and the parabolic rate law is followed. At the highest temperature in Fig. 15.13 (700°C), the oxidation rate is also initially parabolic, but at some point the scale breaks away locally and there is an immediate increase in the scaling rate as a new protective scale is formed.

The volume of the oxide formed, relative to the volume of the metal consumed, is an important parameter in predicting the degree of protection provided by the oxide scale. If the oxide volume is relatively low, tensile stresses will tend to crack the oxide layers. Oxides, which are essentially brittle ceramics, are particularly susceptible to fracture and cracking under such tensile stresses. If the oxide volume is very high, stresses will tend to lead to a break in the adhesion between the metal and oxide. For a high degree

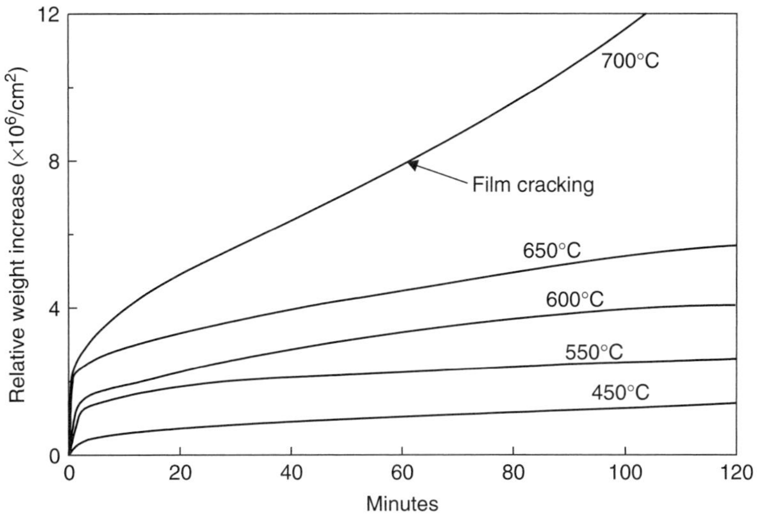

FIGURE 15.13 Oxidation of stainless steel at various temperatures in 0.1 atm oxygen.

of protection, the oxide formed should be similar to that of the metal being oxidized. This argument is the basis for the Pilling-Bedworth ratio (PB):

$$PB = \frac{\text{Volume of oxide}}{\text{Volume of metal consumed}} = \frac{Wd}{nDw} \tag{15.17}$$

where W is molecular weight of oxide
D is density of the oxide
n is number of metal atoms in the oxide molecule
d is density of the metal
w is atomic weight of the metal

PB ratios slightly greater than one are typically expected to indicate "optimal" protection, with modest compressive stresses generated in the oxide layer. Table 15.5 provides the PB ratio of a few metal/oxide systems [4]. In practice, it has been found that PB ratios are generally poor predictors of the actual protective properties of scales. Some of the reasons advanced for deviations from the PB rule include [7]

- Some oxides actually grow at the oxide-air interface, as opposed to the metal-oxide interface.
- Specimen and component geometries can affect the stress distribution in the oxide films.
- Continuous oxide films are observed even when $PB < 1$.

Oxide	Oxide/Metal Volume Ratio
K_2O	0.45
MgO	0.81
Na_2O	0.97
Al_2O_3	1.28
ThO_2	1.30
ZrO_2	1.56
Cu_2O	1.64
NiO	1.65
FeO (on α-Fe)	1.68
TiO_2	1.70–1.78
CoO	1.86
Cr_2O_3	2.07
Fe_3O_4 (on α-Fe)	2.10
Fe_2O_3 (on α-Fe)	2.14
Ta_2O_5	2.50
Nb_2O_5	2.68
V_2O_5	3.19
WoO_3	3.30

Table 15.5 Oxide-Metal Volume Ratios of Some Common Metals

- Cracks and fissures in oxide layers can be "self–healing" as oxidation progresses.
- Oxide porosity is not accurately predicted by the PB parameter.
- Oxides may be highly volatile at high temperatures, leading to nonprotective properties, even if predicted otherwise by the PB parameter.

Figure 15.14 represents, in a schematic fashion, typical ways in which scales have been observed to rupture. If temperature variations occur, a weakly adherent scale is likely to separate locally from the base metal, a phenomenon sometimes termed blistering, and, because it has little intrinsic tensile strength, eventually rupture and spall. The relative thermal expansivity between scale and metal also has been cited as an influential factor when temperature fluctuations are involved. Probably of greater significance is the nature of the interface between the metal and the scale. If the interface is planar, it is easy to conceive of a shear crack initiating at the interface and propagating rapidly across the surface.

FIGURE 15.14 Schematic of some of the hypothetical ways the integrity of film or scale is impaired.

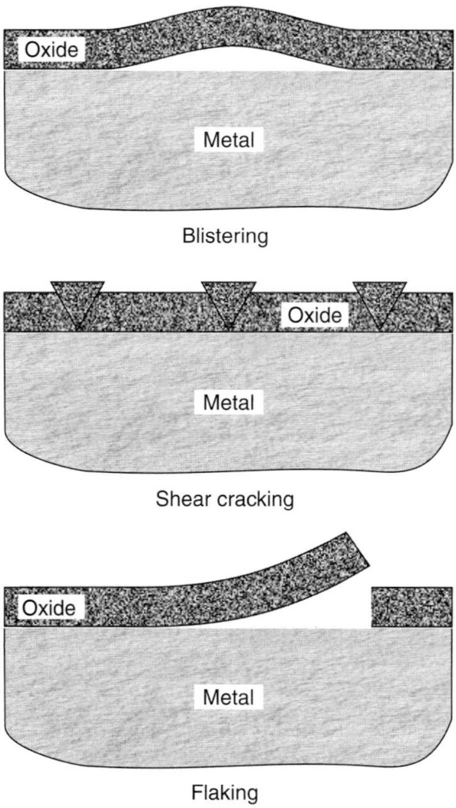

If, on the other hand, the interface is rough, where incipient grain boundary attack beneath the scale is occurring, a keying action may prevent rapid crack propagation and thereby improve scale adherence. It is believed that the major effect of reactive rare earth additions to heat-resistant alloys is not to enter the scale and lower diffusion rates, but to accumulate in grain boundaries and thus ensure a small amount of grain boundary attack to provide the keying action.

A considerable effort is often made in the tubing industry to produce on the interior wall of the tube an initial scale that has optimal properties. Centrifugally cast tubing is often bored to remove porosity near the inner surface, honed to provide a smooth surface profile, and then steamed heavily to produce a thin, continuous, dense, resistant oxide layer on the interior with a remarkably increased life of tubing sections for particular services.

15.4 Practical High-Temperature Corrosion Problems

The oxidation rate laws described earlier are simple models derived from the behavior of pure metals. In contrast, practical high-temperature corrosion problems are much more complex and involve the use of alloys and gas mixtures. For practical problems, both the corrosive environment and the high temperature corrosion mechanism(s) have to be understood. In the introduction, it was pointed out that several high-temperature corrosion mechanisms exist. Although considerable data is available from the literature for high-temperature corrosion in air, for low-sulfur flue gases and for some other common refinery and petrochemical environments, small variations in the composition of a process stream or in operating conditions can cause markedly different corrosion rates. Therefore, the most reliable basis for material selection is operating experience from similar plants and environments or from pilot plant evaluation [8].

When considering specific alloys for high-temperature service, it is imperative to consider other properties besides the corrosion resistance. It would be futile, for example, to select a stainless steel with high-corrosion resistance for an application in which strength requirements could not be met. In general, austenitic stainless steels are substantially stronger than ferritic stainless steels at high temperatures, as indicated by a comparison of stress rupture properties (Fig. 15.15) and creep properties (Fig. 15.16) [3].

Some specific corrosion mechanisms are described in more detail in the following sections. A brief description of the alloys mentioned in these sections is provided in Table 15.6.

15.4.1 Oxidation

Oxidation is generally described as the most commonly encountered form of high-temperature corrosion. However, the oxidation process itself is not always detrimental. In fact, most corrosion and

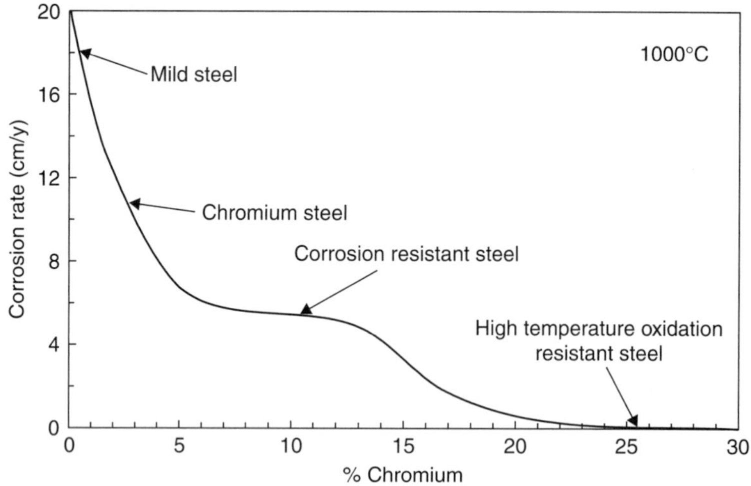

FIGURE 15.15 Effect of chromium on reducing corrosion in air at 1000°C in Fe-Cr alloys.

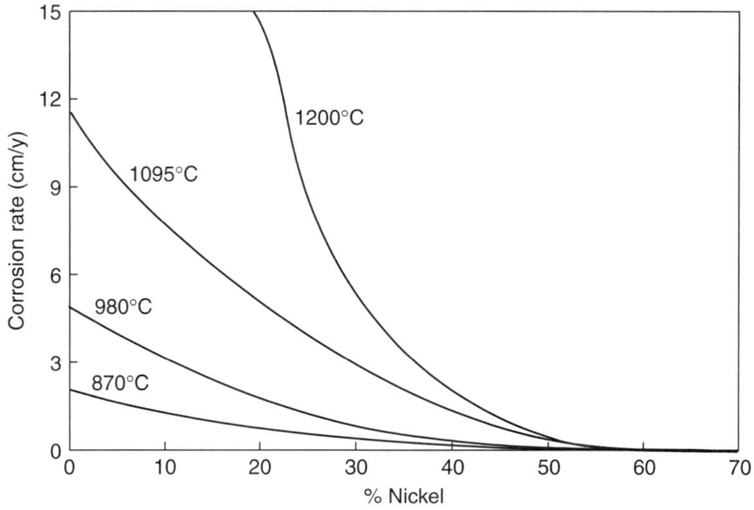

FIGURE 15.16 Effect of adding nickel to a Fe 11-Cr alloy in air at 870, 980, 1095, and 1200°C.

heat-resistant alloys rely on the formation of an oxide film to provide corrosion resistance. Chromium oxide (Cr_2O_3 or chromia) is the most common of such films. In many industrial corrosion problems, oxidation does not occur in isolation and is usually only one of a combination of high-temperature corrosion mechanisms causing material degradation when contaminants (sulfur, chlorine, vanadium, and the like) are present in the atmosphere. Strictly speaking, the oxidation process is only applicable to uncontaminated air and clean combustion atmospheres.

For a given material, the operating temperature assumes a critical role in determining the oxidation rate. As temperature increases, so does the rate of oxidation with important differences in temperature limits between intermittent and continuous service [3]. It has been argued that thermal cycling in the former causes cracking and spalling damage in protective oxide scales, resulting in lower allowable operating temperatures for some alloys such as austenitic stainless steels, whereas others such as the ferritic stainless steels remain protected. Increased chromium content is the most common way of improving oxidation resistance.

Apart from chromium, alloying additions used to enhance oxidation resistance include aluminum, silicon, nickel, and some of the rare earth metals. For oxidation resistance above 1200°C, alloys that rely on protective Al_2O_3 (alumina) scale formation are preferred over those forming chromia [9]. Increasing the nickel content of the austenitic stainless steels up to about 30 percent can have a strong beneficial synergistic effect with chromium.

Common name	UNS Number	Generic Family
6	R30016	Ni-, Ni-Fe-, Co- base alloy
6B	R30016	
25	R30605	..
188	R30188	..
214	N07214	..
230	N06230	..
263	N07041	..
333	N06333	..
556	R30556	..
600	N06600	..
601	N06601	..
617	N06617	..
625	N06625	..
718	N07718	..
800 H	N08810	..
825	N08825	..
Alloy 150(UMCo-50)	N12150	..
Alloy HR-120	N12120	..
Alloy HR-160	N12160	..
Incoloy DS	-	..
Incoloy 801	N08801	..
Incoloy 803	N08803	..
Inconel 602	N06601	..
Inconel 671	–	..
Multimet	R30155	..
Nickel	N02270	..
René 41	N07041	..
RA330	N08330	..
S	N06635	..
WASPALOY	N07001	..
X (Hastelloy X)	N06002	..

Table 15.6 Common Names, UNS Number, and Generic Family of Alloys Used in High Temperature Applications

Common name	UNS Number	Generic Family
304	S30400	austenitic stainless steel
310	S31000	..
316	S31600	..
330	S33000	..
253 MA	S30815	..
410	S41000	martensitic stainless steel
430	S43000	ferritic stainless steel
446	S44600	..
2205	S31803	duplex stainless steel
ACI HK	J94224	cast SS
1Cr-0.5Mo	K11597	Steel
2.25Cr-1Mo	K21590	..
5Cr-0.5Mo	K41545	..
9Cr-1Mo	S50400	..
Carbon Steel	G10200	..
Copper	C11000	copper

Table 15.6 *(continued)*

Fundamental metallurgical considerations impose limits on the amount of alloying additions that can be made in the design of engineering alloys. Apart from oxidation resistance, the mechanical properties must be considered together with processing and manufacturing characteristics. Metallurgical phases that can result in severe embrittlement (such as sigma, Laves, and chi phases) tend to form in highly alloyed materials during high-temperature exposure. A practical example of such problems involves the collapse of the internal heat-resisting lining of a cement kiln. Few commercial alloys contain more than 30 percent chromium. Silicon is usually limited to two percent and aluminum to less than four percent in wrought alloys. Yttrium, cerium, and the other rare earth elements are usually added only as a fraction of a percent [8].

An interesting approach to circumvent the above problems caused by bulk alloying is the use of surface alloying. In this a pproach, a highly alloyed (and highly oxidation resistant) surface layer is produced, whereas the substrate has a conventional composition and metallurgical properties. Bayer has described the formation of a surface alloy containing as much as 50 percent aluminum, by using a pack cementation vapor aluminum diffusion process [10].

The vapor aluminum-diffused surface layer is hard and brittle, but the bulk substrate retains the properties of conventional steels.

Extensive testing of alloys has shown that many alloys establish parabolic time dependence after a minimum time of 1000 hours in air at temperatures above 900°C. If the surface corrosion product (scale) is removed or cracked so that the underlying metal is exposed to the gas, the rate of oxidation can be much faster. The influence of O_2 partial pressure on oxidation above 900°C is specific to each alloy, as illustrated for some common alloys in Fig. 15.17. Most alloys do not show a strong influence of the O_2 concentration upon the total penetration. Alloys such as Alloy HR-120, and Alloy 214 even exhibit slower oxidation rates as the O_2 concentration increases. These alloys are rich

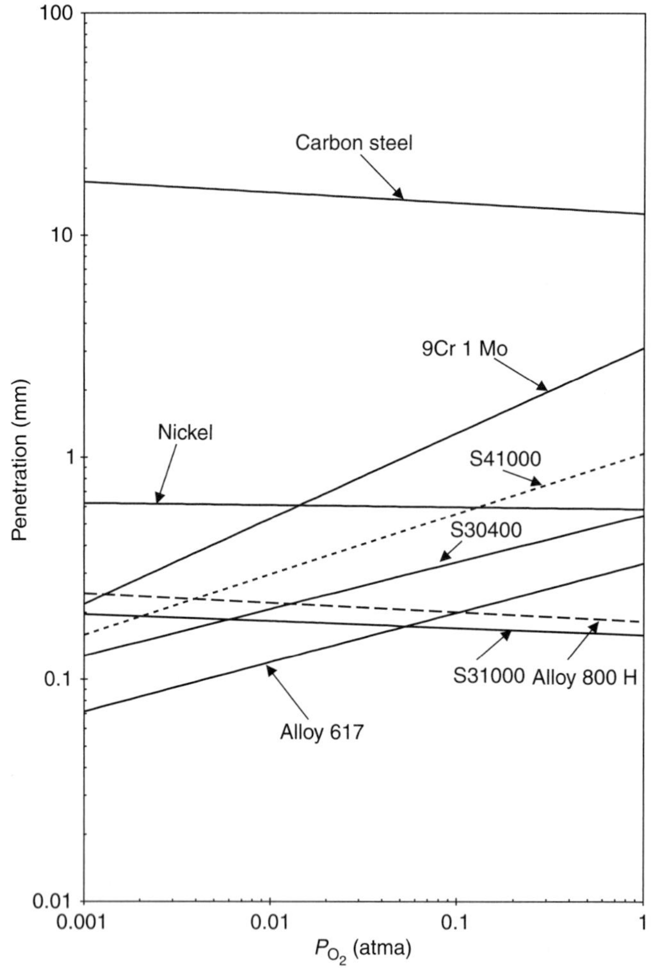

FIGURE 15.17 Effect of oxygen partial pressure upon metal penetration of some common alloys by oxidation after exposure for one year at 930°C.

in Cr or Al, whose oxides are stabilized by increasing O_2 levels. Alloys, which generally exhibit increased oxidation rates as the O_2 concentration increase, are S30400, S41000, and S44600 stainless steels and 9Cr-1Mo, Incoloy DS, alloys 617, and 253MA. These alloys tend to form poor oxide scales [2].

Most alloys tend to have increasing penetration rates with increasing temperature for all oxygen concentrations. Some exceptions are alloys with one to four percent aluminum such as alloy 214. This alloy requires higher temperatures to form Al_2O_3 as the dominant surface oxide, which grows more slowly than the Cr_2O_3 that dominates at lower temperatures. Figure 15.18 summarizes oxidation after one year for some common alloys exposed to air.

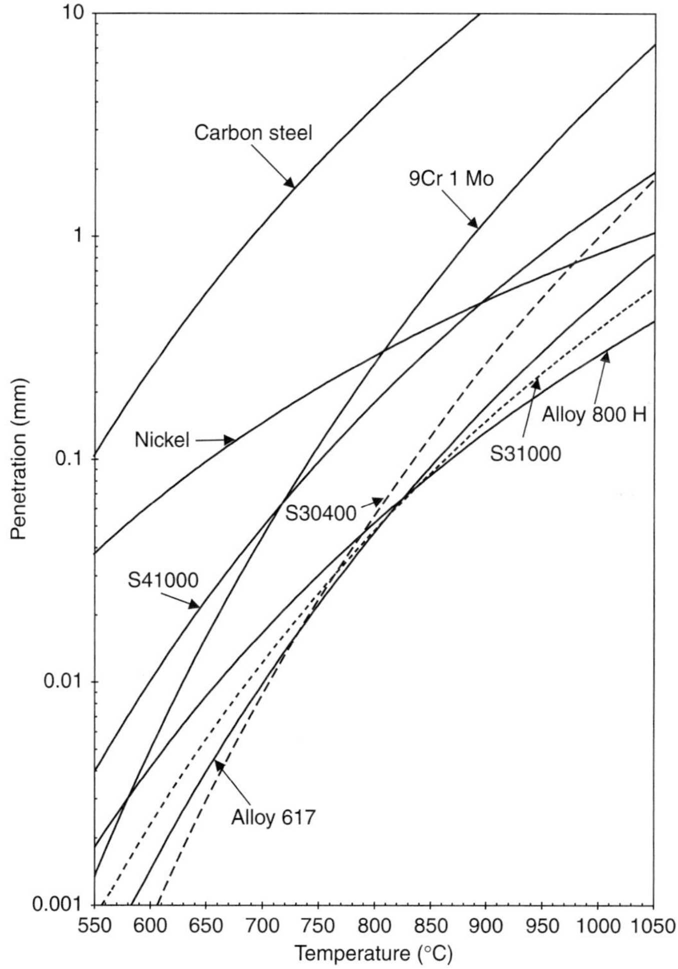

FIGURE 15.18 Effect of temperature upon metal penetration of some common alloys by oxidation after exposure for one year to air.

The alloy composition can influence metal penetration occurring by subsurface oxidation along grain boundaries and within the alloy grains, as schematically shown in Fig. 15.19 [11]. Most of the commercial heat-resistant alloys are based upon combinations of Fe-Ni-Cr. These alloys show about 80 to 95 percent of the total penetration as subsurface oxidation. Some alloys change in how much of the total penetration occurs by subsurface oxidation as time passes even though the corrosion product morphologies may remain constant. Alloys vary greatly in the extent of surface scaling and subsurface oxidation. Tests were conducted in flowing air at 980, 1095, 1150, and 1250°C for 1008 hours. The results of these tests, in terms of metal loss and average metal affected (metal loss and internal penetration), are presented in Table 15.7 [12].

15.4.2 Sulfidation

Sulfidation is a common high-temperature corrosion-failure mechanism. As the name implies, it is related to the presence of sulfur compounds. When examining this form of damage microscopically, a "front" of sulfidation is often seen to penetrate into the affected alloy. Localized pitting-type attack is also possible. A distinction can be made between sulfidation in gaseous environments and corrosion in the presence of salt deposits on corroding surfaces. Only the former is considered in this section, the latter being discussed in

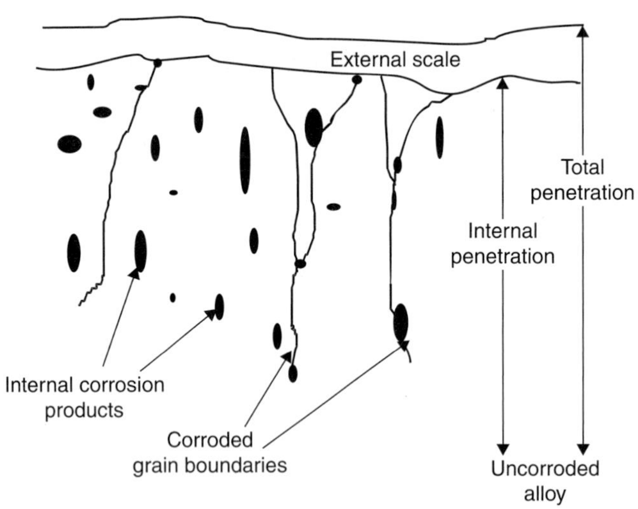

Figure 15.19 Schematic view of total penetration measurement for a typical corrosion product morphology.

	Temperature							
	980 (°C)		1095 (°C)		1150 (°C)		1250 (°C)	
	Loss	Affected	Loss	Affected	Loss	Affected	Loss	Affected
Alloy	(mm)	(mm)	(mm)	(mm)	(mm)	(mm)	(mm)	(mm)
214	0.0025	0.005	0.0025	0.0025	0.005	0.0075	0.005	0.018
601	0.013	0.033	0.03	0.067	0.061	0.135	0.11	0.19
600	0.0075	0.023	0.028	0.041	0.043	0.074	0.13	0.21
230	0.0075	0.018	0.013	0.033	0.058	0.086	0.11	0.20
S	0.005	0.013	3.01	0.033	0.025	0.043	> 0.81	> 0.81
617	0.0075	0.033	3.015	0.046	0.028	0.086	0.27	0.32
333	0.0075	0.025	0.025	0.058	0.05	0.1	0.18	0.45
X	0.0075	0.023	0.038	0.069	0.11	0.147	> 0.9	> 0.9
671	0.0229	0.043	0.038	0.061	0.066	0.099	0.086	0.42
625	0.0075	0.018,	0.084	0.12	0.41	0.46	> 1.2	> 1.2
Waspaloy	0.0152	0.079	0.036	0.14	0.079	0.33	> 0.40	> 0.40
R-41	0.0178	0.122	0.086	0.30	0.21	0.44	> 0.73	> 0.73

Table 15.7 Results of 1008-hours Static Oxidation Tests on Iron, Nickel, and Cobalt Alloys in Flowing Air at Different Temperatures

	Temperature							
	980 (°C)		1095 (°C)		1150 (°C)		1250 (°C)	
	Loss	Affected	Loss	Affected	Loss	Affected	Loss	Affected
Alloy	**(mm)**	**(mm)**	**(mm)**	**(mm)**	**(mm)**	**(mm)**	**(mm)**	**(mm)**
263	0.0178	0.145	0,089	0.36	0.18	0.41	> 0.91	> 0.91
188	0.005	0.015	0.01	0.033	0.18	0.2	> 0.55	> 0.55
25	0.01	0.018	0.23	0.26	0.43	0.49	> 0.96	> 0.96
150	0.01	0.025	0.058	0.097	> 0.68	> 0.68	> 1.17	> 1.17
6B	0.01	0.025	0.35	0.39	> 0.94	> 0.94	> 0.94	> 0.94
556	0.01	0.028	0.025	0.067	0.24	0.29	> 3.8	> 3.8
Multimet	0.01	0.033	0.226	0.29	> 1.2	> 1.2	> 3.7	> 3.7
800H	0.023	0.046	0.14	0.19	0.19	0.23	0.29	0.35
RA330	0.01	0.11	0.02	0.17	0.041	0.22	0.096	0.21
S31000	0.01	0.028	0.025	0.058	0.075	0.11	0.2	0.26
S31600	0.315	0.36	> 1.7	> 1.7	> 2.7	> 2.7	> 3.57	> 3.57
S30400	0.14	0.21	> 0.69	> 0.69	> 0.6	> 0.6	> 1.7	> 1.73
S44600	0.033	0.058	0.33	0.37	>0.55	> 0.55	> 0.59	> 0.59

Table 15.7 *(continued)*

a following section on salt and ash deposit corrosion. Gaseous environments associated with sulfidation have divided into the following categories [12]:

- Hydrogen/hydrogen sulfide mixtures or sulfur vapor of a highly reducing nature
- Moderately reducing mixed gas environments that contain mixtures of hydrogen, water, carbon dioxide, carbon monoxide, and hydrogen sulfide
- Sulfur dioxide-containing atmospheres

In the first category, sulfides rather than protective chromia are thermodynamically stable. Hydrogen/hydrogen sulfide mixtures are typically found in catalytic reformers in oil refining operations. Organic sulfur compounds such as mercaptans, polysulfides, and thiophenes, as well as elemental sulfur, contaminate practically all crude oils in various concentrations and are partially converted to hydrogen sulfide in refining operations. Hydrogen sulfide in the presence of hydrogen becomes extremely corrosive above 260 to 288°C. Sulfidation problems may also be encountered at lower temperatures. Increased temperatures and higher hydrogen sulfide contents generally lead to higher degradation rates.

For catalytic reforming, the 18Cr-8Ni austenitic stainless steels grades are considered to be adequately resistant to sulfidation. However, the use of stabilized grades is advisable. Some sensitization is unavoidable if exposure in the sensitizing temperature range is continuous or long term. Stainless steel equipment subjected to such exposure and to sulfidation corrosion should be treated with a two percent soda-ash solution or an ammonia solution immediately upon shutdown to avoid the formation of polythionic acid, which can cause severe intergranular corrosion and stress cracking [8].

Vessels for high-pressure hydrotreating and other heavy crude fraction upgrading processes (e.g., hydrocracking) are usually constructed of one of the Cr-Mo alloys. To control sulfidation, they are internally clad with one of the 300 series austenitic stainless steels. In contrast, piping, heat exchangers, valves, and other components exposed to high-temperature hydrogen-hydrogen sulfide environments are usually entirely constructed with these austenitic stainless steels. Figure 15.20 illustrates the corrosion behavior of austenitic steels as a function of hydrogen concentration and temperature [3]. In some designs alloy 800H has been used for piping and headers.

The effects of temperature and H_2S concentration upon sulfidation of alloys often used in oil refining services are shown in Figs. 15.21 to 15.25, which represent the metal losses expected after

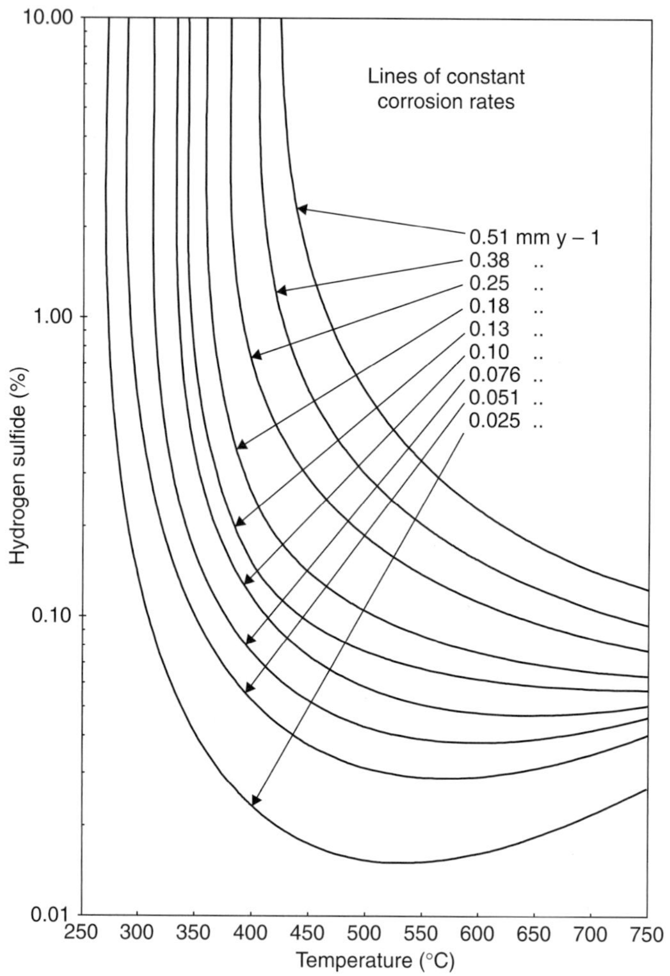

FIGURE 15.20 Effect of temperature and hydrogen sulfide concentration on corrosion rates of austenitic stainless steels for exposure longer than 150 hours.

one year of exposure (note the decreasing corrosion penetration scale in Figs. 15.22 to 15.24). The carbon steel line, in Fig. 15.21, stops for lower concentrations of H_2S because FeS is not stable and the steel does not corrode in such environment [2]. Increasing the temperature and H_2S concentration increases the sulfidation rate.

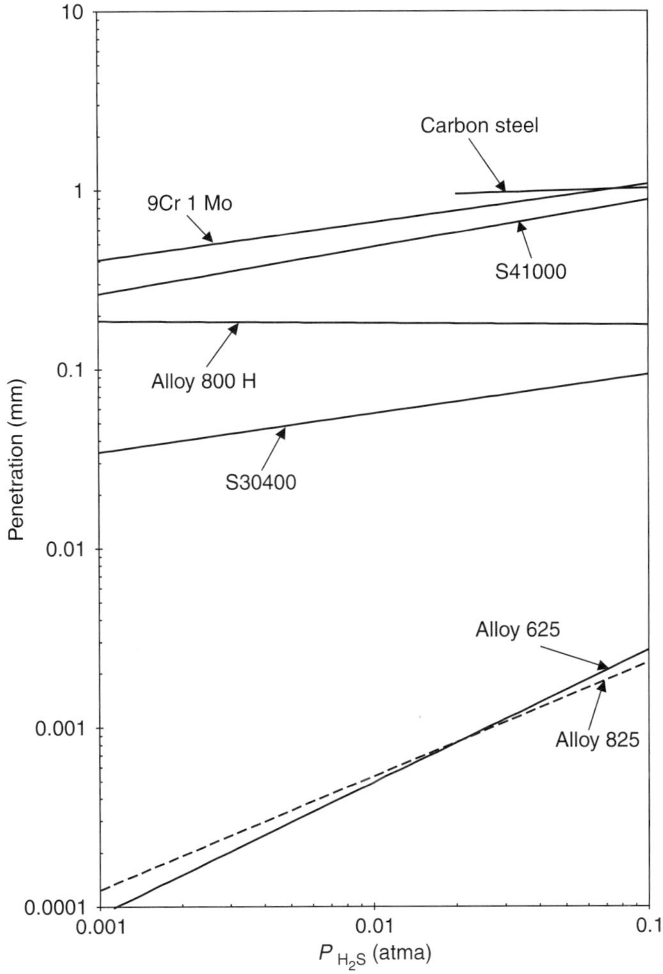

FIGURE 15.21 Effect of H_2S partial pressure upon sulfidation corrosion after one year in H_2-H_2S gases at 34 atma and 540°C.

It is typical that a temperature increase of 55°C will double the sulfidation rate, whereas increasing the H_2S concentration by a factor of 10 may be needed to double the sulfidation rate. Therefore, changes of H_2S concentration are generally less significant than temperature variations.

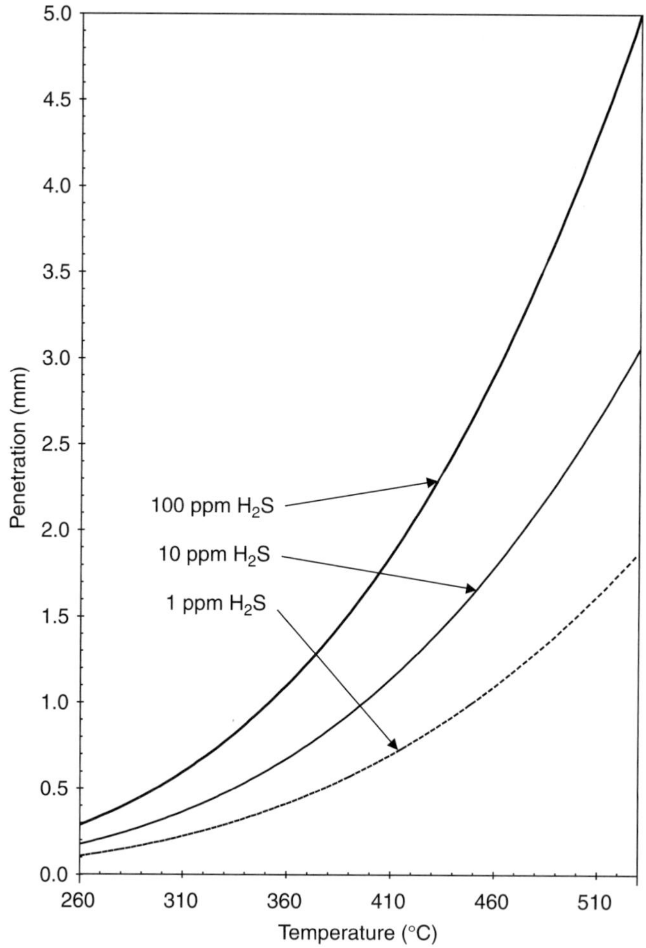

FIGURE 15.22 Effect of temperature upon sulfidation corrosion of 9 Cr 1 Mo after one year in H_2-H_2S gases at 34 atma.

Increasing the Cr content of the alloy greatly slows the sulfidation, as seen in progression from 9Cr-1Mo, S41000, S30400, 800H, 825, and 625 (Fig. 15.25). The ranges of H_2S concentration represented in these figures span the low H_2S range of catalytic reformers to the high H_2S concentrations expected in modern hydrotreaters. A summary of maximum allowable temperatures that will limit the extent of metal

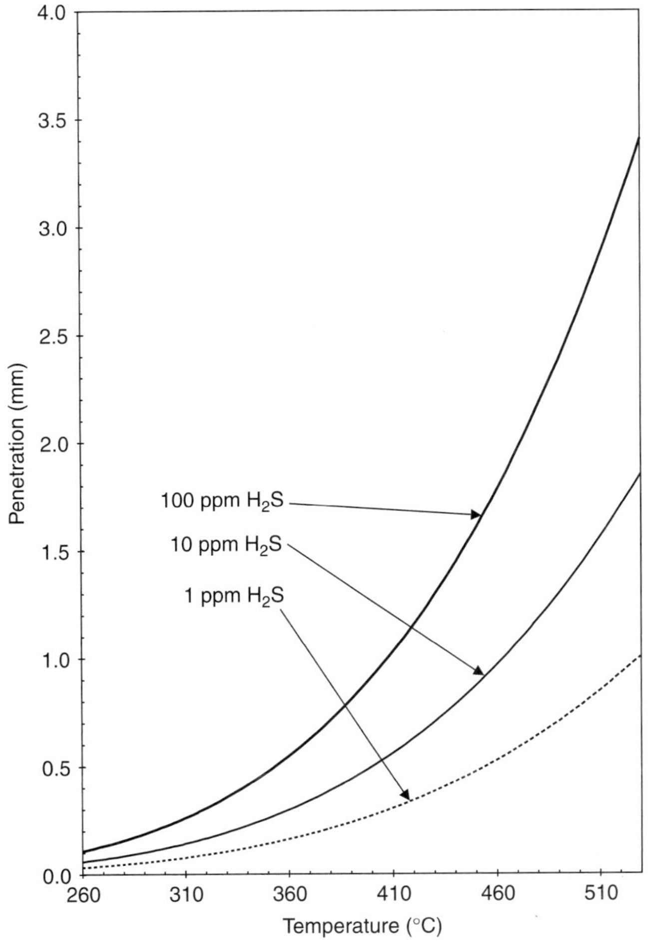

FIGURE 15.23 Effect of temperature upon sulfidation corrosion of S41000 after one year in H_2-H_2S gases at 34 atma.

loss by sulfidation to less than 0.25 mm is shown in Table 15.8 for several gas compositions of H_2S-H_2 at a pressure of 34 atm, which is similar to hydrotreating in an oil refinery. The maximum allowable temperatures for alloys exposed to different gas pressures and compositions can be evaluated with this information.

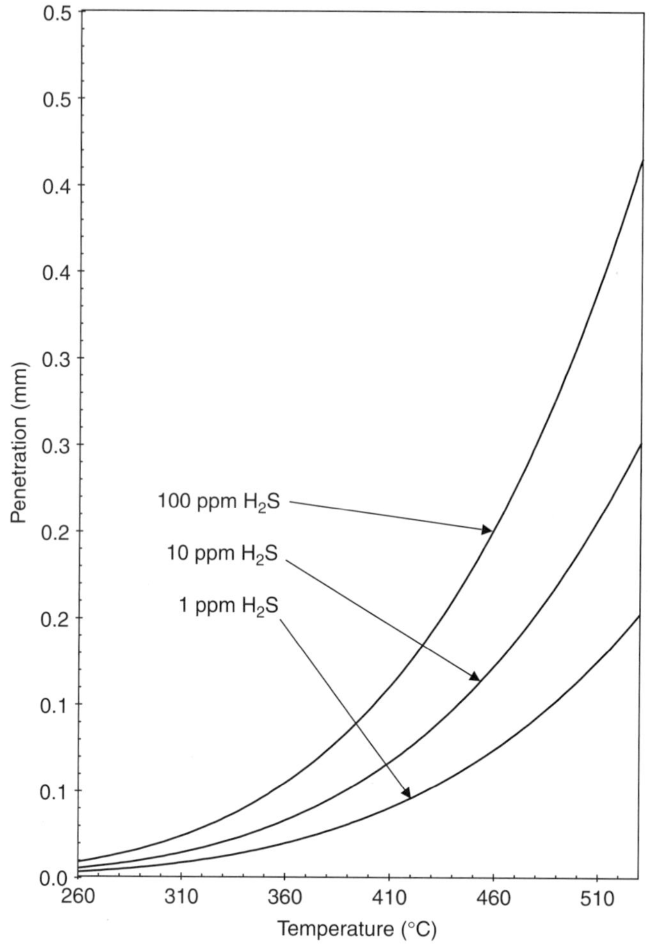

Figure 15.24 Effect of temperature upon sulfidation corrosion of S30400 after one year in H_2-H_2S gases at 34 atm.

In the second category, the presence of oxidizing gases such as H_2O (steam) or CO_2 slow the sulfidation rate below that expected if only the H_2S-H_2 concentrations were considered. This can be important because gases, which are thought to contain only H_2S-H_2, often also contain some H_2O. For example, a gas, which has been well mixed and equilibrated with water at room temperature, may contain up to two percent water vapor in the gas. Sulfidation rate predictions based only upon the H_2S-H_2 concentrations may overestimate the rate of metal loss.

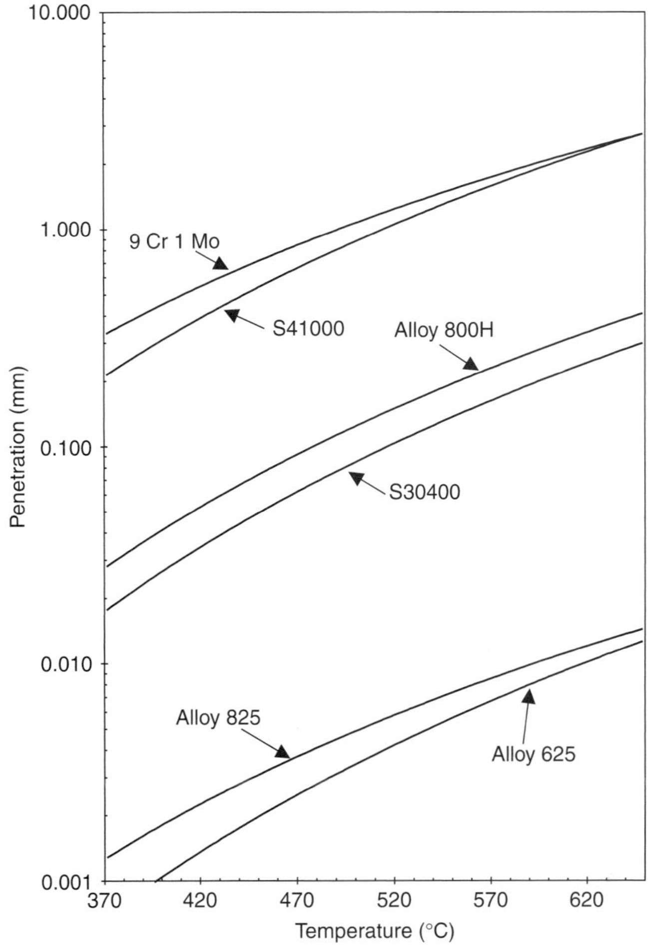

Figure 15.25 Effect of temperature upon sulfidation corrosion after one year in H_2-H_2S gases containing 1 percent vol H_2S at 34 atm.

The precise mechanism of how H_2O slows sulfidation by H_2S is still unclear, although numerous studies have confirmed this effect. This slowed corrosion rate is sometimes called sulfidation/oxidation because it represents a transition between the rapid corrosion of sulfidation and the slow corrosion of oxidation of alloyed metals containing either Cr or Al. Atmospheres high in sulfur dioxide are encountered in sulfur furnaces, where sulfur is combusted in air for manufacturing sulfuric acid. Lower levels of sulfur dioxide are encountered in flue gases when fossil fuels contaminated with sulfur species are combusted.

	Maximum Allowable Temperature (°C)				
Alloy\ H_2S Concentration	(0.001%)	(0.01%)	(0.1%)	(1%)	(10%)
Nickel	395	360	340	310	295
Carbon steel	430	415	405	400	390
9Cr-1Mo	505	445	395	350	310
S41000	570	500	440	390	345
800 H	580	575	575	575	575
430	760	680	615	555	500
S30400	880	790	700	625	565
825	930	630	630	630	630
625	760	630	630	630	630
718	760	630	630	630	630

Table 15.8 Sulfidation Corrosion Temperatures Corresponding to a Maximum Metal Loss of 0.25 mm After One Year in H_2S-H_2 Gases at 34 atm Gas Pressure

15.4.3 Carburization

Carburization can occur when metals are exposed to carbon monoxide and dioxide, methane, ethane, or other hydrocarbons at elevated temperatures. Carbon from the environment combines primarily with chromium but also with any other carbide formers (Nb, W, Mo, Ti, and the like) present in the alloy to form internal carbides. Carbides formed in the microstructure can be complex in composition and structure and may precipitate on the grain boundaries or inside the grains. The main undesirable effect of carbide formation is embrittlement and reduced ductility at temperatures up to 482 to 538°C. By tying up chromium in the form of stable chromium-rich carbides, carburization also reduces oxidation resistance. Creep strength may also be adversely affected, and internal stresses can arise from the volume increase associated with the carbon uptake and carbide formation. This internal pressure represents an additional stress to the operational stresses. Localized bulging, or even cracking, of carburized components is indicative of high internal stress levels that can be generated.

An insidious aspect of carburization is its nonuniform nature. Just as for other forms of localized corrosion, it is extremely difficult to predict and model localized carburization damage. As a rule of thumb, carburization problems only occur at temperatures above 815°C, because of unfavorable kinetics at lower temperatures. Carburization is therefore not a common occurrence in most refining operations because of the relatively low tube temperatures of most refinery-fired heaters.

However, carburization is more common in the petrochemical processing industry. A notable problem area has been the radiant and shield sections of ethylene cracking furnaces, due to high tube temperatures up to 1150°C. Apart from temperature, an increase in carbon potential of the gas mix is responsible for a higher severity of damage. High carbon potentials are associated with the ethane, propane, naphtha, and other hydrocarbons as reactants that are cracked. Carburization has been identified as the most frequent failure mechanism of ethylene furnace tubes.

Less severe and frequent carburization damage has been reported in reforming operations and in other processes handling hydrocarbon streams or certain ratios of $CO/CO_2/H_2$ gas mixtures at high temperature [8]. As it is the case with oxidation and sulfidation, chromium is considered to impart the greatest resistance to carburization [3]. Other beneficial elements include nickel, silicon, columbium, titanium, tungsten, aluminum, and molybdenum. The most important characteristic of a successful alloy is its ability to form and maintain a stable, protective oxide film. Aluminum and silicon alloying additions can contribute positively to this requirement. Unfortunately, the addition of aluminum or silicon to the heat-resistant alloys in quantities to develop full protection involves metallurgical trade-offs in strength, ductility, and/or weldability. Considering fabrication requirements and mechanical properties, viable alloys are generally restricted to about two percent of either element. This is helpful but not a total solution.

The tubes of ethylene-cracking furnaces were originally largely manufactured out of the cast HK-40 alloy (Fe-25Cr-20Ni). Since the mid-1980s, more resistant HP alloys have been introduced, but carburization problems have not been eliminated, probably due to more severe operating conditions at higher temperatures. Some operators have implemented a 35Cr-45Ni cast alloy, with various additions, to combat these conditions. For short residence-time furnaces with small tubes, wrought alloys including HK4M and HPM, Alloy 803, and Alloy 800H have been used. Other wrought alloys (e.g., 85H and HR-160, both with high silicon) have been applied to combat carburization of trays, retorts, and other components used in carburizing heat treatments. However, their limited fabricability precludes broad use in the refining or petrochemical industry [8].

Carburization causes the normally nonmagnetic wrought and cast heat-resistant alloys to become magnetic. The resulting magnetic permeability provides a means for monitoring the extent of carburization damage. Measurement devices range from simple handheld magnets to advanced multifrequency eddy current instruments. Carburization patterns can also reveal uneven temperature distributions that might otherwise have gone undetected.

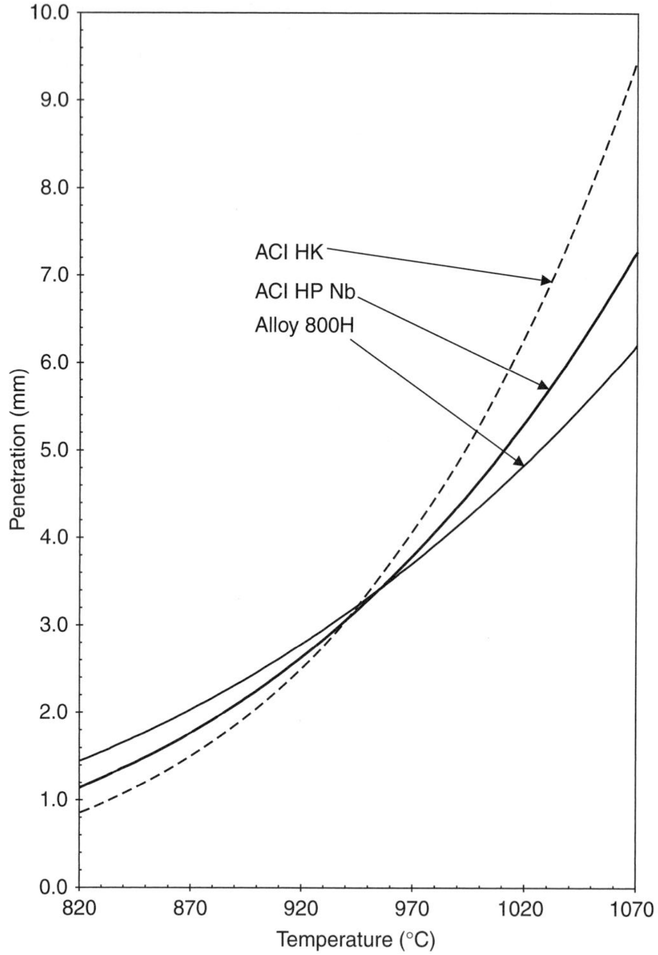

FIGURE 15.26 Effect of temperature upon carburization of several alloys exposed to solid carbon and 200 ppm H_2S at 1 atm.

Most alloys tend to have more carburization penetration with increasing temperatures. Figure 15.26 summarizes carburization after one year for some common alloys exposed to solid carbon and 200 ppm H_2S [2].

The time dependence of carburization has been commonly reported to be parabolic. Removal or cracking of any surface carbide scale will tend to increase the rate of carburization. One thousand hours may be required to establish the time dependence expected for long-term service. Carburization data are properly used when the time dependence is considered. Increasing the concentration of H_2S tends to slow the carburization rate of alloys. Figure 15.27 shows the effect for several alloys widely used in petrochemical equipment. The effect of H_2S is to slow decomposition of the CH_4, which adsorbs onto the metal surface, thus slowing the rate of carburization. Increasing concentrations will slow carburization until the concentrations become high enough to cause sulfidation to become the dominant corrosion mechanism. The conditions for the initiation of sulfidation depend upon the alloy and gas compositions.

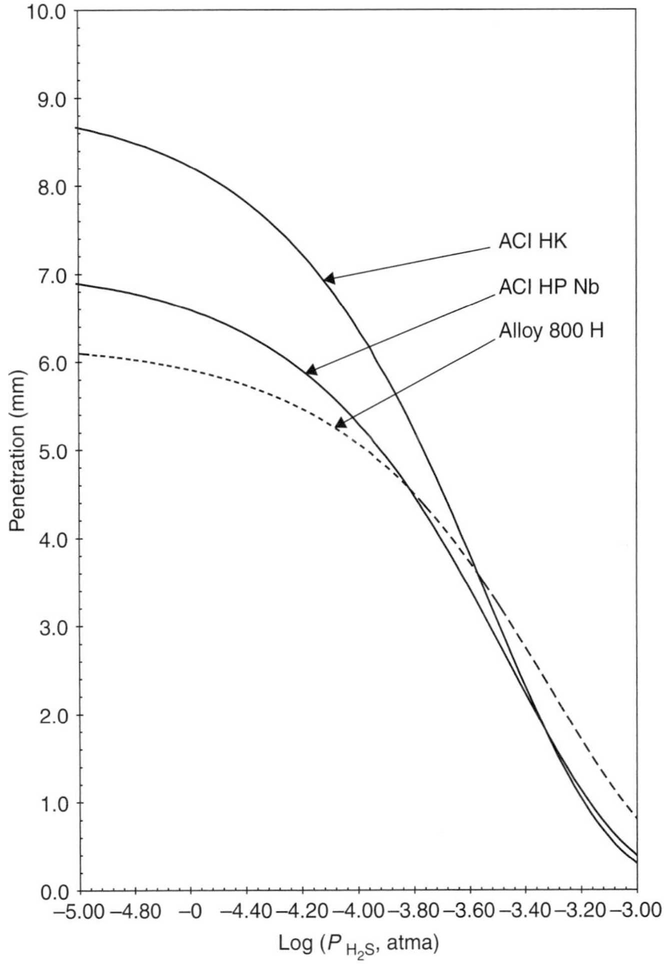

FIGURE 15.27 Effect of H_2S partial pressure upon carburization of several alloys exposed to solid carbon and 200 ppm H_2S at 982°C.

15.4.4 Metal Dusting

Another ultimate effect of carburization can be metal dusting. This phenomenon may occur in process operations in which oxidizing and reducing conditions are cycled. When the environment is on the reducing side (CO predominant), carburization of the metal to a shallow depth can occur at breaks in the protective oxide film. When the exposure then changes to oxidizing, the high-carbon area of the metal is burned out and the metal reacted to the oxide. A depression is left in the metal surface where the carburized area existed (Fig. 15.28), and the metal oxide is swept downstream in the process as metal dust.

Metal dusting first appears as fine powders (hence the term "dusting") consisting of carbides, oxide, and graphite (soot). The morphology of attack can be localized pitting or relatively uniform damage. The underlying alloy may or may not display evidence of carburization in the microstructure. Metal dusting is manifested at lower temperatures than carburization, typically between 425 and 815°C. Maximum rates of metal dusting damage are considered to occur around 650 to 730°C.

Metal dusting is usually associated with gas streams rich in carbon monoxide and hydrogen. Prediction and modeling of metal dusting are difficult, and little relevant quantitative data is available for engineering alloys to assist designers. It appears that most stainless

(a)

Figure 15.28 Two alloys tested in the same H_2-CO-H_2O metal dusting atmosphere: (*a*) 2.25 Cr 1-Mo at 620°C and 1600 hours showing general attack; and (*b*) 9 Cr 1 Mo (P91 steel) at 620°C and 1008 hours, showing metal dusting pitting attack. (Courtesy of Karl-Winnaker-Institute, DECHEMA, e.v.)

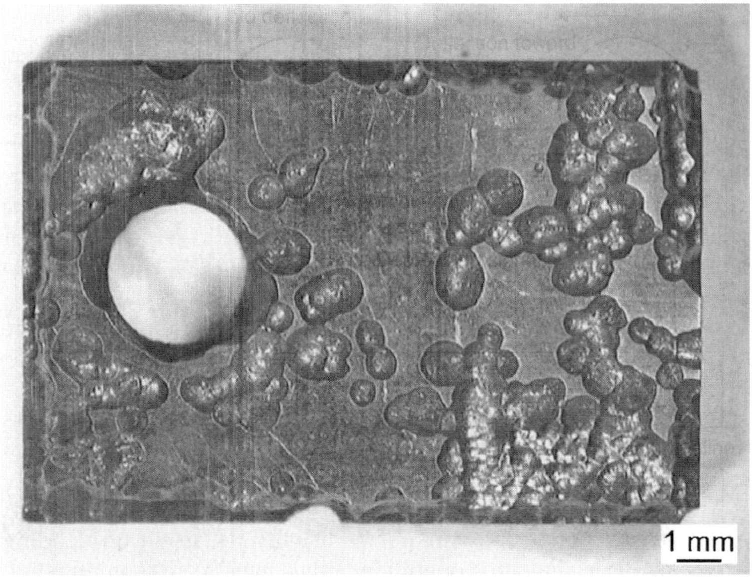

(b)

Figure 15.28 *(continued)*

steels and heat-resistant alloys can be attacked and that the rate of damage can be extremely high. The mechanisms of metal dusting attack are not understood. One remedial measure is adjusting the gas composition by reducing the CO partial pressure [8].

15.4.5 Nitridation

Nitridation usually occurs when carbon, low-alloy, and stainless steels are exposed to an ammonia-bearing environment at elevated temperatures. The production of ammonia, nitric acid, melamine, and nylon generate such conditions. Nitridation can also result from nitrogen atmospheres, especially under reducing conditions and high temperatures. There are many parallels to carburization; nitridation occurs when chromium and other elements combine with nitrogen to form embrittling nitrides in the microstructure.

Although stainless steels may have adequate resistance, high-nickel alloys tend to be more resistant. Increasing nickel and cobalt contents are also considered to be beneficial. However, pure nickel has shown poor resistance. Alloy 600, with 72 percent nickel, is often used in the heat-treating industry and occasionally in refining and petrochemical applications involving ammonia at temperatures above 340°C. Economics and its lower strength, compared with Alloy 800H and cast-modified HP, have limited its applications in the latter industries.

15.4.6 Gaseous Halogen Corrosion

The corrosive effect of halogen on passivating alloys is well known in aqueous media. Chlorides and fluorides also contribute to high-temperature corrosion by interfering with the formation of protective oxides or breaking them down if already formed. The main reason for the reduced corrosion resistance in the presence of halogens is the formation of volatile corrosion products that are nonprotective. The melting points, boiling points, and temperature at which the vapor pressure reaches 10^{-4} atm of selected metal chlorides is presented in Table 15.9 [9]. The high volatility and relatively low melting points of these chloride species should be noted. Clearly these properties are not conducive to establishing an effective diffusion barrier on the corroding alloy surfaces.

In refining operations, chlorides most commonly enter the process operations as salt water or brine. Organic chlorides find their way into crude feed. These are not removed in the desalters but are generally removed in the distillation process. Chlorides can enter the downstream processes. Fluoride contamination is usually the result of blending streams from an alkylation operation. This downstream contamination cannot spill over to petrochemical facilities that take feed from these sources. Halogen contamination during shipment and storage are also of concern to petrochemical facilities [8]. Chlorination processes are used to produce certain metals, as well as in nickel extraction. Calcining operations used in the production of certain rare earth metals and for producing ceramic ferrites for permanent magnets are also associated with high-temperature chlorine-containing environments.

In high-temperature chlorine atmospheres chromium and nickel additions to iron are both regarded as beneficial. Stainless steels are therefore more resistant than the lower-alloyed steels. Austenitic stainless steels tend to outperform the ferritic grades (at equivalent chromium levels). Nickel and nickel-based alloys are widely used under such conditions. The high-nickel alloys are significantly more resistant than the stainless steels to chlorine but not to fluorine, which is more soluble in nickel. When both chlorine and oxygen are present in the environment, essentially a competing situation arises between the formation of stable oxide and volatile chloride species. Therefore, the degradation rate can fluctuate between parabolic, linear, and hybrid behavior. Molybdenum and tungsten alloying additions are considered highly undesirable in such service environments due to the formation of highly volatile oxychlorides (Table 15.9). Aluminum additions are regarded as beneficial, due to the formation of a protective alumina scale at high temperatures [13].

15.4.7 Fuel Ash and Salt Deposits

In many industrial applications, the surfaces undergoing high-temperature corrosion are far from clean and mostly covered with surface deposits of ash and/or salt form on the components.

Chlorides	Melting Point	Temperature at 10^{-4} atm	Boiling Point
	(°C)	(°C)	(°C)
$FeCl_2$	676	536	1026
$FeCl_3$	303	167	319
$NiCl_2$	1030	607	987
$CoCl_2$	740	587	1025
$CrCl_2$	820	741	1300
$CrCl_3$	1150	611	945
CrO_2Cl_2	−95	–	117
CuCl	430	387	1690
$MoCl_5$	194	58	268
WCl_5	240	72	–
WCl_6	280	11	337
$TiCl_2$	1025	921	–
$TiCl_3$	730	454	750
$TiCl_4$	−23	−38	137
$AlCl_3$	193	76	–
$SiCl_4$	−70	−87	58
$MnCl_2$	652	607	1190
$ZrCl_4$	483	146	–
$NbCl_5$	205	–	250
$NbCl_4$	–	239	455
$TaCl_5$	216	80	240
$HfCl_4$	434	132	–
CCl_4	−24	−80	77
NaCl	801	742	1465
KCl	772	706	1407
LiCl	610	665	1382
$MgCl_2$	714	663	1418
$CaCl_2$	772	1039	2000
$BaCl_2$	962	–	1830
$ZnCl_2$	318	349	732
$PbCl_2$	498	484	954

Table 15.9 Melting Points, Temperatures at Which Chloride Vapor Pressure Reaches 10^{-4} atm, and Boiling Points of Various Chlorides

Chemical reactions between these deposits and the protective surface oxide can lead to destruction of the oxide and rapid corrosive attack. In gas turbines, oxidized sulfur contaminants in fuel and chlorides from ingested air (marine atmospheres) tend to react to form salt deposits. The presence of sodium sulfate, potassium sulfate, and calcium sulfate together with magnesium chloride have been reported in such deposits for compressor-stage components [14]. Sodium sulfate is usually regarded as the dominant component of the salt deposits.

Testing has indicated that in commercial nickel- and cobalt-based alloys, chromium additions play an important role in limiting this type of damage. Alloys with less than 15 percent of chromium as alloying addition are considered highly vulnerable to attack.

Refinery heaters and boilers that are fired with low-grade fuels may be vulnerable to such corrosion damage, especially if vanadium, sulfur, and sodium contaminants are present at high levels. Vanadium pentoxide and sodium sulfate deposits assume an important role in this type of environment. The melting point of one of these mixed compound deposits (Na_2SO_4-V_2O_5) can be as low as 630°C, at which point catastrophic corrosion can set in. In these severe operating conditions the use of special high-chromium alloys is required. A 50Ni-50Cr alloy has been recommended over the use of 25Cr-12Ni and 25Cr-20Ni alloys for hangers, tube sheets, and other supports. Ash and salt deposit corrosion is also a problem area in fireside corrosion of waste incinerators, in calcining operations, and in flue gas streams.

15.4.8 Corrosion by Molten Salts

Corrosion damage from molten salts can occur in a wide variety of materials and by different mechanisms. It has been pointed out that although many studies have been performed, quantitative data for materials selection and performance prediction are rarely available [15]. Molten salt corrosion is usually applicable to materials retaining the molten salt, as used in heat treating, solar and nuclear energy systems, batteries, fuel cells, and extractive metallurgical processes. Some factors that can make molten salts extremely corrosive include the following:

- By acting as fluxes, molten salts destabilize protective oxide layers (on a microscopic scale, this effect contributes toward fuel ash corrosion described above).
- High temperatures are typically involved.
- Molten salts are generally good solvents, preventing the precipitation of protective surface deposits.
- Direct chemical reaction between the containment material and the salt.
- The presence of noble metal ions in the molten salt, more noble than the containment material itself.

15.4.9 Corrosion in Liquid Metals

Corrosion in liquid metals is applicable to metals and alloy processing, metals production, liquid metal coolants in nuclear and solar power generation, other nuclear breeding applications, heat sinks in automotive and aircraft valves, and brazing operations. Corrosion damage to containment materials is usually the concern. Again, practical design and performance data are extremely limited. Several possible corrosion mechanisms need to be considered during the materials selection design phase. The most severe problems arise at high temperatures and aggressive melts. Molten steel is typically regarded as a nonaggressive melt, whereas molten lithium is much more corrosive.

Practical problems are complicated by the fact that several of these forms can occur simultaneously. In fact, opposing actions may be required for individual effects that act in combination. The following categories can be used to classify relevant corrosion phenomena [16]:

- Dissolution
- Impurities and interstitial reactions
- Alloying
- Compound reduction

Corrosion reactions can occur by a simple dissolution mechanism, whereby the containment material dissolves in the melt without any impurity effects. Material dissolved in a hot zone may be redeposited in a colder area, possibly compounding the corrosion problem by additional plugging and blockages where deposition has taken place. Dissolution damage may be of a localized nature, for example, by selective dealloying. The second corrosion mechanism is one of reactions involving interstitial (or impurity) elements such as carbon or oxygen in the melt or containment material. Two further subforms are corrosion product formation and elemental transfer. In the former the liquid metal is directly involved in corrosion product formation. In the latter the liquid metal does not react directly with the containment alloy; rather, interstitial elements are transferred to, from, or across the liquid.

Alloying refers to the formation of reaction products on the containment material, when atoms other than impurities or interstitials of the liquid metal and containment material react. This effect can sometimes be used to produce a corrosion-resistant layer, separating the liquid metal from the containment (for example, aluminum added to molten lithium contained by steel). Lastly, liquid metal can attack ceramics by reduction reactions. Removal of the nonmetallic element from such solids by the melt will clearly destroy their structural integrity. Molten lithium poses a high risk for reducing ceramic materials (oxides).

References

1. Elliott P. Practical guide to high-temperature alloys. *Materials Performance* 1989; 28: 57–66.
2. John, RC. Compilation and use of corrosion data for alloys in various high-temperature gases. *Corrosion 99* Paper #99073. Houston, Tex., NACE International, 1999.
3. Sedriks AJ. *Corrosion of Stainless Steels.* New York, N.Y.: John Wiley & Sons, 1979.
4. Birks N, Meier GH. *Introduction to High Temperature Oxidation of Metals.* London, U.K.: Edward Arnold, 1983.
5. Gulbransen EA, Jansson SA. Thermochemical considerations of high temperature gas-solid reactions. In: Belton GR, Worrell WF, eds. *Heterogeneous Kinetics at Elevated Temperatures.* New York, N.Y.: Plenum Press, 1970.
6. Roine A, Outokumpu HSC. *Chemistry for Windows. [3.0].* 1997. Finland, Outokumpu Research Oy.
7. Jones DA. *Principles and Prevention of Corrosion. 2nd edn.* Upper Saddle River, N.J.: Prentice Hall, 1996.
8. Tillack DJ, Guthrie JE. *Wrought and Cast Heat-Resistant Stainless Steels and Nickel Alloys for the Refining and Petrochemical Industries.* NiDI Technical Series 10071. Toronto, Canada: Nickel Development Institute, 1992.
9. Lai GY. High-temperature corrosion: Issues in alloy selection. *Journal of Materials* 1991; 54–60.
10. Bayer GT. Vapor aluminum diffused steels for high-temperature corrosion resistance. *Materials Performance* 1995; 34: 34–8.
11. John RC, Pelton AD, Young AL, Thompson WT, Wright IG, Besmann TM. Managing corrosion engineering information for metals in hot corrosive gases. In Zhao, J. C., Fahrmann, M., and Pollock, TM. 359-382. 2001. TMS. *Materials Design Approaches and Experiences,* November 4–8, 2001, Indianapolis, Ind.
12. Lai GY. *High Temperature Corrosion of Engineering Alloys.* Materials Park, Ohio: American Society for Metals, 1990.
13. Roberge PR. *Handbook of Corrosion Engineering.* New York, N.Y.: McGraw-Hill, 2000.
14. Bornstein NS. Reviewing sulfidation corrosion — yesterday and today. *Journal of Materials* 1996; 37–9.
15. Koger JW. Fundamentals of High-temperature corrosion in molten salts. In: *Metals Handbook: Corrosion.* Metals Park, Ohio: ASM International, 1987; 50–5.
16. Tortorelli PF, Pawel SJ. Corrosion by Liquid Metals. In: Cramer DS, Covino BS, eds. *Vol. 13A: Corrosion: Fundamentals, Testing, and Protection.* Metals Park, Ohio: ASM International, 2003; 129–34.

APPENDIX A

Historical Perspective

Organized studies have been devoted to corrosion for a longer time than most people realize. The British Association for the Advancement of Science appropriated a sum of money for a series of corrosion experiments on the corrosion of cast and wrought iron in the 1830s. Experiments were conducted by Robert Mallet and reported by him in 1838, 1840, and 1843. Mallet's work was done in an era in which gifted scientists were investigating the properties of matter.

Many interesting references and reflections on historical aspects of corrosion can be found in the special symposium that was held to celebrate NACE International's fiftieth anniversary [1]. Table A.1 presents some historical landmarks of discoveries related to the understanding and management of corrosion.

The science of corrosion has had two periods of rapid advancement. One, in the first half of the nineteenth century, was a result of intense and sustained scientific interest and activity aroused by the invention of the galvanic battery, and the controversy over the nature and source of the galvanic current. The other, in the first half of the twentieth century, was stimulated by growing realization of the immense economic cost of corrosion in a rapidly developing industrial age. In the latter period, a number of theories and facts established in the earlier one were rediscovered or elaborated, or both. These include the electrochemical theory of corrosion, proposed by Wollaston in 1801, developed by de La Rive in 1830, confirmed by Ericson-Auren and Palmaer in 1901, and rediscovered by Whitney in 1903 [2].

As early as 1819, Hall demonstrated the necessity of dissolved oxygen for appreciable corrosion of iron in water at ordinary temperatures and Sir Humphrey Davy published results of his work on cathodic protection of copper bottoms for British naval vessels in 1824. These early experiments established a practical base for the application of cathodic protection, which led to the development of galvanized iron.

In 1906, Committee U of the American Society for Testing Materials (ASTM) was formed to promote the development of corrosion tests.

Date	Landmark	Source
1500	Tin plate	
1675	Mechanical origin of corrosiveness and corrodibility	Boyle
1763	Bimetallic corrosion	HMS Alarm report
1788	Water becomes alkaline during corrosion of iron	Austin
1790	Passivity of iron	Keir
1791	Copper-iron electrolytic galvanic coupling	Galvani
1800	Galvanic battery	Volta
1801	Electrochemical theory of acid corrosion	Wollaston
1819	Insight into electrochemical nature of corrosion	Thenard
1824	Cathodic protection of Cu by Zn or Fe	Sir Humphrey Davy
1325	Differential temperature cell	Walcker
1826	Metal/metal oxide and stress cells	Davy
1827	Differential solution concentration cell	Beckerel
1830	Microstructural aspect of corrosion (Zn)	De la Rive
1830	Differential oxygen concentration cell	Marianini
1834–40	Relations between chemical action and generation of electric currents	Faraday
1836	Cement-lined cast iron pipe	Herschel
1837	Galvanized iron	Craufurd
1872	Organic inhibitors (essential and fixed oils)	Marangoni, Stefanelli
1873–78	Thermodynamics of corrosion	Gibbs
1899	Measurements of hydrogen overvoltage	Caspari
1903	Theory of corrosion of iron in water	Whitney
1904	Hydrogen overvoltage as a function of current	Tafel
1905	Carbonic and other acids are not essential for the corrosion of iron	Dunstan, Jowett, Goulding Tilden
1906	Monel alloy	Monell
1906	Phosphate coating	Coslett
1907	Chromate inhibitor	Cushman
1907	Oxygen action as cathodic stimulator	Walker, Cederholm
1908	Zinc chromate pigment	Cushman

Table A.1 Historical Landmarks of Discoveries Related to the Understanding and Management of Corrosion

Date	Landmark	Source
1908–10	Compilation of corrosion rates in different media	Heyn, Bauer
1909	Synthetic resin (Bakelite)	Baekeland
1910	Impressed current cathodic protection	Cumberland
1910	Microbiological corrosion	Gaines
1910	Inhibitive paint	Cushman, Gardner
1912	Austenitic chromium-nickel stainless steel	Maurer, Strauss
1913	Study of high temperature oxidation kinetics of tungsten	Langmuir
1913	High silicon cast iron (Duriron)	
1916	Differential aeration currents	Aston
1917	Mechanism of corrosion fatigue	Haigh
1916	Chromium stainless steel	Brearley
1919	Mechanism of cavitation	Parsons, Cook
1920–3	"Season-cracking" of brass as intergranular corrosion	Moore, Beckinsale
1923	Anodized aluminum	Bengough, Stuart
1923	High-temperature formation of oxides	Pilling, Bedworth
1924	Galvanic corrosion	Whitman, Russell
1926	Chromium-nickel-molybdenum stainless steel	Strauss
1929	Inconel	Merica
1930–31	Subscaling or "Internal corrosion"	Smith
1931–9	Quantitative electrochemical nature of corrosion	Evans
1936	Langelier index	Langelier
1938	Anodic and cathodic inhibitors	Chyzewski, Evans
1938	E-pH thermodynamic diagrams	Pourbaix
1939	Hastelloy C	McCurdy
1950	Autocatalytic nature of pitting	Uhlig
1956	Tafel extrapolation for measurement of kinetic parameters	Stern, Geary
1968	Electrochemical noise signature of corrosion	Iverson
1970	Study of corrosion processes with electrochemical impedance spectroscopy (EIS)	Epelboin

Table A.1 *(continued)*

Shortly afterward, other organizations began to pay attention to corrosion and its control [3].

Among pioneers in studying the effects of corrosion was the American Committee on Electrolysis, which noted in 1921 that its preliminary report had been published in October, 1916. This committee, composed of representatives of the American Institute of Electrical Engineers, American Electric Railway Association, American Railway Engineering Association, National Bureau of Standards, and others, concerned itself with the then serious problem of stray current damage to underground metal structures, especially the protection of communication cable from electrified street and interurban railways.

In England, the Corrosion Committee of the Iron and Steel Institute issued its first report in 1931 and sixth in 1959. An American Coordinating Committee on Corrosion was organized with representatives from 17 technical societies in 1938. This group, which aimed at coordinating the activities of societies to prevent duplicated work was absorbed by the National Association of Corrosion Engineers (NACE) in 1948 and was renamed the Inter-Society Committee on Corrosion Control. It functioned in a semiautonomous manner until it was finally disbanded about 10 years later [3]. This was largely because the growth of abstract publications and numerous other periodicals permitted easy interchange of most information.

Although Germany had a corrosion journal, *Korrosion and Metallschutz,* prior to World War II, which was interrupted during the war and reissued after its end under a new title, *Werkstoffe und Korrosion*, it was not until after NACE began publishing the magazine *Corrosion* in 1945 that journals on corrosion control were started in other countries. Since then, one or more magazines about corrosion control have been started in most industrialized countries of the world.

In addition to the present NACE International association, many other scientific engineering, governmental, and trade organizations are active in corrosion control work. Leaders among the scientific and engineering groups in North America are the American Society for Metals (ASM), American Society for Testing and Materials (ASTM), American Chemical Society (ASC), and The Electrochemical Society (ECS).

There are many other notable groups and associations dealing with some aspects of corrosion prevention and control in the Americas and in the rest of the world. A simple search of the World Wide Web will reveal the busiest of these groups and associations.

References

1. Gundry RD. Corrosion/93 Plenary & Keynote Lectures. Houston, Tex.: NACE International, 1993.
2. Lynes W. Some historical developments relating to corrosion. *Journal of the Electrochemical Society* 1951; 98: 3C–10C.
3. Van Delinder LS. *Corrosion Basics - An Introduction*. Houston, Tex: NACE International, 1984.

APPENDIX B

Periodic Table

Key

Atomic Number → 6 +4 +2 } Combining Capacity

Symbol → C
Carbon
12.0
↑
Atomic Mass*

*Based on C^{12} = 12.00000

1	2	3	4	5	6	7	8	9	10	11	12	13	14	15	16	17	18
1 ±1 H Hydrogen 1.0																	2 He Helium 4.0
3 +1 Li Lithium 6.9	4 +2 Be Beryllium 9.0											5 +3 B Boron 10.8	6 +4 +2 C Carbon 12.0	7 −3 +5 N Nitrogen 14.0	8 −2 O Oxygen 16.0	9 −1 F Fluorine 19.0	10 Ne Neon 20.2
11 +1 Na Sodium 23.0	12 +2 Mg Magnesium 24.3											13 +3 Al Aluminum 27.0	14 +4 +2 Si Silicon 28.1	15 −3 P Phosphorus 31.0	16 −2 +4 +6 S Sulphur 32.1	17 −1 Cl Chlorine 35.5	18 Ar Argon 40.0
19 +1 K Potassium 39.1	20 +2 Ca Calcium 40.1	21 +3 Sc Scandium 45.0	22 +3 +4 Ti Titanium 47.9	23 +2 +5 V Vanadium 50.9	24 +2 +3 +6 Cr Chromium 52.0	25 +2 +7 Mn Manganese 54.9	26 +3 +2 Fe Iron 55.8	27 +2 +3 Co Cobalt 58.9	28 +2 +3 Ni Nickel 58.7	29 +1 +2 Cu Copper 63.5	30 +2 Zn Zinc 65.4	31 +3 Ga Gallium 69.7	32 +4 +2 Ge Germanium 72.6	33 −3 +5 As Arsenic 74.9	34 −2 +4 +6 Se Selenium 79.0	35 −1 Br Bromine 79.9	36 Kr Krypton 83.8
37 +1 Rb Rubidium 85.5	38 +2 Sr Strontium 87.6	39 +3 Y Yttrium 88.9	40 +4 Zr Zirconium 91.2	41 +5 +3 Nb Niobium 92.9	42 +6 +2 Mo Molybdenum 95.9	43 +7 +3 Tc Technetium (99)	44 +3 Ru Ruthenium 101.1	45 +3 Rh Rhodium 102.9	46 +2 +3 +4 Pd Palladium 106.4	47 +1 Ag Silver 107.9	48 +3 Cd Cadmium 112.4	49 +3 +1 In Indium 114.8	50 +2 +4 Sn Tin 118.7	51 −3 +5 Sb Antimony 121.8	52 −2 +4 +6 Te Tellurium 127.6	53 −1 I Iodine 126.9	54 Xe Xenon 131.3
55 +1 Cs Cesium 132.9	56 +2 Ba Barium 137.3	57 +3 La Lanthanum 138.9	72 +4 Hf Hafnium 178.5	73 +5 Ta Tantalum 180.9	74 +6 +2 W Tungsten 183.9	75 +7 +3 Re Rhenium 186.2	76 +2 +3 +4 Os Osmium 190.2	77 +2 +3 +4 Ir Iridium 192.2	78 +2 +4 Pt Platinum 195.1	79 +1 +3 Au Gold 197.0	80 +1 +2 Hg Mercury 200.6	81 +1 +3 Tl Thallium 204.4	82 +2 +4 Pb Lead 207.2	83 −3 +5 Bi Bismuth 209.0	84 ±2 +4 Po Polonium (209)	85 −1 At Astatine (210)	86 Rn Radon (222)
87 +1 Fr Francium (223)†	88 +2 Ra Radium (226)	89 +3 Ac Actinium (227)															

Lanthanide Series	58 +3 +4 Ce Cerium 140.1	59 +3 +4 Pr Praseody-mium 140.9	60 +3 Nd Neodymium 144.2	61 +3 Pm Promethium (145)	62 +2 +3 Sm Samarium 150.4	63 +2 +3 Eu Europium 152.0	64 +3 Gd Gadolinium 157.3	65 +3 +4 Tb Terbium 158.9	66 +3 Dy Dysprosium 162.5	67 +3 Ho Holmium 164.9	68 +3 Er Erbium 167.3	69 −2 −3 Tm Thulium 168.9	70 −2 −3 Yb Ytterbium 173.0	71 −3 Lu Lutetium 175.0
Actinide Series	90 +4 Th Thorium 232.0	91 +4 +5 Pa Protactinium (231)	92 +3 +4 +5 +6 U Uranium 238.0	93 +3 +4 +5 +6 Np Neptunium (244)	94 +3 +4 +5 +6 Pu Plutonium (244)	95 +3 +4 +5 +6 Am Americium (243)	96 +3 Cm Curium (247)	97 +3 +4 Bk Berkelium (247)	98 Cf Californium (251)	99 Es Einsteinium (252)	100 Fm Fermium (257)	101 Md Mendelevium (258)	102 No Nobelium (259)	103 Lr Lawrencium (260)

†Masses in Parentheses are the Mass Numbers of the Most Stable Isotope.

APPENDIX C

SI Units Conversion Table

A.1 How to Read This Table

The table provides conversion factors to SI units. These factors can be considered as unity multipliers. For example,

Length: m/*X*
0.0254 in 0.3048 ft

Means that

$$1 = 0.0254\ (\text{m/in}) \text{ or } 1 \text{ in} = 0.0254 \text{ m}$$
$$1 = 0.3048 \text{ m/ft or } 1 \text{ ft} = 0.3048 \text{ m}$$

The SI units are listed immediately after the quantity; in this case: Length: m/*X*. The m stands for meter, and the *X* designates the non-SI units for the same quantity. These non-SI units follow the numerical conversion factors.

NOTE: In the following table at all locations, ton refers to U.S. Customary (2000 lb) rather than metric ton (1000 Kg).

Area: m^2/X	1.0×10^{-4}	cm^2
	1.0×10^{-12}	μm^2
	0.0929	ft^2
	6.452×10^{-4}	in^2
	0.8361	yd^2
	4047	Acre
	2.59×10^{6}	mi^2

Density: $(kg/m^3)/X$	1000.0	g/cm^3
	16.02	lbm/ft^3
	119.8	lbm/gal
	27,700	lbm/in^3
	2.289×10^{-3}	$grain/ft^3$
Diffusion coefficient: $(m^2/s)/X$	1.0×10^{-4}	cm^2/s
	2.78×10^{-4}	m^2/h
	0.0929	ft^2/s
	2.58×10^{-5}	ft^2/h
Electrical charge: C/X	1	A-s
	10	abcoulomb
	3.336×10^{-10}	statcoulomb
Electrical conductance: [siemens (S)]/X	1	Ω^{-1}
Electrical field strength: $(V/m)/X$	1	Kg m/A s^3
	100	V/cm
	39.4	V/in
Electrical resistivity: $(\Omega\ m)/X$	1	kg $m^5/A^2\ s^3$
	1.0×10^{-9}	abΩ m
	$8.988 \times 10^{+11}$	statΩ m
Energy: J/X	$3.6 \times 10^{+6}$	kWh
	4.187	cal
	4187	kcal
	1.0×10^{-7}	erg
	1.356	ft lbf
	1055	Btu
	0.04214	ft pdl
	$2.685 \times 10^{+6}$	hp h
	$1.055 \times 10^{+8}$	therm
	0.113	in lbf
	$4.48 \times 10^{+4}$	hp min
	745.8	hp s
Energy density: $(J/m^3)/X$	$3.6 \times 10^{+6}$	kWh/m^3
	$4.187 \times 10^{+6}$	cal/cm^3
	$4.187 \times 10^{+9}$	$kcal/cm^3$

	0.1	erg/cm^3
	47.9	ft lbf/ft^3
	$3.73 \times 10^{+4}$	Btu/ft^3
	$1.271 \times 10^{+8}$	kWh/ft^3
	$9.48 \times 10^{+7}$	hp h/ft^3
Energy, linear: (J/m)/*X*	418.7	cal/cm
	$4.187 \times 10^{+5}$	kcal/cm
	1.0×10^{-5}	erg/cm
	4.449	ft lbf/ft
	3461	Btu/ft
	$8.81 \times 10^{+6}$	hp h/ft
	$1.18 \times 10^{+7}$	kWh/ft
Energy per area: (J/m^2)/*X*	41,868	cal/cm^2
	$4.187 \times 10^{+7}$	kcal/cm^2
	0.001	erg/cm^2
	14.60	ft lbf/ft^2
	11,360	Btu/ft^2
	$2.89 \times 10^{+7}$	hp h/ft^2
	$3.87 \times 10^{+7}$	kWh/ft^2
Flow rate, mass: (kg/s)/*X*	1.0×10^{-3}	g/s
	2.78×10^{-4}	Kg/h
	0.4536	lbm/s
	7.56×10^{-3}	lbm/min
	1.26×10^{-4}	lbm/hr
Flow rate, volume: (m^3/s)/*X*	1.0×10^{-6}	cm^3/s
	0.02832	cfs
	1.639×10^{-5}	in^3/s
	4.72×10^{-4}	cfm
	7.87×10^{-6}	cfh
	3.785×10^{-3}	gal/s
	6.308×10^{-5}	gpm
	1.051×10^{-6}	gph
Force: N/*X*	1.0×10^{-5}	dyn
	1	kg m/s

	9.8067	kg(force)
	9.807×10^{-3}	G(force)
	0.1383	pdl
	4.448	lbf
	4448	kip
	8896	ton(force)
Heat transfer coefficient: (W/m-K)/*X*	41868	cal/s cm^2 °C
	1.163	kcal/h m^2 °C
	1.0×10^{-3}	erg/s cm^2 °C
	5.679	Btu/h ft^2 °F
	12.52	kcal/h ft^2 °C
Henry's constant: (N/m^2)/*X*	$1.01326 \times 10^{+5}$	atm
	133.3	mmHg
	6893	lbf/in^2
	47.89	lbf/ft^2
Length: m/*X*	0.01	cm
	1.0×10^{-6}	μm
	1.0×10^{-10}	Å
	0.3048	ft
	0.0254	in
	0.9144	yd
	1609.3	mi
Mass: kg/*X*	1.0×10^{-3}	G
	0.4536	lbm
	6.48×10^{-5}	grain
	0.2835	oz avdp
	907.2	ton (U.S.)
	14.59	Slug
Mass per area: (kg/m^2)/*X*	10	g/cm^2
	4.883	lbm/ft^2
	703.0	lbm/in^2
	3.5×10^{-4}	ton/mi^2
Power: W/*X*	4.187	cal/s
	4187	kcal/s
	1.0×10^{-7}	erg/s

	1.356	ft lbf/s
	0.293	Btu/h
	1055	Btu/s
	745.8	hp
	0.04214	ft pdl/s
	0.1130	in lbf/s
	3517	ton refrigeration
	17.6	Btu/min
Power density: $(W/m^3)/X$	$4.187 \times 10^{+6}$	cal/s cm^3
	$4.187 \times 10^{+9}$	kcal/s cm^3
	0.1	erg/s cm^3
	47.9	ft lbf/s ft^3
	$3.73 \times 10^{+4}$	Btu/s ft^3
	10.36	Btu/h ft^3
	$3.53 \times 10^{+4}$	kW/ft^3
	$2.63 \times 10^{+4}$	hp/ft^3
Pressure, stress: (Pa)/X	0.1	dyn/cm^2
	1	N/m^2
	9.8067	Kg(f)/m^2
	$1.0 \times 10^{+5}$	Bar
	$1.0133 \times 10^{+5}$	std. atm
	1.489	pdl/ft^2
	47.88	lbf/ft^2
	6894	lbf/in^2 (psi)
	$1.38 \times 10^{+7}$	ton(f)/in^2
	249.1	in H_2O
	2989	ft H_2O
	133.3	torr, mmHg
	3386	in Hg
Resistance: Ω/X	1	Kg m^2/A^2 s^3
	1	V/A
	1.0×10^{-9}	abΩ
	$8.988 \times 10^{+11}$	StatΩ

Specific heat, gas constant: (J/kg-K)/*X*	1	M^2/s^2 K
	4187	cal/g °C
	1.0×10^{-4}	erg/g °C
	4187	Btu/lbm °F
	5.38	ft lbf/lbm °F
Surface tension: (N/m)/*X*	1.0×10^{-3}	dyn/cm
	14.6	lbf/ft
	175.0	lbf/in
Temperature: K/*X* (difference)	0.5555	°R
	0.5555	°F
	1.0	°C
Thermal conductivity: (W/m-K)/*X*	418.7	cal/s cm °C
	1.163	kcal/h m °C
	1.0×10^{-5}	erg/s cm °C
	1.731	Btu/h ft °F
	0.1442	Btu in/h ft^2 °F
	2.22×10^{-3}	ft lbf/h ft °F
Time: s/*X*	60.0	min
	3600	hour
	86,400	day
	$3.156 \times 10^{+7}$	year
Torque: N-m/*X*	1.0×10^{-7}	dyn cm
	1.356	lbf ft
	0.0421	pdl ft
	2.989	kg(f) ft
Velocity: (m/s)/*X*	0.01	cm/s
	2.78×10^{-4}	m/h
	0.278	Km/h
	0.3048	ft/s
	5.08×10^{-3}	ft/min
	0.477	mi/h
Velocity, angular: (rad/s)/*X*	0.01667	rad/min
	2.78×10^{-4}	rad/h
	0.1047	rev/min

Viscosity, dynamic: (Kg/m-s)/*X*	1	N s/m^2
	0.1	P
	0.001	cP
	2.78×10^{-4}	Kg/m h
	1.488	lbm/ft s
	4.134×10^{-4}	lbm/ft h
	47.91	lbf s/ft^2
(g/cm-s)/*X*	1	P
Viscosity, kinematic: (m^2/s)/*X*	1.0×10^{-4}	St
	2.778×10^{-4}	M^2/h
	0.0929	ft^2/s
	2.581×10^{-5}	ft^2/h
(cm^2/s)/*X*	1	St
Volume: m^3/*X*	1.0×10^{-6}	cm^3
	1.0×10^{-3}	L
	1.0×10^{-18}	μm^3
	0.02832	ft^3
	1.639×10^{-5}	in^3
	3.785×10^{-3}	gal (U.S.)
Voltage, electrical potential: V/*X*	1.0	kg m^2/A s^3
	1	W/A

A.2 Using the Table

The quantity in braces { } is selected from the table.

Example 1 To calculate how many meters are in 10 ft, the table provides the conversion factor as 0.3048 m/ft. Hence multiply 10 ft × {0.3048 m/ft} = 3.048 m

Example 2 Convert thermal conductivity of 10 kcal/h m °C to SI units. Select the appropriate conversion factor for these units, 10 (kcal/h m °C) × {1.163 (w/m K)/(kcal/h m °C)} = 11.63 w/m K

Index

B

C

F

G

M

N

Q

R

S

T

Z